272
Advances in Polymer Science

Aims and Scope

The series Advances in Polymer Science presents critical reviews of the present and future trends in polymer and biopolymer science. It covers all areas of research in polymer and biopolymer science including chemistry, physical chemistry, physics, material science.

The thematic volumes are addressed to scientists, whether at universities or in industry, who wish to keep abreast of the important advances in the covered topics.

Advances in Polymer Science enjoys a longstanding tradition and good reputation in its community. Each volume is dedicated to a current topic, and each review critically surveys one aspect of that topic, to place it within the context of the volume. The volumes typically summarize the significant developments of the last 5 to 10 years and discuss them critically, presenting selected examples, explaining and illustrating the important principles, and bringing together many important references of primary literature. On that basis, future research directions in the area can be discussed. Advances in Polymer Science volumes thus are important references for every polymer scientist, as well as for other scientists interested in polymer science - as an introduction to a neighboring field, or as a compilation of detailed information for the specialist.

Review articles for the individual volumes are invited by the volume editors. Single contributions can be specially commissioned.

Readership: Polymer scientists, or scientists in related fields interested in polymer and biopolymer science, at universities or in industry, graduate students.

Special offer:
For all clients with a standing order we offer the electronic form of Advances in Polymer Science free of charge.

More information about this series at http://www.springer.com/series/12

Karl Leo
Editor

Elementary Processes in Organic Photovoltaics

Springer

Editor
Karl Leo
Technische Universität Dresden
Dresden, Germany

ISSN 0065-3195 ISSN 1436-5030 (electronic)
Advances in Polymer Science
ISBN 978-3-319-28336-4 ISBN 978-3-319-28338-8 (eBook)
DOI 10.1007/978-3-319-28338-8

Library of Congress Control Number: 2016945019

Printed on acid-free paper

This Springer imprint is published by Springer Nature
The registered company is Springer International Publishing AG Switzerland

Preface

The transition to renewable energies is one of the most important and challenging tasks for humankind: On the one hand, the emission of carbon dioxide induces a climate change with negative consequences that are already visible today. On the other hand, the limited reserve of fossil fuels will not be able to provide sufficient supplies for the world's growing energy demand.

One of the most elegant principles of renewable energy generation is photovoltaics: They convert the most abundant energy source, light from the sun, directly into electrical energy. The last two decades have seen a rapid development in photovoltaics, leading to an already significant worldwide deployment of the technology. The current market leader is crystalline silicon technology, which has many advantages such as its high efficiency and stability. Nevertheless, there is a strong drive for novel technologies that could open up new application areas.

One technology with excellent potential is organic photovoltaics: This technology is based on carbon compounds; thus, the raw material is virtually inexhaustible. Furthermore, organic photovoltaics allow flexible modules on plastic foil substrates, which can be transparent and have adjustable color. The flexible substrates and the easy deposition of organic materials by methods such as printing offer the potential for low-cost roll-to-roll production. The main challenges for organic photovoltaics are their still comparably low conversion efficiencies and lack of stability.

This book presents a collection of chapters summarizing a large collaborative research effort of many groups: In the *Schwerpunktprogramm* (priority program) 1355 "Organic Photovoltaics" of the Deutsche Forschungsgemeinschaft (German Research Foundation), more than 150 researchers during the years 2008–2015 investigated the chemistry, physics, and engineering aspects of this technology. The collaborative effort within this research program led to substantial improvements and to a much better understanding of the working principles of the technology.

At the beginning of the program, the best efficiencies reported for organic solar cells were in the range of 6 %. During the course of the program, the efficiencies approximately doubled. The currently reported best value of 13.2 % is partially based on work performed in the framework of the program reported here.

One of the key aspects of research on organic photovoltaics is its interdisciplinarity: The work starts with the design of new molecules, which is a task for theoretical physics and chemistry, and continues with the synthesis of the materials by organic chemistry. Characterization and device principles are subjects of physics, while device design and module integration are engineering tasks. Accordingly, most of the individual projects reported in the chapters of this book are based on the close collaboration of researchers from different disciplines, and the new results reported address diverse subjects such as new molecules, novel characterization techniques, new device designs, and integration aspects such as novel electrodes.

I wish to thank the many persons who made the *Schwerpunktprogramm* and its publication in this book a pleasure: First, I am grateful to the many researchers who worked in a highly collaborative style on this challenging subject. I am glad that many of the younger researchers (and in particular the females among them) used this work as a stepping stone in their career to academic positions. I thank those colleagues who helped to successfully apply for the grant proposal. Finally, I thank Dr. Annette Polte for her competent coordination work, Dr. Michael Mößle and Dr. Stefan Krückeberg from the DFG, and Dr. Claus Ascheron from the publisher for the excellent collaboration.

Dresden, Germany Karl Leo

Contents

In situ Studies of Morphology Formation in Solution-Processed Polymer–Fullerene Blends

Esther Barrena, Felix Buss, Ana Perez-Rodriguez, Monamie Sanyal, Benjamin Schmidt-Hansberg, Michael F.G. Klein, Philip Scharfer, Wilhelm Schabel, and Uli Lemmer

Contents

E. Barrena (✉) • A. Perez-Rodriguez
Instituto de Ciencia de Materiales de Barcelona (ICMAB CSIC), Campus UAB, Bellaterra, Barcelona 08193, Spain
e-mail: ebarrena@icmab.es

F. Buss • B. Schmidt-Hansberg • P. Scharfer • W. Schabel
Institute of Thermal Process Engineering, Thin Film Technology, Karlsruhe Institute of Technology, Kaiserstrasse 12, 76131 Karlsruhe, Germany
e-mail: wilhelm.schabel@kit.edu

M. Sanyal
Max Planck Institut für Metallforschung, Heisenbergstrasse 3, 70569 Stuttgart, Germany

M.F.G. Klein • U. Lemmer
Light Technology Institute, Karlsruhe Institute of Technology, Engesserstraße 13, 76131 Karlsruhe, Germany
e-mail: uli.lemmer@kit.edu

K. Leo (ed.), *Elementary Processes in Organic Photovoltaics*, Advances in Polymer Science 272, DOI 10.1007/978-3-319-28338-8_1

Abstract Control of the blend nanomorphology in bulk heterojunctions (BHJs) is still a challenge that demands more fundamental knowledge of the mechanism of phase separation and crystallization during solvent drying. In this review we show that in situ studies using combined laser reflectometry and grazing-incidence wide-angle X-ray scattering provide a fundamental understanding on how the nanomorphology develops dynamically during film drying. We identify influencing parameters for controlled film formation in order to obtain optimized solar cell performance. We review here our results on BHJs of poly(3-hexylthiophene)–[6,6]-phenyl-C61-butyric acid methyl ester and poly{[4,40-bis(2-ethylhexyl)dithieno(3,2-b;20,30-d)silole]-2,6-diyl-alt-(2,1,3 benzothidiazole)-4,7-diyl} with [6,6]-phenyl-C71-butyric acid methyl ester.

Keywords Bulk heterojunction • Crystallization • Drying kinetics • GIWAXS • P3HT • PCBM • PSBTBT • Real-time studies • Solution processing

1 Introduction

Morphological control in bulk heterojunction (BHJ) photovoltaics is one pillar of the recent remarkable progress in the power conversion efficiency (PCE) of organic photovoltaic (OPV) devices [1–3]. The performance of solution-processed polymer solar cells is highly dependent on the phase-separated nanomorphology that dynamically forms during the drying process. Whether the polymeric component is in a crystalline or an amorphous state, pure or intermixed with the fullerene, with domain sizes on the order of magnitude of the exciton diffusion length or larger, are factors that largely affect the mechanisms involved in the energy-conversion process, that is, charges carrier photogeneration, dissociation, conduction, and collection.

In general, the final film structure that results upon solvent evaporation is not at thermodynamic equilibrium and, in addition, is very material dependent. The more quickly the drying step is performed, the less time remains for the solid components to arrange into their thermodynamically favored order. Consequently, quickly dried as-cast thin films fabricated into a functional OPV device generally exhibit a low PCE, requiring the use of an additional posttreatment step such as thermal annealing [4] or solvent annealing [5].

A more promising approach is to foster the formation of a more optimized nanomorphology in a solution, as recently exploited with the use of additives [1, 6]. However, predictive control of the blend nanomorphology remains a challenge that demands more fundamental knowledge of how the interplay between thermodynamic and kinetic factors (miscibility, solubility, blend ratio, solvent evaporation rate, etc.) triggers the evolution of the BHJ during film drying. To gain a full understanding of the mechanisms of phase separation and organization of the polymer and fullerene components, it is necessary to characterize the dynamic process of film formation from its earliest stage.

With this purpose, we designed a drying setup to monitor in real time the evolution of the thickness and structure in the wet film during solvent evaporation by combining laser reflectometry and grazing-incidence wide-angle X-ray scattering (GIWAXS). We performed extensive investigation of the blend system poly(3-hexylthiophene) (P3HT):[6,6]-phenyl-C61-butyric acid methyl ester (PCBM) and gained an understanding of the film structure formation [7, 8], the effect of the drying temperature [9], the role of PCBM [10], and the impact of using nonhalogenated solvent mixtures [11]. We also extended our studies to other material systems, such as poly{[4,40-bis(2-ethylhexyl)dithieno(3,2-b;20,30-d)silole]-2,6-diyl-alt-(2,1,3 benzothidiazole)-4,7-diyl} (PSBTBT) [11] and poly[2,6-(4,4-bis-(2-ethylhexyl)-4H-cyclopenta[2,1-*b*;3,4-*b*0]dithiophene)-*alt*-4,7(2,1,3-benzothiadiazole)] (PCPDTBT).

Here we review some of our results of the structural evolution during the drying of BHJ films of poly(3-hexylthiophene)–[6,6]-phenyl-C_{61}-butyric acid methyl ester (P3HT:PCBM) and poly{[4,40-bis(2-ethylhexyl)dithieno(3,2-*b*;20,30-d)silole]-2,6-diyl-*alt*-(2,1,3 benzothidiazole)-4,7-diyl}–[6,6]-phenyl-C_{71}-butyric acid methyl ester (PSBTBT:$PC_{71}BM$). We also provide detailed information on the experimental setup and the measurements performed in situ.

2 In situ Characterization of Film Formation During Drying

2.1 *Drying Channel and Experimental Setup*

Investigations of the drying behavior of polymer–fullerene films require a setup that facilitates the adjustment of all experimentally accessible drying parameters. In a drying channel the gas-phase drying conditions, that is, gas temperature T_g and velocity u_g, and the temperature of the substrate holder T_S, can be defined. An automated knife coater (doctor blade) allows the deposition of thin films in the range of several micrometers and subsequent drying. Immediately after coating, the drying process is monitored using two measurement techniques simultaneously: Laser reflectometry gives insight into the drying kinetics by providing information on the actual film thickness and composition, and simultaneously collected GIWAXS data reveal the appearance of structural features as the solid components begin to aggregate or crystallize. This setup enables the solubility limits and general thermodynamic properties that are known from supplementary measurements to be correlated to the onset and development of structure formation during the drying process (Fig. 1).

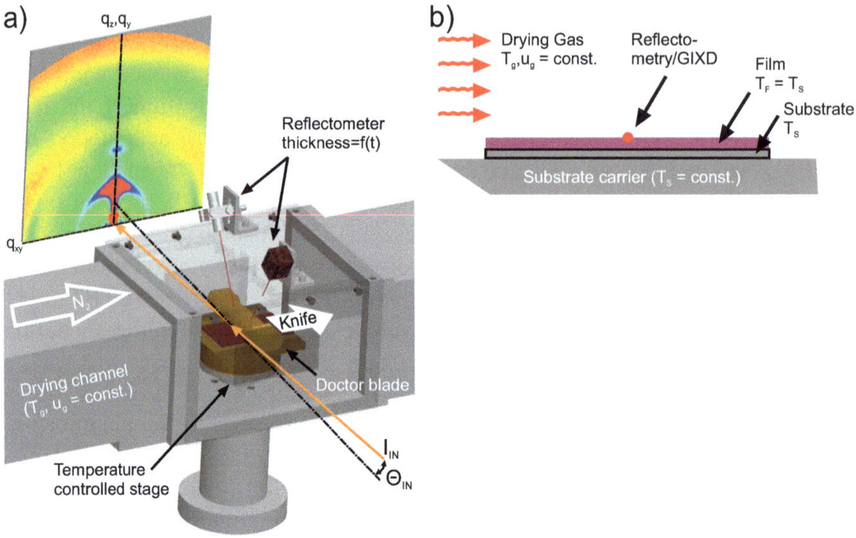

Fig. 1 Schematic of the experimental setup for simultaneous laser reflectometry and grazing-incidence wide-angle X-ray spectroscopy under controlled drying conditions [4]

2.2 *Basics of Laser Reflectometry*

Laser reflectometry is used in our experimental setup as a nondestructive method to determine the thickness of a transparent film [12–14]. In case of a drying film it is based on the optical interference of a laser beam that is reflected at the air–film and the film–substrate interfaces. As the film thickness decreases, constructive and destructive interference conditions of the reflected light are detected by a photodiode. With the knowledge of the refractive index and the dry film thickness, the thickness evolution over time can be calculated. Figure 2 illustrates the geometric conditions that need to be considered for the analysis of the reflectometry signal.

Using Snell's law of diffraction, basic geometric rules, and assuming a refractive index $n_0 = 1$ for the ambient, the difference in optical path [Eq. (1)] and the criterion for constructive and destructive interference [Eq. (2)] can be written as [8, 15]

$$\Delta = 2d\sqrt{n_1^2 - \sin^2\alpha} \tag{1}$$

$$\Delta = \frac{m}{2}\cdot\lambda, \tag{2}$$

where $m = [1, 2, 3, \ldots]$ with even (odd) values for constructive (destructive) interference and λ is the laser wavelength. Due to solvent evaporation, the refractive index n_1 of the solution is not a constant, but a function of the volume fractions φ_i

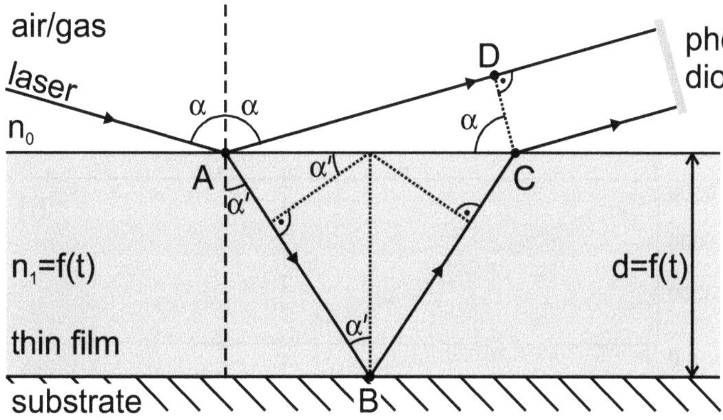

Fig. 2 Optical path of a laser beam through a drying film. The laser light hits and is partly reflected from the film surface with an angle α. At the same time light penetrates into the wet film and experiences a change in direction caused by the higher refractive index n_1 of the solution. The difference in optical path (Δ) of these two beams leads to the appearance of interference

(s = solid, l = liquid) of the components:

$$n_1 = \sum\nolimits_{i=1}^{K} \varphi_i \cdot n_i, \tag{3}$$

$$\varphi_s = 1 - \varphi_l = d_{\text{dry}}/d, \tag{4}$$

Insertion of Eqs. (2)–(4) into Eq. (1) yields an expression for the calculation of the film thickness:

$$d = \frac{-B + \sqrt{B^2 - 4AC}}{2A}, \tag{5}$$

$$A = n_l^2 - \sin^2\alpha;\ \ B = 2d_{\text{dry}} n_l (n_s - n_l);\ \ C = d_{\text{dry}}^2 (n_s - n_l)^2 - m^2\lambda^2/4,$$

$$x_S = \left(1 + \frac{d_{\text{dry}}}{d - d_{\text{dry}}} \cdot \frac{\rho_s}{\rho_l}\right)^{-1}, \tag{6}$$

The thickness can be converted into mass fractions using Eq. (6), where x_s is the solvent mass fraction and ρ_i are the densities of the liquid and the solid components.

Figure 3 shows the reflectometer signal of a drying P3HT:PCBM in ortho-dichlorobenzene (o-DCB) solution, the associated thickness evolution over time, i.e., the drying curve, and—deduced from the latter—the composition of the solution. The drying process is divided into a constant-rate period, governed by

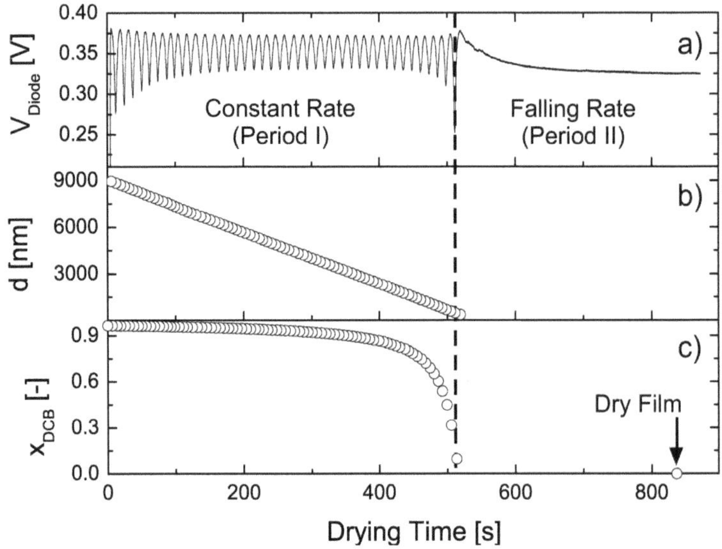

Fig. 3 (**a**) Voltage signal of the reflectometer photodiode. (**b**) Thickness and (**c**) solvent mass fraction evolution over drying time derived from the reflectometer signal. The film is dry when there is no further change of photovoltage [15]

the mass transport of solvent in the gas phase. This phase is characterized by a regular appearance of interference fringes and a linear decrease in film thickness. At low solvent content, the solvent diffusion in the film dominates the drying process and solvent removal is slowed down by orders of magnitudes, defined as the falling rate period. The thickness evolution in this range cannot be resolved, since the thickness change between two interference fringes is about 100 nm for the given setup ($\lambda = 650$ nm, $\alpha = 35°$). The duration of this period can be estimated by the time interval between the last interference fringe and the point where a constant photovoltage is reached. Other methods, such as spectroscopic ellipsometry [16] or quartz crystal microbalances [17, 18], are feasible techniques to resolve changes during this period. In order to predict the drying behavior at given drying parameters, a model for the constant-rate period can be used [12].

2.3 Experimental Details of the GIWAXS Measurements

The real-time GIWAXS studies were performed at beamline ID10B in the European Synchrotron Radiation Facility (ESRF) (Grenoble, France) using the experimental setup shown in Fig. 1. Depending on the experiment, the photon energy varied between 10–12.3 keV. Two-dimensional (2D) diffraction patterns were acquired by an X-ray area detector (MarCCD). The incident angle was set below the critical angle of the substrate but above the critical angle of the blend. It should be taken into account that the intensity collected with a flat detector in grazing-incidence geometry is not a direct map of reciprocal space and misses a part of the reciprocal space along the vertical direction of the sample [19–21]. Consequently, out-of-plane profiles obtained from the 2D pattern contain an in-plane component of the scattering wave vector transfer (q).

We give values as a function of the magnitude of the scattering vector q determined from the total scattering angle. In our work, complementary measurements with a point detector were performed to obtain specular (along q_z) and in-plane scans (along q_{xy}). The point detector was mounted in a second arm of the diffractometer and was used to align the sample and to collect X-ray reflectivity data before and after each drying experiment. Two-dimensional frames were acquired with an integration time of a few seconds. After each frame, the sample was moved horizontally in the direction of coating by 0.6 mm (the horizontal beam size was controlled by the horizontal slit width, which was 0.5 mm) to measure a fresh region of the film not previously exposed to direct radiation.

As a measure of the crystallinity, the intensity of the (100) was azimuthally integrated, accounting for the orientation distribution of crystallites. The coherence correlation length was measured from the full width at half maximum (FWHM) of the peaks. This is a convolution of the coherence length of the lamellar stacking and crystalline quality.

3 Drying of P3HT/P_{61}CBM

3.1 Binary Phase Diagram and Morphology Development

Before addressing the structure formation of the BHJ in the ternary polymer–fullerene–solvent system (P3HT-PCBM-DCB), we have first determined the phase diagrams of the binary systems P3HT-DCB and PCBM-DCB [7]. Figure 4a depicts the composition of the two coexisting phases of a P3HT solution in DCB for each temperature at thermodynamic equilibrium, namely, the solid-poor sol phase (half-filled circles) and the solid-rich gel phase (filled circles). These two states form the binodal curve which surrounds the thermodynamically unstable two-phase region. If the P3HT solution is situated in this unstable region it tends to phase separate, leading to the aggregation of a swollen polymer with solvent ($x_s = 10.2$ wt% solid fraction) and a residual sol phase (5.1 wt% solid fraction) [22, 23]. The solution is stable at room temperature in the concentration range of 20–40 mg/ml [$x_s = 1.5$–3 wt%; $x_s = m_{\mathrm{solid}}/(m_{\mathrm{DCB}} + m_{\mathrm{solid}})$] usually used for organic solar cells. With solvent evaporation at a constant temperature (the arrow indicates evaporation at 25 °C), the P3HT solution reaches the solubility limit at a solid mass fraction of $x_s = 5.1$ wt% (interpolated for 25 °C) and enters the unstable binodal area. The width of the unstable two-phase region is almost zero at 50 °C (overlapping error

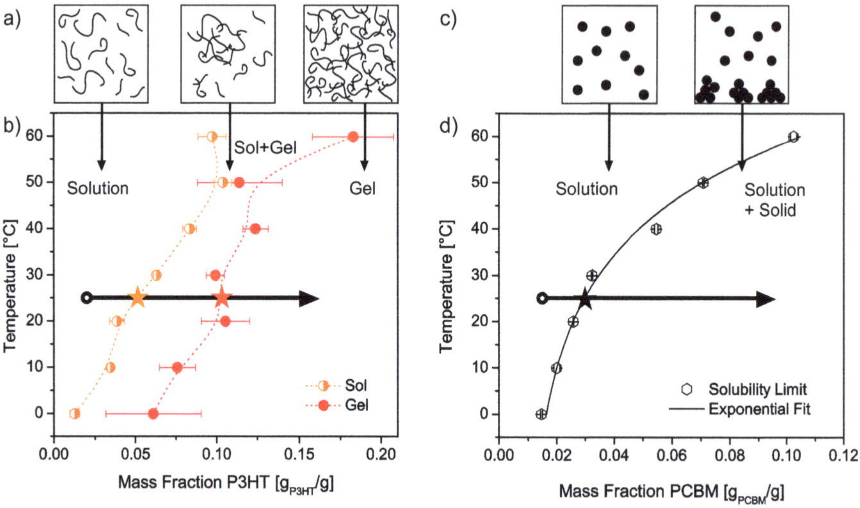

Fig. 4 Binary phase diagram of P3HT and PCBM solutions in dichlorobenzene (DCB). (**a**) Schematic of the transition from solution to gel state with polymer aggregation in the unstable two-phase region, before the solution turns into a higher viscous gel during solvent evaporation. (**b**) Phase diagram of P3HT–DCB solution with unstable two-phase region between the solution and gel state. *Dashed lines* are a visual guide. (**c**) Schematic of fullerene aggregation after crossing the solubility limit. (**d**) Solubility of PCBM in DCB as a function of temperature. The *arrows* indicate the drying process at 25 °C, and the *stars* represent the instant of phase transition [7]

bars), indicating an almost continuous increase of viscosity without phase separation at this drying temperature.

PCBM does not show a gelation behavior and exhibits a single solubility limit, as depicted in Fig. 4d, which increases with temperature, in contrast to P3HT. By crossing the solubility limit caused by solvent evaporation, solid PCBM aggregates precipitate, maintaining the solubility limit as shown schematically in Fig. 4c. Depending on the evaporation kinetics of pure PCBM solution, this can lead to macroscopic crystals of several micrometers [24].

We compared the pathway through the superposition of the binary phase diagrams with the evolution of the crystallization of the ternary blend in thin films, studied in situ using GIWAXS and laser reflectometry. Using the experimental setup explained in Sect. 2, we obtained 2D GIWAXS diffraction patterns immediately after doctor-blading the solution on a glass/poly(3,4-ethylenedioxythiophene) polystyrene sulfonate (PEDOT:PSS) substrate. Figure 5a shows five snapshots of the structural evolution of a P3HT:PCBM (1:0.8) blend during the evaporation of DCB. The solvent mass fraction (x_{DCB}) at each time instant can be obtained from the optical thickness measurement based on interference patterns.

The time evolution of the film composition obtained from laser reflectometry is plotted in Fig. 5b. To study the entire evolution of the solvent fraction, a model of the constant gas-phase mass-transfer coefficient for the film drying kinetics was adapted

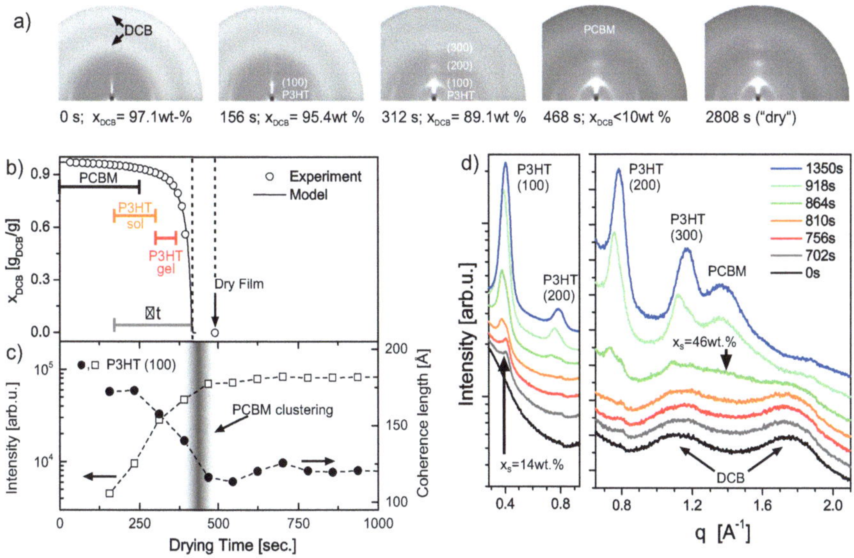

Fig. 5 (**a**) Two-dimensional grazing-incidence wide-angle X-ray spectroscopy diffraction patterns acquired during different stages in the evolution of the film during drying. Below each image, the drying time and actual solvent fraction are mentioned. (**b**) Evolution of film composition (solvent mass fraction x_{DCB}) during film drying (data, symbols; fit from model, line). The labeled bars indicate ranges of phase transitions as expected from the superposition of the binary phase diagrams. The constant-rate period time after crossing P3HT solubility is denoted as Δt. (**c**) The integrated area of the P3HT (100) Bragg peak and the correlation length are plotted vs. drying time. (**d**) Selected profile plots at increasing drying times showing the emergence of the diffraction peaks of PCBM, P3HT (h00) Bragg peaks, and the disappearance of the diffraction peaks of the evaporating solvent (dichlorobenzene) on a silicon dioxide/silicon substrate [7]

to the experimental data [8]. The bars in Fig. 5b labeled PCBM, P3HT (sol), and P3HT (gel) indicate the instants on the timescale when the solubility limits/phase transitions of each component are reached for the binary cases. The length of each bar originates from two cases of transferring the binary phase diagrams to the ternary system as described in detail in [7]. Because of the low solid fraction in highly diluted solutions, the decrease in the solvent mass fraction x_{DCB} proceeds slowly in the beginning. This leads to a long residence time at solvent fractions $x_{\mathrm{DCB}} > 85$ wt%. As the solid fraction becomes significant, the solvent fraction x_{DCB} reduces rapidly and a wide range of solvent fractions in the phase diagram is crossed quickly in a small window of drying time, offering little time for further molecular ordering. In the last stage of drying, at high solid fractions $x_{\mathrm{DCB}} < 5$ wt%, the evaporation rate decreases because of the limited diffusion of solvent molecules in the film.

We compared the expected ranges of transitions in Fig. 5b, which are proposed according to the binary phase diagrams, with the observed evolution of P3HT crystallinity (Fig. 5c) for the ternary system under investigation. For P3HT, we initially expect a delay in crystallization until the solubility limit is reached. P3HT nucleation at the solubility limit is expected for a solid mass fraction of $x_s = 5.1$–9.1 wt%, which is confirmed by the appearance of the P3HT (100) lamellar peak in the early stage of film drying ($x_s = 4.6$ wt%). During the constant-rate period the intensity increases; near the range of complete gelation, the second- and third-order Bragg peaks appear. PCBM crystallizes at a later stage of drying.

On PEDOT:PSS-covered substrates it is difficult to decipher the appearance of the PCBM scattering ring ($q = 1.37$ Å^{-1}) following the disappearance of DCB broad diffraction scattering rings (at $q = 1.12$ Å^{-1} and 1.75 Å^{-1}). Hence, a drying experiment was performed without PEDOT:PSS and using a silicon substrate with native silicon oxide because of its low surface roughness. Profile lines at different stages of drying are plotted in Fig. 5d. The drying experiment confirmed the development of the PCBM diffraction ring after crystallization of P3HT at $x_s = 46$ wt%. Even for the highest published PCBM solubility of about 100 mg/ml [24] ($x_s = 7.7$ wt% at room temperature), PCBM would be expected to crystallize at an earlier stage. This shows that polymer crystallization occurs first, followed by PCBM clustering at a later stage of drying. These observations of oppressed PCBM crystallization suggest a strong interaction between fullerene and polymer.

Additional insight in the drying process is obtained from the analysis of the evolution of the (100) lamellar peak with time. The spot-like P3HT (100) Bragg peak ($q \approx 0.4$ A^{-1}) at the earliest stage of drying indicates that the initially formed P3HT nuclei are predominantly oriented with an edge-on configuration. This implies that P3HT nucleation is induced by a planar interface, either at the film–substrate or at the film–air interface. During the solidification there is a progressive shift in the (100) peak position, which indicates a densification of the lamella packing (with a reduction of the spacing from 17.4 to 16.3 Å). The correlation length of the lamellae with the drying time decreases slightly with time (Fig. 5c). The fact that the crystalline correlation length does not increase during the film formation indicates that the crystallization of P3HT occurs as a result of the nucleation of new crystals in the bulk of the wet film rather than as a result of further growth of the initially formed nuclei at the interface. The arcing of the diffracted intensity extending from the (h00) P3HT peaks reveals a rapid increase in the fraction of crystallites tilted with respect to the edge-on orientation during P3HT crystallization. It is worth noting that the decreased correlation length cannot be explained by the decrease in the lamella spacing and suggests an overall smaller lamellar coherence length of P3HT bulk crystallized aggregates than those nucleated at the interface.

The evolution of the blend's structure is schematically illustrated in Fig. 6. Blend solidification proceeds in three steps: (1) the earliest stage of P3HT nucleation at the interface (substrate or air); (2) crystallization of disordered P3HT crystallites out of the wet film; and (3) late formation of PCBM aggregates in the last stage of the drying. We show in Sect. 3.3 that the stage of bulk crystallization is crucial for the development of π–π ordering in P3HT.

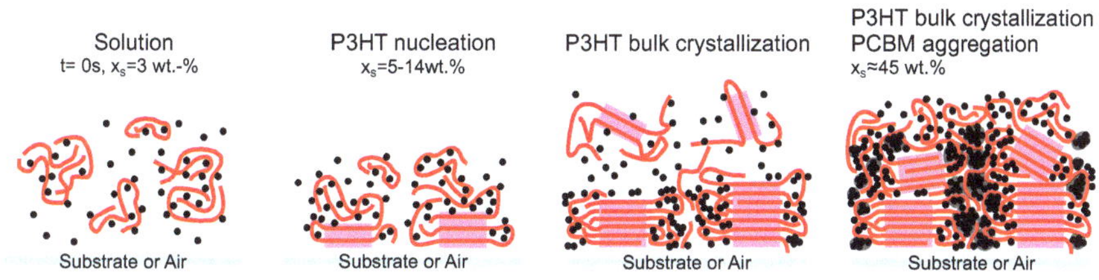

Fig. 6 Scheme of the stages of P3HT:PCBM film formation during solvent (dichlorobenzene) evaporation

In contrast to a low PCBM solubility of only 3–4 % in crystalline P3HT, Collins et al. [25, 26] observed a solubility of about 16 % in amorphous P3HT regions. This confirms the image of three coexisting phases: (1) mainly pure crystalline P3HT domains comprising very few PCBM inclusions; (2) an intermixed phase formed by amorphous P3HT regions containing higher amounts of PCBM molecules; and (3) PCBM-rich clusters partly intermixed in the amorphous P3HT regions. PCBM-solving additives may selectively reduce the interaction between P3HT and PCBM and promote polymer crystallization; this strategy has been used to improve the PCE with the use of mixed solvents and additives [1, 26–28].

3.2 Effect of the Blend Ratio

Previous works have reported that the optimum P3HT:PCBM weight ratio composition for the best device performance is close to 1:0.8 [29–35]. To understand the role of PCBM in the crystallization of P3HT during film drying, we investigated the structural evolution for P3HT:PCBM blends in weight ratios of 1:0.5, 1:0.8, and 1:2 on glass/PEDOT:PSS [10]. Figure 7c displays four 2D X-ray scattering frames illustrating different stages of drying for each of the investigated blend weight ratios. The appearance of the P3HT (100), (200), and (300) is indicated in the drying paths of the phase diagram of different blending ratios (Fig. 7a). The structural evolution with drying time observed for the blends with ratios 1:0.5 and 1:0.8 is similar and matches the picture of film formation illustrated in Fig. 6: an initial interface nucleation of edge-on P3HT crystallites followed by P3HT bulk crystallization, with a consequent increase in the orientation distribution, and late aggregation of PCBM. The evolution is remarkably different for the blend with a 1:2 ratio, that is, with a larger amount of PCBM. First, the appearance of the P3HT (100) Bragg peak ($x_{\mathrm{DCB}} = 92$ wt%) is delayed [as were the development of the (200) and (300) peaks], which can be attributed to the increased intermixing of P3HT and PCBM because of the larger amount of PCBM. Another striking difference is that for the blend with a 1:2 ratio, the (h00) peaks maintain a spot-like shape during drying, that

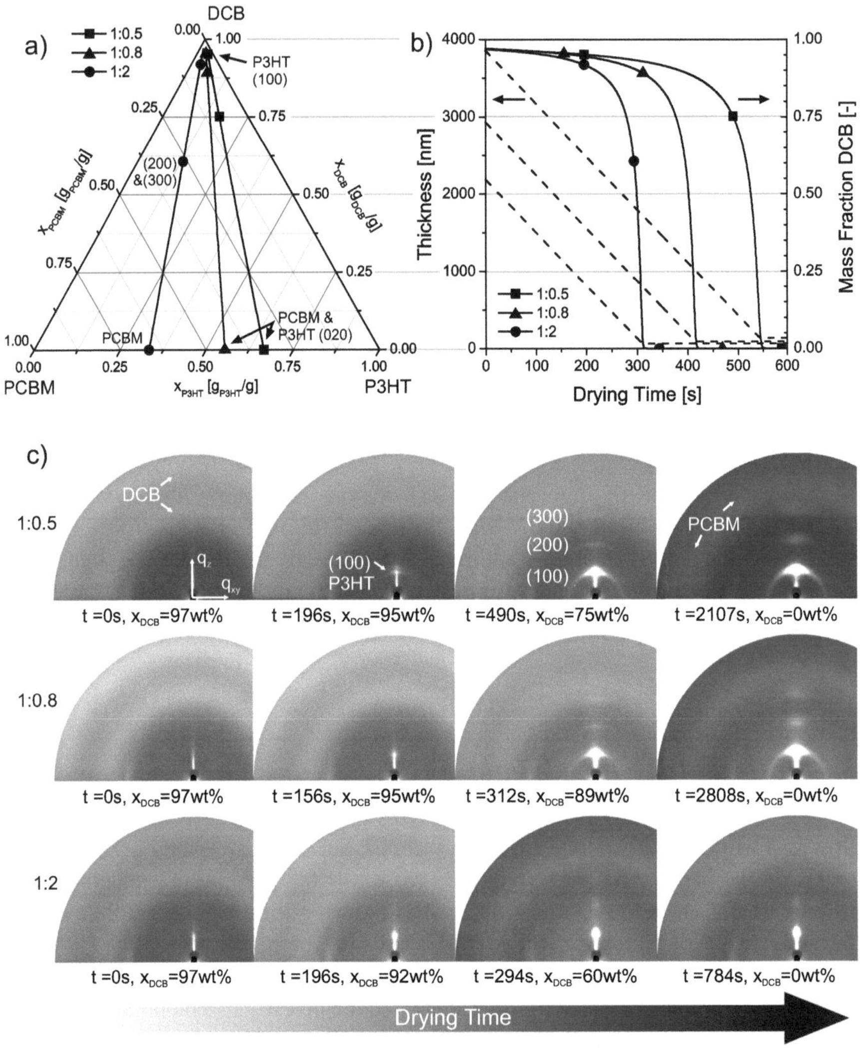

Fig. 7 (**a**) Drying path through the phase diagram of different P3HT:PCBM blending ratios at 25 °C. *Symbols* indicate the appearance of P3HT (100) Bragg peak, (200) and (300), and the PCBM diffraction ring in order of appearance from *top* to *bottom*. (**b**) Drying kinetics of the investigated films. Each symbol represents a grazing-incidence X-ray diffraction measurement. *Filled symbols* are related to the instants of structural changes indicated in the ternary phase diagram. (**c**) Two-dimensional X-ray diffraction patterns at different instants of evaporation of the solvent (dichlorobenzene) for three blend ratios [36]

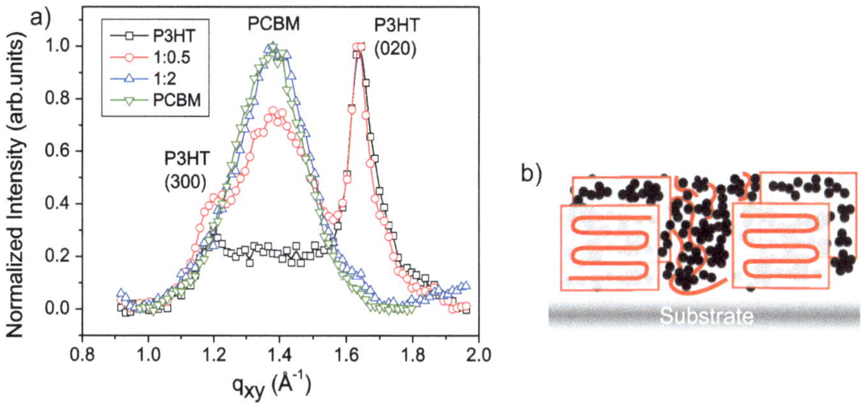

Fig. 8 (**a**) In-plane data using a point detector of films consisting of pure P3HT (*black squares*), P3HT:PCBM blend with 1:0.5 ratio (*red circles*), P3HT:PCBM blend with 1:2 ratio (*blue triangles*), and pure PCBM (*green inverted triangles*). (**b**) Schematic picture of the disruption of π–π stacking of P3HT polymer chains by PCBM [10]

is, lamellae with an edge-on orientation. In spite of the lamella organization with an edge-on orientation, in-plane GIWAXS data scans reveal that the π–π stacking of the polymer backbones is suppressed by PCBM: The scans lack the characteristic in-plane (020) peak ($q = 1.64$ Å) [37] that is observed for films of pure P3HT and for blends with ratios of 1:0.5 and 1:1.8 (Fig. 8). This observation is confirmed by optical absorption spectra data [10].

Our results confirm that increased loading of PCBM in the solution significantly inhibits the P3HT ordering during the bulk crystallization stage. Interface-induced nucleation of edge-on crystallites with lamellar ordering seems not to be hindered although the intermixing of P3HT and PCBM at a molecular level hinders the π–π stacking of P3HT polymer chains. This is an important issue because the π–π ordering is considered beneficial for the delocalization and mobility of charge carriers and polaron excitations over neighboring chains. An alternative explanation recently proposed by Richter et al. [28] attributes the suppression of the crystallization of P3HT by PCBM to polymer vitrification caused by the increased glass transition temperature (T_g) of the mixed amorphous phase formed by P3HT and PCBM.

3.3 *Effect of the Drying Temperature*

As we already mentioned, the structural development during drying can be influenced with solvent mixtures or with additives. The substrate temperature is another parameter that can be used to influence the nanomorphology formation during drying. The drying temperature influences several parameters, including the solubility

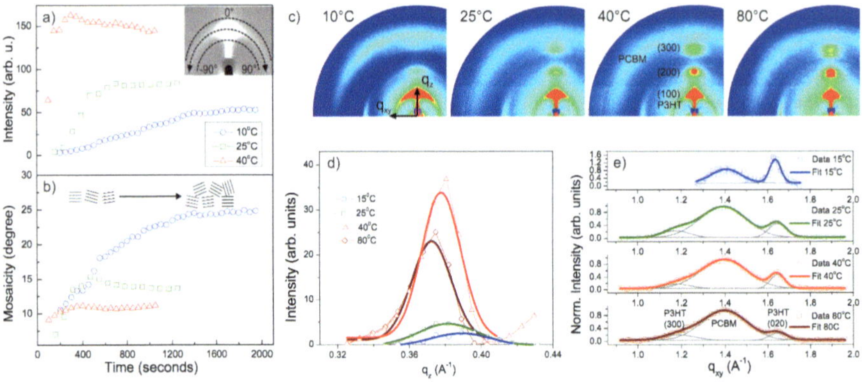

Fig. 9 Evolution with drying time of (**a**) P3HT (100) integrated intensity and (**b**) orientation distribution (mosaicity) for drying at substrate temperatures of 10, 25, and 40 °C. (**c**) Two-dimensional diffraction pattern measured for the dried P3HT:PCBM films upon drying at the indicated temperatures. (**d**) X-ray specular data with Gaussian fits (*thick continuous line*) of P3HT (100) Bragg peak for films dried at different substrate temperatures (15–80 °C). (**e**) Grazing-incidence X-ray diffraction data with Gaussian fits of the same dried films (normalized to maximum) [9]

and the diffusion kinetics in gas and liquid phases. To study the structural evolution during drying, we doctor-bladed a 1:0.8 solution of P3HT and PCBM dissolved in DCB on PEDOT:PSS-coated glass substrates at substrate temperatures of 10, 25, 40, and 80 °C. The thickness decreased as the processing temperature increased because of the lower viscosity of the wet film for a constant solid fraction. The solid fraction of the blend–solvent film was kept constant in order to start the drying process at the same position in the phase diagram.

The evolution with drying time of the overall integrated (100) intensity and of the orientation distribution is depicted in Fig. 9a and b, respectively. The drying proceeds faster at higher temperatures. In the earliest stage of P3HT nucleation, a similar orientation distribution is observed at all substrate temperatures. During film drying at lower temperatures, a larger fraction of misaligned crystallites develops than at higher-temperature drying. The larger orientation distribution of P3HT crystallites for the films dried at lower temperatures can be related to the slower kinetics of film formation. Similar observations were reported for solvent annealed spin-coated films in comparison with faster drying [33, 38]. Nevertheless the results reveal better (100) crystallinity for the films dried at higher temperatures (and with faster drying), that is, a larger fraction of aggregates with lamellar stacking. Drying at higher temperatures also leads to better alignment of the polymer chains with an edge-on orientation. The 2D diffraction patterns obtained for dried films are shown in Fig. 9c.

Further quantitative structural information on the ordering perpendicular and parallel to the substrate surface was obtained by a point detector. X-ray specular and in-plane data are plotted in Fig. 9d and e, respectively. The higher specular

X-ray intensity of the (100) Bragg peak for P3HT:PCBM blends dried at higher drying temperatures confirms better lamellar packing. The mean coherence length of the P3HT (100) peak increases with the drying temperature from 194 Å at 15 °C to 308 Å at 80 °C. It is difficult to hypothesize any major effect on devices from differences in lamellae organization since this direction of the crystalline structure is less relevant for charge generation or transport. Of higher relevance is the π–π organization of the polymer. In-plane GIWAXS data in Fig. 9e show the P3HT (020) in-plane peak, demonstrating the π–π assembly of P3HT chains for all four drying temperatures. We note, however, that the π–π crystallinity increases significantly with lower drying temperatures, as revealed by the larger intensity of the P3HT (020) peak. Additionally, the coherence length of the π–π stacking within the P3HT layers was largest for the sample dried at the lowest temperature: 118 Å at a drying temperature of 15 °C vs. 77 Å at 80 °C. Increased π–π ordering for the films dried at lower temperatures has been corroborated by the optical absorption spectra. The morphology of the films measured by atomic force microscopy upon PCBM removal reveals that the size of PCBM clusters increase as the drying temperature increases, most likely as a result of the stronger increase in PCBM solubility at higher temperatures [7].

These results suggest that the drying time after the solubility limit (Δt) is crossed, that is, during P3HT bulk crystallization, is crucial for π–π ordering (see Fig. 5b). Drying at a higher temperature leads to a larger lamellar organization; however, the higher solvent evaporation rate reduces the time that interchain P3HT π–π stacking can develop.

We measured how the solar cell efficiency is affected by drying at different temperatures; Fig. 10 summarizes the results. Further information can be found in [7]. We performed experiments with two drying gas velocities with two different nitrogen flow rates for each of the drying temperatures. The time after crossing solubility (Δt) decreased for higher-temperature drying. The short-circuit current density and PCE (Fig. 10b) are plotted as a function of Δt in Fig. 10. The solar cell efficiency increased with lower-temperature—and thus slower—drying, according to the longer crystallization time after crossing the P3HT solubility. This trend still existed after postannealing the samples for 5 min at 150 °C.

In terms of high-throughput fabrication, a long drying process is not desirable. Our in situ crystallization studies of film drying have shown that the initial drying period prior to crossing the P3HT solubility does not contribute to the polymer crystallization. One strategy to decrease the overall drying time is altering the thermodynamics of phase separation by adding an "unfriendly" solvent for both components, reducing the solubility of P3HT and PCBM [39]. For instance, if one adds the "unfriendly" solvent cyclohexanone (CHN) to chlorobenzene (CB) solution of P3HT:PCBM, the solubility can be crossed prior to the drying process. We performed a real-time GIWAXS study of P3HT:PCBM from a solvent mixture of CHN:CB (1:1) and demonstrated the formation of a P3HT (100) diffraction ring directly after coating ($t = 0$), confirming the crystallization of randomly oriented

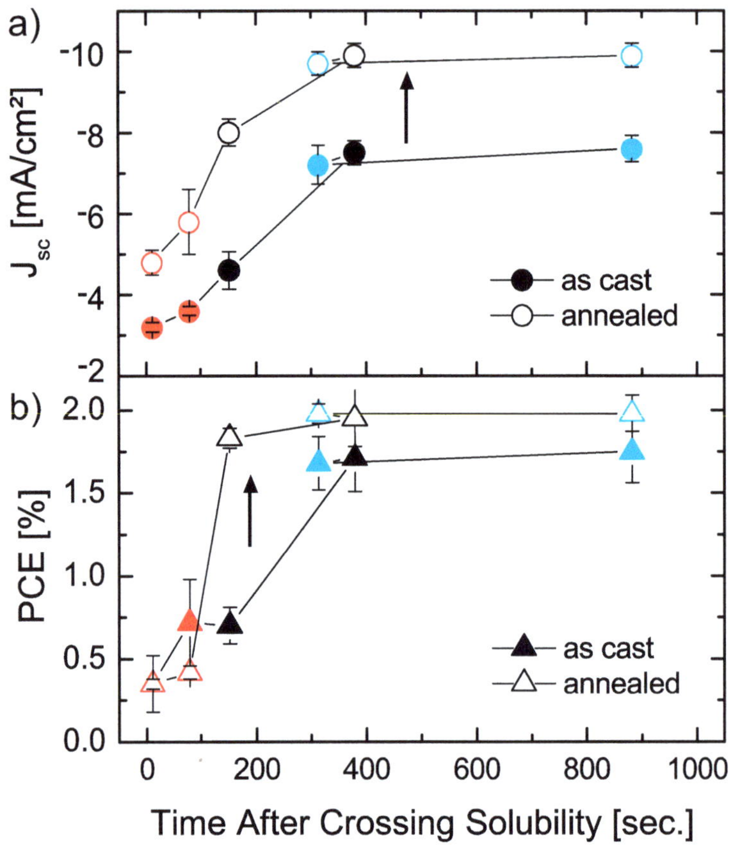

Fig. 10 (**a**) Short-circuit current density and (**b**) power conversion efficiency of the untreated (*filled symbols*) and for 5 min at 150 °C annealed (*open symbols*) solar cells plotted against the drying time after crossing P3HT solubility for films dried at 15, 25, 40 °C drying temperatures at a nitrogen flow of 0.15 and 0.5 m/s [7]

solid-like P3HT aggregates in the dispersion and, therefore, the acceleration of polymer aggregation [7].

4 Drying of PSBTBT:$PC_{71}BM$

The PCE of P3HT:PCBM solar cells is limited to maximum values of approximately 4–5 % [40]. This limit can be overcome by introducing low-bandgap polymers that are able to absorb light also at higher wavelengths above 600 nm and therefore cover a higher fraction of the solar spectrum [41]. The use of the asymmetric fullerene derivative $PC_{71}BM$ further promotes the efficiency because of an increased

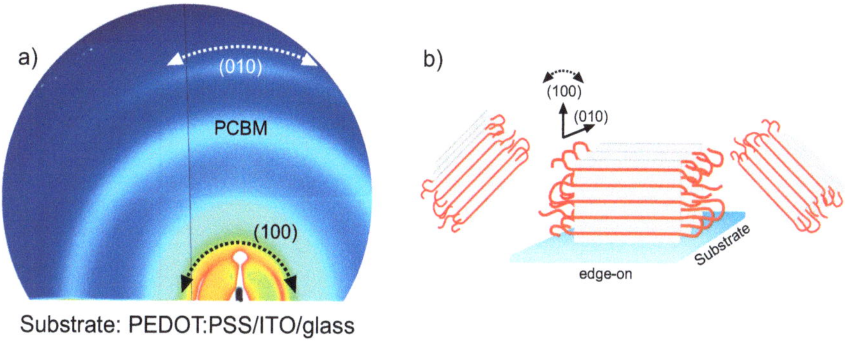

Fig. 11 (**a**) Grazing-incidence X-ray diffraction pattern of PSBTBT:$PC_{71}BM$ cast from dichlorobenzene. (**b**) Schematic representation of the edge-on configuration of PSBTBT crystallites [36]

absorption in the visible region compared to PCBM. For PSBTBT, one of the earlier synthesized low-bandgap polymers for photovoltaics, PCEs of 5.5 % have been reported at a blending ratio of 1:2 PSBTBT:$PC_{71}BM$ in single devices [42–46]. The polymer is characterized by a neutral-color absorption in the visible spectrum, making it an ideal candidate for the manufacturing of semitransparent solar cells [47]. In tandem devices, which comprise two active layers harvesting sunlight in complementary regions of the solar spectrum, efficiencies of 5.8 % have been reported with the use of PSBTBT [48, 49]. This concept was expanded to ternary blends consisting of two photoactive polymers and a fullerene component. Consequently, the interplay between processing and morphology in these blends is even more complex [50–52].

Figure 11 depicts a 2D GIWAXS diffraction pattern of PSBTBT:$PC_{71}BM$ cast from o-DCB showing the diffraction features commonly observed for this system: the (010) diffraction scattering ring attributed to the π–π stacking, the (100) peak indicative of the lamellar organization, and the prominent broad diffraction scattering ring from amorphous $PC_{71}BM$. In contrast to P3HT, blends with PSBTBT exhibit a broader orientation distribution, being beneficial for charge transport into all directions of the photoactive layer. This chapter highlights similarities of the P3HT:PCBM and PSBSBT:$PC_{71}BM$ systems, as we try to present a more generalized description of the drying process of photovoltaic polymer:fullerene blends.

4.1 *In situ Study of the Crystallization of PSBTBT:$PC_{71}BM$*

The structural evolution during drying in o-DCB has been investigated for PSBTBT:$PC_{71}BM$ with the reportedly optimal weight ratio of 1:2. As in the preparation of a P3HT:PCBM casting solution, PSBTBT:$PC_{71}BM$ is dissolved

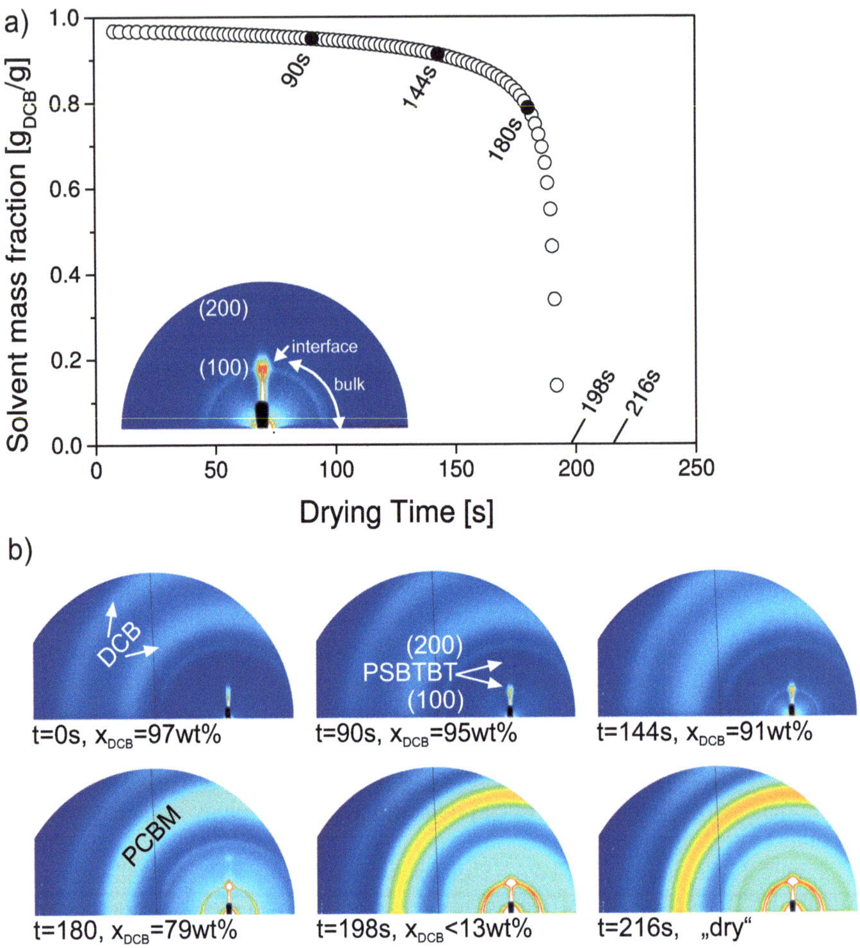

Fig. 12 Evolution of (**a**) solvent mass fraction and (**b**) grazing-incidence X-ray diffraction (GIXD) patterns during the drying of PSBTBT:PC_{71}BM (1:2) film cast from dichlorobenzene at 40 °C. The *filled symbols* in the drying curve and the stated time correspond to the GIXD images [36]

at a low solid content (PSBTBT:PC_{71}BM:DCB = 1:2:97 wt%). This leads to comparable drying kinetics during the initial constant-rate period, where only the properties of the solvent govern the drying process. Similarly, the drying kinetics are characterized by a long residence time at a high solvent mass fraction with a steep decrease at the end toward a low solvent content, as depicted in Fig. 12a. After this decrease the solution enters the falling-rate period, where the diffusional mass transport is the dominating mechanism.

Figure 12b summarizes the structural evolution of the drying blend. At $X_{DCB} = 95$ wt% the onset of (100) polymer aggregation is observed. The spot-like shape of the (100) peak indicates that crystallites lie parallel to the substrate surface with an edge-on orientation. As for P3HT:PCBM, this suggests the onset of nucleation in the vicinity of an interface, either the substrate or the surface of the drying solution. In contrast to P3HT:PCBM, this material system shows the concomitant appearance of a sharp powder-like diffraction ring with (100) spacing after 108 s, indicating that a significant portion of the polymer component crystallizes under a random orientation (see caption in Fig. 12a). Thus, different from P3HT, during the initial constant-rate period of drying, PSBTBT also crystallizes in the bulk solution, giving evidence of strong intermolecular interactions between PSBTBT molecules as reported in previous studies [46]. During the drying, the (100) crystallinity increases (from both the bulk and the interface). The crystallization rate significantly increases after 180 s ($X_{DCB} \approx 79$ wt%), when the strong $PC_{71}BM$ diffraction signals appear, indicating the aggregation of $PC_{71}BM$ [36]. This suggests a certain molecular miscibility of $PC_{71}BM$ with PSBTBT, as $PC_{71}BM$ is expected to aggregate earlier [53]. A (010) diffraction ring also emerges, revealing π–π-stacked aggregates of PSBTBT randomly oriented in the film. Figure 13 summarizes the crystallization process.

4.2 *Effect of the Drying Temperature*

A broad orientation distribution of polymer crystallites and the development of π–π stacking turned out to be beneficial for photoactive layers consisting of P3HT:PCBM. For PSBTBT the π–π stacking is visible by the appearance of a (010) ring (Fig. 12) and a pronounced absorption shoulder at 760 nm in the optical absorption spectrum (Fig. 14b). Studying the effect of the drying temperature on the film formation [36] has shown that the drying temperature does not significantly influence either the crystallinity or the orientation distribution of PSBTBT. The independence from the casting temperature is visible in the absorption spectra

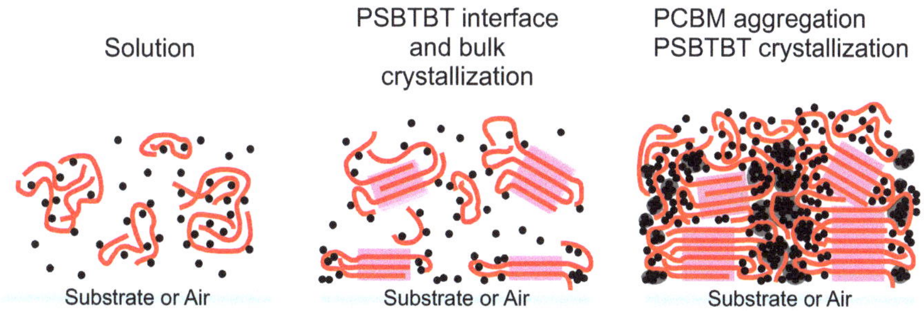

Fig. 13 Schematic of the drying process for PSBTBT:$PC_{71}BM$. Unlike P3HT, there is coexisting nucleation of PSBTBT in solution and at the interface ($X_{DCB} \approx 95$ wt%), followed by a stage of $PC_{71}BM$ aggregation ($X_{DCB} \approx 80$ wt%) and rapid crystallization of PSBTBT

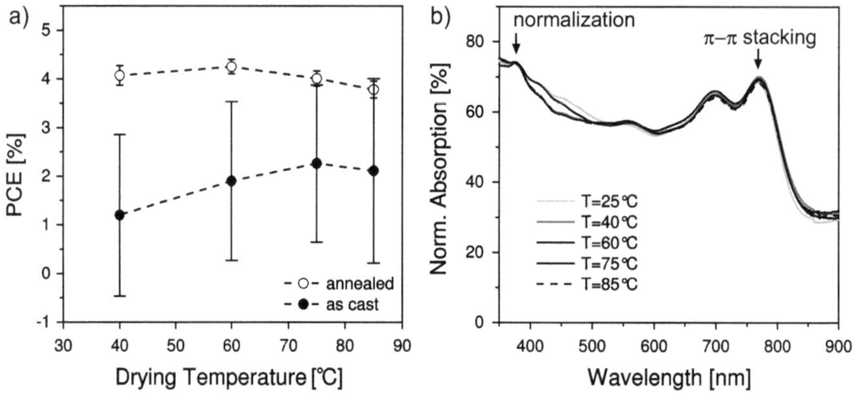

Fig. 14 Effect of the drying temperature and annealing on PSBTBT:$PC_{71}BM$ (**a**) device performance and (**b**) absorption spectra. The latter are normalized to the $PC_{71}BM$ absorption peak at 375 nm [36]

(Fig. 14a). Similarly, postannealing treatment does not cause appreciable changes in the film structure [36].

The comparison of PCEs of as-cast and annealed solar cells is shown in Fig. 13a for different drying temperatures ranging from 40–85 °C. The PCE of the as-casted films is not significantly affected by the drying temperature. There is a large dispersion in the PCE value obtained for OPV devices before annealing, which is mainly caused by interface effects; these are reduced after thermal treatment.

As a consequence, PSBTBT:$PC_{71}BM$ proves to be a photoactive blend whose film formation is largely independent of the drying kinetics or processing conditions. This is attributed to the early polymer aggregation in solution, plausibly giving rise to a rapid π–π assembly. The relative insensitivity of the nanomorphology formation to the drying conditions is possibly a result of the strong interaction of PSBTBT molecules in solution as proposed earlier by Chen et al. [45]. This enables manufacturing of cells at a constant efficiency, even for inhomogeneous drying conditions, making PSBTBT:$PC_{71}BM$ a good material for the large-scale production of 4.5–5 % PCE solar cells [2, 5, 25, 28, 29].

5 Conclusions

We have shown that real-time studies of the film formation during drying by combined laser reflectometry and GIWAXS provide fundamental knowledge of how the components may self-organize at different stages in the ternary phase diagram (i.e., the solvent, donor material, and acceptor material). Furthermore, the influencing parameters or kinetic factors for controlled film morphology during drying can be identified to improve the solar cell performance.

In this review we have shown that the structure formations for P3HT:PCBM and PSBTBT:$PC_{71}BM$ are remarkably different. For P3HT:PCBM blends, P3HT crystallization starts at the interface and is followed by crystallization in the bulk of the wet film and by the late aggregation of PCBM in the final stage of drying. We identify the intermediate stage of bulk crystallization during drying to be particularly significant with regard to the π–π self-organization of the polymer. We have found evidence of strong polymer–fullerene interactions that impede the crystal growth of PCBM and disrupt the π–π stacking of P3HT. Sufficiently slow drying of P3HT–PCBM at low temperatures allows the polymer molecules to assemble and reach a favorable nanomorphology driven by the slow organization of P3HT in solution and the spatially limited aggregation of $P_{61}CBM$. For PSBTBT, the predominant mechanism for BHJ nanoscale morphology seems to be crystallization of the polymer, which takes place in the bulk of the solution from the earliest stage. There is coexisting nucleation of PSBTBT in solution and at the interface, followed by a stage of $PC_{71}BM$ aggregation ($X_{DCB} \approx 80$ wt%) and rapid crystallization of PSBTBT. The final nanomorphology of PSBTBT:$PC_{71}BM$ is not significantly influenced by drying conditions such as temperature and drying gas velocity. Since short drying times are not detrimental for the optoelectronic device properties of PSBTBT, these materials should be advantageous for fast production speeds.

Although the number of in situ studies represents a small percentage of the overall literature about the characterization of BHJs, the importance of gaining an understanding of the dynamics of crystallization can be recognized in the increasing number of in situ studies published in recent years [16, 28, 53–62]. It is expected that a combination of complementary techniques capable of gaining insight into the nanomorphology formation at different length- and timescales will offer a better understanding of the nature of the different phases formed (a mixture of polymer-rich, fullerene-rich, and mixed polymer–fullerene domains) and their implications for charge transport and charge generation. Moreover, dynamic studies are of general applicability to a wide range of solution-processed materials and systems beyond OPV research.

6 Acknowledgments

We thank the German Research Foundation (DFG) for funding our project within the Priority Program 1355, "Elementary Processes of Organic Photovoltaics" (BA 3772/1-1, RU 342/25-2, BA 3772/1-2, BA 3772/1-3). We acknowledge the European Synchrotron Radiation Facility (ESRF) for providing synchrotron radiation facilities and Alexei Vorobiev and Oleg Konovalov for their support. We thank Ralf Weigel for his support in the MPI–MF beamline of the Synchrotron Radiation Facility ANKA (Karlsruhe). We thank all the coworkers involved in this project, in particular, Prof. Dr. Helmut Dosch, Dr. Erik Ahlswede, Prof. Dr. Sigurd Höger, Dr.

Alexander Colsmann, Dr. Felix Pasker, Dr. Jonas Hanisch, Dr. Andreas Bauer, Dr. Aina Quintilla, and Dr. Carmen Munuera.

References

1. Liao H-C et al (2013) Additives for morphology control in high-efficiency organic solar cells. Mater Today 16(9):326–336
2. Liu F et al (2013) Characterization of the morphology of solution-processed bulk heterojunction organic photovoltaics. Prog Polym Sci 38(12):1990–2052
3. Treat ND, Chabinyc ML (2014) Phase separation in bulk heterojunctions of semiconducting polymers and fullerenes for photovoltaics. Annu Rev Phys Chem 65:59–81
4. Clarke TM et al (2008) Free energy control of charge photogeneration in polythiophene/fullerene solar cells: the influence of thermal annealing on P3HT/PCBM blends. Adv Funct Mater 18(24):4029–4035
5. Chen F-C et al (2010) Morphological study of P3HT:PCBM blend films prepared through solvent annealing for solar cell applications. Sol Energy Mater Sol Cells 94(12):2426–2430
6. Liu X et al (2012) Solvent additive control of morphology and crystallization in semiconducting polymer blends. Adv Mater 24(5):669–674
7. Schmidt-Hansberg B et al (2011) Moving through the phase diagram: morphology formation in solution cast polymer-fullerene-blend films for organic solar cells. ACS Nano 5(11):8579–8590
8. Schmidt-Hansberg B et al (2009) In situ monitoring the drying kinetics of knife coated polymer-fullerene films for organic solar cells. J Appl Phys 106(12):124501
9. Sanyal M et al (2011) In situ X-ray study of drying-temperature influence on the structural evolution of bulk-heterojunction polymer-fullerene solar cells processed by doctor-blading. Adv Energy Mater 1(3):363–367
10. Sanyal M et al (2011) Effect of photovoltaic polymer/fullerene blend composition ratio on microstructure evolution during film solidification investigated in real time by X-ray diffraction. Macromolecules 44(10):3795–3800
11. Schmidt-Hansberg B et al (2012) Investigation of non-halogenated solvent mixtures for high throughput fabrication of polymer–fullerene solar cells. Sol Energy Mater Sol Cells 96:195–201
12. Schmidt-Hansberg B et al (2011) Spatially resolved drying kinetics of multi-component solution cast films for organic electronics. Chem Eng Process Process Intensif 50(5–6):509–515
13. Bergqvist J et al (2013) In situ reflectance imaging of organic thin film formation from solution deposition. Sol Energy Mater Sol Cells 114:89–98
14. Heriot SY, Jones RA (2005) An interfacial instability in a transient wetting layer leads to lateral phase separation in thin spin-cast polymer-blend films. Nat Mater 4(10):782–786
15. Schmidt-Hansberg B (2012) Process-structure-property relationship of polymer-fullerene bulk heterojunction films for organic solar cells: drying process, film structure and optoelectronic properties. Cuvillier, Göttingen
16. Wang T et al (2010) The development of nanoscale morphology in polymer:fullerene photovoltaic blends during solvent casting. Soft Matter 6(17):4128–4134
17. Vogt BD et al (2005) Moisture absorption into ultrathin hydrophilic polymer films on different substrate surfaces. Polymer 46(5):1635–1642
18. Eastman SA et al (2012) Effect of confinement on structure, water solubility, and water transport in Nafion thin films. Macromolecules 45(19):7920–7930
19. Baker JL et al (2010) Quantification of thin film crystallographic orientation using X-ray diffraction with an area detector. Langmuir 26(11):9146–9151

20. Muller-Buschbaum P (2014) The active layer morphology of organic solar cells probed with grazing incidence scattering techniques. Adv Mater 26(46):7692–7709
21. Rivnay J et al (2012) Quantitative determination of organic semiconductor microstructure from the molecular to device scale. Chem Rev 112(10):5488–5519
22. van Krevelen DW (1990) Properties of polymers: their correlation with chemical structure: their numerical estimation and prediction from additive group contributions. Elsevier, Amsterdam
23. Yang X et al (2004) Crystalline organization of a methanofullerene as used for plastic solar-cell applications. Adv Mater 16(9–10):802–806
24. Yao Y et al (2008) Effects of solvent mixtures on the nanoscale phase separation in polymer solar cells. Adv Funct Mater 18(12):1783–1789
25. Westacott P et al (2013) On the role of intermixed phases in organic photovoltaic blends. Energy Environ Sci 6(9):2756
26. Collins BA, Tumbleston JR, Ade H (2011) Miscibility, crystallinity, and phase development in P3HT/PCBM solar cells: toward an enlightened understanding of device morphology and stability. J Phys Chem Lett 2(24):3135–3145
27. DeLongchamp DM et al (2011) Molecular characterization of organic electronic films. Adv Mater 23(3):319–337
28. Richter LJ et al (2015) In situ morphology studies of the mechanism for solution additive effects on the formation of bulk heterojunction films. Adv Energy Mater 5(3):1400975
29. Dennler G, Scharber MC, Brabec CJ (2009) Polymer-fullerene bulk-heterojunction solar cells. Adv Mater 21(13):1323–1338
30. Ma W et al (2005) Thermally stable, efficient polymer solar cells with nanoscale control of the interpenetrating network morphology. Adv Funct Mater 15(10):1617–1622
31. Li G et al (2005) High-efficiency solution processable polymer photovoltaic cells by self-organization of polymer blends. Nat Mater 4(11):864–868
32. Mihailetchi VD et al (2006) Origin of the enhanced performance in poly(3-hexylthiophene): [6,6]-phenyl C61-butyric acid methyl ester solar cells upon slow drying of the active layer. Appl Phys Lett 89(1):012107
33. Li G et al (2007) "Solvent annealing" effect in polymer solar cells based on poly(3-hexylthiophene) and methanofullerenes. Adv Funct Mater 17(10):1636–1644
34. Shin M et al (2010) Abrupt morphology change upon thermal annealing in poly(3-hexylthiophene)/soluble fullerene blend films for polymer solar cells. Adv Funct Mater 20(5):748–754
35. Campoy-Quiles M et al (2008) Morphology evolution via self-organization and lateral and vertical diffusion in polymer:fullerene solar cell blends. Nat Mater 7(2):158–164
36. Schmidt-Hansberg B et al (2012) Structure formation in low-bandgap polymer:fullerene solar cell blends in the course of solvent evaporation. Macromolecules 45(19):7948–7955
37. Brinkmann M et al. (2014) Understanding the structure and crystallization of regioregular poly (3-hexylthiophene) from the perspective of epitaxy, vol 265. Springer, Heidelberg, pp 83–106
38. Chu C-W et al (2008) Control of the nanoscale crystallinity and phase separation in polymer solar cells. Appl Phys Lett 92(10):103306
39. Moulé AJ, Meerholz K (2008) Controlling morphology in polymer–fullerene mixtures. Adv Mater 20(2):240–245
40. Dang MT, Hirsch L, Wantz G (2011) P3HT:PCBM, best seller in polymer photovoltaic research. Adv Mater 23(31):3579–3602
41. Bundgaard E, Krebs FC (2007) Low band gap polymers for organic photovoltaics. Sol Energy Mater Sol Cells 91(11):954–985
42. Scharber MC et al (2010) Influence of the bridging atom on the performance of a low-bandgap bulk heterojunction solar cell. Adv Mater 22(3):367–370
43. Maurano A et al (2010) Recombination dynamics as a key determinant of open circuit voltage in organic bulk heterojunction solar cells: a comparison of four different donor polymers. Adv Mater 22(44):4987–4992

44. Hou J et al (2008) Synthesis, characterization, and photovoltaic properties of a low band gap polymer based on silole-containing polythiophenes and 2,1,3-benzothiadiazole. J Am Chem Soc 130(48):16144–16145
45. Chen HY et al (2010) Silicon atom substitution enhances interchain packing in a thiophene-based polymer system. Adv Mater 22(3):371–375
46. Morana M et al (2010) Nanomorphology and charge generation in bulk heterojunctions based on low-bandgap dithiophene polymers with different bridging atoms. Adv Funct Mater 20(7):1180–1188
47. Colsmann A et al (2011) Efficient semi-transparent organic solar cells with good transparency color perception and rendering properties. Adv Energy Mater 1(4):599–603
48. Chou C-H et al (2011) A metal-oxide interconnection layer for polymer tandem solar cells with an inverted architecture. Adv Mater 23(10):1282–1286
49. Sista S et al (2010) Highly efficient tandem polymer photovoltaic cells. Adv Mater 22(3):380–383
50. Ameri T et al (2013) Organic ternary solar cells: a review. Adv Mater 25(31):4245–4266
51. Koppe M et al (2010) Near IR sensitization of organic bulk heterojunction solar cells: towards optimization of the spectral response of organic solar cells. Adv Funct Mater 20(2):338–346
52. Ameri T et al (2012) Performance enhancement of the P3HT/PCBM solar cells through NIR sensitization using a small-bandgap polymer. Adv Energy Mater 2(10):1198–1202
53. Pearson AJ et al (2014) Morphology development in amorphous polymer:fullerene photovoltaic blend films during solution casting. Adv Funct Mater 24(5):659–667
54. Agostinelli T et al (2011) The role of alkane dithiols in controlling polymer crystallization in small band gap polymer:fullerene solar cells. J Polym Sci B 49(10):717–724
55. Rogers JT et al (2011) Structural order in bulk heterojunction films prepared with solvent additives. Adv Mater 23(20):2284–2288
56. Rogers JT et al (2012) Time-resolved structural evolution of additive-processed bulk heterojunction solar cells. J Am Chem Soc 134(6):2884–2887
57. Pearson AJ, Wang T, Lidzey DG (2013) The role of dynamic measurements in correlating structure with optoelectronic properties in polymer:fullerene bulk-heterojunction solar cells. Rep Prog Phys 76(2):022501
58. Engmann S et al (2015) Real-time X-ray scattering studies of film evolution in high performing small-molecule-fullerene organic solar cells. J Mater Chem A 3(16):8764–8771
59. Chou KW et al (2013) Spin-cast bulk heterojunction solar cells: a dynamical investigation. Adv Mater 25(13):1923–1929
60. Schmidt K et al (2014) A mechanistic understanding of processing additive-induced efficiency enhancement in bulk heterojunction organic solar cells. Adv Mater 26(2):300–305
61. Smilgies D-M et al (2013) Look fast: crystallization of conjugated molecules during solution shearing probed in-situ and in real time by X-ray scattering. Phys Status Solidi RRL 7(3):177–179
62. Wei Chou K et al (2014) Late stage crystallization and healing during spin-coating enhance carrier transport in small-molecule organic semiconductors. J Mater Chem C 2(28):5681–5689

Organic and Hybrid Solar Cells Based on Well-Defined Organic Semiconductors and Morphologies

Amaresh Mishra, Volker Schmidt, René A.J. Janssen, and Peter Bäuerle

Contents

A. Mishra (✉) • P. Bäuerle
Institute of Organic Chemistry II and Advanced Materials, University of Ulm, Albert-Einstein-Allee 11, 89081 Ulm, Germany
e-mail: amaresh.mishra@suniv.ac.in; peter.baeuerle@uni-ulm.de

V. Schmidt
Institute of Stochastics, Ulm University, Helmholtzstr. 18, 89069 Ulm, Germany
e-mail: volker.schmidt@uni-ulm.de

R.A.J. Janssen
Molecular Materials and Nanosystems, Eindhoven University of Technology, P.O. Box 513, 5600 MB Eindhoven, The Netherlands
e-mail: r.a.j.janssen@tue.nl

K. Leo (ed.), *Elementary Processes in Organic Photovoltaics*,
Advances in Polymer Science 272, DOI 10.1007/978-3-319-28338-8_2

Abstract Organic and hybrid bulk heterojunction (BHJ) solar cells are investigated. We describe the synthesis and solar cell characteristics of well-defined functional thiophene dendrimers. Three-dimensional morphologies of the polymer–metal oxide BHJs are analyzed with electron tomography and stochastic models, and are simulated with the latter, to establish the effect of processing on morphology. Device models based on stochastically simulated data of morphologies and transport parameters are used to provide accurate descriptions of solar cell performance.

Keywords Bimolecular recombination • Bulk heterojunction • Charge carrier mobility • Charge transport • Energy gap • External quantum efficiency • Internal quantum efficiency • Modeling • Oligothiophene • Phase separation • Photoactive layer • Protecting group • Self-organization • Simulation • Stochastic • Tomography

1 Introduction

Polymer-based bulk heterojunction (BHJ) solar cells are considered one of the most promising photovoltaic technologies in producing large-area and low-cost photovoltaic systems. The significant improvement in performance witnessed in recent years mainly resulted from the continuous development and improvement of new organic semiconducting materials that provide better overlap with the solar spectrum, higher charge carrier mobility, and improved film morphologies for charge generation, separation, and collection. Solar cells based on polymer–fullerene blend junctions have recently reached power conversion efficiencies (PCEs) over 10 % [1].

In parallel to developing conjugated polymers for solar cell applications, interest in developing solution-processable, small-molecule–based solar cells has increased over the past 5 years. In comparison to polymeric materials, small molecules have distinct advantages in terms of straightforward synthesis, defined structures, and purity, which greatly improve the fabrication reproducibility and avoid batch-to-batch variations. The best single-junction, solution-processed, small-molecule–based solar cells also reach PCEs close to 10 % [2].

It is widely accepted that the performance of organic solar cells critically depends on the three-dimensional (3D) nanomorphology of their photoactive layers, which is influenced by various process parameters. Presently, the choice of appropriate process parameters required to balance phase separation for charge generation and, at the same time, for charge transport is a challenging multiparameter optimization problem that is solved in a purely empirical way. This is unsatisfactory, but quantitative approaches based on a systematic understanding of the influence of the 3D nanostructure on performance are missing. One reason for this situation is that the experimental determination of the 3D nanostructure of photoactive layers with two organic (donor–acceptor) semiconductors via electron tomography is hampered by the low contrast between the two organic phases.

Progress in PCE compared with the present state of the art is possible by further reducing energetic and recombination losses that presently limit performance. For single-junction solar cells, two main goals can be identified:

1. development of pure, well-defined organic semiconductors
2. rational design of donor–acceptor morphologies

In this chapter we provide an overview of recent developments toward these goals, with an emphasis on results that were obtained in our collaborative research project in which chemistry, materials science, physics, and mathematics were combined to investigate organic solar cells based on blends of well-defined thiophene dendrimers with fullerenes and hybrid solar cells based on π-conjugated polymers and metal oxides.

2 All-Thiophene Dendrimers

Functional oligothiophenes have attracted comprehensive attention among researchers all over the world and have actually been advanced to be among the most frequently used π-conjugated materials, in particular as active components in organic electronic devices and molecular electronics [3].The traditional linear π-conjugated oligothiophenes have recently been extended to 3D dendritic and hyperbranched structures. In this respect, a multitude of various functional molecular architectures have been established and characterized.

Thiophene chemistry is well established, and there are numerous methods to modify the core molecule [4]. Most importantly, thiophenes are ideal building blocks in transition metal-catalyzed cross-coupling reactions, which have been enormously developed in the past 30 years and nowadays provide the basis for the synthesis of the great variety of conjugated π-systems. Oligothiophene-based materials possess outstanding chemical and physical properties, and their electronic properties can be tuned by enormous structural variations. They are typically stable in various oxidation states and can be readily characterized by many methods. Their unique electronic, optical, and redox properties are intriguing, as are their unique self-assembling properties on solid surfaces or in the bulk. Finally, the high polarizability of sulfur atoms in thiophene rings leads to a stabilization of the conjugated chain and to excellent charge-transport properties, which are two of the most crucial assets for applications in organic and molecular electronics.

Three-dimensional branched architectures, generally termed dendrimers, represent a class of synthetic macromolecules that have dramatically impacted the field of organic and polymer chemistry and created a new branch in synthetic and material chemistry. These dendritic macromolecules can be synthesized by either divergent or convergent approaches. The divergent method comprises the sequential addition of building blocks or repeating units to the central multifunctional core. The buildup of the dendrimers goes toward the periphery via protection and deprotection techniques. In contrast, in the convergent method the building blocks are first

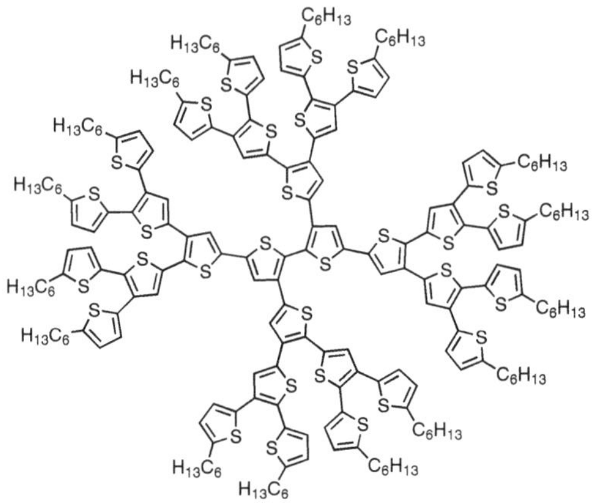

Fig. 1 Third-generation *all*-thiophene dendrimers reported by Advincula et al. [10]

constructed stepwise with different generations and are then attached to a specific functional core unit. Functionalization of dendrimers with oligothiophenes at the core or periphery as well as purely thiophene-based dendrimers was materialized just recently.

2.1 *Thiophene Dendrimers Functionalized at the Periphery*

Over the past years, a variety of dendrimers have been constructed with different core, periphery, and branching units, with flexible and rigid structures. Conjugated, rigid, and shape-persistent dendritic structures were constructed comprising phenylacetylene [5, 6], phenylenevinylene [7], or exclusively phenylene units or phenylene–thiophene systems [8, 9]. Purely thiophene-based dendrimers and dendrimers with oligothiophenes at the core or periphery are of a more recent date. The first examples of *all*-thiophene dendrimers were reported by Advincula et al. [10] in 2002. They synthesized dendritic oligothiophenes (DOTs) consisting of 30 thiophene units [up to a third generation (*G3*), Fig. 1] by a convergent approach using transition metal-catalyzed Kumada- and Stille-type couplings. The primary building block used was an α,β-branched terthiophene, in which hexyl side chains are attached at the periphery to improve solubility in common organic solvents.

Bäuerle's group recently developed an effective approach to novel functionalized DOTs with the aim of creating 3D semiconducting materials that can be further functionalized at the periphery with, for instance, dyes, redox active, self-organizing, or biological groups. These functional dendrons can also be easily attached to core structures of interesting geometric and electronic properties. As a basic building block, trimethylsilyl (TMS)-protected branched 2,2′:3′,2″-terthiophene was used to directly allow selective reactions at the free α-position

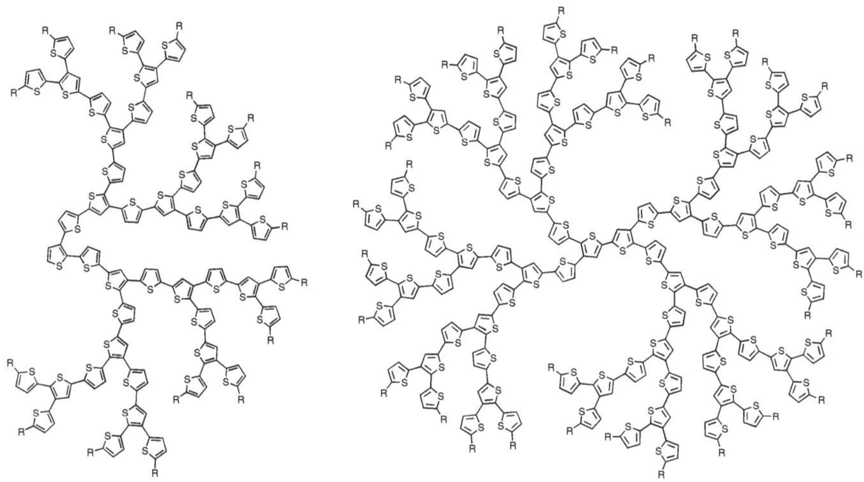

45T-Si (R = Si)
45T (R = H)

90TSi (R = $SiMe_3$)
90T (R = H)

Fig. 2 Structures of dendron with 45 thiophenes and dendrimer with 90 thiophenes

to build up higher-generational DOTs [11]. In contrast to the system developed by Advincula, in our case the TMS-protecting groups allowed further transformations at the other α-positions and can be either cleaved off to give the "pure" oligothiophenes or converted to other functional groups by electrophilic *ipso*-reactions.

Via repetition of the sequential divergent–convergent synthetic protocols, the first series of protected and nonsubstituted DOTs up to the fourth generation (*G4*) were synthesized. The *G*4 dendron **45T** and dendrimer **90T** are monodisperse and highly soluble in common organic solvents (Fig. 2).

The ultraviolet–visible spectroscopy absorption profiles of the dendrimers with different generations measured in dichloromethane are shown in Fig. 3. All of these thiophene dendrimers showed intense broad and structureless absorption spectra covering 300–550 nm. With increasing generation the molar extinction coefficient increases, while the optical energy gaps (E_g) decrease from 2.5 to 2.2 eV and are within the range of linear semiconducting oligo- and polythiophenes [11–13]. In these dendritic systems the absorption profile revealed conjugation paths along the α–β connections of the chromophoric subunits. It has also been shown that the absorption spectra were a superimposition of multiple chromophores that correlate to α-conjugated oligothiophene subunits. Emission typically came from the longest chromophoric α-conjugated pathway and was invariant to the excitation wavelength. This behavior, together with the very low-fluorescence quantum yields, clearly indicated intramolecular energy transfer from shorter chromophores to the longest one, which then emits in these systems.

Solution-processed BHJ solar cells were fabricated (device architecture indium tin oxide/poly(3,4-ethylenedioxythiophene):polystyrene sulfonate/dendritic oligothiophene:[6,6]-phenyl-C_{61}-butyric acid methyl ester/lithium fluoride/

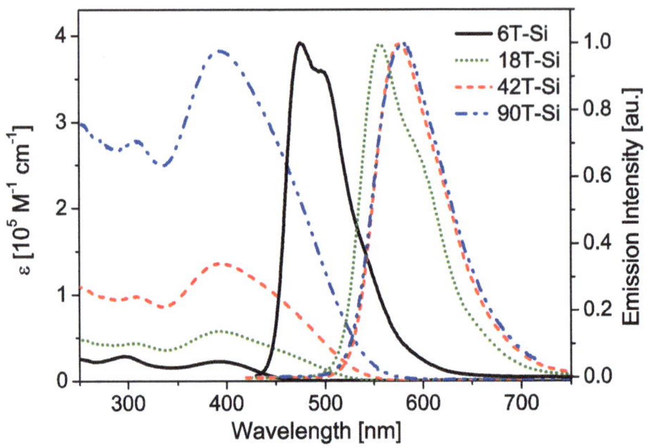

Fig. 3 Optical absorption and emission spectra of dendritic oligothiophenes with increasing generation. 6T-Si = *G*1; 18T-Si = *G*2; 42T-Si = *G*3; and 90T-Si = *G*4 (*G* = generation)

aluminum) (ITO/PEDOT:PSS/DOT:$PC_{61}BM$/LiF/Al) using these dendrimers as donor materials in combination with $PC_{61}BM$ as the acceptor [12]. BHJ solar cells for dendrimer **D1** (R = H) (Fig. 4) in a D:A ratio of 1:2 generated a PCE of 1.7 % with a high open-circuit voltage (V_{OC}) of 0.97 V and a short-circuit current (J_{SC}) of 4.19 mA/cm^2. It is important to note that for optimal performance there should be five to six thiophene units in these monodispersed DOTs per $PC_{61}BM$. Tapping-mode AFM investigation on the BHJ layers revealed smooth surfaces for the different generations with nanoscale phase separation of the dendrimer and the $PC_{61}BM$ [12]. \vspace*{-6pt}

A hole mobility around 10^{-5} cm^2/Vs for these pristine dendrons and dendrimers with 21 or 42 thiophene units measured via the charge extraction by linearly increasing voltage (CELIV) technique was found to be relatively invariant, indicating that the shape and size of the studied dendrimers have little influence on their charge-transport properties. The hole mobility was decreased by a factor of 3–10 for the TMS-protected molecules, which was attributed to the insulating effect of the peripheral TMS groups reducing charge-transfer rates between neighboring dendritic molecules. The decreased hole mobility in the blend and a rather short charge carrier lifetime measured by photo CELIV are thought to be the limiting factors for the overall PCE in these devices [14].

2.2 *Thiophene Dendrimers Functionalized in the Branches or at the Core*

We have also functionalized the dendrimers by introducing ethynyl units in the branching points. Investigations of the optical properties revealed that, in comparison to *all*-thiophene dendrimer **D1**, the introduction of ethynyl groups in the

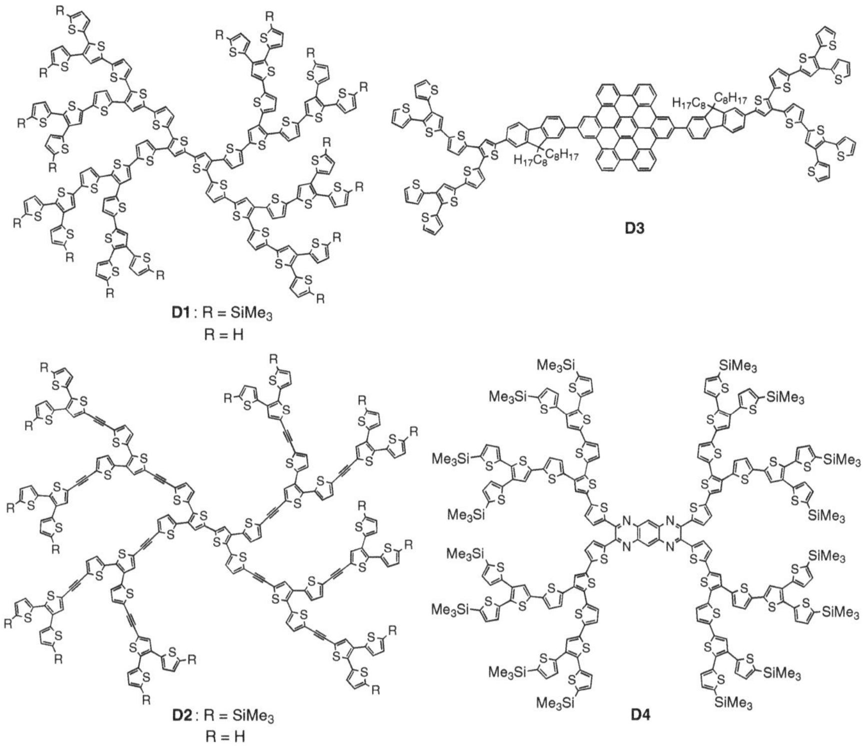

Fig. 4 Structures of **D1**, **D2**, **D3**, and **D4**

branching units in **D2** caused a hypsochromic shift of the low-energy absorption band (Fig. 4) [15]. The highest occupied molecular orbital (HOMO) energy level of **D2** ($R = SiMe_3$) comprising ethynyl units was around −5.6 eV, which is about 0.3 eV lower than **D1** ($R = SiMe_3$) (−5.3 eV). BHJ solar cells were prepared with dendrimer **D2** ($R = SiMe_3$) as donor and $PC_{61}BM$ as acceptor in a D:A blend ratio of 1:4, giving rise to a PCE of 0.4 % and a V_{OC} of 0.81 V. In a similar D:A ratio, dendrimer **D1** ($R = SiMe_3$) generated a higher PCE of 1.0 % because of higher J_{SC} and V_{OC} values. The lower J_{SC} for the device based on ethynylated dendrimer **D2** compared to **D1** could be caused by unbalanced charge carrier mobility of the former, which was also reflected in the lower external quantum efficiency (EQE) value of 17 % for **D2** compared to 45 % for **D1**.

Wong et al. [16] and our group synthesized a series of hexa-*peri*-hexabenzocoronene (HBC) derivatives comprising dioctylfluorenyl moieties end-capped with oligothiophene dendrons and implemented as donor material in the active layer of BHJ solar cells. The HBC derivatives showed self-association behavior into ordered structures in solution and in the solid state. BHJ solar cells fabricated using HBC derivative **D3** (Fig. 4) as electron donor and $PC_{61}BM$ as electron acceptor (1:2, w/w) gave PCEs of up to 1.5 %. This study clearly demonstrated the positive effect of molecular self-organization in the solid state on device performance. The PCE of devices based on **D3** was further increased to 2.5 % using [6,6]-phenyl-C_{71}-butyric acid methyl ester ($PC_{71}BM$) as acceptor because of an increase in the J_{SC} value to 6.4 mA/cm^2.

In order to extend the absorption range of DOTs, we introduced a strong electron-accepting pyrazino[2,3-*g*]quinoxaline core into the dendritic structure in **D4** (Fig. 4) [17]. Compared to thiophene dendrimer **D1**, **D4** showed a broad absorption band covering 300–700 nm and a reduced bandgap of 1.7 eV. In BHJ solar cells, low-bandgap dendrimer **D4** gave a PCE of 1.3 % in combination with $PC_{61}BM$ as acceptor, whereas a PCE of 1.0 % was obtained with all-thiophene dendrimer **D1** (R=$SiMe_3$). Using $PC_{71}BM$ as acceptor, devices based on **D4** gave a PCE of 1.7 % and a high V_{OC} of 0.98 V. This study showed that the longest-wavelength intramolecular charge-transfer band contributes to the EQE spectrum, which was extended up to 750 nm, and consequently to the photocurrent. Recently, desilylated dendron **D4** containing a pyrazino[2,3-*g*]quinoxaline core was also synthesized, exhibiting PCEs of 0.9 % by blending with $PC_{61}BM$ and 1.6 % by blending with $PC_{71}BM$ as acceptor, respectively [18].

Furthermore, dendrons were functionalized with pyridinium units. For example, dendron **D5** (Fig. 5) showed absorption bands at 390 and 535 nm that were assigned to the oligothiophene dendron and to the corresponding charge-transfer absorption arising from the oligothiophene donor and pyridinium acceptor, respectively [19]. A pyridine unit was used to form complexes with phthalocyanines (Pc), as realized in the case of **D6** (Fig. 5) [20]. It has been shown that the absorption band of the Pc–dendron complexes increased at higher energies when the dendron generation increased. However, the HOMO energies (~ -5.1 eV) were found to be independent of the dendron size. Implementation of the functional dendrons **D5** and **D6** in BHJ solar cells generated PCEs of 0.45 % and 0.7 %, respectively, in combination with $PC_{61}BM$ as acceptor. The devices based on Pcs, which comprise a 3T-dendron, gave PCEs of 1.0 % with $PC_{61}BM$ and 1.5 % with $PC_{71}BM$ as acceptor. The contribution of the Pc core in the photocurrent generation, about 40 % at 630 nm, can be clearly seen in the EQE spectrum, which corresponds to the absorption band of the Pc unit.

Ma and coworkers synthesized a dendron functionalized with squaraine dye **D7** (Fig. 5) that showed strong absorption around 680 nm and a HOMO energy level in the range of −5.11 to −5.16 eV. The bandgap of dendrimer **D7** determined by cyclic voltammetry was 1.73 eV. In planar heterojunction solar cells, PCEs up to 1.6 % have been measured at low-light condition (50 mW/cm^2).

2.3 Thiophene Dendrimers with Extended Branches

Zhang [21] designed and synthesized dendron **D8** and dendrimer **D9** functionalized in the periphery with benzothiadiazole as acceptor units (Fig. 6). The absorption feature of these molecules showed two absorption bands—one at higher energy around 330 nm that is assigned to the π–π^* transition and one at lower energy around 500 nm assigned to an intramolecular charge-transfer (ICT) band. In comparison

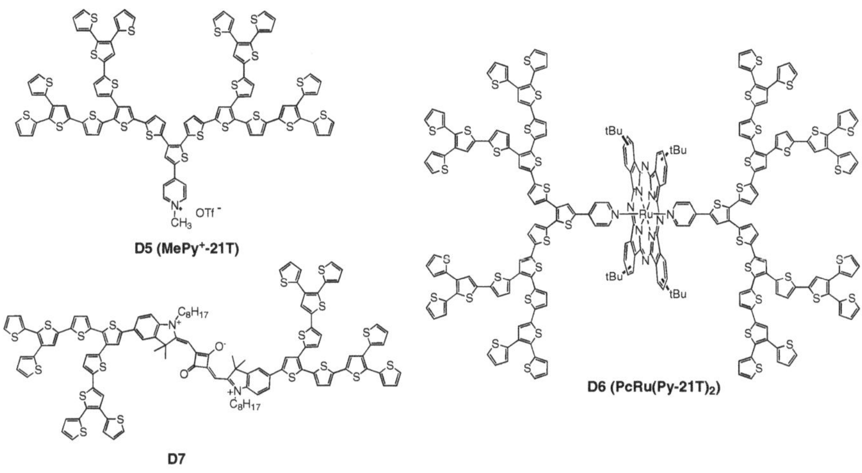

Fig. 5 Structures of **D5**, **D6**, and **D7**

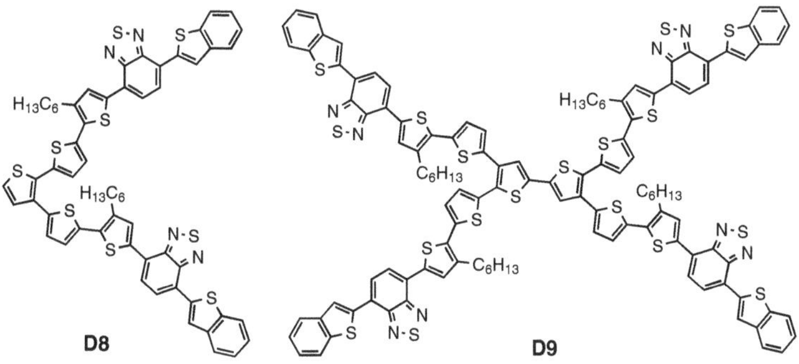

Fig. 6 Structures of **D8** and **D9**

to solution spectra, a redshift of 25 nm was observed in thin films. The optical energy gap for **D8** and **D9** was 2.1 and 2.03 eV in solution and 2.0 and 1.82 eV in thin films, respectively. The HOMO energy levels were determined to be about −5.6 and −5.3 eV, respectively. The devices [ITO/PEDOT:PSS/ DOT:$PC_{61}BM$ (1:3) (w/w)/LiF/Al] based on **D8** and **D9** as donor and $PC_{61}BM$ as acceptor gave a high V_{OC} of up to 1.04 V, J_{SC} up to 3.7 mA/cm^2, and PCEs of 1.2 % and 1.3 %, respectively [21].

Insertion of additional thiophene units into the branches of a terthiophene dendron to prepare a quinquethiophene (5T) dendron was further developed by Zhang [21]. She prepared higher-generational dendrimers up to the third generation (**D10**, **D11**, Fig. 7). The dendrimers were synthesized starting from a 5T-dendron following the similar synthetic strategies applied for the terthiophene-based dendrimers **D1**. Both **D10** and **D11** showed a broad and structureless absorption

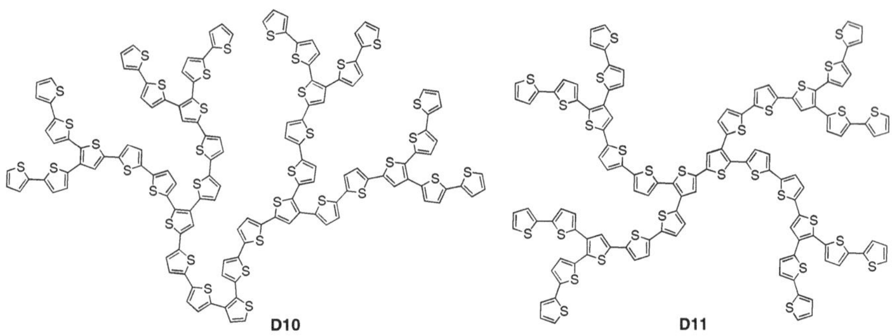

Fig. 7 Structures of **D10** and **D11**

band with a maximum around 438 nm and an optical energy gap of 2.3 eV. BHJ solar cells prepared from these dendrons and dendrimers as donor and $PC_{71}BM$ as acceptor in a weight ratio of 1:3 generated high V_{OC}s of up to 0.95 V and PCEs of 2.2 % and 3.1 %, respectively. Dendron **D10** comprising 35 thiophene units showed an EQE maximum of 53 % at 470 nm. Postprocessing by thermal annealing or solvent additives did not lead to a further improvement of the device performance.

3 Morphology and Performance of Hybrid and Polymer Solar Cells

The morphology of the photoactive layer in donor–acceptor BHJ solar cells is undoubtedly one of the most crucial parameters in determining the photovoltaic performance. The size and extent of the phase separation determine the distance over which excitons have to travel to reach the donor–acceptor interface, where they can be dissociated into charge carriers. As a result, exciton dissociation is favored when the two materials are intimately mixed. In contrast, charge separation involves diffusion and drift of holes and electrons away from the interface and benefits from delocalized wave functions in large semicrystalline domains and is therefore favored in phase-separated blends. Finally, charge collection requires that holes and electrons move via unobstructed percolation pathways to the electrodes, where they can be collected. This requires not only the presence of almost pure phases, but also the absence of cul-de-sacs in the morphology in which charges may be physically trapped. If charge separation or charge transport is sluggish, the photogenerated holes and electrons recombine via monomolecular (or geminate) and bimolecular recombination, resulting in a loss of photocurrent, fill factor, and PCE.

Koster et al. [22] developed a straightforward method to quantify the fraction of charge carriers that recombines bimolecularly under operating conditions. The experimental technique is based on measuring the differential current density ΔJ induced by a small modulation in light intensity (ΔI) in the presence of a background light intensity I. This steady-state differential current technique easily resolves nonlinear behavior that is obscured in conventional J–I measurements and enables the direct experimental determination of intensity-dependent losses. Koster et al. [22] also derived an analytical model to quantitatively relate these losses to bimolecular recombination. This model offers a direct link between the bimolecular recombination loss efficiency (η_{BR}) and the intensity dependence of the short-circuit current as $\eta_{BR} = \alpha^{-1} - 1$, where α is the exponent of the intensity (I) dependence of the short-circuit current $J_{SC} \propto I^{\alpha}$. Remarkably, it was found that bimolecular losses in BHJ solar cells can amount to tens of percentage points under solar illumination while J_{SC} vs. I seemingly remains linear.

To reach optimum performance, the typical domain size of phase-separated domains should be on the order of 10 nm as a compromise between efficient charge generation and collection. In reality, the actual morphology is often hierarchical in nature in the sense that apart from pure phases, mixed domains also exist and charge collection occurs via a network of short, narrow filaments that interconnect larger domains, much like the artery system. The desired ~10-nm size of the features presently precludes the use of top-down fabrication of the ideal morphology. Instead, morphology formation involves autonomous processes such as self-assembly, crystallization, and phase separation using the natural tendency of polymers not to mix as a consequence of a low mixing entropy. In general, morphology formation from a solution of donor–acceptor materials is a nonequilibrium process that can occur within a fraction of a second, governed by the evaporation rate of the solvent. It is often assumed that the resulting morphology is a kinetically trapped state and that postdeposition techniques such as thermal or solvent vapor annealing result in thermodynamically more stable states.

Although the importance of morphology is very well recognized as a crucial parameter in organic solar cells, many questions and challenges remain. For example, few techniques are available that allow one to determine the 3D morphology on a nanoscale. If the morphology has been measured, is it then possible to predict the device performance with a suitable physical device model? A more daunting challenge is to rationalize the relationship between processing conditions and the resulting morphology. In the next sections we address these issues for two examples: polymer–metal oxide and polymer–fullerene blends.

3.1 Polymer–Metal Oxide Solar Cells

An elegant way of fabricating hybrid polymer–metal oxide solar cells is via depositing a semiconducting polymer and a metal oxide precursor from solution in which the precursor is converted by reacting with water from the ambient to form a metal oxide network inside the polymer. An example of this method is using

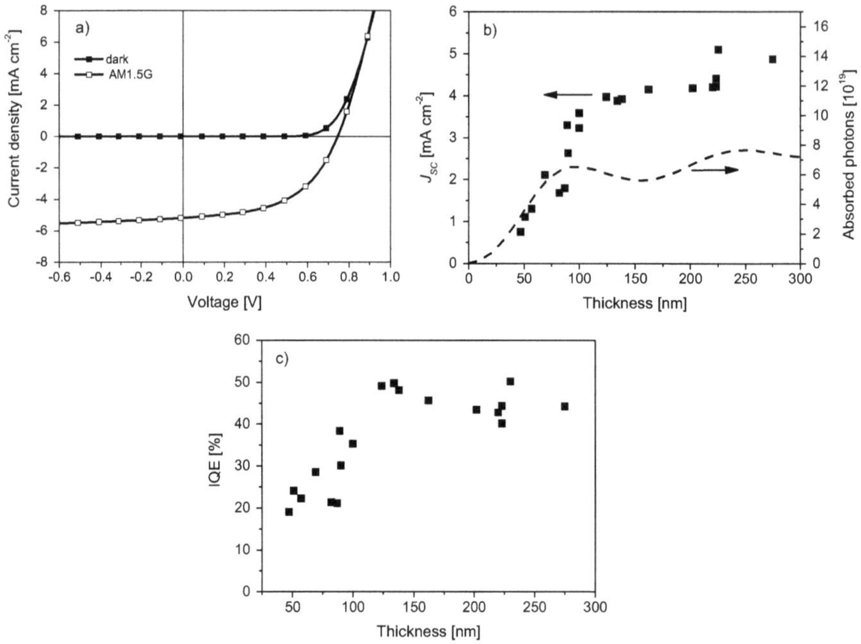

Fig. 8 Performance of P3HT:ZnO solar cells. (**a**) *J–V* characteristics. (**b**) *J* and absorbed photon flux vs. thickness. (**c**) IQE vs. thickness. The thickness was varied by altering the spin speed. (Reprinted from [23] with permission from Nature Publishing Group)

poly(3-hexylthiophene) (P3HT) and diethyl zinc as precursor for ZnO. In this way it is possible to make P3HT:ZnO solar cells with a PCE of about 2 % at an active layer thickness of 225 nm [23]. In these cells the EQE maximizes at 44 % at 520 nm.

Thicker P3HT:ZnO layers gave a higher internal quantum efficiency (IQE) than thin cells (Fig. 8). The number of photons absorbed by the P3HT:ZnO layer, calculated by optical modeling of the entire stack of layers, revealed that the IQE (Fig. 8) increases significantly with layer thickness [23]. Hence, somewhat counterintuitively, in thicker layers photons are also more efficiently transformed into collected charges. To understand this behavior, we spatially resolved the morphology of P3HT:ZnO layers in the nanometer range using electron tomography (Fig. 9). ZnO provides a good contrast with the polymer in electron microscopy, enabling a nanometer resolution. To quantify the effect of the coarseness of the experimental morphologies on the charge carrier generation of P3HT:ZnO solar cells, we modeled the exciton diffusion in the active layer. The diffusion of excitons in the polymer phase was described by the diffusion equation for excitons [23] in the experimentally determined morphologies. Figure 9 shows the calculated exciton dissociation efficiency for different layer thicknesses and reveals that exciton dissociation is far more efficient in thicker layers as a consequence of their

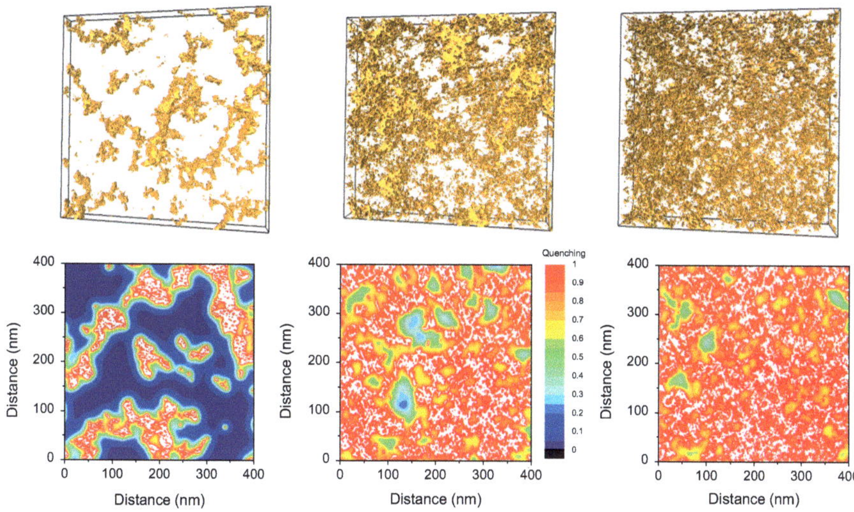

Fig. 9 *Top row*: electron micrographs of typical P3HT:ZnO blends; from left to right, the thickness is 57, 100, and 167 nm. The size of all samples is 700 × 700 nm; ZnO appears *yellow*, P3HT transparent. *Bottom row*: modeled exciton dissociation efficiency cross sections in the same blends. (Reprinted from [23] with permission from Nature Publishing Group)

finer morphology, giving the first explanation of why the IQE increases with layer thickness.

While the influence of morphology on charge generation is relatively easy to model, charge transport in disordered donor–acceptor blends is more difficult to describe. In a first study, Koster theoretically investigated the charge carrier mobility in disordered organic blends for photovoltaics by numerically solving the Pauli master equation [24]. He found that the well-known dependence of mobility on charge carrier concentration is even more pronounced in blends than in pure materials and that it is influenced by the electric field strength. At low charge carrier densities in the blend, the mobility was found to decrease with increasing field. For strongly disordered materials, charge transport is favored by relatively large domains.

In a subsequent study, Koster combined optical and charge-transport models to establish the relationship between morphology and performance in more detail and create a quantitative link between efficiency and morphology using the experimental 3D morphology of P3HT:ZnO blends and a 3D optoelectronic device model [25]. This model includes the effects of exciton diffusion and dissociation; space-charge; recombination, generation, drift, and diffusion of charge carriers; and the injection/extraction of carriers at the contacts. The observed trend in IQE as a function of layer thickness as shown in Fig. 8 was reproduced with a single set of parameters. The model reveals that isolated clusters of ZnO that are more than a few times the exciton diffusion length away from the main ZnO phase do not affect the overall device efficiency. The impact of cul-de-sacs, however, is more complex

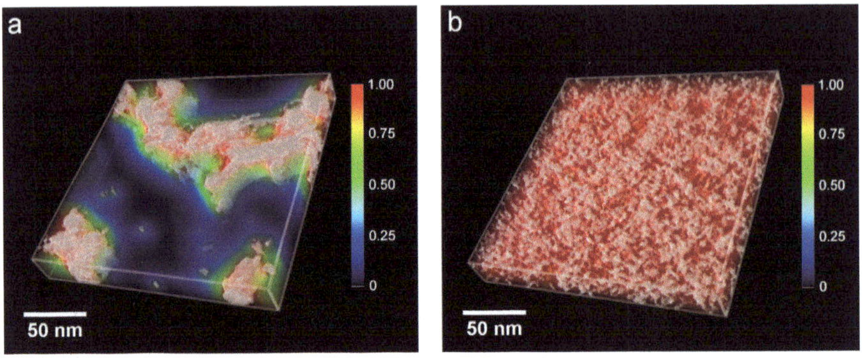

Fig. 10 Calculated charge carrier generation efficiency in experimentally determined morphologies for (**a**) P3HT:ZnO and (**b**) P3HT-E:ZnO. The ZnO appears *white*; the efficiency is indicated by the *color*. (Reprinted from [26] with permission from Wiley)

and depends on the local electric field and the cul-de-sac depth. This first direct use of morphological data in an optoelectronic device model shows that it is indeed possible to quantitatively link morphology to device performance.

In an attempt to control the morphology of conjugated polymer:ZnO solar cells by the chemical structure, we have investigated the effect of side chain functionalization [26]. Using an ester-functionalized side chain poly(3-hexylthiophene-2,5-diyl) derivative (P3HT-E), we found that the nanoscale morphology of polymer:ZnO solar cells is significantly more intimately mixed compared to P3HT:ZnO, as evidenced experimentally from a 3D reconstruction of the phase separation using electron tomography and the modeling of the charge-generation efficiency (Fig. 10). In the P3HT-E:ZnO blends, nearly quantitative charge transfer occurs. For thin P3HT-E:ZnO active layers (~50 nm) this yields a significant improvement in the solar cell performance. For thicker cells, however, the reduced hole mobility and a reduced percolation of ZnO pathways hinder charge carrier collection, limiting the PCE.

In summary, using a combination of electron tomography and electro-optical device modeling has enabled a coherent and consistent picture to emerge for the device physics of polymer–metal oxide solar cells in relation to the morphology. Although these cells are unable to compete in terms of device performance with the most efficient polymer–fullerene solar cells, the characteristics of these studies seem generally applicable. In Sect. 4.1 we will come back to these P3HT:ZnO blends and describe a stochastic morphology model and a stochastic multiscale model that are then used to create virtual 3D microstructures for different processing conditions.

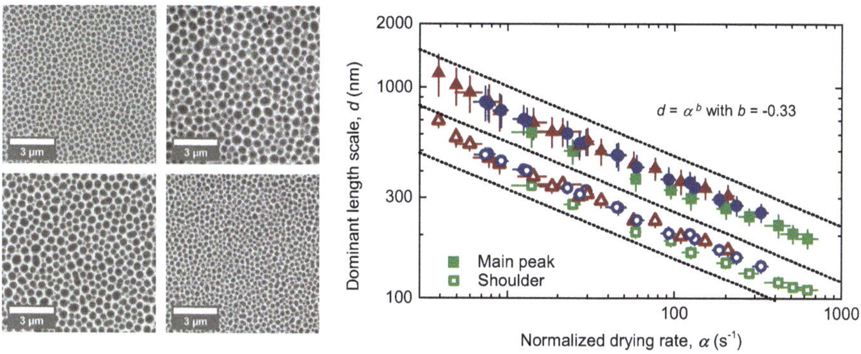

Fig. 11 *Right*: examples of morphologies obtained for different total concentrations of polymer and fullerene in chloroform. *Left*: relationship between the dominant length scale and normalized drying rate. Different initial polymer concentrations are indicated: *green squares* are 3 mg/ml, *dark blue circles* are 6 mg/ml, and *dark red triangles* are 9 mg/ml. The scaling law is valid for all concentrations. *Solid symbols* indicate the main peak position from the Fourier transform, while *open symbols* represent the location of a shoulder in the Fourier transform. (Reprinted from [28] with permission from Wiley)

3.2 Polymer–Fullerene Solar Cells

The active layer morphology in polymer–fullerene blends has been extensively studied, but relatively little information is available on the active layer formation. Van Franeker et al. [27] recently developed in situ optical techniques that allow one to determine the active layer thickness, the formation of phase separation, and the aggregation of polymer chains during spin coating with a millisecond time resolution. One of the interesting phenomena that quite often occur during spin coating of the photoactive layer in polymer–fullerene blends is a process called liquid–liquid phase separation, which results in the formation of large droplet-like fullerene domains in a matrix that consists of polymer and fullerene. An example of this morphology is shown in Fig. 11 for a mixture of a diketopyrrolopyrrole–quinquethiophene (PDPP5T) copolymer with $PC_{71}BM$ in a 1:2 weight ratio [28]. Depending on the spin rate and total concentration of polymer and fullerene, different sizes of domains can be obtained. Although these strongly phase-separated morphologies are less relevant for efficient solar cells, and they can be circumvented by the use of a cosolvent [27], it is important to understand their formation and the parameters that control length scales in phase-separated films if we want to understand morphology formation from evaporating solvents in more detail. The droplet-like features are a consequence of spinodal demixing, and we were interested in the parameters that determine the size of these domains. We therefore analyzed the drying process of spin-coated inks that undergo liquid–liquid phase separation by determining the drying rate in the evaporation phase of the spin coating. Light-scattering experiments showed that during this phase liquid–liquid demixing occurs. As such we were able to show that the size of these droplets scales with the normalized drying rate; that is, the drying rate normalized to the final thickness of the film. The scaling is valid over two orders of magnitude and has been verified for several material combinations. The normalized drying rate can be related to two possible "determinants" of domain size: (1) the quench rate, which would determine length scales by the amplification of an initial length scale by spinodal decomposition (early stage); or (2) to the coarsening time, in which the growth of already existing domains can take place (late stage, for example, Ostwald ripening).

4 Stochastic Modeling and Predictive Simulation of Morphology and Functionality of Organic Semiconductor Materials at Various Length Scales

The section focuses on stochastic modeling and predictive simulation of the morphology of organic semiconductors and its coherence with functional properties of these materials, such as the dissociation efficiency of excitons at microscopic scale or the mobility of charge carriers at molecular scale.

To begin, morphological characteristics of microscopic image data of the underlying materials are computed using tools of stochastic geometry and spatial statistics [29]. Then parametric stochastic morphology models are developed and the parameters are fitted to experimental data. Whenever a material can be produced using different processing parameters, the model can be fitted to a whole set of materials processed under different conditions, and a relationship between processing parameters and model parameters can be found. This allows so-called predictive simulations, that is, creating virtual structures on the computer that display processing parameters that have not been tested in the laboratory so far. Analyzing the efficiency of those virtual materials is a much cheaper way to investigate the results of various processing parameters than creating those materials in the laboratory. Doing this allows processing parameters to be optimised and improves functional properties of organic devices. This process is called virtual materials design.

This section discusses the stochastic morphology modeling and virtual materials design of three different types of materials. In Sect. 4.1, hybrid polymer–ZnO solar cells (to be more precise, P3HT:ZnO solar cells) are considered and their 3D morphology is described by a suitably chosen stochastic model. The model is fitted to various experimental data sets that have been produced using different spin-coating velocities. This makes it possible to use the model for predictive simulations for further spin-coating velocities. Then, in Sect. 4.2, thin polymer–fullerene films (PDPP5T:$PC_{71}BM$ in a 1:2 ratio) are considered. This material is also produced using spin coating. A stochastic 2D model is fitted to transmission electron microscopy images of polymer–fullerene films produced with seven different spin-coating velocities. The model accounts for the droplet-like fullerene parts in the films. Again, we perform regression in parameter space and show that the model is able to realize realistic morphologies for arbitrary spin-coating velocities using cross validation. Finally, in Sect. 4.3 we briefly describe results we obtained regarding morphological and electronic properties of networks of dicyanovinyl-substituted oligothiophene (DCV4T) molecules.

Additionally, in [30] we investigated the nanostructure of a further kind of organic–inorganic composites that can be used as photoactive layers in hybrid–polymer solar cells. These materials consist of a polymeric (OC1C10-PPV) phase with CdSe nanoparticles embedded therein. On the basis of 3D image data with high spatial resolution, gained by electron tomography, we developed an algorithm in [30] to automatically extract the CdSe nanoparticles from grayscale images, where we assume them to be spheres. The algorithm is based on a modified version of the Hough transform, where a watershed algorithm is used to separate the image data into basins such that each basin contains exactly one nanoparticle. Moreover, in [31] we applied techniques of spatial statistics for the characterization of the nanoscale 3D morphology of a hybrid solar cell based on a novel hierarchical nanostructured photoanode, where a cross section of the solar cell device has been prepared by focused ion beam milling in a micropillar geometry. This allows a detailed 3D reconstruction of the titania photoanode by electron tomography. It turns out that the hierarchical titania nanostructure facilitates polymer infiltration, thus favoring intermixing of the two semiconducting phases, essential for charge separation. Again, the respective 3D nanoparticle network has been analyzed using stochastic geometry tools to extract information related to charge transport in the hierarchical solar cell. In particular, the experimental data set considered in [31] allows direct visualization of the percolation pathways that contribute to the photocurrent.

4.1 Hybrid Polymer–ZnO Solar Cells

In this section, we consider the case of hybrid polymer–ZnO solar cells. In a first study [23], the relationship between the 3D morphology and functionality of thin photoactive films of this type was investigated via a descriptive statistical analysis using electron tomography image data; see Fig. 9. This includes the analysis of spherical contact distances and percolation pathways as well as solving the 3D exciton diffusion equation. Furthermore, in [26] the effect of side chain functionalization was studied. Due to this analysis, a good understanding of the complex morphology could be gained. This knowledge has been used to develop a spatial stochastic model of the 3D morphology of hybrid polymer–ZnO solar cells, which was first introduced in [32]. An overview with further mathematical background information can also be found in [33]. The stochastic morphology model is then used to perform virtual materials design and to analyze the resulting simulated 3D structures; see [34]. A more detailed analysis of the dependency between morphology and efficiency has been provided in [25]. In the following, the parametric morphology model of hybrid polymer–ZnO solar cells and the ideas

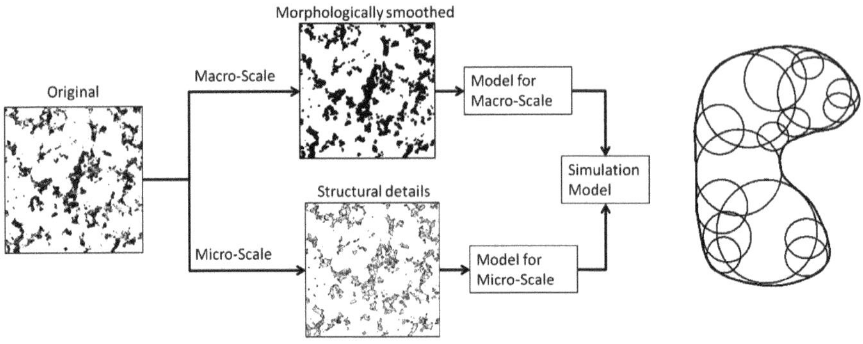

Fig. 12 *Left*: distinction between macro- and microscale morphology; *right*: representation of ZnO phase by union of overlapping spheres. (Reprinted from [35])

of predictive simulation of such 3D morphologies for virtual materials design are described in more detail.

The stochastic morphology model is built on a multiscale approach to the representation of 3D image data, which has been introduced and applied to polymer solar cells in [35]. The basic idea is sketched in Fig. 12 (left). First, the binarized image is split into two parts—a morphologically smoothed one, representing the macroscale morphology, and a second image containing structural details, which are referred to as microscale morphology and are caused, for instance, by thin ZnO branches or isolated ZnO particles. Both parts are modeled separately and combined to form a stochastic model of the whole 3D structure.

Modeling of the macroscale morphology is based on a representation of the ZnO phase by a union of overlapping spheres; see Fig. 12 (right) for an illustrative planar section. This representation of the experimental data is achieved using an efficient stochastic segmentation algorithm; see [35] for details. Note that because of this representation, the macroscale morphology can be interpreted as a marked point pattern, where the points are the midpoints of the spheres and the marks are the corresponding radii.

Modeling of such structures is possible using stochastic geometry tools. In this particular case, the point pattern is modeled using a Markov chain of planar Poisson cluster processes, where clusters of points occur in ellipsoidal form and the centers of these clusters form a two-dimensional Poisson process; that is, they do not interact with each other but possess the property of complete spatial randomness [32]. Given the point pattern, the radii are simulated. To account for the structure of the radii from the representation of the experimental data derived in [35], the distribution of radii and the so-called mark correlation function are analyzed, where the mark correlation function of a marked point process describes the correlation of pairs of marks with a given distance between the corresponding points. A positive correlation for small distances is observed. Therefore, radii are simulated using a so-called moving average procedure, where radii are first simulated independently of each other using a gamma distribution. Note that the gamma distributions form a

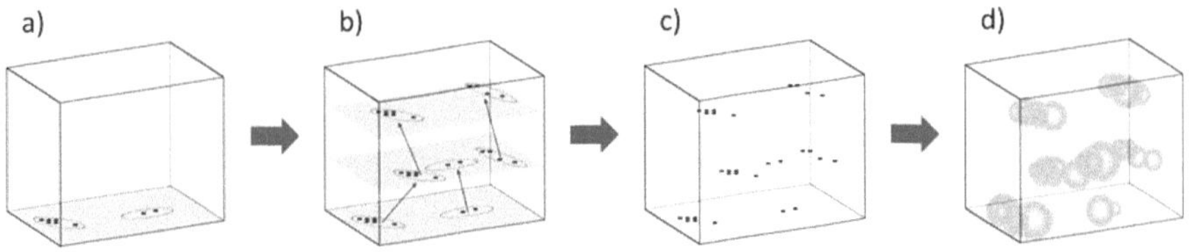

Fig. 13 Simulation algorithm. (**a**) First two-dimensional realization of cluster process. (**b**) Markov chain of two-dimensional cluster processes. (**c**) Final point pattern. (**d**) Final sphere system. (Reprinted from [34] with permission from Wiley)

two-parameter family of continuous probability distributions, of which the common exponential distribution and chi-squared distribution are special cases. Then to each point's mark the marks of a certain number of neighboring points are added. This leads to a positive correlation of radii, as desired. Furthermore, because the family of gamma distributions is closed with respect to convolution, the resulting distribution of radii is again gamma distributed, which fits the observations in the experimental data. The whole simulation algorithm for the macroscale morphology is summarized in Fig. 13.

Finally, the microscale morphology, which has been left out so far, is modeled given the macroscale morphology. As an example, thin ZnO branches and isolated ZnO particles are simulated using marked Cox processes, where the intensities and marks depend on the distance to the ZnO phase of the macroscale morphology. Again, the marks correspond to the radii of respective spheres. Detailed information can be found in [32].

The model is fitted to experimental data and validated using various morphological image characteristics. A good fit can be found; see [32] for details. In total, six different experimental data sets with different layer thicknesses that result from different spin-coating velocities have been considered. The same model type can be fitted to all six data sets, and the parameter vector of the morphology model can be interpreted as a function of the spin-coating velocity. This function is parametrized, such that using regression in the parameter space allows predictive simulations for arbitrary spin-coating velocities to be made. Doing so enables the simulation of virtual microstructures of polymer–ZnO solar cells which realistically mimic the dependence of the morphology on the spin-coating velocity. For an impression of this dependence, see Fig. 14. The stochastic multiscale model is then used to create virtual 3D microstructures for a large number of spin-coating velocities, which are subsequently analyzed with respect to exciton dissociation efficiency and connectivity. It turns out that the microstructure tends to coarsen with increasing spin-coating velocity, where larger domains of polymer and ZnO are created. As a consequence, the exciton dissociation efficiency decreases significantly with increasing spin-coating velocity.

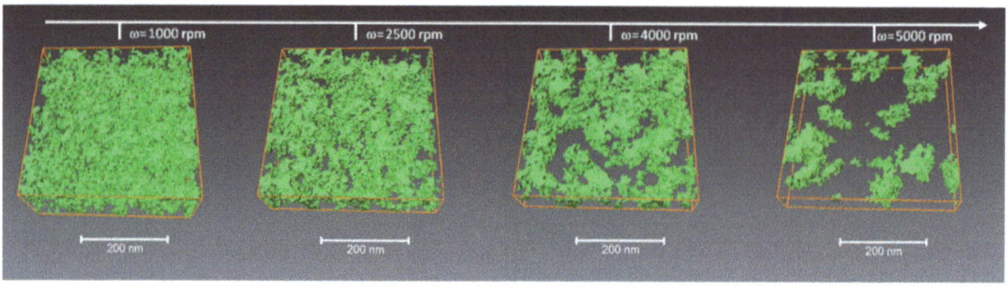

Fig. 14 Simulated realizations of the model for different spin-coating velocities

4.2 *Polymer–Fullerene Semiconductor Films*

We now turn to thin (50–500-nm) polymer–fullerene semiconductor films. In [28], stochastic geometry and spatial statistics tools helped to correlate the dominant length scale of phase separation in this material to a normalized drying rate. Again, after comprehensive statistical analysis of the data, a parametric stochastic model has been developed; see [36].

The polymer–fullerene films consist of droplet-like fullerene agglomerates [28], which are called fullerene particles or simply particles in the following. The stochastic morphology model consists of three steps. First, the midpoints of fullerene particles are simulated using tools of point-process theory [29]. Then, based on the simulated point pattern, a so-called Voronoi tessellation is considered that divides the space into disjoint cells, where each point generates exactly one cell. Note that a Voronoi cell consists of all those points that are closer to the generator point of the considered cell than to other generator points. Particles are simulated according to the size and shape of the cell corresponding to their midpoint. In the model, particles are described by either circles or the union of two overlapping circles, which fits the shape of the particles observed in the experimental data. As a third step, a slight modification of particle positions is performed using a collective rearrangement algorithm. This is necessary to ensure a minimum distance between particles, as observed in the experimental data. An overview of the simulation algorithm is given in Fig. 15.

The model parameters are fitted to experimental data processed using seven different spin-coating velocities. In each case, the model is determined by a surprisingly small number of parameters but still describes the experimental data very well. This is seen from a good visual impression as well as morphological image analysis of experimental and simulated data; see [36] for details.

Because of the huge amount of different data sets, each of the model parameters could reliably be described as a function of the spin-coating velocity. As we did with the hybrid polymer–ZnO solar cells, this allowed us to perform regression in the parameter space. In order to show the viability of this approach, we performed cross validation. We fitted the model only to six of the seven data sets and used the information gained from that to estimate the model parameters for the seventh data set using the functional correlation between the spin-coating velocity and model parameters. Realizations using the model parameters estimated in that way were compared to the corresponding experimental data set, and a good coherence was found [36]. This proves that the model is able to capture the main morphological features of the thin polymer–fullerene films for arbitrary spin-coating velocities.

4.3 Networks of DCV4T Molecules

In [37] we proposed a novel framework to study charge-transport properties of materials via so-called random walkers in disordered random media. This framework combines a stochastic 3D model for the fast generation of random environments that realistically models material morphology with an algorithm for the efficient estimation of key properties of the resulting random walk. This algorithm, called aggregate Monte Carlo (AMC), was originally introduced in [38] and is a method for estimating the velocity of a random walker as it passes through a realization of the random environment; see also [39], where a prototype of the model to estimate the charge carrier velocity in one dimension is introduced. The algorithm works by first identifying and then aggregating problem regions in the random environment. We show that we are able to do this in such a way that our estimator is strongly consistent. Our stochastic morphology model represents the disordered medium as a random spatial graph with directed edge weights, where the edge weights represent the transition rates of a Markov jump process modeling the motion of the random walker and extends the spatial graph model introduced in [40]. It makes use of spatial statistics tools and the theory of random geometric graphs. A particular strength is that it allows the efficient simulation of large-scale molecular systems.

In particular, in [37] we describe our modeling framework via an example that illustrates an amorphous mesophase of an organic dye used as an electron donor in organic solar cells; see Fig. 16. Here, the random environment represents a molecular morphology, and the random walker describes the movement of a charge

Fig. 15 *Left*: point pattern of particle midpoints. *Center*: Voronoi cells (*colored*) and corresponding particles. *Right*: final particle system after collective rearrangement

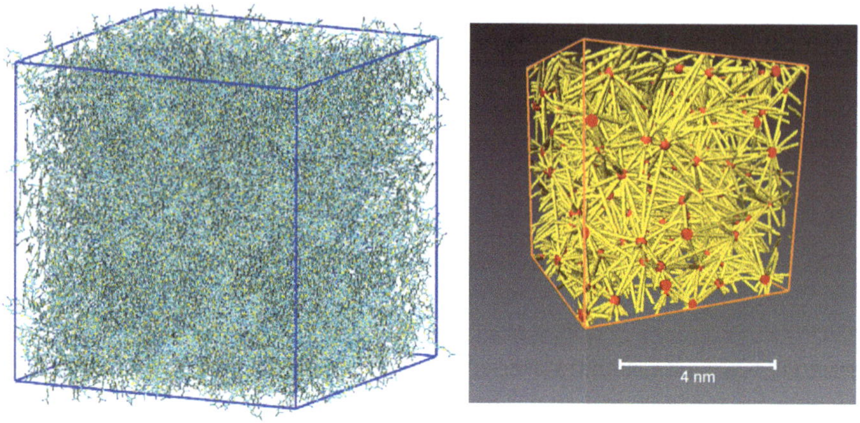

Fig. 16 *Left*: large-scale morphology of DCV4T gained by microscopic simulations. *Right*: corresponding 3D graph extracted from DCV4T morphology (cutout), where vertices are displayed in *red* and edges are marked in *yellow*. (Reprinted from [37] with permission from the Society for Industrial and Applied Mathematics)

(hole) through the morphology. More specifically, we adjust the parameters of the model to the electronic properties of DCV4T molecules in a small (microscopic) system. The fitted model can be used to predict the mobility of holes in a morphology of DCV4T molecules; see also [41]. Such a model can be of use to better understand solar cells, as it allows for larger-scale (experimentally relevant) modeling of such microscopic structures. For specific details on the microscopic simulation of DCV4T, we refer to [42, 43].

In the framework proposed in [37], the motion of the charge carrier is modeled using a Markov jump process, where the state space is a set of points. The transition probabilities depend on an underlying energy landscape that is modeled using a Gaussian random field. Due to this procedure, there are regions with very low energy in which the charge carrier becomes stuck; see Fig. 17 for a sketch. Therefore, the Monte Carlo estimation of charge carrier velocity has high variances and takes a lot of time. To reduce the number of simulation steps and decrease the variance, the ACM method is introduced. It includes a sening of the state space based on a stochastic watershed algorithm.

The stochastic approach presented in [37] is not limited to the particular organic semiconductor (DCV4T) we have used as a test system. In fact, the majority of host materials in organic light-emitting diodes have a large energetic disorder (deep traps) and are therefore challenging to model using small systems and the conventional variable step-size method; see [44–46]. The modeling of network morphology using stochastic geometry tools helps to increase the size of the system, while AMC provides an efficient way of evaluating its properties (e.g., charge carrier mobility). Moreover, the techniques developed are not restricted to organic semiconductors. Similar problems are encountered, for example, when studying

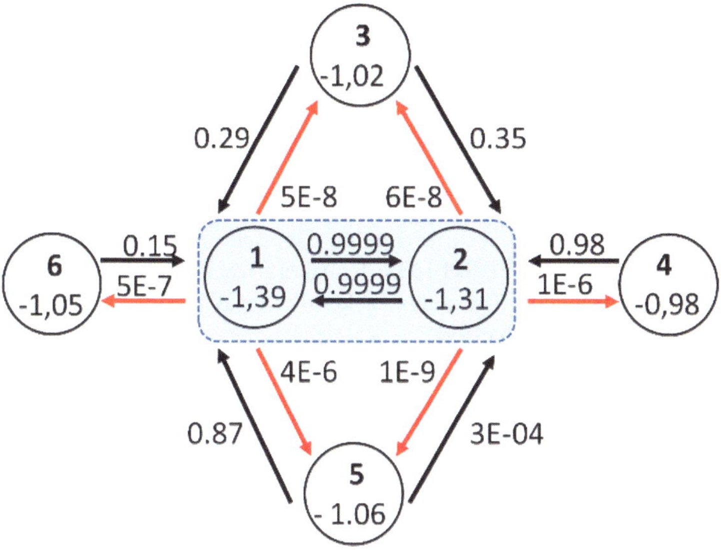

Fig. 17 Energy at different positions and corresponding transition probabilities. The carrier becomes stuck in the center with a very high probability. (Reprinted with permission from [38] from Springer Science and Business Media)

surface reactions (catalysis) [47] or hydrogen and oxygen transport in biological systems [48].

5 Conclusions

In this chapter we have addressed the design and synthesis of well-defined all-thiophene dendrons and dendrimers of different generations that can be functionalized at the periphery, in the branches, and at the core with units that extend the absorption spectrum and influence the intermolecular interactions. These functionalized thiophene dendrimers were used to fabricate BHJ organic solar cells in which fullerenes act as acceptor, reaching PCEs up to 3 %.

We have further described the relationship between morphology and device performance using experimental electron tomography data on polymer–metal oxide solar cells, using P3HT–ZnO as an example. The remarkable increase in IQE and the device characteristics of the cells can be quantitatively described using a 3D electron optical model that uses the actual nanometer-scale morphology.

In the last section we have highlighted the use of stochastic methods to analyze the morphology of the polymer–metal oxide and polymer–fullerene BHJ blends. We have further shown that with statistical models it is possible to create virtual blend morphologies that have the same characteristics as the experimentally determined

BHJs. Stochastic models have also been used to efficiently model the charge transport in Monte Carlo simulations of device performance.

6 Acknowledgments

The authors would like to acknowledge the Deutsche Forschungsgemeinschaft in the framework of the DFG SPP1355, Collaborative Research Center 569 (SFB 569), for financial support. The authors acknowledge all the coworkers who actively participated in this research effort.

References

1. Liu Y, Zhao J, Li Z, Mu C, Ma W, Hu H, Jiang K, Lin H, Ade H, Yan H (2014) Nat Commun 5:5293
2. Kan B, Li M, Zhang Q, Liu F, Wan X, Wang Y, Ni W, Long G, Yang X, Feng H, Zuo Y, Zhang M, Huang F, Cao Y, Russell TP, Chen Y (2015) J Am Chem Soc 137:3886
3. Müllen K, Wegner G (eds) (1998) Electronic materials: the oligomer approach. Wiley, Weinheim
4. Gronowitz S, Hörnfeldt A-B (2004) Thiophenes. Elsevier, San Diego
5. Moore JS (1997) Acc Chem Res 30:402
6. Gong L, Hu Q, Pu L (2001) J Org Chem 66:2358
7. Dev SK, Maddux TM, Yu L (1997) J Am Chem Soc 119:9079
8. John H, Bauer R, Espindola P, Sonar P, Heinze J, Müllen K (2005) Angew Chem Int Ed 44:2447
9. Berresheim AJ, Müller M, Müllen K (1999) Chem Rev 99:1747
10. Xia C, Fan X, Locklin J, Advincula RC (2002) Org Lett 4:2067
11. Ma CQ, Mena-Osteritz E, Debaerdemaeker T, Wienk MM, Janssen RA, Bäuerle P (2007) Angew Chem Int Ed 46:1679
12. Ma C-Q, Fonrodona M, Schikora MC, Wienk MM, Janssen RAJ, Bäuerle P (2008) Adv Funct Mater 18:3323
13. Ma C-Q, Mena-Osteritz E, Wunderlin M, Schulz G, Bäuerle P (2012) Chem – Eur J 18:12880
14. Mozer AJ, Ma C-Q, Wong WWH, Jones DJ, Bäuerle P, Wallace GG (2010) Org Electron 11:573
15. Mishra A, Ma C-Q, Janssen RAJ, Bäuerle P (2009) Chem – Eur J 15:13521
16. Wong WWH, Ma C-Q, Pisula W, Yan C, Feng X, Jones DJ, Müllen K, Janssen RAJ, Bäuerle P, Holmes AB (2010) Chem Mater 22:457
17. Mastalerz M, Fischer V, Ma C-Q, Janssen RAJ, Bäuerle P (2009) Org Lett 11:4500
18. Schulz GL, Mastalerz M, Ma C-Q, Wienk M, Janssen RAJ, Bäuerle P (2013) Macromolecules 46:2141
19. Fischer MKR, Ma C-Q, Janssen RAJ, Debaerdemaeker T, Bäuerle P (2009) J Mater Chem 19:4784
20. Fischer MKR, López-Duarte I, Wienk MM, Martínez-Díaz MV, Janssen RAJ, Bäuerle P, Torres T (2009) J Am Chem Soc 131:8669
21. Zhang W-S (2014) Dendritic oligothiophenes: synthesis, characterization and application in organic photovoltaics, University of Ulm
22. Koster LJA, Kemerink M, Wienk MM, Máturova K, Janssen RAJ (2011) Adv Mater 23:1670

23. Oosterhout SD, Wienk MM, van Bavel SS, Thiedmann R, Koster LJA, Gilot J, Loos J, Schmidt V, Janssen RAJ (2009) Nat Mater 8:818
24. Koster LJA (2010) Phys Rev B 81:205318
25. Koster LJA, Stenzel O, Oosterhout SD, Wienk MM, Schmidt V, Janssen RAJ (2013) Adv Energy Mater 3:615
26. Oosterhout SD, Koster LJA, van Bavel SS, Loos J, Stenzel O, Thiedmann R, Schmidt V, Campo B, Cleij TJ, Lutzen L, Vanderzande D, Wienk MM, Janssen RAJ (2011) Adv Energy Mater 1:90
27. van Franeker JJ, Turbiez M, Li W, Wienk MM, Janssen RAJ (2015) Nat Commun 6:6229
28. van Franeker JJ, Westhoff D, Turbiez M, Wienk MM, Schmidt V, Janssen RAJ (2015) Adv Funct Mater 25:855
29. Chiu SN, Stoyan D, Kendall WS, Mecke J (2013) Stochastic geometry and its applications. Wiley, Chichester
30. Thiedmann R, Spettl A, Stenzel O, Zeibig T, Hindson JC, Saghi Z, Greenham NC, Midgley PA, Schmidt V (2012) Image Anal Stereol 31:23
31. Divitini G, Stenzel O, Ghadirzadeh A, Guarnera S, Russo V, Casari CS, Bassi AL, Petrozza A, di Fonzo F, Schmidt V, Ducati C (2014) Adv Funct Mater 24:3043
32. Stenzel O, Hassfeld H, Thiedmann R, Koster LJA, Oosterhout SD, van Bavel SS, Wienk MM, Loos J, Janssen RAJ, Schmidt V (2011) Ann Appl Stat 5:1920
33. Schmidt V, Gaiselmann G, Stenzel O (2015) Stochastic geometry, spatial statistics and random fields: models and algorithms. Lecture notes in mathematics. In: Schmidt V (ed), vol 2120. Springer, Cham, p 95
34. Stenzel O, Koster LJA, Thiedmann R, Oosterhout SD, Janssen RAJ, Schmidt V (2012) Adv Funct Mater 22:1236
35. Thiedmann R, Hassfeld H, Stenzel O, Koster LJA, Oosterhout SD, van Bavel SS, Wienk MM, Loos J, Janssen RAJ, Schmidt V (2011) Image Anal Stereol 30:19
36. Westhoff D, van Franeker JJ, Brereton T, Kroese DP, Janssen RAJ, Schmidt V (2015) Model Simul Mater Sci Eng 23:045003
37. Stenzel O, Hirsch C, Brereton T, Baumeier B, Andrienko D, Kroese D, Schmidt V (2014) Multiscale Model Simul 12:1108
38. Brereton T, Stenzel O, Baumeier B, Andrienko D, Schmidt V, Kroese DP (2014) Methodol Comput Appl Probab 16:465
39. Brereton TJ, Stenzel O, Baumeier B, Kroese DP, Schmidt V (2012) Proceedings of the 2012 winter simulation conference. In: Laroque C, Himmelspach J, Pasupathy R, Rose O, Uhrmacher AM (eds), IEEE, Piscataway
40. Baumeier B, Stenzel O, Poelking C, Andrienko D, Schmidt V (2012) Phys Rev B 86:184202
41. Kordt P, Stenzel O, Baumeier B, Schmidt V, Andrienko D (2014) J Chem Theory Comput 10:2508
42. Schrader M, Fitzner R, Hein M, Elschner C, Baumeier B, Riede M, Leo K, Bäuerle P, Andrienko D (2012) J Am Chem Soc 134:6052
43. Elschner C, Schrader M, Fitzner R, Levin AA, Bäuerle P, Andrienko D, Leo K, Riede M (2013) RSC Adv 3:12117
44. Rühle V, Lukyanov A, May F, Schrader M, Vehoff T, Kirkpatrick J, Baumeier B, Andrienko D (2011) J Chem Theory Comput 7:3335
45. May F, Al-Helwi M, Baumeier B, Kowalsky W, Fuchs E, Lennartz C, Andrienko D (2012) J Am Chem Soc 134:13818
46. May F, Baumeier B, Lennartz C, Andrienko D (2012) Phys Rev Lett 109:136
47. Jansen APJ (2012) An introduction to kinetic Monte Carlo simulations of surface reactions. Springer, New York
48. Wang PH, Best RB, Blumberger J (2011) J Am Chem Soc 133:3548

Dicyanovinylene-Substituted Oligothiophenes for Organic Solar Cells

Christian Koerner, Hannah Ziehlke, Roland Fitzner, Moritz Riede, Amaresh Mishra, Peter Bäuerle, and Karl Leo

Contents

Abstract We investigate dicyanovinyl-substituted oligothiophene derivatives as absorber materials in organic solar cells. We determine structure–property relationships, which are important for materials design. We demonstrate the influence of those structural changes on the processing ability, energy levels, optical properties, thin-film morphology, and charge transport. Furthermore, we give a detailed picture of the microscopic processes between photon absorption and charge carrier generation, in particular, the importance of triplet exciton losses and a relationship between the yield of charge carrier generation and macroscopic charge-transport properties.

C. Koerner (✉) • H. Ziehlke • R. Fitzner • M. Riede • A. Mishra • P. Bäuerle • K. Leo
Institut für Angewandte Photophysik, Technische Universtität Dresden, Dresden, Germany

Institut für Organische Chemie II und Neue Materialien, Universität Ulm, Ulm, Germany
e-mail: leo@iapp.de; peter.baeuerle@uni-ulm.de

K. Leo (ed.), *Elementary Processes in Organic Photovoltaics*,
Advances in Polymer Science 272, DOI 10.1007/978-3-319-28338-8_3

Keywords Charge carrier generation • Multijunction • Oligothiophene • Photoinduced absorption • Small molecule • Structure-property relationships • Triplet exciton

1 Introduction

In recent years, organic solar cell research has increased power conversion efficiencies (PCEs) to market-relevant values of more than 10 %. Notwithstanding this success, some of the basics are still not completely understood. Two of the most important topics in organic solar cell research are the following: "How exactly are free-charge carriers generated and how can this process be optimized?" and "What is the structure of the perfect absorber material?" Both questions have been addressed by our research on acceptor–donor–acceptor (A–D–A)-type oligothiophenes, representing an efficient family of donor materials for vacuum-processed small-molecule organic solar cells (SMOSCs). The A–D–A concept allows us to separately influence the highest occupied molecular orbital (HOMO) or the lowest unoccupied molecular orbital (LUMO) energy level by changing the acceptor or donor. The acceptor $2,2'$-dicyanovinyl (DCV) proved to be ideal to achieve the optimal optoelectronic properties for donor materials in combination with the commonly used acceptor C_{60}. Furthermore, alkyl side chains are attached to influence molecular packing in thin films. The general structure of those molecules is depicted in Fig. 1. In the first very promising solar cell application, a butyl-substituted quinquethiophene was deployed in SMOSC devices, yielding a PCE of 3.4 %[1]. Meanwhile, the extensive exploration of this material class in synthesis and characterization led to excellent certified PCE values of 8.3 % in single-junction and 9.7 % in multijunction cells using a centrally methylated quinquethiophene derivative [2]. Those achievements were enabled by targeted materials design, following the structure–property relationships determined from variations of the oligothiophene chain length [3, 4] and the length of the alkyl side chains [5, 6]. In general, in addition to the high variability of this material class for in-depth investigations, the success of those molecules is attributed to strong absorption, finely adjusted energy levels, well-ordered thin films, and good charge-transport properties.

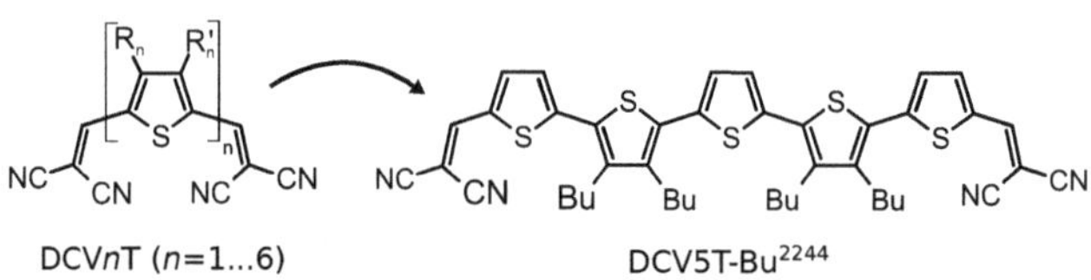

Fig. 1 (*Left*) General structure of DCV*n*T derivatives. R_n denote alkyl side chains of various lengths attached at different positions to the oligothiophene core. (*Right*) The nomenclature used in this work is explained according to the given example: DCV5T describes the length of the backbone, followed by the length of the side chains (butyl = Bu) and the number of respective rings to which they are attached

In this chapter, we demonstrate the potential of this material class for determining structure–property relationships and give some examples of the obtained results. We show the influence of structural variations to the molecule on thermal properties, energy levels, optical properties, molecular packing, and charge transport. Furthermore, we discuss detailed photophysical investigations offering insight into the microscopic processes between photon absorption and charge carrier generation, in particular, the interplay between charge separation and recombination from the charge-transfer (CT) state to triplet excitons. Finally, we discuss the most important achievements in device physics and optimization.

2 Synthesis of Dicyanovinyl-Functionalized Oligothiophenes (DCV*n*Ts)

In recent years, we have developed various series of dicyanovinyl-substituted oligothiophenes (DCVnTs) by varying the conjugated backbone or the length of the alkyl side chains. The series of oligomers with different alkyl substituents at the oligothiophene backbone were synthesized by using two different synthetic strategies (see Scheme 1) [1, 7]. As an example, the synthesis of DCV5T-Bu^{2244} is shown in Scheme 1. In the first approach (A), the quinquethiophene **3** was prepared by a Pd^0-catalyzed Suzuki-type cross-coupling reaction of butyl-substituted diiodoterthiophene **1** and boronic ester **2** in 70 % yield. For the introduction of the DCV acceptor groups, quinquethiophene **3** was formylated in both terminal

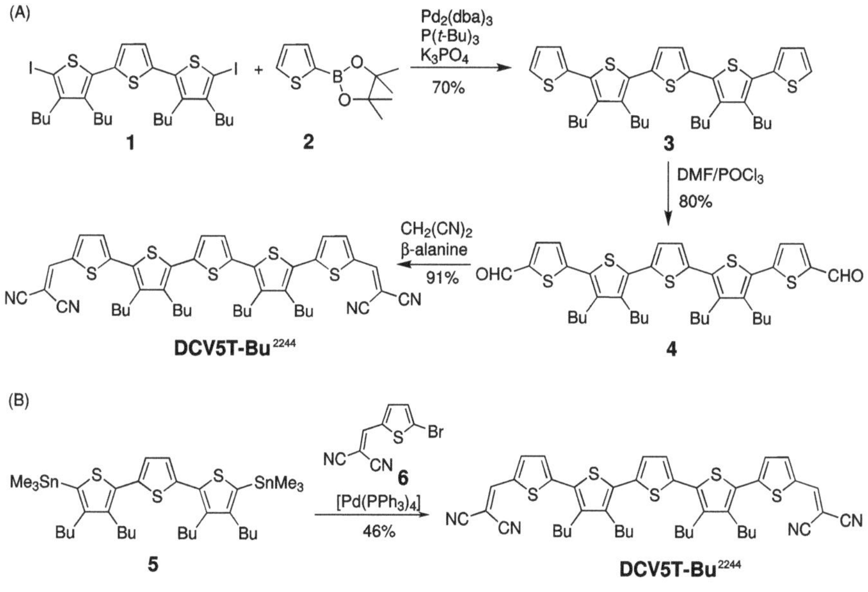

Scheme 1 Synthesis of dicyanovinyl-substituted quinquethiophene DCV5T-Bu^{2244} using two different approaches

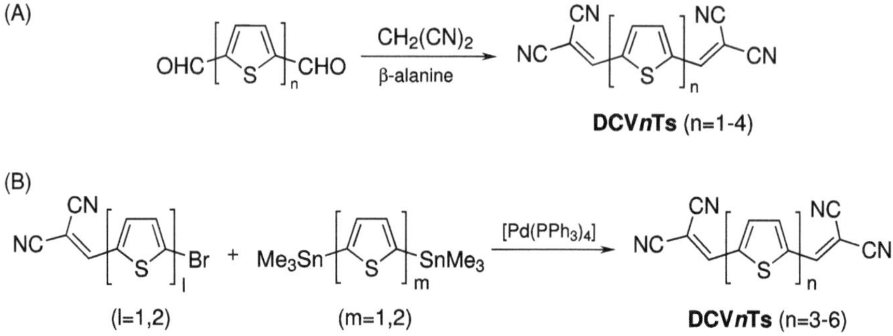

Scheme 2 Synthetic strategies for the development of dicyanovinyl-substituted oligothiophenes DCV*n*T **1–6**

α-positions under Vilsmeier–Haack conditions to yield dialdehyde **4**, which was subsequently transformed to the target molecule DCV5T-Bu^{2244} by a Knoevenagel condensation with malononitrile using β-alanine as catalyst. The compound was obtained as a deep-purple to black solid. In the second strategy (B), DCV-functionalized bromothiophene **6** was reacted with distannylated butyl–substituted terthiophene **5** in a twofold Pd^0-catalyzed Stille-type coupling reaction to obtain DCV5T-Bu^{2244} in 46 % yield. Similar synthetic approaches were also implemented in the synthesis of nonalkylated DCVnTs [4]. In this case, the smaller members of the series, DCV1T–DCV4T, method (A) was used. The oligomers were prepared by the reaction of the corresponding dialdehydes, which were synthesized by Vilsmeier–Haack formylation of the parent oligothiophenes [8], with malononitrile and β-alanine as catalyst in near-quantitative yields and high purity (see Scheme 2). However, because of a dramatic decrease in solubility, method (A) was not successful for the preparation of DCV4T in high purity. The reaction resulted in an inseparable mixture of the target DCV4T and the singly reacted intermediate with a DCV group as an impurity. For this reason, method (B) was used for the preparation of higher homologues DCV4T to DCV6T. DCV-substituted bromothiophene [9] and bithiophene [10] were reacted with central distannylated mono- and bithiophene [11, 12] in a Stille-type coupling under Pd[PPh$_3$]$_4$-catalysis in DMF as solvent to obtain DCVnTs ($n = 4 \ldots 6$). After washing and extracting with various solvents, the materials were finally purified by gradient vacuum sublimation to obtain the product in high purity. Using a similar Pd-catalyzed Stille-type cross-coupling approach, we have further developed various series of methylated and ethylated DCV4Ts and methylated DCV5Ts (see Figs. 7 and 12) [5, 6].

3 Structure–Property Relationships

The question of how to design functional materials for specific needs is most important for organic electronics. Knowledge about structure–property relationships is desired to gain a deeper understanding of how to tune molecular properties in a desired fashion. The material class of DCV*n*Ts is ideal to study those correlations. Because of the A–D–A architecture, the HOMO of the molecule is mainly determined by the oligothiophene backbone, whereas the LUMO is mainly influenced by the DCV acceptor unit and hardly changes upon donor variations. Therefore, it is possible to separately study the impact of changes in the HOMO or the LUMO energy on optoelectronic and/or device properties by changing either donor or acceptor.

The morphology in thin films, which is an important tuning parameter for organic solar cells, is changed by attaching alkyl side chains of varying lengths and numbers at different positions of the molecule backbone. Those variations allow one to alter the morphology without significantly impacting the energy level in most cases. Finally, the strong absorptivity and reasonable hole mobilities are important ingredients to grant this material class high relevance in the field of SMOSCs.

3.1 Thermal Properties

All synthesized nonalkylated and alkylated DCV*n*Ts showed high thermal stability as investigated using differential scanning calorimetry (DSC) and thermogravimetric analysis. Thermal stability is of utmost importance as we use thermal evaporation in vacuum for the deposition of thin films. Furthermore, thermally stable materials allow purification by thermal gradient sublimation before using the materials for investigations. Depending on the purity following the synthesis, one or more consecutive steps can be necessary to fully purify the obtained powder, which was demonstrated for the unsubstituted DCV4T derivative by an increasing melting temperature T_M obtained from DSC measurements [4]. However, not all the materials showed high enough sublimation yields for practical use, meaning that they were not stable enough to withstand this step without significant decomposition. In those cases, the materials were used unsublimed, demanding high standards in the purification and separation of side products after synthesis.

We have monitored the fragmentation of the molecules during evaporation using a mass spectrometer directly attached to the vacuum chamber, showing that particularly the DCV acceptor part breaks during evaporation. For longer molecules, other fragments can also be observed and the peak intensities, that is, the degree of decomposition, increase with the length of the molecule [13].

The melting temperatures were used above to qualify the material purity. Furthermore, these values can already give preliminary information about, for instance, molecular packing or intrinsic dipole moments [4]. Figure 2 shows the melting temperatures for the investigated DCV*n*T compounds. Upon comparing molecules with similar substitution pattern, one observes a systematic odd–even effect, meaning that even-numbered compounds generally have higher melting temperatures than odd-numbered analogues [4, 14]. We attribute this finding to the symmetry of the molecules. In contrast to the even-numbered oligothiophenes, which exhibit a C_{2h} (point) symmetry, the odd-numbered compounds show C_{2v}

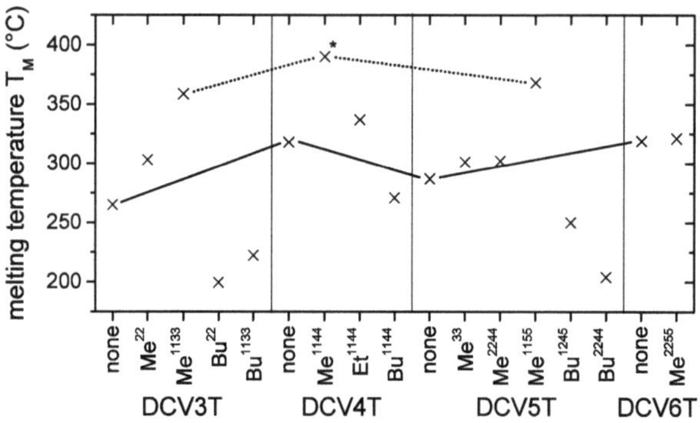

Fig. 2 Melting temperatures T_M determined using differential scanning calorimetry for several DCV*n*T molecules with different side chain substitution patterns. The odd–even effect is indicated by lines for the unsubstituted (*solid*) as well as for methylated derivatives with four methyl groups at the terminating thiophene units (*dotted*). For DCV4T-Me^{1144}, a melting temperature could not be detected. The point in the graph (marked by *) denotes the decomposition temperature of the compound. The melting temperatures shown here are taken in part from [4–6]

(mirror-plane) symmetry. As a result, the even-numbered molecules do not have a dipole moment. For the thin-film deposition, the higher melting temperatures allow very high sublimation yields for DCV4Ts, whereas the higher-mass DCV6T compounds cannot benefit from high melting temperatures. Figure 2 furthermore shows that the side chains have a significant effect on T_M. Throughout the whole series, methylated compounds have the highest values, whereas those with butyl side chains show the lowest T_M, indicating decreased molecular interaction in the solid state for the latter. Moreover, there are subtle differences with the position of the side chains attached to the backbone: For both DCV3Ts and DCV5Ts, endstanding methyl groups lead to the highest melting temperatures observed in the whole series [6].

Additional information on thermal properties and vacuum processing of those materials is given in [13].

3.2 Energy Levels

It was already mentioned that the A–D–A architecture allows one to separately address the HOMO or the LUMO upon molecular changes. An elongation of the molecule leads to an increase in the HOMO level, whereas the LUMO is practically unchanged at least for $n \geq 3$. For $n < 3$, the mutual influence of the donor and the acceptor parts leads to a lowering of both the HOMO and LUMO [4]. In contrast to DCV as acceptor, the LUMO can be increased by using weaker acceptor moieties such as benzothiadiazole or related thiadiazolopyridine units [15] or further lowered using the even stronger acceptor tricyanovinyl [16]. However, the DCV acceptor turned out to be highly suitable for using the compounds in solar cells as donor in combination with C_{60} as acceptor, as the LUMO of C_{60} is slightly lower,

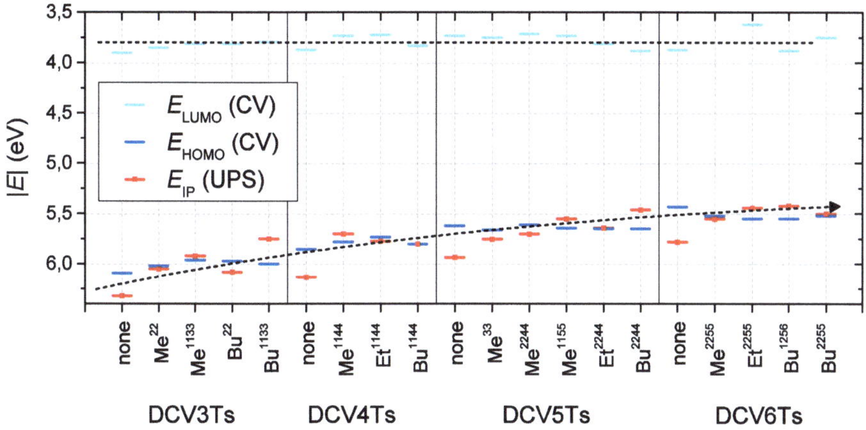

Fig. 3 The displayed HOMO/LUMO levels were taken in part from [4–6, 17]. Some values are corrected for the different ferrocene reference potentials used (4.8 eV in older publications vs. 5.1 eV in newer publications). The E_{IP} values were taken in part from [1, 4, 13, 17–20]

allowing for efficient charge transfer without significant voltage losses relative to the optical gap.

Figure 3 shows the energy levels measured in solution with cyclic voltammetry and the corresponding ionization potential (E_{IP}) in thin film measured using ultraviolet photoelectron spectroscopy. As expected from the A–D–A architecture, an extension of the molecular backbone steadily increases the HOMO level, whereas the LUMO is practically constant. The side chains are not strongly coupled to the π system of the molecule core and therefore have only a minor influence on the HOMO level in solution.

In contrast, larger differences are observed in thin films, particularly for DCV3Ts. The highest E_{IP} of 6.3 eV is obtained for the unsubstituted compound, whereas the twice- and four-times–substituted derivatives show much smaller values of 6.1 eV and 5.8 eV, respectively. The length of the side chains does not play a significant role here [17]. A similarly high E_{IP} was observed for the other unsubstituted DCV*n*Ts as well. As particularly the short-chain derivatives and the unsubstituted derivatives are prone to strong crystallization [17, 21], we attribute those larger deviations between solution and thin film to morphologic effects. Furthermore, as observed above in the discussion of the melting temperatures, the derivatives bearing alkyl chains at the endstanding thiophene rings (cp. DCV3Ts and DCV5T-Mes) exhibit special properties, showing the smallest ionization potentials within their respective series.

3.3 Optical Properties

DCVnTs show a very strong absorptivity of 40,000–80,000 L/mol cm [4, 6], allowing for high photocurrent densities already for very thin photoactive layers of 40–60 nm. The extinction coefficient of $k = 1.5$ [13] determined for DCV5T-Me[33] is one of the highest values published for small-molecule donor materials for OSCs [22, 23] and twice as much as for the well-known dye zinc phthalocyanine (ZnPc) [13].

The energy-level discussion already indicates that the optical properties, that is, absorption and emission, are determined to a large extent by the molecular backbone, at least regarding the wavelength range. Because of the increasing HOMO level, the absorption redshifts when the number of thiophene units increases in the molecule core. This effect was nicely demonstrated, for instance, for the series of butylated ($n = 3 \ldots 6$) [3] and unsubstituted derivatives ($n = 1 \ldots 6$) [4]. As a first approximation, the excitation energy is inversely proportional to the number of double bonds in the conjugated backbone [3]. The alkyl chains hardly alter the absorption range, as they do not significantly contribute to the π system. However, they determine the thin-film growth of the molecules, which is reflected in the optical properties as well. Figure 4 shows the normalized absorption and emission spectra of four DCV4T derivatives in solution compared to the respective thin films. The molecules exhibit an identical substitution pattern but different lengths of the alkyl side chains, ranging from 0 (H) to 4 (Bu) carbon atoms. In addition to the typical redshift between solution and thin-film spectra, the completely unstructured absorption spectra in solution evolve into substructured vibrational modes particularly for the shorter side chains. The appearance of such modes (which is also reflected in the luminescence spectra) together with the strongest redshift for the unsubstituted DCV4T compared to the butylated derivative is frequently attributed to a more crystalline growth of the molecules [24, 25].

As already observed in the discussion of the melting temperatures (and in part also the energy levels), the methylated compounds are somewhat special. The photoluminescence (PL) of DCV4T-Me[1144] in Fig. 4b is strongly redshifted compared to the other derivatives, and this is not reflected in the absorption spectra. Taking the structureless shape of the spectrum into account, we attribute this PL to emission from an extended intermolecular state. This assignment requires a close packing of the molecules, which is supported by the determined crystal structure (see Sect. 3.4). Further support for this assignment is obtained from temperature-dependent PL measurements, which are discussed in more detail in [13].

3.4 Structural Properties and Morphology in Thin Films

In the following, the influence of the side chains on the crystal packing and the implications for OSCs are discussed in more detail. The importance of side chains is emphasized using two examples: a series of DCV4T [5] and a series of DCV5T molecules [6].

In the first series, we compare the aforementioned DCV4T molecules with different side chain lengths (H, Me, Et) [4]. From the discussion of the energy levels (Fig. 3) as well as the optical properties (Fig. 4), we concluded that there is only a minor effect in solution but a stronger effect in thin films. A first glance at the different packing of the molecules is obtained from single-crystal analysis. The crystal structure for all three materials is shown in Fig. 5 as a top view (looking at the molecular plane) and a side view (looking at the π–π-stacking direction). When we compare all three derivatives, the DCV4T-Me shows the highest degree of interconnection[1] in-plane as well as out-of-plane. In comparison to the compound without side chains, the molecules are shifted parallel to the molecular axis by half of the molecular length, giving a brickstone-like packing. The advantage of this two-dimensional (2D)–like stacking compared to the nearly perfect columnar one-dimensional (1D) stacking is revealed by simulations of the charge carrier mobility [26]. In spite of the higher electronic couplings for the 1D stacked DCV4T

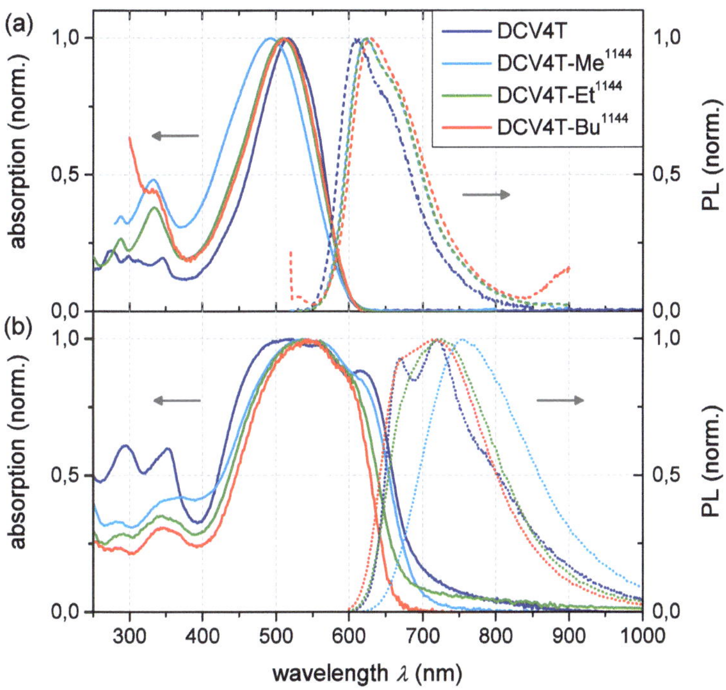

Fig. 4 Absorption and emission spectra of DCV4T derivatives with different side chain lengths measured in (**a**) solution and (**b**) thin films. All spectra are normalized to their respective maximum

[1]Number of nonbonding short contacts between molecules.

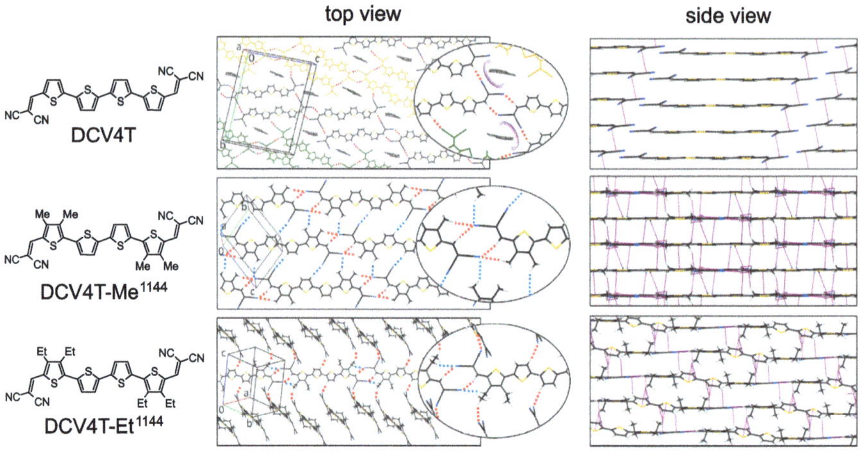

Fig. 5 Crystal structures of DCV4T derivatives with different side chain lengths (unsubstituted, methyl, and ethyl): (*left*) molecular structure; (*middle*) crystal structure in the molecular plane (*top view*); (*right*) crystal structure perpendicular to the molecular plane. Short contacts between the molecules are indicated by connecting lines and magnified in the insets. [Reprinted from Fitzner et al., Advanced Materials, 24(5), 675, (2012), copyright 2012, with permission from John Wiley and Sons]

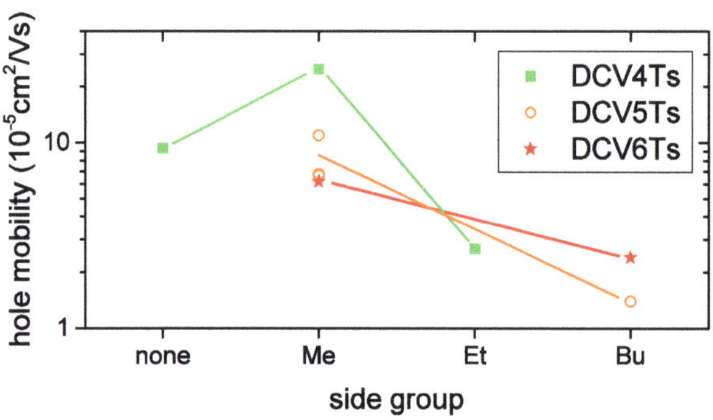

Fig. 6 Hole mobilities (determined in organic field-effect transistor geometry) of different DCV*n*T derivatives that depend on the length of the side chains attached. The numbers were taken in part from [5, 20, 28]. [Details on sample preparation, measurement, and evaluation are given in [13]]

molecules, the higher interconnection of DCV4T-Me gives higher mobility values because of the increased number of possible pathways for a charge traveling through the film. The differences are even stronger when site energy differences (energetic disorder) are taken into account as in real systems: As a result of the restriction to one transport direction, the mobility is reduced by two orders of magnitude in the 1D stacked compound, whereas the mobility drops only by one order of magnitude in the case of the highly networked DCV4T-Me. Qualitatively, the same results are obtained for the respective terthiophenes [26]. Comparing the hole mobilities obtained from OFET measurements, the conclusions from the structure

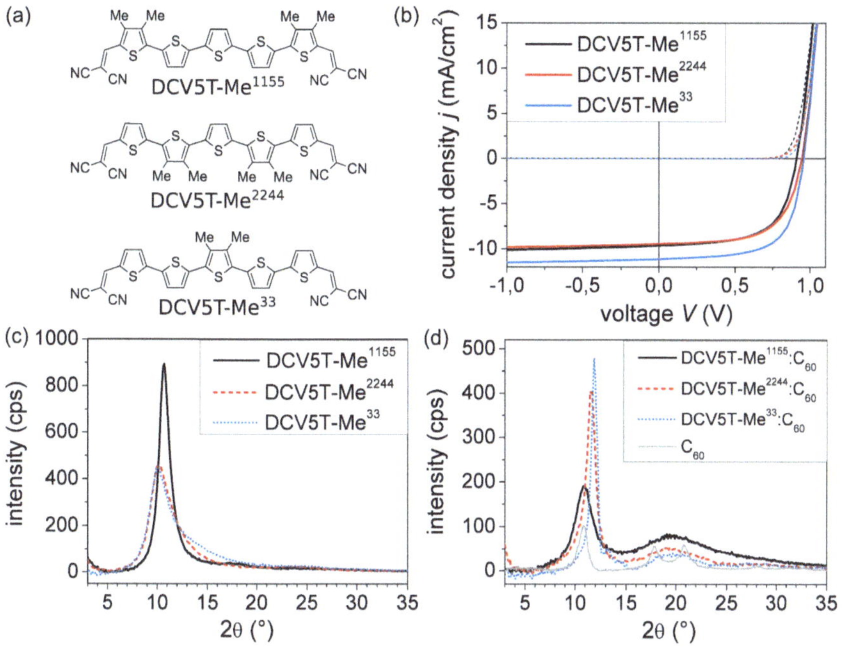

Fig. 7 (**a**) Molecular structures of the methylated DCV5T derivatives. (**b**) *J–V* curves of test solar cells with all DCV5T-Me derivatives. The basic stack is identical and contains a 30-nm-thick blend layer of donor: C_{60} mixed at a volume ratio of 2:1 and deposited at a substrate temperature of 90 °C. (**c**) Grazing-incidence X-ray diffraction (GIXRD) measurements on pure films of all DCV5T-Me derivatives deposited on glass substrates. (**d**) GIXRD measurements on DCV5T-Me:C_{60} blends as used in the solar cells in (a), but higher thickness of 75 nm. A diffraction pattern of C_{60} is included for comparison. [Adapted from [6]]

analysis and the predictions from simulation are validated in the DCV4T series. The highest hole mobility of $2.4 \cdot 10^{-4}\,\mathrm{cm^2/V\,s}$ is obtained for the methylated compound, followed by the unsubstituted ($9.4 \cdot 10^{-5}\,\mathrm{cm^2/V\,s}$) and the ethylated DCV4T derivative ($2.7 \cdot 10^{-5}\,\mathrm{cm^2/V\,s}$). In the series of investigated molecules, the methylated derivatives consistently obtain the highest hole mobilities independent of the backbone length (see Fig. 6). The decreasing mobility for longer molecules is attributed to an increased reorganization energy and higher energetic disorder stemming from the less crystalline packing in films [27].

Another example of the importance of side chains for solar cell optimization is obtained from a comparison of three DCV5T derivatives [6]. Each of these molecules has methyl side chains but differs in their number and the substitution pattern (see Fig. 7). In fact, for those molecules the side chains make the difference between a good- and a high-efficiency absorber molecule. Solar cells with an identical stack besides the donor material used give a 4.8 % efficiency for both DCV5T-Me^{1155} and DCV5T-Me^{2244}, whereas a PCE of 6.1 % is obtained for DCV5T-Me^{33}. The difference in PCE mainly comes from the larger photocurrent (despite similar absorption) and a higher fill factor (FF), which could be the result of differences in the crystallinity in thin film. The crystallinity in pure films is investigated by grazing-incidence X-ray diffraction (GIXRD) measurements, indicating the highest degree of crystallinity[2] for DCV5T-Me^{1155}. However, for high-efficiency solar cells, mixing with C_{60} in substrate-heated blend layers is essential. Comparing mixed films of those three donor molecules, we find that the picture is upside down, with the highest crystallinity obtained for DCV5T-Me^{33}. This enhanced crystallinity is expected to be the main reason for the enhanced FF and photocurrent. Although the crystal structures of those blends and the mutual orientation of the molecules are not yet revealed, those measurements show that the miscibility between donor and acceptor in the blend—determining phase separation, crystallinity, and phase purity—decides the race.

Those two examples of DCV4T and DCV5T derivatives demonstrate the necessity for a precise tuning of the side chains to yield the best morphology in thin films both for the pure phase as well as for the mixed phase with the acceptor. In the extreme case when side groups are completely omitted—as for the unsubstituted derivatives—a strong tendency to crystallization was observed, leading to either very high roughness upon processing or a metastable thin-film morphology [13, 21]. Such properties negatively influence the performance of the devices (see Sect. 5.1) and their long-term stability.

4 Free-Charge Carrier Generation

4.1 Energy Versus Charge Transfer

In Sect. 3.2, we showed that upon increasing the number of thiophene units in the backbone, the HOMO level increases. This variation not only influences the optical properties with the observed redshift but also determines the charge photogeneration yield in combination with an appropriate acceptor. A CT between the donor and acceptor is only efficient if the energy of the resulting CT state is lower in energy than the singlet states of both the donor and acceptor.[3] Taking 6.0 eV for the ionization potential of DCV3T (see Fig. 3) and 4.0 eV for the electron affinity of C_{60} [29], the CT state energy is estimated at 1.7 eV,[4] which is close to the singlet energy of C_{60}. Therefore, CT will not be efficient for this D–A combination. Instead, as the emission of DCV3T and the absorption of C_{60} slightly overlap, singlet energy transfer is feasible and was demonstrated by efficient quenching of

[2]Characterized comparing the full width at half maximum and the scattering intensity for identical samples with chemically very similar materials and identical measurement conditions.

[3]Usually, it is expected that a driving force of 0.1 eV between the excitonic and the CT state is sufficient for CT with a near-quantitative yield [31].

[4]Assuming a binding energy for the CT exciton of 0.3 eV.

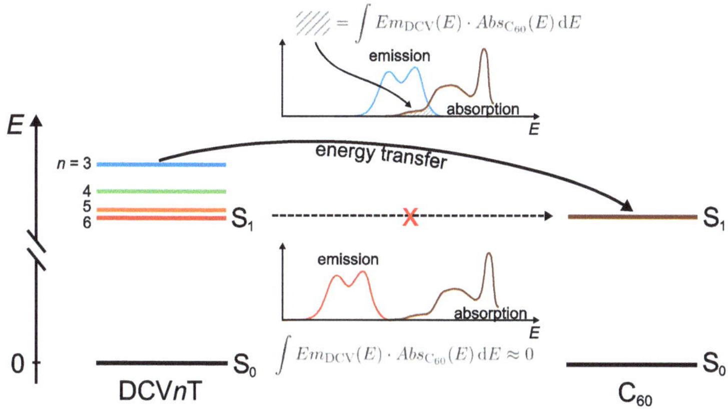

Fig. 8 Sketch of the singlet exciton S_1 levels for a DCV*n*T series with different backbone lengths ($n = 3 \ldots 6$). The energy-transfer rate is determined by the overlap integral between donor emission, Em_{DCV}, and C_{60} absorption, $Abs_{C_{60}}$, which is illustrated as the shaded area in the upper example spectrum for $n = 3$, respectively. For the longer compounds ($n = 4 \ldots 6$), the overlap is zero because of the redshifted absorption and emission spectrum (lower example spectrum)

the PL of DCV3T and sensitized emission from C_{60} [30]. The feasibility of this energy transfer is sketched in Fig. 8. After exciton transfer, the singlet excitons on C_{60} are in part converted to triplet excitons and back-transferred to the lower-lying oligothiophene triplet state[5] [30]. In summary, this so-called excitonic ping-pong effect, which is observed independent of the side chains attached [17], prevents free-charge carrier generation, rendering the DCV3T/C_{60} couple very ineffective for solar cell application.[6]

In contrast, upon extending the oligothiophene chain, the rising HOMO and the likewise redshifted emission prevent the excitonic energy transfer, as sketched in Fig. 8. Simultaneously, the CT state energy is decreased below the singlet energy of C_{60} enabling CT instead of energy transfer, as indicated by the near-quantitative PL quenching without the observation of sensitized fullerene emission. Still, DCV4T molecules in particular show a rather low photogeneration yield in combination with C_{60}, which is discussed in more detail below.

[5]See [3] for calculated energies of the triplet exciton T_1 state for DCV*n*Ts with $n = 3 \ldots 6$.

[6]Still, it may be possible to use DCV3T derivatives in combination with other acceptors or itself as acceptor, as shown, for example, in [18, 34] using ZnPc or DIP as donor material.

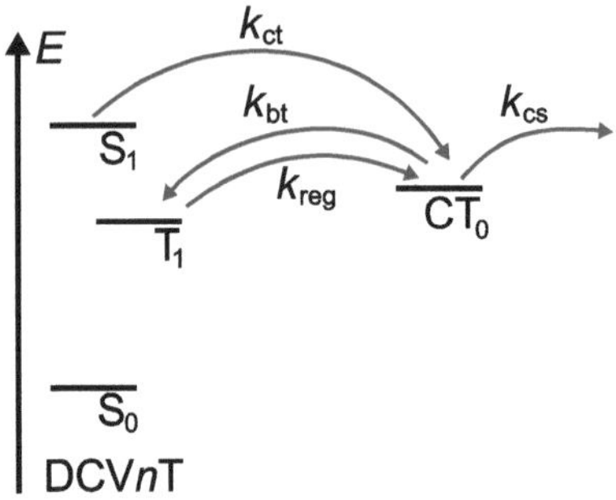

Fig. 9 Sketch of the energy levels at the donor/C_{60} interface. The photogeneration yield is determined by the rates for charge transfer (CT), k_{ct}, back transfer to the triplet state, k_{bt}, (thermal) regeneration of the CT state from triplet excitons, k_{reg}, and the final separation of CT states into free carriers, k_{cs}

4.2 Influencing the Charge Photogeneration Yield

Intense photophysical studies using near-steady-state photoinduced absorption (PIA) spectroscopy revealed that the population of the oligothiophene triplet exciton state is increased when DCV*n*Ts ($n = 4 \ldots 6$) are mixed with the acceptor C_{60} [3, 32, 33]. The result is identical to the DCV3Ts although the pathway is different. In this case, instead of an excitonic transfer, an electron is transferred from the donor to the acceptor, leading to a still coulombically bound CT state at the interface. However, instead of the desired further separation of electron and hole, the CT state can recombine into the lower-lying triplet state on the donor, as this is still energetically the lowest state in the system. A sketch of the different energy levels is shown in Fig. 9. As a consequence, back transfer from the CT state at the D–A interface leads to a strong population of the donor triplet state, acting as a loss mechanism in solar cells. Therefore, the final yield of photogenerated charge carriers sensitively depends on the respective rates for charge transfer, k_{ct}, back transfer to the triplet state, k_{bt}, (thermal) regeneration of the CT state from triplet excitons, k_{reg}, and the final separation of CT states into free carriers, k_{cs} (see Fig. 9). Some examples are discussed in the following.

4.2.1 Side Chain Spacer Effect

In theory, the first step, namely the CT from the donor to the fullerene acceptor, could be influenced by side chains attached to the molecule backbone acting as a spacer between both molecules. The reduced wavefunction overlap would then decrease the rate for the forward CT as well as the subsequent triplet back transfer, or as well for excitonic energy transfer between DCV3T and C_{60}. Such effects were already observed comparing tetracene to its sterically inaccessible counterpart rubrene [35]. However, a detailed study of several DCV4T derivatives with different alkyl side chain length did not reveal a clear indication of a negative influence of long side chains on the CT rates [13] although previously Ziehlke et al. claimed such a spacer effect for the energy transfer from certain DCV3T derivatives to C_{60} [17]. The main difference between the above-mentioned example of tetracene and rubrene and the alkylated DCV*n*Ts is the rather extended side group for the latter in contrast to the compact but bulky phenyl group for the former. Therefore, as long as C_{60} can directly access the donor backbone for CT, a spacer effect is not reasonable. Finally, the reduced recombination rates would also be visible in the open-circuit voltages of the devices. For the DCV4T derivatives, V_{oc} just increases slightly with increasing side chain length along with an increasing E_{IP} [13], again suggesting the absence of a spacer effect.

4.2.2 Triplet Exciton State as Loss Mechanism

The main difference between the longer ($n = \{5, 6\}$) and the shorter oligothiophenes ($n = 4$) is the higher photocurrent of the former. For DCV5T and DCV6T derivatives, high external quantum efficiency (EQE) values between 50–80 % were measured, corresponding to internal quantum efficiencies (IQEs) above 80 % [2, 19, 36]. In contrast, the photocurrents obtained using DCV4Ts as donors are rather low, with the highest EQE values reported ranging below 50 % [5]. As this trend is rather general and independent of side chains or device architecture (planar or mixed heterojunction), differences in the extraction efficiency because of varying thin-film morphologies are excluded as the main reason. Taking into account the similar absorption spectra and strength, the photogeneration yield, that is, the efficiency of creating a free-charge carrier from an exciton, must be different. PIA measurements revealed that at low temperatures of $T = 10\,K$[7] triplet excitons are generated by back transfer from the CT state independent of the backbone length [3]. As this temperature is far from solar cell operating temperatures, we investigated the influence of temperature on the excited state properties to see whether triplet excitons still play a role at realistic temperatures [13, 33]. We found that triplet generation and the generation of free-charge carriers are competing processes, with the latter dominating at temperatures above 150 K (see Fig. 10). Furthermore, temperature activation of singlet exciton diffusion also plays a role, initially increasing the generation of triplet excitons at lower temperatures. Figure 10a also shows that the triplet generation rate at room temperature is still at the same level compared to 10 K, explaining the low quantum yields observed in solar cells.[8]

Although the overall behavior was not observed to be different for longer molecules, in contrast to the DCV4Ts, the generation rate of triplet excitons is one order of magnitude lower for the longer derivatives ($n = \{5, 6\}$). This decrease can be explained by a regeneration of the CT state from the triplet state depending on the energetic distance between both states. For $n = \{5, 6\}$, the CT state energies

---

[7]The PIA signal is proportional to the product of generation rate and lifetime of the respective excited state. Therefore, signals are much higher at low temperatures due to longer lifetimes, facilitating measurements because of a high signal-to-noise ratio.

[8]For more detailed information, refer to [13, 33].

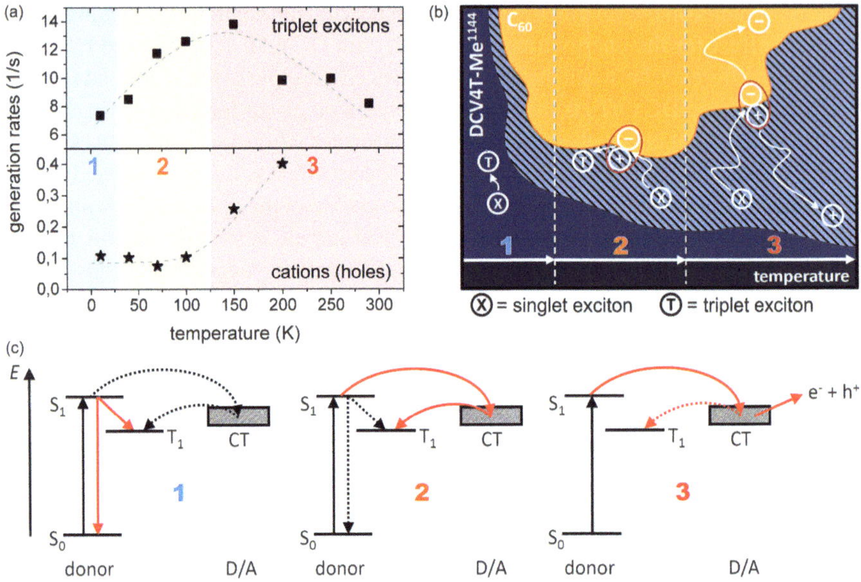

Fig. 10 (**a**) Temperature dependence of the triplet exciton and cation (i.e., holes on the donor, giving the spectral signature of a positively charged donor molecule, i.e., a donor cation) generation rates (The displayed generation rates are effective values that still contain sample and setup-specific terms. For more information, see [13, 33]) in a DCV4T-Me^{1144}:C_{60} mixed film obtained from photoinduced absorption spectroscopy analysis. (**b**) Sketch of the mixed phase of donor and acceptor, showing a magnified region around the mutual interface. The *shaded area* depicts the part of the donor phase that is within the exciton diffusion length from the acceptor phase. As a result of the temperature activation of exciton diffusion, this area increases with increasing temperature (to the *right* in the figure). Three characteristic regimes are indicated according to (**a**). (**c**) The important processes in each regime are depicted in more detail as red lines (dominant) and *black/dashed lines* (secondary), respectively. [Adapted from [33] and [13]]

come closer to the triplet T_1 energy because of the higher HOMO level. Therefore, the regeneration rate from the triplet exciton to the CT state is increased, explaining the reduced photocurrent loss for those materials. Overall, the success for DCV5Ts and DCV6Ts over the shorter DCV4Ts in terms of solar cell efficiencies is clearly attributed to differences in the triplet recombination loss.

4.2.3 Correlation Between Free-Charge Carrier Generation and Hole Mobility

As a first approximation, the exponential-like increase in the charge carrier generation rate g with increasing temperature (see Fig. 10a) can be modeled with an exponential function of the form $g = g_0 + g_1 \exp(-E_A/kT)$ with an offset generation rate g_0, a scaling parameter g_1, and a characteristic activation energy E_A [20, 33, 37]. The data in Fig. 10a give an activation energy of roughly 40 meV,

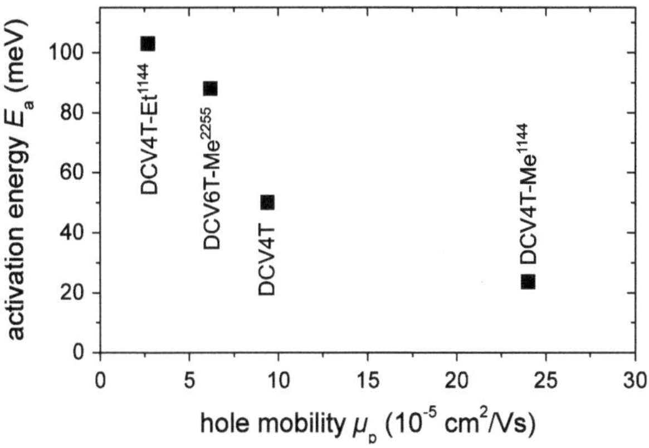

Fig. 11 Activation energy for charge carrier generation determined from temperature-dependent photoinduced absorption spectroscopy measurements on blends of C_{60} with different oligothiophene donor materials depending on their hole mobility. For further information and data on this topic, refer to [37]

which is much less than the difference between the triplet exciton and the CT state energy.[9] In principle, to separate the CT state, the hole and the electron have to overcome their mutual Coulomb attraction by moving away from the interface. The higher the probability to escape from the interface, the higher the charge-separation yield. Macroscopically, the main parameter to describe charge motion is the charge carrier mobility, which in turn is determined by parameters such as the reorganization energy or the energetic disorder. The charge carrier mobility itself is influenced by temperature because of the hopping process in a disordered energy landscape. Therefore, a temperature activation of charge carrier transport should be directly reflected in the temperature dependence of the charge carrier generation. Finally, in a first approximation, a reduced temperature activation of charge carrier transport should be correlated to less energetic disorder, which in turn should result in a higher charge carrier mobility. Putting all the pieces together, the activation energy we determine from PIA spectroscopy should inversely correlate with the charge carrier mobility [37]. Figure 11 shows the activation energies determined for blends of C_{60} with different oligothiophene derivatives depending on their hole mobility. As derived above, the activation energies, ranging between 20–100 meV, decrease with increasing hole mobility. This correlation supports the assumption that charge-transport properties influence the charge-photogeneration yield. To

[9]The CT state energy can again be estimated from the IP of the donor and the electron affinity of the acceptor to $E_{CT} = 5.75\,eV - 4.0\,eV - 0.3\,eV = 1.45\,eV$, assuming a binding energy of the CT exciton of 0.3 eV. Triplet exciton T_1 energies were calculated in [3] and range between 1.23–1.27 eV. Because of a commonly observed redshift between density functional theory (DFT) calculations and experiment, we estimate an energy difference of 0.2–0.4 eV between the triplet exciton and the CT exciton.

strengthen the proposed link between the determined activation energies and energetic disorder, we perform temperature-dependent charge extraction by linearly increasing voltage (CELIV) measurements on the DCV6T-Me compound. The measurements reveal an energetic disorder parameter of 109 meV [37], matching the activation energy found in PIA measurements. Similar values for the energetic disorder and trends with increasing backbone length were also found in charge-transport simulations with simulated morphologies based on measured crystal structures [27].

Finally, we can correlate these results back to the morphology discussion from the previous chapter. The methylated DCV4T compound, which showed the highest connectivity with a closely packed 2D network and the highest hole mobility, exhibits the lowest activation energy for charge separation. Furthermore, although a direct spacer effect was not observed, the side chains do indirectly influence the charge-generation yield because of the lower charge carrier mobility of those compounds.

5 Solar Cell Properties and Optimization

Vacuum-deposited SMOSCs exhibit a large variability in and high control over the processing conditions and especially the layer stack details. Planar and bulk heterojunction devices, normal and inverted structure, and exact variations of layer thicknesses are easily achieved. Using the p-i-n concept with doped transport layers [38], we can additionally tune the thicknesses of these layers to shift the absorber layer into the maximum of the optical field inside the device to obtain the highest photocurrents [39], which is especially relevant for tandem devices [40, 41]. Planar heterojunction devices (PHJ) serve as test devices to assess the potential of new molecules, because the performance is independent of the mixing behavior with the acceptor. Figure 12 compares PHJ solar cells with C_{60} as acceptor and DCVnTs with varying backbone lengths ($n = 4 \ldots 6$) as donor. The open-circuit voltage (V_{oc}) is highest for the DCV4T compound, whereas the FF and the short-circuit current density (j_{sc}) increase with higher n. The decrease in V_{oc} follows directly from the increasing HOMO level, which decreases the effective gap (the CT state energy) at the interface to C_{60}. The increasing j_{sc} is a result of the reduced triplet recombination rate and the redshifted absorption, which leads to a broader coverage of the solar spectrum. The FF can also be attributed to the higher photogeneration yield and reduced recombination. The lower hole mobility of larger molecules does not play a role for those ultrathin donor films (6 nm thickness). These trends are rather general and were described before for compounds with different side chain pattern as well [3, 4].

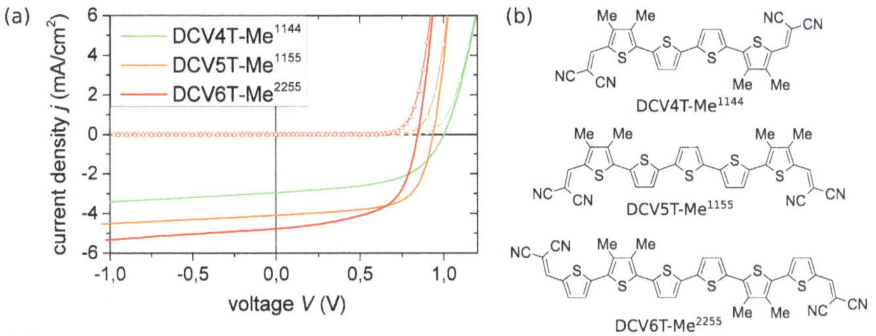

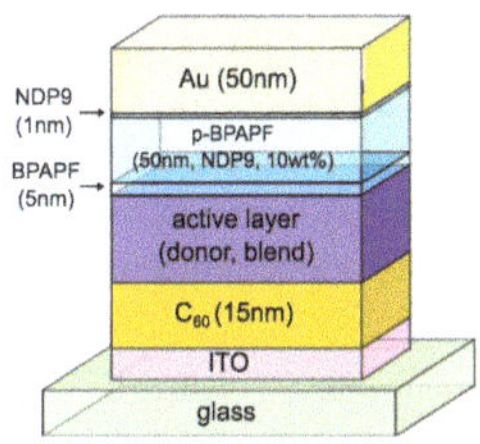

(d)

	$j_{sc,100}$ (mA/cm^2)	V_{oc} (V)	FF (%)	PCE (%)	I_L (mW/cm^2)
DCV4T-Me^{1144}	2.9	1.00	54.9	1.6	107
DCV5T-Me^{1155}	4.1	0.94	65.1	2.6	99
DCV6T-Me^{2255}	4.8	0.84	58.9	2.4	100

Fig. 12 (**a**) *jV*-characteristics of planar heterojunction solar cells with a donor film thickness of 6 nm. The molecular structures of the donor materials are given in (**b**). The device stack is given in (**c**). The device parameters are given in the table (**d**). As the illumination intensities vary slightly, the given j_{sc}s are scaled to 100 mW/cm^2 for better comparability [13]

To achieve higher efficiencies, bulk heterojunction (BHJ) devices are necessary because of the limited exciton diffusion length.[10] Control over the morphology of the blend is achieved by changing the deposition rates and particularly the temperature of the substrate during evaporation of the blend, which will be discussed in the next section in more detail.

5.1 *On the Influence of Substrate Heating on Morphology*

Figure 13 shows an example of how the performance changes from a PHJ to a BHJ to a substrate-heated BHJ device. PHJ devices lack photocurrent as a result of the limited donor thickness but exhibit high FF caused by the high mobility and low recombination in pure layers. Blend films with higher thicknesses deliver higher photocurrent but suffer from high recombination and lower V_{oc} because small-molecule blends are usually too finely mixed. To increase the separation of donor and acceptor and increase their respective phase sizes, the substrate is heated during the deposition of the blend layer, which leads to even higher j_{sc} and a high FF, which is similar to the PHJ device.

[10] For these oligothiophene derivatives, values between 5–10 nm were measured [42, 43].

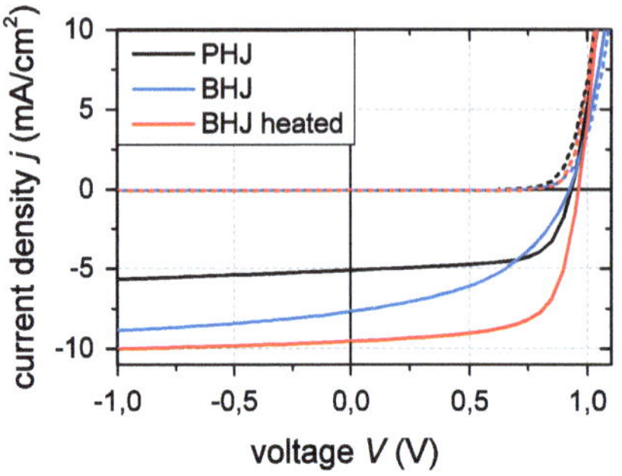

	j_{sc} (mA/cm²)	V_{oc} (V)	FF (%)	PCE (%)
PHJ	5.0	0.93	67.3	3.1
BHJ	7.6	0.93	46.0	3.2
BHJ heated	9.5	0.96	67.6	6.1

Fig. 13 Comparison of the solar cell performance of planar heterjunction (PHJ), bulk heterjunction (BHJ), and substrate-heated BHJ devices. The $J - V$ characteristics are shown to the *left*, whereas the device parameters are shown in the table to the *right*. All devices are made with the same stack, shown in Fig. 12b, differing only in their active layers (6 nm donor in the PHJ and 30 nm of donor:C60 blend in the BHJ). The molecular structure of the donor molecule DCV5T-Me33 is shown in Fig. 7

The technique of substrate heating has been applied to DCV*n*T materials by Wynands et al. [19]. The increasing phase separation was indicated by a decrease in the donor PL quenching, whereas X-ray diffraction, optical absorption, and atomic force microscopy measurements demonstrated an increase in the crystallinity of the film. Overall, the improved charge collection efficiency led to an increase in the IQE from 50 % to 80 % and a concomitant efficiency increase from 1.6 % to 3.8 %. Using PIA and impedance spectroscopy, we also observed an increase in the charge carrier lifetime for a substrate-heated blend film using a DCV5T derivative as donor material [32].

However, when the temperature becomes too high, the crystallinity and phase separation evolve to unfavorable dimensions, resulting in rough films and high exciton recombination. A particularly interesting example is the unsubstituted DCV4T, which shows the best performance without substrate heating (see Fig. 14a) [21]. When the substrate is heated during deposition, the molecules agglomerate in elongated worm-like structures on the surface of the film (d). The crystallinity of both phases is strongly increased (e), but because of the lower coverage of the surface, the absorption is decreased (c), leading to a reduction in the photocurrent.[11] This behavior is still persistent when short side groups are attached to the core molecule. DCV5T-Me33 shows a similarly strong decrease in the PCE for substrate temperatures beyond 90 °C (see Fig. 14f) and a detrimental increase in both crystallinity and phase separation [13, 44]. For longer side chains, this overly strong morphology transition is smoother or happens at temperatures that are irrelevant in practice [28].

[11]The EQE is exclusively lowered in the range where the donor absorbs; see Fig. 14b.

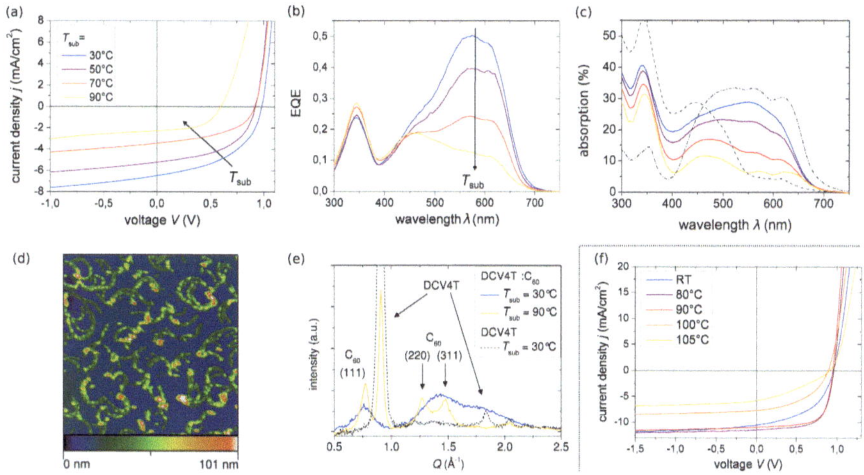

Fig. 14 (**a**) $J - V$ characteristics of solar cells with the DCV4T:C_{60} mixed active layer deposited onto a heated substrate at $T = T_{sub}$. (**b**) External quantum efficiency spectra corresponding to the devices in (**a**). (**c**) Absorption spectra of the blends deposited on glass at $T = T_{sub}$. The spectra are corrected for increased scattering by a linear approximation [13]. For comparison, absorption spectra of pure films of DCV4T and C_{60} are shown. (**d**) Atomic force microscopy topography ($5 \times 5\,\mu m^2$) of the DCV4T:C_{60} blend film deposited at $T_{sub} = 50\,°C$. (**e**) Grazing-incidence X-ray diffraction patterns of the blend deposited at $T_{sub} = 30\,°C$ and $90\,°C$. For comparison, the diffraction pattern of a pure donor film deposited on a room-temperature substrate is given. The pattern of a pure C_{60} film can be reviewed in Fig. 7. (**f**) Substrate temperature dependence of equivalent solar cells with DCV5T-Me^{33} as donor material. [Parts (**a**)–(**e**) are reprinted/adapted from Koerner et al., Organic Electronics, 13(4), 623, (2013), copyright 2013, with permission from Elsevier. The data of part (**c**) was reprocessed as detailed in [13]]

6 Summary

Oligothiophene derivatives have demonstrated outstanding properties for application in SMOSCs. After the first successful devices with a PCE of 3.4 % with DCV5T-Bu^{2244} [1], we continuously improved this value to 4.9 % with DCV6T-Bu^{1256} [36], and 5.2 % with DCV5T [4] by side chain variation and device optimization. Finally, the highest efficiency of 6.1 % was obtained with DCV5T-Me^{33} in a test device [6], which was then further optimized to a certified record value of 8.3 % [2]. Those results were achieved in close collaboration with the Heliatek company, taking advantage of their experience in this field. We also used DCV5T-Me^{33} in homo-multijunction (tandem and triple) structures, giving up to 9.7 % PCE [2]. The optical simulation for the optimization of transport and active layer thicknesses is in good agreement with the EQE measurement on the final devices. Furthermore, those examples nicely demonstrate the advantages of multijunction devices: In a comparison of lab-size devices with an increased active area of $1\,cm^2$, the reduced photocurrent density at higher voltage in multijunction devices decreases resistive losses, giving higher FFs. Beyond those attractive efficiency values, the presented DCV*n*T molecule class is ideal to study structure–property relationships resulting from the A–D–A architecture. Those results are important to guide the way for further improvements versus targeted materials design.

Here, we show that the temperature stability and, therefore, the processing ability decrease with increasing length of the molecular backbone. The melting temperatures are furthermore dependent on the side chains, which determine the packing in solid state. Furthermore, the pronounced odd–even effect in the melting temperatures mirrors the symmetry of the molecules.

The HOMO of the molecules is determined by the backbone length, the LUMO by the acceptor, whereas alkyl side chains show only a minor effect on the energy levels. However, because they influence the solid-state packing, their properties influence optical properties and determine the charge transport. Overall, short side chains (Me) show the best performance, with good processing properties, closest packing, and highest hole mobility, finally giving the highest PCE values in solar cells.

The presented photophysical investigations reveal that back transfer of charge carriers from the CT state to the donor triplet exciton is the reason for the low external quantum yield in DCV4T materials. Furthermore, we determined activation energies for the generation of charge carriers in D–A blend layers, which can be correlated to the hole mobility of the donor, indicating that charge-transport properties are closely related to the process of charge carrier generation.

Further optimization could be performed in several directions: Increasing the stability of the molecules is necessary with respect to production, because easy purification and processing is mandatory for reliability and reproducibility. Increasing the charge carrier mobility would allow thicker films to enhance the absorption in the device. The blend layer of DCV5T-Me[33] and C_{60} in the above-mentioned high-efficiency single-junction device is only 40 nm thick, corresponding to less than 30 nm of oligothiophene material. Furthermore, according to the above discussion about activation energies for charge carrier generation, an enhanced mobility could further improve the IQE for higher photocurrents and FFs. The latter might also be achieved by reducing the triplet exciton losses by reducing the gap between CT and triplet states. Finally, complementary absorber materials in the blue and near-infrared region are necessary for high-efficiency multijunction devices beyond the realized homojunctions.

In summary, both structure–property relationships and a deeper understanding of microscopic processes are necessary for further progress in the field of (small-molecule) organic photovoltaics. Oligothiophene derivatives have proven to allow for both, demonstrating their potential and importance in this field.

Acknowledgments The authors gratefully acknowledge the following people for their contribution to this work: S. Olthof, C. Uhrich, M. Tietze, and M. Schwarze for UPS measurements; M. P. Hein and J. Jankowski for mobility measurements; C. Elschner, N. Cates-Miller, and M. McGehee (+team) for thin-film X-ray measurements; F. Selzer and D. Schütze and the IAPP Lesker team for preparation of samples; A. Petrich for material purification; M. Pfeiffer and C. Uhrich from Heliatek for our fruitful collaboration; E. Mena-Osteritz and R. Gresser for DFT calculation and discussion; M. Weil for single-crystal X-ray analysis. The authors gratefully acknowledge the DFG for funding this project within the priority program 1355 (LE747-38 and BA 880/5).

References

1. Schulze K, Uhrich C, Schüppel R, Leo K, Pfeiffer M, Brier E, Reinold E, Bäuerle P (2006) Efficient vacuum-deposited organic solar cells based on a new low-bandgap oligothiophene and fullerene C60. Adv Mater 18(21):2872–2875
2. Meerheim R, Körner C, Leo K (2014) Highly efficient organic multi-junction solar cells with a thiophene based donor material. Appl Phys Lett 105(6):063306
3. Schueppel R, Schmidt K, Uhrich C, Schulze K, Wynands D, Brédas J, Brier E, Reinold E, Bu HB, Baeuerle P, Maennig B, Pfeiffer M, Leo K (2008) Optimizing organic photovoltaics using tailored heterojunctions: a photoinduced absorption study of oligothiophenes with low band gaps. Phys Rev B 77(8):085311
4. Fitzner R, Reinold E, Mishra A, Mena-Osteritz E, Ziehlke H, Körner C, Leo K, Riede M, Weil M, Tsaryova O, WeißA, Uhrich C, Pfeiffer M, Bäuerle P (2011) Dicyanovinyl-substituted oligothiophenes: structure-property relationships and application in vacuum-processed small molecule organic solar cells. Adv Funct Mater 21(5):897–910
5. Fitzner R, Elschner C, Weil M, Uhrich C, Körner C, Riede M, Leo K, Pfeiffer M, Reinold E, Mena-Osteritz E, Bäuerle P (2012) Interrelation between crystal packing and small-molecule organic solar cell performance. Adv Mater 24(5):675–680
6. Fitzner R, Mena-Osteritz E, Mishra A, Schulz G, Reinold E, Weil M, Körner C, Ziehlke H, Elschner C, Leo K, Riede M, Pfeiffer M, Uhrich C, Bäuerle P (2012) Correlation of n-conjugated oligomer structure with film morphology and organic solar cell performance. J Am Chem Soc 134(27):11064–11067
7. Schulz GL, Urdanpilleta M, Fitzner R, Brier E, Mena-Osteritz E, Reinold E, Bäuerle P (2013) Optimization of solution-processed oligothiophene: fullerene based organic solar cells by using solvent additives. Beilstein J Nanotechnol 4(1):680–689
8. Tamao K, Kodama S, Nakajima I, Kumada M, Minato A, Suzuki K (1982) Nickel-phosphine complex-catalyzed Grignard coupling–II. Tetrahedron 38(22):3347–3354
9. Qi T, Liu Y, Qiu W, Zhang H, Gao X, Liu Y, Lu K, Du C, Yu G, Zhu D (2008) Synthesis and properties of fluorene or carbazole-based and dicyanovinyl-capped n-type organic semiconductors. J Mater Chem 18(10):1131–1138
10. Raposo MMM, Fonseca AMC, Kirsch G (2004) Synthesis of donor–acceptor substituted oligothiophenes by Stille coupling. Tetrahedron 60(18):4071–4078
11. van Pham C, Macomber RS, Mark HB, Zimmer H (1984) Lithiation reaction of 2,5-dibromothiophene. Carbon-13 NMR spectra of 3-substituted derivatives. J Org Chem 49(26):5250–5253
12. Asawapirom U, Güntner R, Forster M, Farrell T, Scherf U (2002) Dialkylfluorene-oligothiophene and dialkylfluorene-dithienylvinylene alternating copolymers. Synthesis 2002(09):1136–1142
13. Körner C (2013) Oligothiophene materials for organic solar cells—photophysics and device properties. Dissertation, TU Dresden
14. Fitzner R, Mena-Osteritz E, Walzer K, Pfeiffer M, Bäuerle P (2015) A-D-A-type oligothiophenes for small molecule organic solar cells: extending the n-system by introduction of ring-locked double bonds. Adv Funct Mater 25:1845–1856
15. Steinberger S, Mishra A, Reinold E, Levichkov J, Uhrich C, Pfeiffer M, Bäuerle P (2011) Vacuum-processed small molecule solar cells based on terminal acceptor-substituted low-band gap oligothiophenes. Chem Commun 47(7):1982–1984
16. Pappenfus TM, Burand MW, Janzen DE, Mann KR (2003) Synthesis and characterization of tricyanovinyl-capped oligothiophenes as low-band-gap organic materials. Org Lett 5(9):1535–1538

17. Ziehlke H, Fitzner R, Koerner C, Gresser R, Reinold E, Bäuerle P, Leo K, Riede MK (2011) Side chain variations on a series of dicyanovinyl-terthiophenes: a photoinduced absorption study. J Phys Chem A 115(30):8437–8446
18. Uhrich C, Schueppel R, Petrich A, Pfeiffer M, Leo K, Brier E, Kilickiran P, Baeuerle P (2007) Organic thin-film photovoltaic cells based on oligothiophenes with reduced bandgap. Adv Funct Mater 17(15):2991–2999
19. Wynands D, Levichkova M, Riede M, Pfeiffer M, Baeuerle P, Rentenberger R, Denner P, Leo K (2010) Correlation between morphology and performance of low bandgap oligothiophene:C60 mixed heterojunctions in organic solar cells. J Appl Phys 107(1):2–7
20. Ziehlke H (2012) Excited state properties in dicyanovinyl-oligothiophene donor materials for small molecule organic solar cells. Dissertation, TU Dresden
21. Koerner C, Elschner C, Miller NC, Fitzner R, Selzer F, Reinold E, Bäuerle P, Toney MF, McGehee MD, Leo K, Riede M (2012) Probing the effect of substrate heating during deposition of DCV4T:C60 blend layers for organic solar cells. Org Electron 13(4):623–631
22. Gommans H, Cheyns D, Aernouts T, Girotto C, Poortmans J, Heremans P (2007) Electro-optical study of subphthalocyanine in a bilayer organic solar cell. Adv Funct Mater 17:2653–2658
23. Griffith OL, Forrest SR (2014) Exciton management in organic photovoltaic multidonor energy cascades. Nano Lett 14(5):2353–2358
24. Brown P, Thomas D, Köhler A, Wilson J, Kim JS, Ramsdale C, Sirringhaus H, Friend R (2003) Effect of interchain interactions on the absorption and emission of poly(3-hexylthiophene). Phys Rev B 67(6):064203
25. Gierschner J, Cornil J, Egelhaaf HJ (2007) Optical bandgaps of n-conjugated organic materials at the polymer limit: experiment and theory. Adv Mater 19(2):173–191
26. Schrader M, Fitzner R, Hein M, Elschner C, Baumeier B, Leo K, Riede M, Bäuerle P, Andrienko D (2012) Comparative study of microscopic charge dynamics in crystalline acceptor-substituted oligothiophenes. J Am Chem Soc 134(13):6052–6056
27. Schrader M, Körner C, Elschner C, Andrienko D (2012) Charge transport in amorphous and smectic mesophases of dicyanovinyl-substituted oligothiophenes. J Mater Chem 22(41):22258–22264
28. Wynands D (2011) Strategies for optimizing organic solar cells. Dissertation, TU Dresden
29. Yoshida H (2013) New experimental method to precisely examine the LUMO levels of organic semiconductors and application to the fullerene derivatives. MRS Proc 1493:1
30. Schueppel R, Uhrich C, Pfeiffer M, Leo K, Brier E, Reinold E, Baeuerle P (2007) Enhanced photogeneration of triplet excitons in an oligothiophene-fullerene blend. Chem Phys Chem 8(10):1497–1503
31. Veldman D, Meskers SCJ, Janssen RAJ (2009) The energy of charge-transfer states in electron donor–acceptor blends: insight into the energy losses in organic solar cells. Adv Funct Mater 19(12):1939–1948
32. Ziehlke H, Burtone L, Koerner C, Fitzner R, Reinold E, Bäuerle P, Leo K, Riede M (2011) Increase of charge carrier lifetime in dicyanovinyl–quinquethiophene: fullerene blends upon deposition on heated substrates. Org Electron 12(12):2258–2267
33. Koerner C, Ziehlke H, Gresser R, Fitzner R, Reinold E, Bäuerle P, Leo K, Riede M, Bäuerle P (2012) Temperature activation of the photoinduced charge carrier generation efficiency in quaterthiophene: C60 mixed films. J Phys Chem C 116(47):25097–25105
34. Scholz R, Luschtinetz R, Seifert G, Jägeler-Hoheisel T, Körner C, Leo K, Rapacioli M (2013) Quantifying charge transfer energies at donor-acceptor interfaces in small-molecule solar cells with constrained DFTB and spectroscopic methods. J Phys Condens Matter 25:473201
35. Schlenker CW, Thompson ME (2011) The molecular nature of photovoltage losses in organic solar cells. Chem Commun 47(13):3702–3716
36. Wynands D, Levichkova M, Leo K, Uhrich C, Schwartz G, Hildebrandt D, Pfeiffer M, Riede M (2010) Increase in internal quantum efficiency in small molecular oligothiophene: C_{60} mixed heterojunction solar cells by substrate heating. Appl Phys Lett 97(7):073503

37. Koerner C, Hein MP, Kažukauskas V, Sakavičius A, Janonis V, Fitzner R, Bäuerle P, Leo K, Riede M (2014) Correlation between temperature activation of charge-carrier generation efficiency and hole mobility in small-molecule donor materials. ChemPhysChem 15:1049–1055
38. Walzer K, Maennig B, Pfeiffer M, Leo K (2007) Highly efficient organic devices based on electrically doped transport layers. Chem Rev 107(4):1233–1271
39. Falkenberg C, Leo K, Riede MK (2011) Improved photocurrent by using n-doped 2,3,8,9,14,15-hexachloro-5,6,11,12,17,18-hexaazatrinaphthylene as optical spacer layer in p-i-n type organic solar cells. J Appl Phys 110(12):124509
40. Schueppel R, Timmreck R, Allinger N, Mueller T, Furno M, Uhrich C, Leo K, Riede M (2010) Controlled current matching in small molecule organic tandem solar cells using doped spacer layers. J Appl Phys 107(4):1–6
41. Riede M, Uhrich C, Widmer J, Timmreck R, Wynands D, Schwartz G, Gnehr WM, Hildebrandt D, Weiss A, Hwang J, Sundarraj S, Erk P, Pfeiffer M, Leo K (2011) Efficient organic tandem solar cells based on small molecules. Adv Funct Mater 21(16):3019–3028
42. Wynands D, Männig B, Riede M, Leo K, Brier E, Reinold E, Bäuerle P (2009) Organic thin film photovoltaic cells based on planar and mixed heterojunctions between fullerene and a low bandgap oligothiophene. J Appl Phys 106(5):054509
43. Uhrich C (2008) Strategien zur Optimierung organischer Solarzellen: Dotierte Transportschichten und neuartige Oligothiophene mit reduzierter Bandlücke. Dissertation, TU Dresden [German]
44. Moench T, Friederich P, Holzmueller F, Rutkowski B, Benduhn J, Strunk T, Koerner C, Vandewal K, Czyrska-Filemonowicz A, Wenzel W, Leo K (2016) Adv Energy Mater 1501280. doi: 10.1002/aenm.201501280

Charge Separation at Nanostructured Molecular Donor–Acceptor Interfaces

Andreas Opitz, Rupak Banerjee, Stefan Grob, Mark Gruber, Alexander Hinderhofer, Ulrich Hörmann, Julia Kraus, Theresa Linderl, Christopher Lorch, Andreas Steindamm, Anna Katharina Topczak, Andreas Wilke, Norbert Koch, Jens Pflaum, Frank Schreiber, and Wolfgang Brütting

Contents

A. Opitz (✉) • A. Wilke
Department of Physics, Humboldt-Universität zu Berlin, Newtonstraße 15, 12489 Berlin, Germany
e-mail: Andreas.Opitz@hu-berlin.de

N. Koch
Department of Physics, Humboldt-Universität zu Berlin, Newtonstraße 15, 12489 Berlin, Germany

Helmholtz-Zentrum Berlin für Materialien und Energie GmbH, Albert-Einstein-Straße 15, 12489 Berlin, Germany
e-mail: Norbert.Koch@physik.hu-berlin.de

R. Banerjee • A. Hinderhofer • C. Lorch • F. Schreiber
Institut für Angewandte Physik, Eberhard-Karls Universität Tübingen, Auf der Morgenstelle 10, 72076 Tübingen, Germany
e-mail: Frank.Schreiber@uni-tuebingen.de

S. Grob • M. Gruber • U. Hörmann • J. Kraus • T. Linderl • W. Brütting
Institute of Physics, University of Augsburg, Universitätsstraße 1, 86135 Augsburg, Germany
e-mail: Wolfgang.Bruetting@physik.uni-augsburg.de

A. Steindamm • A.K. Topczak
Experimental Physics VI, Julius Maximilian University of Würzburg, Am Hubland, 97074 Würzburg, Germany

K. Leo (ed.), *Elementary Processes in Organic Photovoltaics*,
Advances in Polymer Science 272, DOI 10.1007/978-3-319-28338-8_4

Abstract Planar and bulk heterojunctions of organic donor and acceptor molecules are used to understand elementary processes in photovoltaic cells. The electronic structure, interface and film morphology, excitonic behavior, device characteristics, and correlations between these properties are reviewed here using a wide range of material combinations.

Keywords Device performance • Electronic structure • Exciton transport • Interface morphology

1 Introduction to Molecular Donor–Acceptor Solar Cells

Organic photovoltaic cells (OPVCs) have recently surmounted the psychologically important barrier of 10 % efficiency for the conversion of incoming solar to electrical power, and the current record value for the power conversion efficiency of molecular OPVCs has reached more than 11 % [1, 2]. The rapid progress over recent years poses the question of what the fundamental efficiency limits of these materials are? For inorganic photovoltaics, this question was answered more than 50 years ago by the well-known Shockley–Queisser (SQ) limit [3]. Based on very few fundamental assumptions, the SQ theory is able to predict an upper limit for the power conversion efficiency of single-junction cells with the energy gap of the active semiconductor material being the only input parameter. Because of their excitonic nature, however, an important modification of this concept is required for OPVCs, where a donor–acceptor (D–A) heterojunction is used to enable charge separation. This is related to the existence of a charge-transfer (CT) state (see Fig. 1) as an intermediate in the dissociation process from the photoexcited (excitonic) state on one molecular species toward free charge-separated states on both molecular species [4–7].

The overall charge-generation process can be split into four steps [9], as illustrated in Fig. 1:

1. absorption of light with photon energy above the optical gap E_{opt} and generation of excitons (ideally in both materials, but here shown for simplicity only in the donor);

J. Pflaum
Experimental Physics VI, Julius Maximilian University of Würzburg, Am Hubland, 97074 Würzburg, Germany

Bavarian Center for Applied Energy Research (ZAE Bayern e.V.), Magdalene-Schoch-Straße 3, 97074 Würzburg, Germany
email: Jens.Pflaum@physik.uni-wuerzburg.de

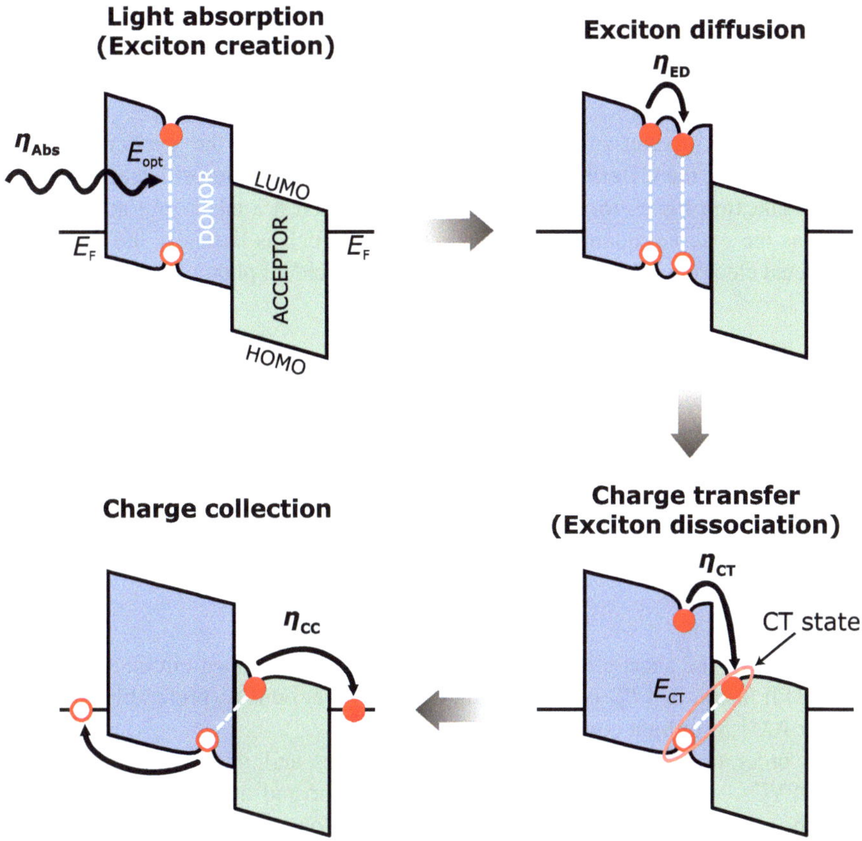

Fig. 1 Basic processes in organic solar cells visualized in a schematic energy diagram of a D–A cell under short-circuit conditions. E_F are the Fermi energies of the cathode and anode contacts, respectively. LUMO is the lowest unoccupied molecular orbital, and HOMO is the highest occupied molecular orbital of the organic film. *Filled circles* represent electrons and *open circles* holes. A *line* between an electron and a hole symbolizes an exciton, with a dip in the energy levels depicting the lowering of energy by Coulomb interaction between electron and hole. [Adapted from [8]]

2. exciton diffusion from the bulk of the film to the D–A interface, where the exciton diffusion length determines the fraction of the layer that is actually relevant;
3. exciton dissociation and formation of a CT state at the interface, which is still coulombically bound;
4. charge separation and carrier collection at the electrodes to produce electrical power in the external circuit.

Accordingly, the internal quantum efficiency is given by [9]

$$\eta_{\text{int}} = \eta_{\text{Abs}} \cdot \eta_{\text{ED}} \cdot \eta_{\text{CT}} \cdot \eta_{\text{CC}} , \tag{1}$$

which is the product of the absorption efficiency η_{Abs}, the exciton diffusion efficiency η_{ED}, the CT efficiency η_{CT}, and the charge-collection efficiency η_{CC}.

If reflection losses for coupling light into the cell are taken into account, one obtains the external quantum efficiency (EQE), which is basically the number of collected electrons with respect to the number of incident photons,

$$\eta_{\text{ext}} = (1 - R) \cdot \eta_{\text{int}}, \tag{2}$$

with R being the reflectivity of the device. η_{ext} is experimentally determined by the measured current density at short-circuit conditions divided by the incident light intensity at a given wavelength and is also termed the incident photon-to-current efficiency (IPCE).

The overall power conversion efficiency of a solar cell is given by

$$\eta_{\text{P}} = \frac{J_{\text{SC}} \cdot V_{\text{OC}} \cdot \text{FF}}{P_{\text{in}}} , \tag{3}$$

where J_{SC} is the short-circuit current density, V_{OC} is the open-circuit voltage, FF is the fill factor, and P_{in} is the incident optical power density, preferably measured under AM1.5 sunlight conditions.

In order to realize different nanomorphologies and investigate their impact on OPVCs, we chose a series of molecular donor and acceptor materials from which we prepared different D–A heterojunctions. The chemical structures of α-sexithiophene (6T), copper phthalocyanine (CuPc), pentacene (PEN), tetraphenyldibenzoperiflanthene (DBP), diindenoperylene (DIP), Buckminster fullerenes C60 and C70, perfluorinated CuPc (F16CuPc), and perfluorinated PEN (PFP), which were all deposited as thin films by vacuum sublimation, are shown in Fig. 2. For all of these materials the energy levels corresponding to the highest occupied molecular orbital (HOMO) and the lowest unoccupied molecular orbital (LUMO) are given as the solid-state ionization potential (IP) and electron affinity (EA) in Fig. 3. A wide range of energy levels is covered. As long as the HOMO and LUMO levels of the acceptor are lying lower than the respective levels of the donor (staggered or type II heterojunction [21]) and the energy offset between the HOMO and LUMO levels of the two materials is large enough [22], photovoltaic action appears and the difference between the IP of the donor and the EA of the acceptor—the CT energy or photovoltaic gap—is an upper limit for V_{OC} [22, 23].

Another material parameter that is important for solar cell performance is the optical absorption spectrum, which is obtained from transmission measurements

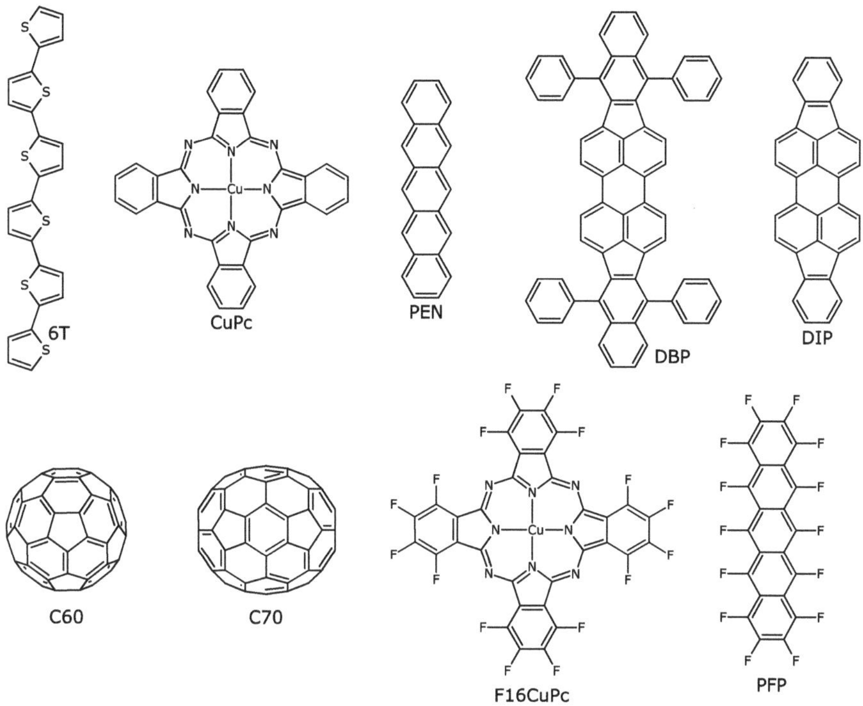

Fig. 2 Chemical structures of molecular materials (6T α-sexithiophene, CuPc copper phthalocyanine, PEN pentacene, DBP tetraphenyldibenzoperiflanthene, DIP diindenoperylene, C60 and C70 Buckminster fullerenes, F16CuPc perfluorinated CuPc, PFP perfluorinated pentacene)

and is shown in Fig. 4 for the above-mentioned materials. Spectral complements as for CuPc and C60 [24] or a higher absorption strength of DBP and C70 in comparison to DIP and C60, respectively, [25] are advantageous. However, the interplay between optical absorption, energy levels, morphology, exciton diffusion, and transport has to be balanced to improve the overall efficiency of a solar cell.

Furthermore, the architecture of a solar cell—the arrangement of donor and acceptor molecules in a device—is important. Figure 5 shows different examples. In all of these architectures an interface to separate the photogenerated excitons is present. The simplest case, as also used by Tang in 1986 [26], is the planar heterojunction (PHJ), where the acceptor layer is deposited on top of the donor film (Fig. 5a). A rougher D–A interface increases the area where charge separation can take place (Fig. 5b). The ideal case is an interdigitated interface with a separation of the columns in the range of the exciton diffusion lengths (Fig. 5c). Often this occurs in a self-organized manner in crystalline materials because of the presence of an intermediate scenario between layer and island growth. The interface area can be increased on purpose also by glancing angle deposition, as shown for PEN and DIP [27, 28]. Strong phase separation caused by the structural incompatibility of the two partners, as in the case of PEN or 6T codeposited with C60 [8, 29],

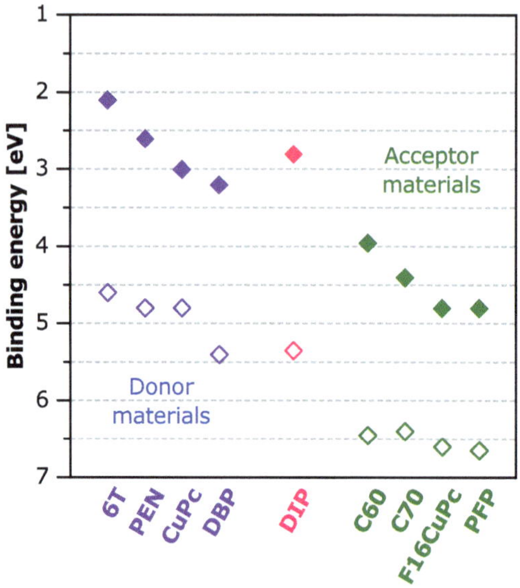

Fig. 3 Energy levels of molecular materials in the solid state (*closed symbols*: electron affinity, or EA, *open symbols*: ionization potential, or IP). The IPs were determined by ultraviolet photoelectron spectroscopy [10–15]. The EAs were taken from inverse photoelectron spectroscopy [16–19] or by adding the exciton binding energy [20] to the optical gap

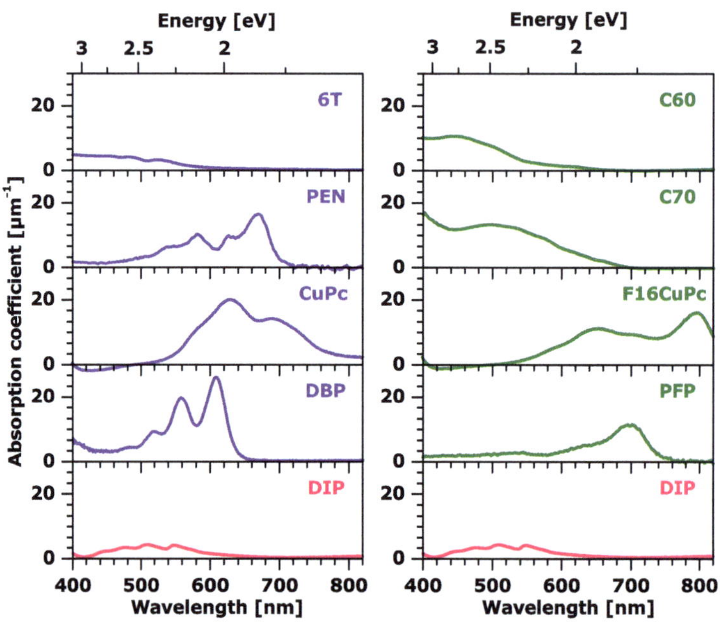

Fig. 4 Absorption spectra of the investigated materials (*left column*: donors; *right column*: acceptors). The spectra were obtained from transmission measurements on transparent substrates

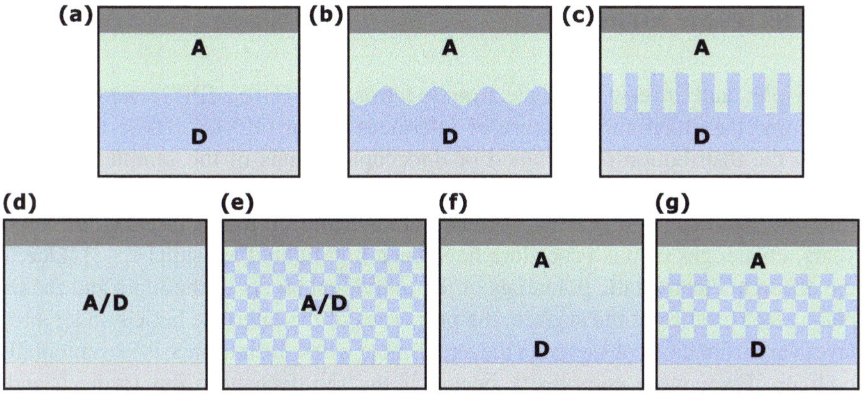

Fig. 5 Different architectures for solar cells: (**a**) planar donor–acceptor (D–A) heterojunction; (**b**) D–A heterojunction with rough interface; (**c**) D–A layers with interdigitated interface; (**d**) bulk heterojunction (molecular mixture); (**e**) phase-separated bulk heterojunction; (**f**) gradient heterojunction; (**g**) planar heterojunction with mixed interface. [Adapted from [8]]

can also lead to a morphology close to an interdigitated interface. However, the formation of leakage pathways between the electrodes by one of the materials or very high islands [8, 29] is unfavorable for solar cell performance. To cope with low exciton diffusion lengths, the D–A interface can be increased by mixing the materials in bulk heterojunctions (BHJ) [30], as illustrated for a molecular mixture and a phase-separated mixture in Fig. 5d and e, respectively [31]. Molecular mixtures are possible for the codeposition of compounds with similar shapes such as CuPc/F16CuPc [32], PEN/PFP [33], or DIP/PFP [34]. Phase separation takes place for the codeposition of molecules with different shapes such as CuPc/C60 [35] or DIP/C60 [12, 36]. The advantages of both basic structures, PHJ and BHJ, can be combined by surrounding the mixed film by neat D and A layers as shown in Fig. 5f for a graded heterojunction and in Fig. 5g for a planar-mixed heterojunction (PM-HJ). In such a case the neat layers provide carrier-selective contacts and to some extent also serve as an exciton-blocking layer to eliminate exciton quenching at the metal electrodes [25].

To maximize the overall efficiency it is not enough to improve only one of the four processes. In addition to balancing the exciton diffusion length with the optical absorption length for the optimal film thickness of absorbing materials, the morphology of the films and the respective alignment of the energy levels also influence the cell performance. In this chapter we start with an overview of the different aspects of photocurrent generation (electronic structure, interface morphology and phase separation, exciton diffusion and dissociation, CT, and device performance) and end with a more conclusive picture that includes several of these aspects. DIP/C60 serves as a model system throughout this review, but other D–A combinations are mentioned as well.

2 Electronic Structure

Ultraviolet and inverse photoelectron spectroscopy (UPS, IPES) were used to determine the electronic structure of interfaces in our OPVCs. These techniques assess the distribution of occupied or unoccupied states of the samples [16, 37–41]. Typically, a stepwise deposition enables the investigation of the underlying film, the interface as it is being formed, and the film on top as in PHJs. In recent years, challenges of this procedure have been highlighted as well[14, 42]. Due to the different electrostatic potentials of the individual layers of the films and the in-plane electrostatics of the surface, the properties of the interface buried under a top layer vary from those observed in the experiment when the interface is incrementally built up. Even more complexity occurs if the top-deposited film forms islands instead of closed layers, for the same reasons. The combination of electrostatic simulations [42] and measurements of energy levels lead to an improved and reliable understanding of the energy-level alignment at heterojunctions [43, 44]. Furthermore, the detrimental effects of the measurement itself have to be considered. The bombardment of organic materials either with UV photons or with low-energy electrons can induce damage to the molecules and thus leads to irreversible changes of the energy-level alignment [14].

In this section the focus lies on the interface between DIP and C60 in both layer sequences. Clevios™ P AI4083 (designated as PEDT:PSS) and Clevios™ HIL1.3 (designated as HIL1.3) both purchased from Heraeus Clevios GmbH (Leverkusen, Germany) are used as the hole injection layers (HILs) and contact materials.

Figure 6a and b shows the secondary electron cutoff (SECO) and the valence region UPS spectra for the layer sequence C60 on DIP on top of HIL1.3 [12, 46]. The shift of the SECO position $\Delta\Phi$ indicates a change of the vacuum level when DIP is deposited on this high-work-function HIL. In contrast, when PEDT:PSS is used instead of HIL1.3, no change in work function Φ is observed (not shown) even if the work functions of the thick films are the same. The Fermi level is pinned for both HILs inside the DIP only 0.4 eV above the HOMO onset, which represents the hole injection barrier (HIB) magnitude for PEDT:PSS as HIL. The high Φ HIL1.3 additionally generates a doped contact layer with a reduced HIB of about 0.3 eV. The SECO position and with it the work function stay constant after depositing C60 on top of DIP. Nonetheless, the offset between the HOMO onsets of the DIP and the C60 layers ΔE_{HOMO} is equal for both of the HILs used.

To analyze the unoccupied states as well, we performed combined IPES and UPS measurements on pristine DIP and C60 layers as well as at the interface of a thin C60 film deposited on top of a DIP layer [17]. The results are displayed in

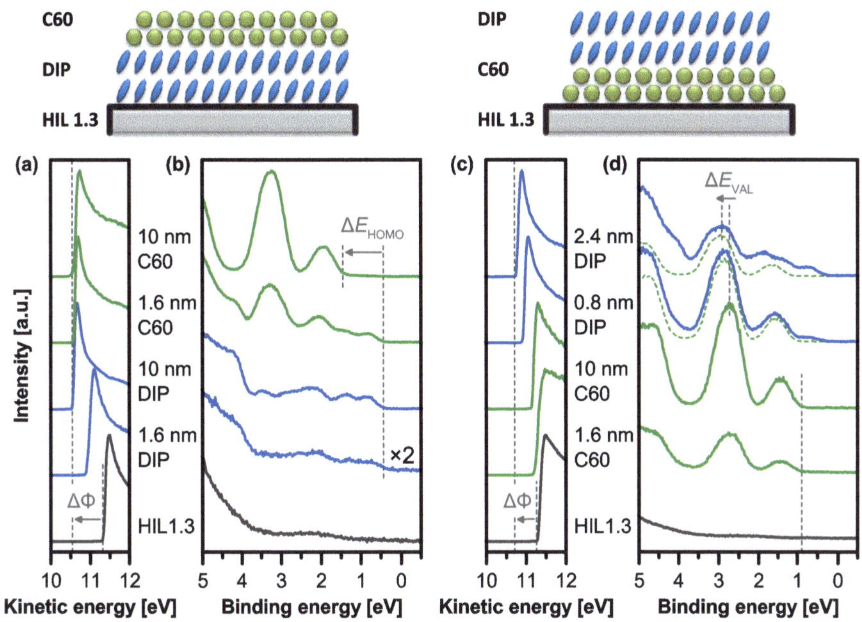

Fig. 6 Summary of the secondary electron cutoff and the valence region ultraviolet photoelectron spectroscopy measurements for the two stepwise deposited layer sequences C60 on DIP (**a** and **b**) and DIP on C60 (**c** and **d**) with a high work function HIL1.3 as the underlying contact layer. Vacuum-level shifts are marked as $\Delta\Phi$, shift of the valence region as ΔE_{VAL}, and HOMO-level offset as ΔE_{HOMO}. The *dashed lines* in part (**d**) represent the C60 content of the spectra. [Adapted from [12, 45]]

Fig. 7. Deconvolution of the interface spectra is done as a linear superposition of the pristine layer spectra, corrected for a shift of the Fermi level. This shows that the transport gaps of the DIP and C60 are 2.55 eV and 2.50 eV, respectively, and that they have the same values in pristine films and at the interface between the two materials.

The observed energy-level alignment of this layer sequence is summarized in Fig. 8a and b [12, 46]. Because of the contact doping, the reduced HIB when using HIL1.3 results in an improved fill factor of solar cells. S-shaped behavior is present in the solar cell characteristics if the work function is too low and contact doping is absent [12, 46]. In a detailed study it was shown that the S-shaped behavior can also be linked to the work function when the substrate is heated during the deposition of the DIP layer, essentially because of a lowered water content of the HIL [46]. The injection barrier is relevant at voltages around open-circuit conditions because the injection and photocurrents (of opposite direction) are balanced at this point. The offset between the HOMO of the donor and the LUMO of the acceptor, the photovoltaic gap, is an upper limit for the open-circuit voltage [22, 23], which is discussed further in the summary, Sect. 6. The vacuum level at the interface is constant in the case presented here, but it can vary by hundreds of

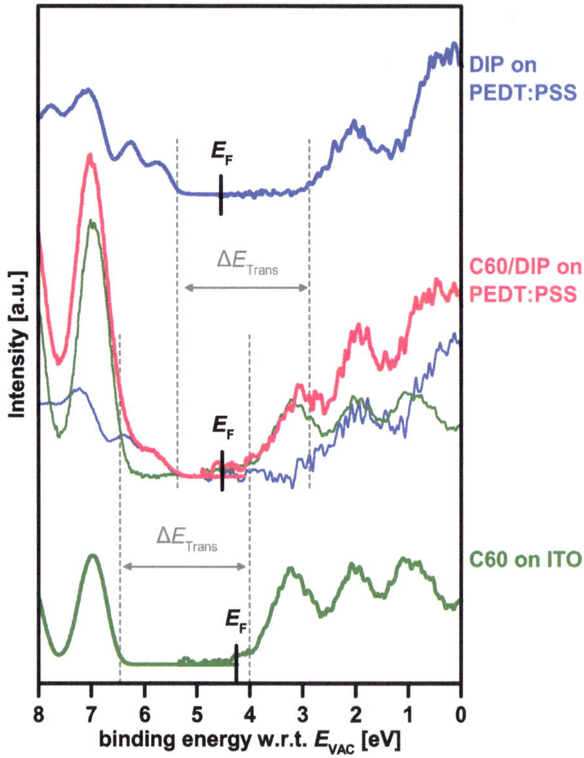

Fig. 7 Combined ultraviolet photoelectron and inverse photoelectron spectroscopy analyses of pristine layers of DIP and C60 together with the analysis of the planar heterojunction (PHJ) containing a thin C60 layer on top of a diindenoperylene film. The spectra of the PHJ can be well described as the linear superposition of the pristine film spectra. The transport gaps ΔE_{Trans}, which are invariant at the interface, have been determined for DIP 2.55 eV and for C60 2.50 eV, including broadening correction. E_F is the Fermi level for each measurement. [Adapted from [17]]

milli-electron volts (meV) in general [14]. Any difference changes the photovoltaic gap and therefore the open-circuit voltage. The correct value has to be considered when analyzing the losses in open-circuit voltage in comparison to those in the photovoltaic gap. To date this energy-level alignment could not reliably be predicted by calculations alone and thus had to be measured.

The UPS measurements for the SECO and valence regions of the layer sequence DIP on C60 on top of HIL1.3 are displayed in Fig. 6c and d [45]. The SECO and the HOMO onset positions stay constant during the deposition of C60 on HIL1.3. Upon the addition of DIP on top, the SECO shifts, indicating a lower work function of the sample. Additionally, the features of the valence region attributed to the fullerene move to a higher binding energy. The onset of the HOMO level of DIP is very close to the Fermi level. The Fermi level within the DIP layer is pinned again close to the HOMO level. As in the case of DIP in direct contact with HIL1.3, the DIP layer is contact-doped, which results from the pinning behavior. As a result of the absence of pinning for the C60 layer itself no charges are induced, giving an intrinsic interlayer. Therefore, the transfer of charges into the DIP layer has to originate from the underlying HIL1.3 (see Fig. 8c). This long-range CT happens when the layer in between the contact and the doped layer is vacuum-level aligned with the substrate, which was observed for organic as well as inorganic interlayers [43, 47]. It has

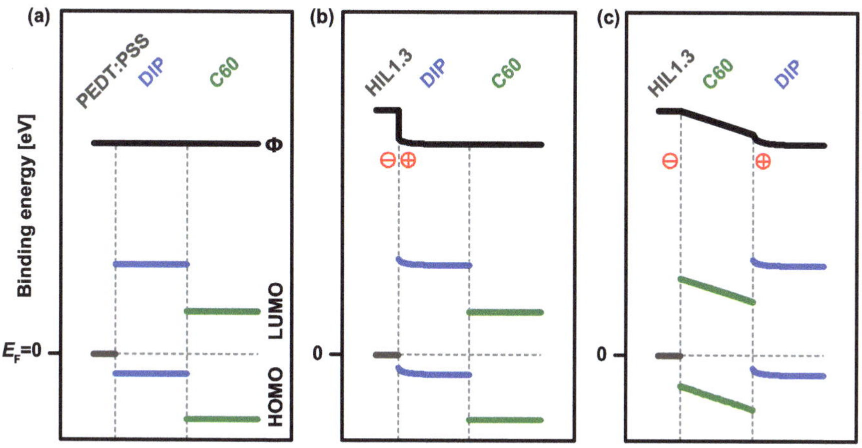

Fig. 8 Schematic energy-level alignment for three different cases of DIP/C60 planar heterojunctions. (**a**) Vacuum-level alignment at the contact and the organic–organic interface with the layer sequence PEDT:PSS/DIP/C60 [12, 14, 46]. (**b**) Contact doping of the DIP layer with the layer sequence HIL1.3/DIP/C60 [12, 14, 46] and vacuum-level alignment at the organic–organic interface. (**c**) Long-range charge transfer across the C60 layer with the layer sequence HIL1.3/C60/DIP [45]

to be noted that a CT between the two organic layers would give similar shifts in the valence region of the underlying material to higher binding energies, which, however, was ruled out for the present case.

Generally, it can be stated that most organic heterojunctions of relevance for OPVCs exhibit no charge transfer in the ground state and thus they feature a constant electrostatic potential across the interface. However, the contact to an electrode may induce Fermi-level pinning of one of the organic materials and, depending on the layer sequence, may result in long-range CT and potential shifts. We demonstrated this for a high-work-function electrode (HILs) here, but this also holds true for low-work-function electrodes, where pinning at the LUMO of one of the heterojunction materials may then occur.

As an example, the full electronic structure of an OPVC made of a C60/DIP heterojunction sandwiched between PEDT:PSS and samarium electrodes and including an exciton blocking layer of bathocuproine (BCP) is displayed in Fig. 9 [14]. Blocking layers made of BCP or bathophenanthroline (BPhen) were reported to reduce exciton quenching at the metal top contact and they simultaneously serve to protect the C60 layer from metal penetration and damage during evaporation of the top contact [46, 48, 49]. Energy-level shifts at the C60/BCP interface and inside the BCP layer are caused by a preferential orientation of BCP dipoles. The electrostatic field drop across DIP/C60 is therefore reduced compared to the full effective work-function difference of the contact materials. Even if the photovoltaic gap at the active interface is unaffected by the electrostatic potential landscape, changes in

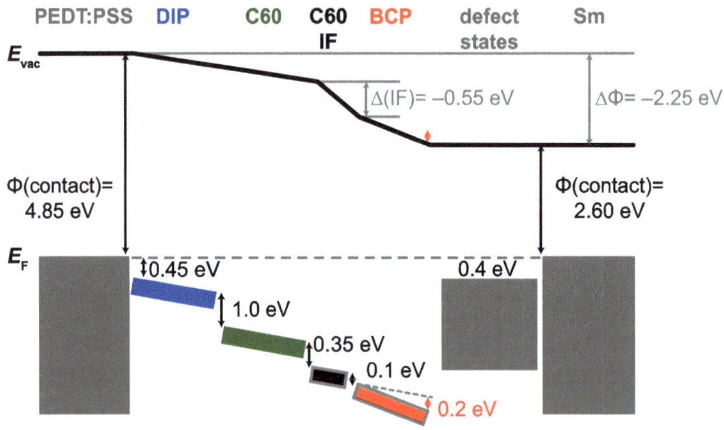

Fig. 9 Schematic energy-level diagram of the electronic structure across a PEDT:PSS/DIP/C60/BCP/Sm heterojunction photovoltaic cell. [Adapted from [14]]

the electric field distribution throughout the devices influence the charge collection efficiency [50].

3 Interface Morphology and Mixing Behavior

The morphology and crystal structure of the active layer are crucially important for the efficiency of OPVCs. The orientation of the molecules in the film determines the coupling strength of the transition dipole to the incoming light and hence strongly affects the absorption of the active layer [51, 52]. The crystal structure and the defect density in the organic films strongly influence the exciton diffusion length and also the charge-transport properties of the films, which are vital for the efficiency of the photovoltaic cell. In addition, the interface morphology, especially in PHJ cells, and the degree of phase separation in BHJ cells play a significant role for the efficiency since ideally all of the created excitons should reach the D–A interface to be separated.

There are various methods to study the interface and structure of organic thin films. One of the most prominent and widely used techniques is atomic force microscopy (AFM). However, AFM is limited to the top surface; it is impossible to obtain information on the structure further below in the film, close to the substrate. Also, it is difficult to distinguish between different materials, namely, donor and acceptor, because of the lack of contrast and hence to determine the phase separation, if any. Another suitable technique to investigate thin organic films is X-ray scattering. Different experimental geometries allow the investigation of the out-of-plane structure via X-ray reflectivity (XRR), including the extraction of the electron density profile along the surface normal, the estimation of the surface roughness, and the out-of-plane crystallinity of the film. Grazing-incidence X-ray diffraction (GIXD) can be used to reveal the in-plane crystal structure of the film, and grazing-incidence small-angle X-ray scattering allows us to determine morphological features like the surface island distance and their shape [53, 54]. One of the benefits of X-ray scattering is that it is possible to use it directly in situ during the preparation of the films and follow the film growth in a noninvasive way in real time, giving a more detailed understanding of the film growth.

In general, there are several interfaces in OPVCs, all of which are important for the efficiency. We focus on the D–A, so the organic–organic interface. For details of the interfaces between the active layer and the (metal) electrodes, which is in itself a very complex and interesting topic, we refer to the literature [55, 56]. In the following we first focus on different interfaces occurring in PHJ geometries and afterward turn to effects observed in mixtures used in BHJ cells.

3.1 Interface in PHJ Geometries

As shown in Fig. 5a–c, the interface in PHJ can have different geometries. Obviously, the schematics represent only ideal cases. In real thin films, the interfaces are usually not as well defined, and even the distinction between PHJ and BHJ is often not as clear as that depicted in Fig. 5.

The growth of an organic material on top of another one can differ from the observed growth of this material on a bare substrate [57]. The growth of the overlayer can strongly depend on the film properties of the materials underneath. The interactions between different organic materials might drastically differ from those between a bare substrate and an organic compound and can have a strong influence on the growth [31]. These interactions depend *inter alia* on the type of energetic arrangement, that is, whether dipole interactions or higher-order electronic interactions dominate the van der Waals interaction, on the sterical compatibility of the materials, which can lead to strain in the growth of the overlayer, and on the orientation of the bottom material, for instance, a lying-down or standing-up orientation of rod-like molecules.

As an example, we show how real-time XRR can be used to characterize the growth of an organic heterostructure consisting of a PFP overlayer on top of DIP [58]. From the real-time data the roughness evolution of the film was extracted (Fig. 10). The PFP first fills the voids between the DIP islands, leading to a smoothing of the DIP layer. The resulting structure is very similar to the interface sketched in Fig. 5b.

The growth of an organic compound on top of another can be different from the growth on a bare substrate. The structural order of C60 thin films is shown to be significantly improved by inserting a templating layer of DIP between the SiO2 substrate and C60. In contrast to growth on an amorphous substrate like SiO2, C60 grown on DIP exhibits alignment of the fcc-domains with the (1 1 1) plane parallel

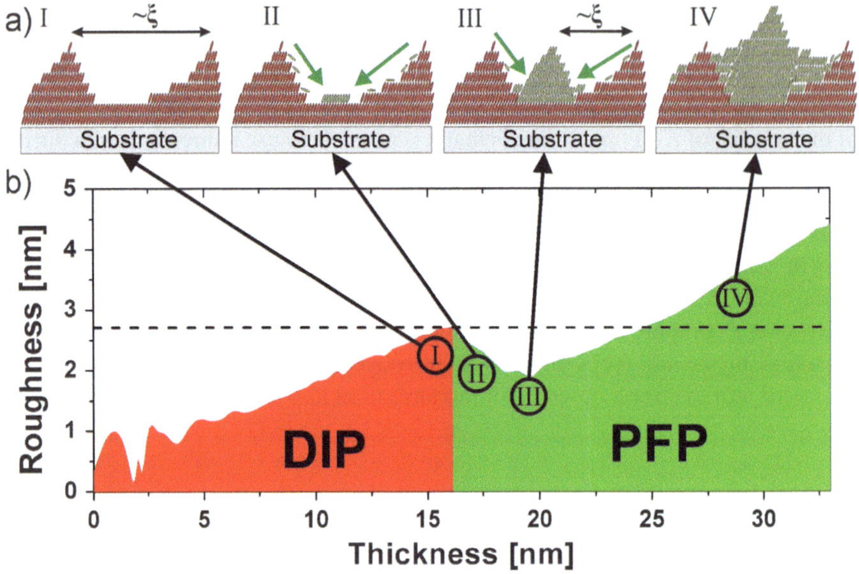

Fig. 10 (**a**) Growth of PFP islands in the voids between the DIP islands, leading to a filling of the latter. (**b**) Roughness evolution of a DIP–PFP heterostructure. [Reprinted with permission [58]]

to the substrate and a significant increase of the coherent in-plane island size by a factor of ∼4 (Fig. 11) [57].

3.2 *Phase Separation in BHJ Geometries*

Mixtures of at least two organic compounds show interesting effects during the growth of thin films. The mixing scenario of two organic compounds depends on several aspects, *inter alia* on the sterical compatibility of the two materials, the interaction energies between the two materials and between the materials and the substrate, and also on the preparation conditions, such as substrate temperature and deposition rate. According to [31, 59], the free energy of a mixture following a mean-field approach [60] can be written as

$$\frac{F_{\text{mix}}}{k_B T} = x_A \ln x_A + x_B \ln x_B + \chi x_A x_B. \tag{4}$$

Here x_A and x_B are the relative concentrations of the two materials and χ is a dimensionless interaction parameter depending on the interaction energies between molecules of the same species W_{AA}, W_{BB} and of the different species W_{AB}, respectively, and can be expressed as

$$\chi = \frac{1}{k_B T}\left[W_{AA} + W_{BB} - 2W_{AB}\right]. \tag{5}$$

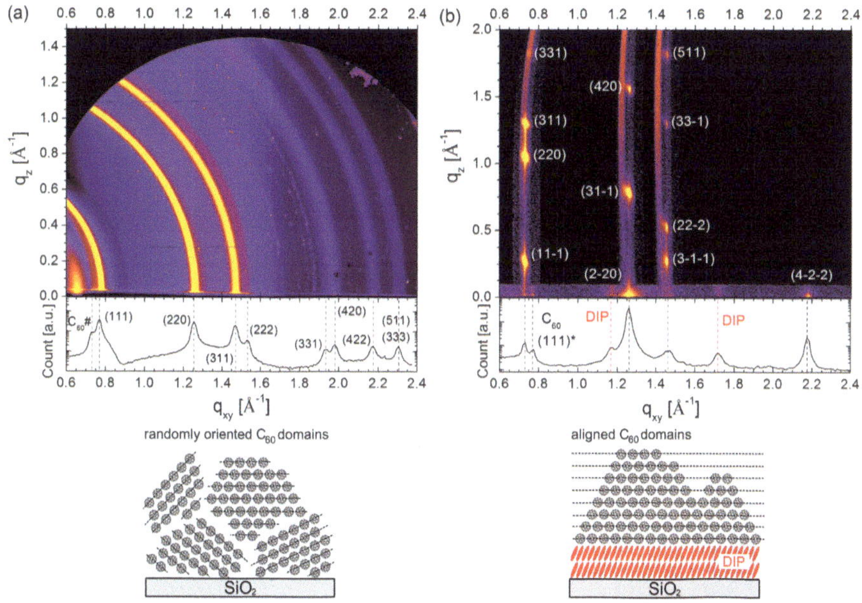

Fig. 11 (**a**) Reciprocal space map from a 60 nm C60 film recorded with a MARCCD area detector. At the bottom, additional grazing-incidence X-ray diffraction (GIXD) data measured with a point detector at $q_z = 0.02$ Å^{-1} are shown. (**b**) Two-dimensional GIXD data from a 15 nm C60 film grown on a DIP templating layer ($d = 4$ nm) indexed according to the C60 fcc structure. The data consist of four detector scans at a fixed angle of incidence at 0.1° performed with a PILATUS II area detector. Images from each data point were transformed into q coordinates and then assembled into one image. One scan was performed parallel to the substrate plane, for which the integrated GIXD data are shown at the bottom. Three scans were performed along the C60 crystal truncation rods. The schematics below show the growth of C60 on bare SiO2 and on a DIP templating layer, respectively. [Adapted from [57]]

Please note that this equation does not directly depend on the shape and dimensions, namely, the sterical compatibility, of the two molecules. As a guideline, for good intermixability, sterical compatibility can be assumed to be necessary. Depending on the value of the interaction parameter χ, three different mixing scenarios can occur (see Fig. 12):

- **Solid solution:** If the interaction energy between the two species is similar to that within one species (e.g., $\chi \approx 0$), the formation of a solid solution is favored. In this case a molecule of one species can be substituted by one of the other species. Such a mixing can usually be observed in systems of compounds with a similar shape and an energetically similar arrangement, like PEN and DIP [61].
- **Phase separation:** If the two materials are sterically incompatible [62] or the interaction energies between equal compounds are significantly more attractive than the interaction energy between different compounds ($\chi > 2$), usually phase separation of both materials is observed.

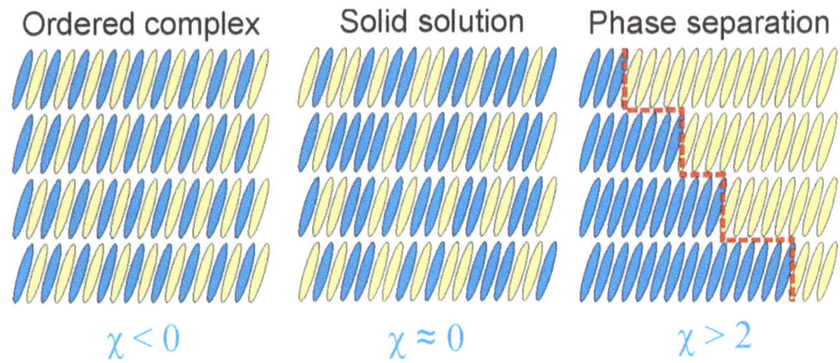

Fig. 12 Sketch of the different mixing scenarios usually observed in binary mixtures of organic semiconductors. Depending on the intermixing parameter χ, a mixed crystal, a solid solution, or phase separation can occur. [Adapted from [31]]

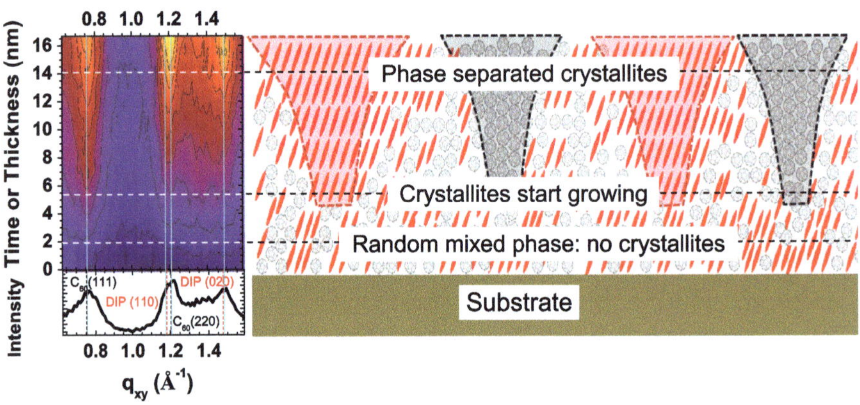

Fig. 13 The *left-hand side* shows at the top real-time grazing-incidence X-ray diffraction data from the film growth, indicating stronger and sharper peaks with increasing film thickness. On the *right-hand side* a schematic of the cross section parallel to the surface normal of a DIP:C60 mixture, shows initially a random mixed phase and then the formation and follow-up extension of domains of the pure materials, as the growth progresses in time. [Reprinted with permission from [62]]

- **Ordered complex:** An ordered complex is usually formed if the interaction energy between the two different materials is higher than the interaction energy within one species and the materials are sterically compatible [61, 63, 64].

A prototypical phase-separating system is a mixture of the two sterically incompatible materials DIP and C60. Real-time, in situ GIXD measurements revealed that in this system first a randomly mixed phase is formed. Then, after a certain film thickness the phase separation begins [62]. As the film thickness increases, the size of the crystalline domains and hence the magnitude of phase separation increase (Fig. 13). This effect shows that thin film growth is far away from equilibrium and the formation of the domains is kinetically limited. The resulting bottom-to-top asymmetry might have a significant impact on the charge and exciton transport both parallel and perpendicular to the substrate surface.

4 Exciton Diffusion and Dissociation

As indicated in the introductory section, the exciton diffusion toward the separating D–A interface constitutes an essential step in the power conversion of PHJ solar cells. Therefore, the individual molecular layer thickness has to be optimized to guarantee a maximum absorption of incident light ($\nearrow$ thickness) while simultaneously enabling photogenerated excitons to reach the D–A junction within their lifetime τ ($\searrow$ thickness). The one-dimensional diffusion relation $L_D = \sqrt{D\tau}$ translates the latter into an effective exciton diffusion length L_D. Accounting for the exponential thickness dependence of the absorption (Lambert–Beer's law) and the one-dimensional diffusion behavior (Fick's law), we illustrated in Fig. 14a the mutual relation of the different contributions and their variation with layer thickness.

Whereas the spectral absorption is mainly determined by the individual molecules and their next-neighboring coupling in (poly-)crystalline thin films, the exciton diffusion length L_D shows a pronounced dependence on the respective film morphology, in particular, on the crystalline phase stabilized for a given substrate and its structural coherence length [66]. The susceptibility of the exciton transport on the underlying morphology becomes immediately obvious considering

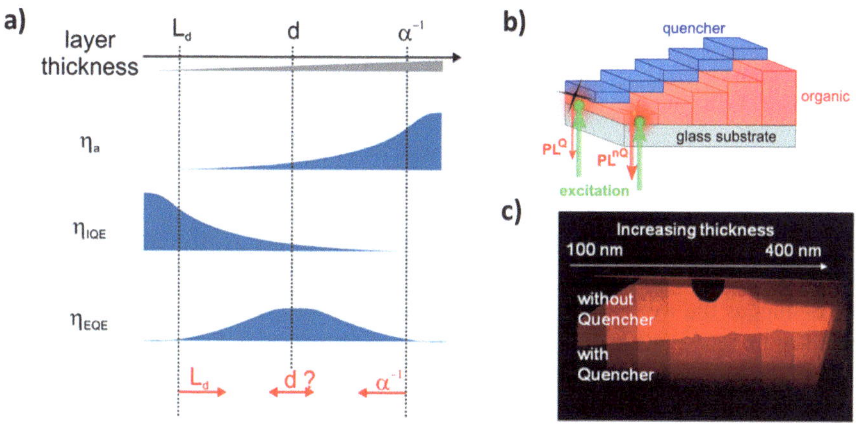

Fig. 14 (**a**) Thickness dependence of absorption depth, α^{-1}, and exciton diffusion length L_D, determining the optimum film thickness, d, of stacked photovoltaic cells. (**b**) Schematic of the photoluminescence (PL) quenching measurements. [Adapted from [65]] (**c**) Photograph of the thickness-dependent PL of DIP with and without CuPc quenching layer atop. With increasing DIP thickness, less excitons are able to reach the DIP/CuPc interface within their lifetime, effectively increasing the PL

the broad variety of L_D values reported in the literature, even for the same molecular compound on similar substrates [67].

Hence, a key challenge to access the inherent mechanisms governing exciton transport and its influence by structural or chemical inhomogeneities is the reliable determination of this property on well-defined molecular reference systems. In addition to various techniques, photoluminescence (PL) quenching measurements facilitate a straightforward approach on this parameter and, via complementary structural information and time-resolved spectroscopy, establish its relationship to sample morphology and excitonic state dynamics, respectively. The general concept of this technique is highlighted in Fig. 14 and has been discussed in detail in the literature [67, 68]. An exciton-transport layer of a defined thickness is partially covered by a quenching material, which by its energy-level positions allows for quenching of the excitons reaching the interface. This effect is visualized in Fig. 14c for a wedged DIP layer of 100–400 nm thickness, capped by a 10 nm-thick CuPc quencher layer (see schematic in Fig. 14b). Apparently, as the DIP film thickness increases, the PL intensities measured on the bare and CuPc-covered areas resemble each other, indicating that excitons are no longer able to reach the DIP/CuPc quencher interface. This quenching can appear either by a complete exciton transfer, such as by a Förster or Dexter mechanism as in the case of DIP/CuPc [65], or by converting the Frenkel-type exciton into a long-lived CT state or free-charge carriers as for DIP/C60 or ZnPc/C60 [8, 69]. As a figure of merit, the relative quenching Q defined by the PL intensity ratio with and without quencher

$$Q = \frac{I_{\mathrm{PL,Q}}}{I_{\mathrm{PL,nQ}}} \tag{6}$$

can be utilized and provides detailed information on the diffusion length and quenching quality at the quencher interface. In addition, more elaborate models can account for interference effects by the incident light as well as for the microscopic morphology of real interfaces [65, 70]. For many molecular semiconductors, growth on weakly interacting substrates such as ITO/PEDT:PSS or MoO_3 proceeds via island formation (Vollmer–Weber mode), leading to a pronounced interface roughness as indicated in Fig. 10. In the case of nonperfectly smooth interfaces in long-range ordered crystalline films, that is, where the exciton diffusion length might reach the absorption length, care has to be taken on a comprehensive modeling, as the well-established Feng–Ghosh model [71] in its initial form is not applicable.

Representative PL quenching studies on prototypical molecular semiconductors are displayed in Fig. 15 together with the structural properties as determined by X-ray diffraction. Despite the X-ray amorphous film structure, thickness-dependent PL quenching measurements on tris-(8-hydroxyquinoline)aluminum (Alq3) revealed an exciton diffusion length of about 22 nm at room temperature, by far exceeding that of many disordered organic materials [67]. Remarkably, a similar trend has been reported in the literature for C60 thin films ($L_D = 25$ nm), indicating

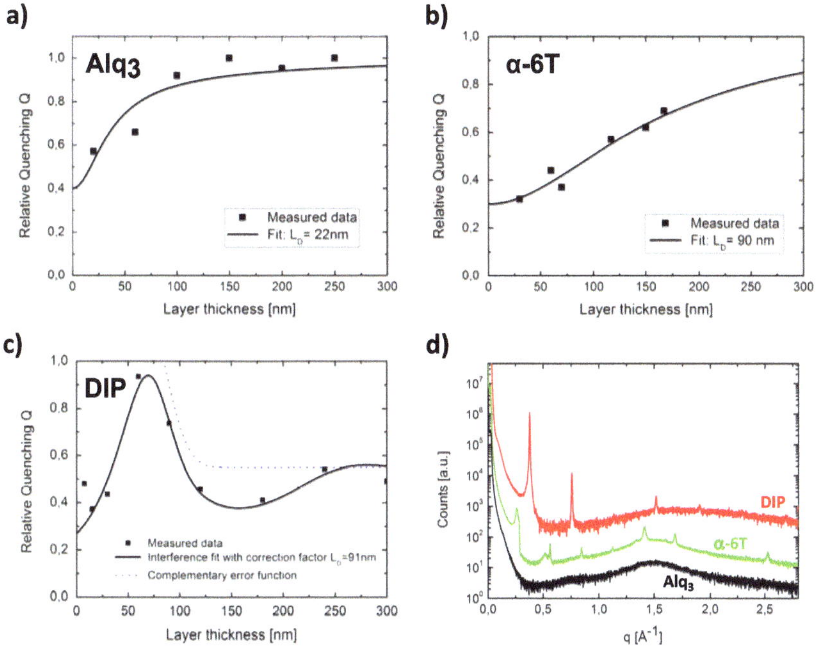

Fig. 15 (**a**)–(**c**) Measured and simulated photoluminescence quenching on various prototypical molecular semiconductors. (**d**) X-ray diffraction analyses of the corresponding thin-film structures evidencing the long-range ordering in case of DIP and 6T and the X-ray amorphous film structure of Alq3

a support of exciton motion by the spherical shape of the molecules via dense packing of the homogeneously distributed transition dipoles.

In contrast, DIP and 6T layers show a pronounced crystalline ordering in the X-ray diffraction spectra that coincides with exceptional long singlet diffusion lengths of about 90 nm for both compounds at room temperature (see Fig. 15b and c). In analogy to charge carrier motion, this might be related to a distinct wavefunction overlap along preferred directions of the crystalline lattice. However, as it turned out by comprehensive time-dependent DFT calculations (SCS-CC2 method) on a dimer level, the relative motion between adjacent molecules upon electronic excitation is essential for supporting or impeding exciton diffusion [72, 73]. Depending on the respective unit cell packing, this relative motion can lead to a continuous downward shift of the excited state on the related potential energy landscape (PEL) until a conical intersection with the PEL of an energetically lower-lying state is reached and the exciton is efficiently transferred to this dark state within its lifetime. Typically, this process occurs within several hundred femtoseconds after photoexcitation and thus defines an intrinsic localization time [73]. Whereas for many materials, including perylene derivatives, this process leads to rather short exciton diffusion lengths e.g., 22 nm for perylenetetracarboxylic dianhydride

(α-PTCDA), in agreement with the literature [66], the situation differs remarkably with DIP. Because of the specific alignment of molecular pairs within the thin-film unit cell, the intermolecular motion upon electronic excitation imposes an energy barrier on the related energy manifold, which prevents excitons from reaching the conical intersection and thereby extends their effective diffusion length. As a consequence of this steric hindrance, L_D is mainly governed by the radiative lifetime of the excited state and the spatial extension of crystalline grains. This was evidenced by temperature-dependent PL quenching studies on DIP as well as 6T rendering exciton transport to be nonthermally activated below a critical temperature and film thickness. This observation was explained by a reduced exciton–phonon scattering at low temperatures and hints at a coherent exciton motion within the crystalline grains [65]. Furthermore, in such long-range ordered crystalline systems efficient exciton trapping might occur at grain boundaries, where the structural constraints are lifted, as reported for rubrene microcrystals [74].

Motivated by the enhanced exciton diffusion length observed in DIP, various derivatives have been investigated and are shown with respect to their thickness-dependent PL quenching behavior in Fig. 16. According to simulations, the related exciton diffusion lengths amount to 5, 13, and 25 nm for tetrapropyltetraphenyldiindenoperylene (P4-Ph4-DIP), DBP (B2-Ph4-DIP), and tetraphenyldiindenoperylene (Ph4-DIP), respectively. Moreover, as can be concluded from the constant quenching offset for B2-Ph4-DIP and Ph4-DIP thicknesses above 50 nm, a roughening of the films occurs, which constitutes a precondition for a self-organized interdigitated film structure, the latter considered to be preferential for PHJ solar cells (see Fig. 5c).

Even though, because of the amorphous film structure upon deposition on unheated substrates, the diffusion lengths in these samples are inferior to that of long-range ordered DIP (L_D = 90 nm), this disadvantage is partially compensated by the much better absorption characteristics because of the randomly distributed transition dipole moments (see Fig. 4). Hence, the overall power conversion efficiencies for planar Ph4-DIP/C60 or bulk Ph4-DIP:C60 heterojunction cells resemble

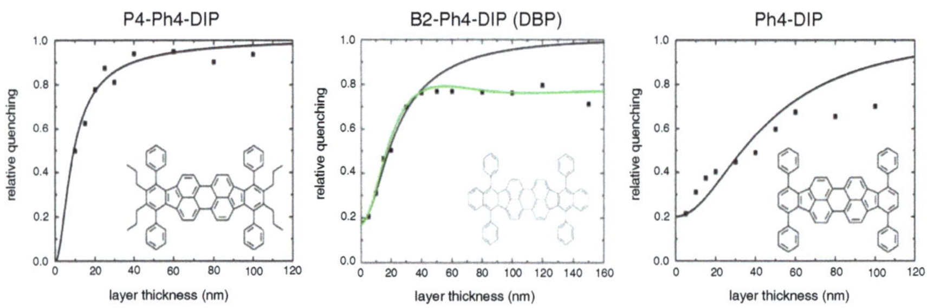

Fig. 16 Relative photoluminescence quenching measurements on various DIP derivatives (provided by courtesy of IAPP, TU Dresden). The respective exciton diffusion lengths amount to 5, 13, and 25 nm for P4-PH-DIP, B2-Ph4-DIP (DBP), and Ph4-DIP, respectively. All layers form an X-ray amorphous film structure

those of the corresponding DIP devices, but with the additional possibility of further enhancing the crystalline order by, for instance, thermal treatment [75].

Finally, addressing the nonradiative losses by exciton–polaron interaction in thin-film solar cells, we propose a new approach, allowing not only for discrimination between effects by free and localized, namely, trapped, charge carriers but also for distinction between electrons and holes. For this purpose, standard thin-film transistor (TFT) geometry has been used, enabling photogenerated excitons to interact with an adjustable amount of charge carriers within a spatially well-defined sample region of monolayer thickness. The general principle of this method is illustrated in Fig. 17a and relies on the voltage-controlled accumulation of the desired charge carrier density at the boundary between the organic transport layer and a chemically passivated gate dielectric.

Applying this technique on crystalline DIP thin films, we can measure the relative quenching Q, defined by the PL intensity ratio at various gate voltages V_G with respect to that at $V_G = 0$ V, as shown in Fig. 17b. A distinct linear variation of the relative quenching occurs up to a certain threshold voltage of about -1.7 V, whereas above this value, the quenching remains almost constant. According to TFT theory, this low-voltage regime can be attributed to gradual trap filling at the interface of the organic transport layer by injected holes (negative gate voltage), while the high-voltage regime is assigned to free-charge carriers within the conduction channel. Therefore, the exciton losses, indicated by an increase in the relative quenching, can be related to the interaction with localized polarons rather than with free holes. Moreover, the maximum quenching of about 4.5 % indicates the minor importance of nonradiative recombination losses in crystalline DIP thin films, as has been confirmed by spectrally resolved photocurrent measurements. Accordingly, the recombination rate of nonradiative exciton–hole processes can be estimated to 1.1×10^{-12} cm^3/s [76, 77], evidencing the potential of the described approach for identifying the microscopic mechanisms of exciton–charge carrier interaction as well as quantifying their respective cross sections.

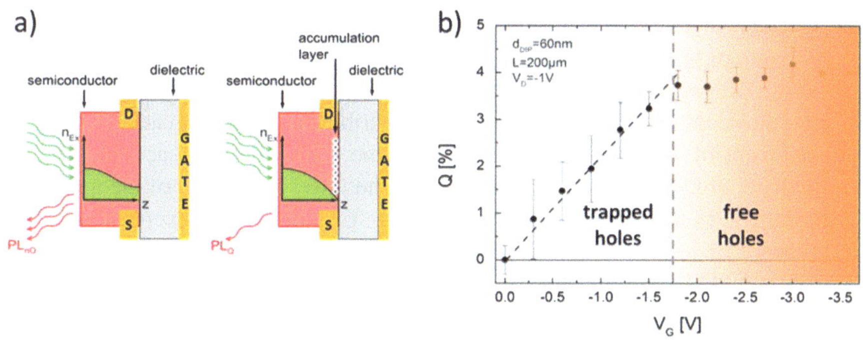

Fig. 17 (**a**) Principle of thin-film transistor–assisted determination of loss mechanisms by exciton–polaron interaction. (**b**) Relative PL quenching Q at different gate voltages, indicating the impact of trapped and free-charge carriers on nonradiative exciton losses. [Adapted from [76]]

Based on the experimental findings and theoretical results, one can develop general strategies toward a rational design of molecular compounds, further improving existing device concepts by an optimized PHJ architecture with respect to exciton generation and transport.

5 Charge Transport and Device Performance

Solar cells have been investigated for various D–A combinations shown in Figs. 2–4 (see [12, 25, 36, 78–80]). Among them DIP/C60 turned out to be a model system, because in-depth studies of the electronic structure, morphology, and exciton transport were available. Remarkably, DIP grows in a crystalline fashion on many technical substrates and, in particular, on ITO covered with PEDT:PSS or HIL1.3 used as anode for OPVCs in this study. Even without substrate heating, the molecules form crystalline domains with sizes of a few tens of nanometers and for heated substrates (with $T_S \approx 100\,°C$) the lateral domain size reaches almost micron dimensions. This high degree of structural order is reflected in good charge-transport properties of DIP, reaching carrier mobilities of $10^{-1}\,cm^2/Vs$ for holes as well as electrons in field-effect transistors [81]. This ambipolar transport behavior, together with its energy levels being intermediate between many classical donor and acceptor materials (see Fig. 3), make DIP a unique candidate for implementation as both donor or acceptor, depending on the partner material [78]. On the one hand, it is a donor with respect to fullerenes, but with significantly larger V_{OC} than CuPc, 6T, or PEN. On the other hand, it can also be used as an acceptor for 6T and its polymeric analogue P3HT, yielding a V_{OC} of more than 1.3 V.

Figure 18 shows the photoelectrical characteristics of DIP and its related analogue DBP combined with C60 or C70 in different architectures. DIP/C60 cells yield a high open-circuit voltage exceeding 0.9 V and high fill factors (up to 75 %); however, they lack in current because as a result of the upright-standing DIP growth, there is basically no photon absorption and therefore no significant contribution of DIP to the photocurrent (see Fig. 18c). This deficit can be improved by using DBP with a lying orientation as the donor material. As the EQE spectra clearly show, there is now an equivalent contribution of both materials. Further improvement of J_{SC} is possible if the fullerene C70 is used as acceptor, which has a higher absorption than C60 because of the lower symmetry of the molecule. Nevertheless, as the photocurrent spectra show, part of the improvement comes at the expense of less photocurrent from DBP. The latter can be enhanced if a thin layer of α-NPD is introduced as an exciton-blocking layer to prevent quenching at the interface to the highly conducting PEDT:PSS. The biggest increase in short-circuit current, however, is achieved when the PHJ architecture is replaced by a PM-HJ—in this case DBP mixed with C70 in a 1:2 ratio by weight. As Fig. 18 shows, the current almost doubles and an EQE between 70–80 % is achieved over a wide range. From the point of view of exciton quenching, the α-NPD layer

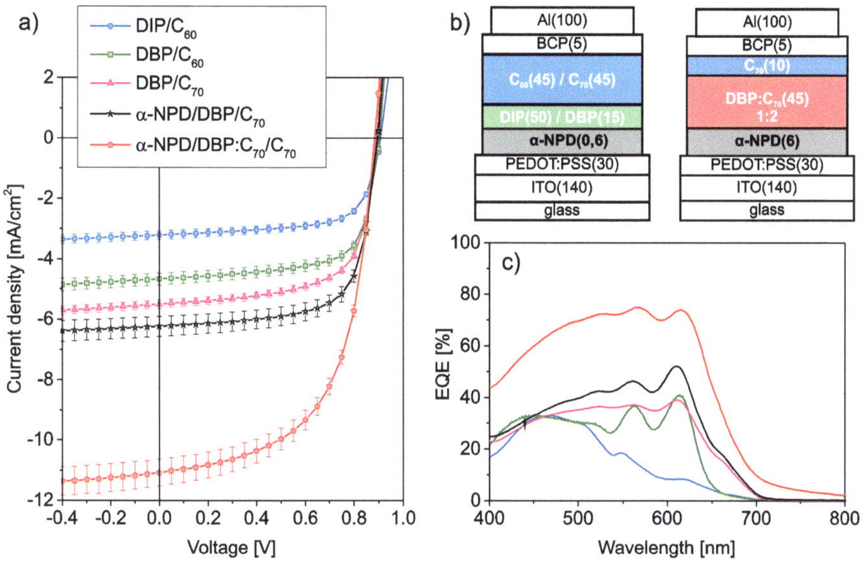

Fig. 18 (**a**) Comparison of J–V characteristics under simulated AM1.5g illumination of PHJ and PM-HJ devices with DIP and DBP as donors and C60 as well as C70 as acceptors, together with (**b**) photocurrent spectra of the same devices. The inset shows the schematic layer structure of the cell. Four nominally identical pixels with an active area of 4 mm^2 each were placed on a 20 mm × 20 mm substrate with a common ITO electrode. J–V measurements were performed without masking the illuminated pixel; the illumination spot for the photocurrent spectra was smaller than the pixel size. By comparing both values—the directly measured J_{SC} and the spectrally integrated photocurrent—we obtained consistent values. [Adapted from [25]]

Table 1 Summary of device data for DIP/C60 planar heterojunctions and DBP:C70 planar-mixed heterojunctions under simulated AM1.5g illumination. The values were averaged over four nominally identical pixels on the same substrate. A sample J–V curve for each type is shown in Fig. 18

	DIP/C60	α-NPD/DBP:C70/C70
j_{SC} [mA/cm^2]	3.16 ± 0.18	11.07 ± 0.44
V_{OC} [mV]	916 ± 1	884 ± 1
FF [%]	65.8 ± 0.9	58.9 ± 0.7
η_{PCE} [%]	1.90 ± 0.03	5.76 ± 0.19

is no longer required, but it is necessary to keep the effective work function of the anode high enough to maintain the high open-circuit voltage. Overall, in going from a PHJ of DIP/C60 to a PM-HJ with DBP and C70, the power conversion efficiency has been increased from about 2 % to almost 6 % (see Table 1).

To understand the origin and magnitude of the open-circuit voltage in OPVCs, we performed detailed studies of the temperature-dependent electrical device characteristics (see also [79, 80, 82]). As an example, Fig. 19 shows J–V characteristics of a

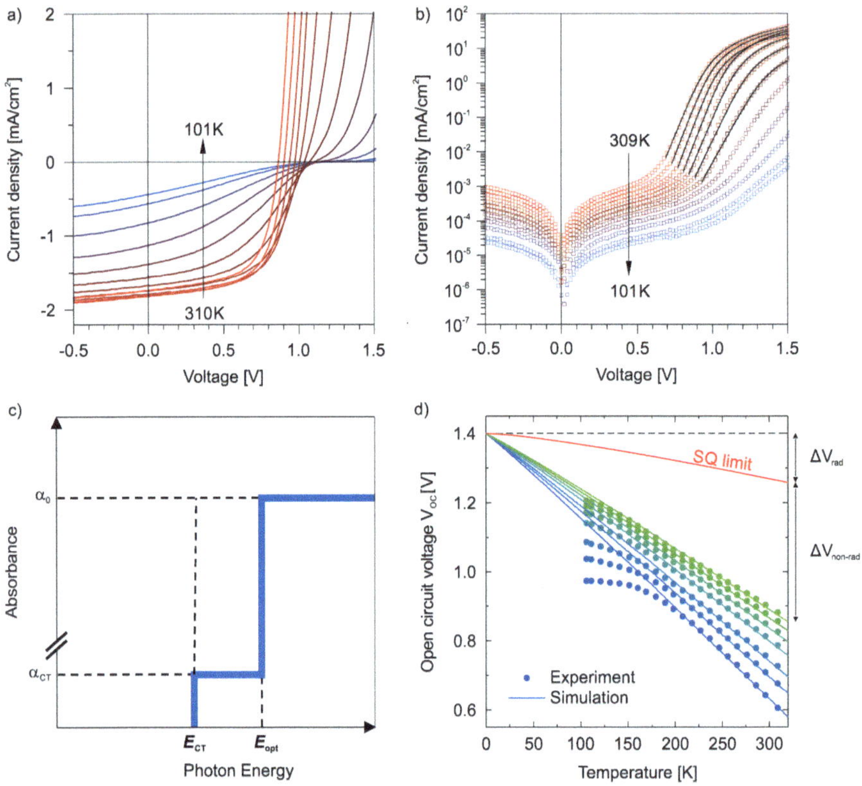

Fig. 19 Temperature-dependent *J*–*V* characteristics for a DIP/C60 PHJ under with light illumination (**a**) and in the dark (**b**). Part (**d**) shows the temperature dependence of the V_{OC} under different illumination intensities together with simulations according to a modified Shockley–Queisser limit, where the schematic absorption spectrum displayed in (**c**) was used. The temperature-dependent V_{OC} data are well described assuming a charge-transfer (CT) energy $E_{CT} = 1.4\,\text{eV}$, an absorption strength of the CT state of $\alpha_{CT} = 1.5 \times 10^{-4}$ (relative to the main absorption α_0 at $E_{opt} = 1.9\,\text{eV}$), and the nonradiative recombination to be higher by a factor $\beta = 2 \times 10^6$ than the radiative recombination. [Adapted from [79]]

DIP/C60 PHJ device measured in dark and under different illumination conditions. In general, under white light illumination (with intensity comparable to AM1.5g; Fig. 19a), one observes an increase in V_{OC} upon cooling, which is accompanied by a decrease in photocurrent and the formation of a so-called S-shape. Both of these unwanted effects are related to thermally activated injection and transport of carriers. The dark and light characteristics can be described with a modified SQ model as explained in [82]. The description is based on a simplified absorption spectrum displayed in Fig. 19c, where in addition to the fundamental absorption across the optical gap E_{opt}, a second absorption step at the CT energy E_{CT} with an absorption strength α_{CT} is introduced. According to this model, the temperature dependence of V_{OC} follows a relation [7, 83]:

$$V_{OC} = \frac{E_{CT}}{e} - \Delta V_{rad} - \Delta V_{non\text{-}rad}\,, \tag{7}$$

where both, the radiative and nonradiative loss terms depend approximately linearly on the temperature [79]. Thus, from the temperature and intensity-dependent V_{OC} data shown in Fig. 19d, one obtains a CT energy of 1.4 eV, which sets the upper limit for the chemical potential of photogenerated charge carriers at $T = 0$ K. Additionally, the graph shows that the major loss at room temperature is caused by nonradiative recombination ($\Delta V_{non\text{-}rad} \approx 0.4$ V)—in agreement with the nonradiative recombination current being more than six orders of magnitude higher than the radiative one, whereas the thermodynamically unavoidable radiative recombination across the CT gap amounts to a loss ΔV_{rad} of only 0.1 V. This exemplifies that the energy loss between the CT state and the open-circuit voltage is of the order of 0.5 eV and thus in the same range as in many polymer/fullerene OPVCs [7]. However, one has to keep in mind that an additional energy loss occurs by the formation of the CT state in the course of exciton dissociation at the D–A interface (see Fig. 1).

The apparently linear extrapolation of the measured $V_{OC}(T)$ toward $T = 0$ K also deserves some comment. Since the introduction by Vandewal et al. [7], this approach to determining the relevant photovoltaic gap has been frequently used in the literature [82, 84]. However, as we have found with the system 6T/DIP, it can overestimate the actual CT energy at the D–A interface if the CT gap is close to the optical gap of one of the two materials and if the electronic coupling between both of them is extremely weak, as is the case for the standing orientation of both molecules on top of each other [80]. In this case, a linear extrapolation of measured $V_{OC}(T)$ data yields the optical gap of DIP, which indicates that this is the dominant recombination pathway at higher temperatures.

Direct spectroscopic access to CT states in OPVCs can be obtained from electroluminescence (EL) and photocurrent spectra measured with high sensitivity below the optical gap of the main absorber [85]. Figure 20a shows EL spectra of diodes with neat layers of DIP and C60 each, which are compared to EL spectra of PHJ and PM-HJ of both materials, where the OPVCs are operated as light-emitting devices in the forward bias regime. For both D–A heterojunctions one observes a broad emission band in the near-infrared region (850–1000 nm), where none of the neat materials shows luminescence. In particular for the PM-HJ, the EL measured close to V_{OC} stems exclusively from light emission from CT states. A quantitative analysis combining the reduced emission spectrum with the EQE spectrum as shown in Fig. 20b reveals a CT energy of 1.45 eV, in excellent agreement with the temperature-dependent electrical measurements discussed above. What the spectra also show is a fairly large Stokes shift between CT absorption (modeled as a Gaussian in the low-energy onset of the EQE spectrum) and CT emission, reflected in a shift between both maxima of twice the reorganization energy $\lambda_0 \approx 0.3$ eV. This large shift together with the low EL quantum efficiency can be seen as major

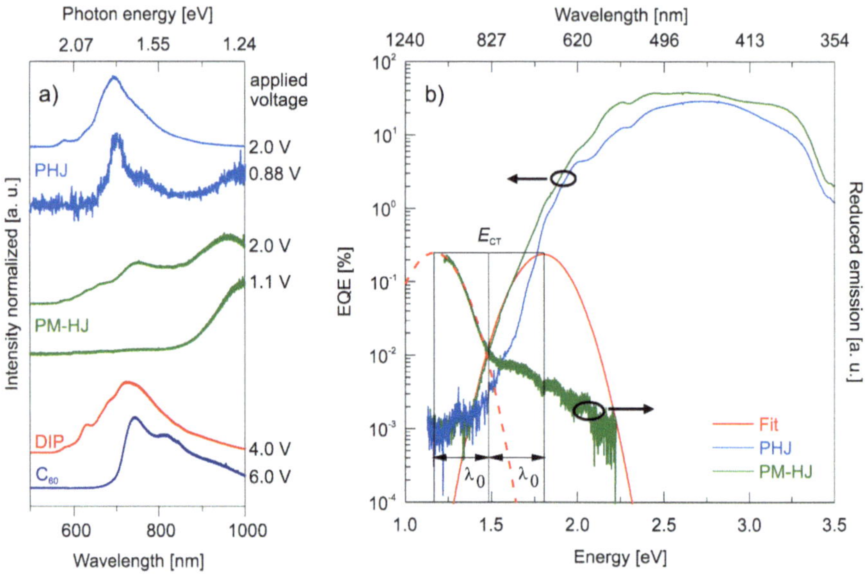

Fig. 20 (**a**) Electroluminescence spectra of single-layer diodes with DIP and C60 as active layers together with the spectra of planar heterojunction and planar-mixed heterojunction structures of both materials clearly showing an additional emission band in the near-infrared, where none of the two neat materials shows luminescence. (**b**) Reduced electroluminescence and sensitively measured external quantum efficiency spectra in the range of the charge-transfer (CT) state absorption and emission. From Gaussian fits with identical reorganization energy $\lambda_0 \approx 0.3$ eV, a CT energy of 1.45 eV is obtained

sources of the relatively large energy losses in OPVCs compared to, for example, hybrid inorganic–organic perovskite cells [86]. Understanding the detailed role of CT states in the process of charge generation in OPVCs, however, is the subject of ongoing research.

6 Conclusions

Concerted studies over recent years have revealed important design rules for improving organic solar cell performance. Using well-defined model systems of molecular donor–acceptor pairs, we have been able to understand their growth behavior in heterostructures with different architecture and, thus, to a certain extent tune their nanomorphology toward favorable conditions for OPVCs. This is reflected, for instance, in highly crystalline layers allowing coherent exciton transport with diffusion lengths limited only by the film thickness. The interface energetics have been studied for all relevant interfaces under device-relevant conditions and were correlated to solar cell performance data.

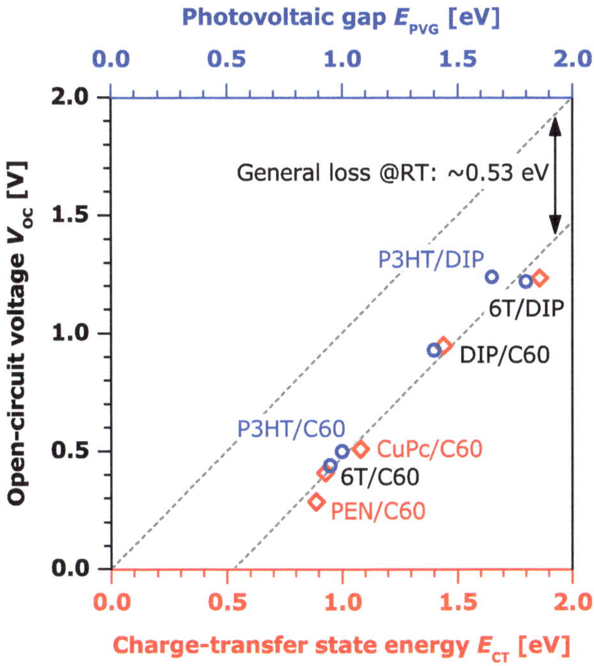

Fig. 21 Comparison of open-circuit voltage dependence on charge-transfer energy (*red diamonds*) and photovoltaic gap (*blue circles*). The difference for material pairs analyzed with both methods (*black labels*) are within the uncertainties because of variations of material batches and sample preparation. The values for V_{OC} and E_{CT} are measured on the same sample [79]. For the other set V_{OC} is taken from the literature [78, 80] in order to show small fluctuations between experimental runs

As a synopsis, Fig. 21 shows a compilation of the open-circuit voltages of all investigated D–A combinations with CT energies determined from temperature-dependent photoelectrical characterization and the relevant photovoltaic gap from photoelectron spectroscopy. In both cases a linear dependence of V_{OC} on the respective interfacial gap is observed, yielding the so-called bandgap-voltage offset of 0.53 eV. Compared to most inorganic pn-junction solar cells [87] as well as the emerging perovskite cells [86], this offset is significantly larger and one of the reasons for the lower performance of OPVCs. Future research will therefore, *inter alia*, have to find strategies to reduce energy losses in organic photovoltaics.

References

1. http://www.nrel.gov/ncpv/ (2015). Accessed 9 September
2. Green MA, Emery K, Hishikawa Y, Warta W, Dunlop ED (2015) Solar cell efficiency tables (Version 46). Prog Photovolt Res Appl 23:805–812. doi:10.1002/pip.2637

3. Shockley W, Queisser HJ (1961) Detailed balance limit of efficiency of p-n junction solar cells. J Appl Phys 32:510–519. doi:10.1063/1.1736034
4. Veldman D, Meskers SCJ, Janssen RAJ (2009) The energy of charge-transfer states in electron donor-acceptor blends: insight into the energy losses in organic solar cells. Adv Funct Mater 19:1939–1948. doi:10.1002/adfm.200900090
5. Clarke TM, Durrant JR (2010) Charge photogeneration in organic solar cells. Chem Rev 110:6736–6767. doi:10.1021/cr900271s
6. Deibel C, Dyakonov V (2010) Polymerfullerene bulk heterojunction solar cells. Rep Prog Phys 73:096401. doi:10.1088/0034-4885/73/9/096401
7. Vandewal K, Tvingstedt K, Gadisa A, Inganäs O, Manca JV (2010) Relating the open-circuit voltage to interface molecular properties of donor:acceptor bulk heterojunction solar cells. Phys Rev B 81:125204. doi:10.1103/PhysRevB.81.125204
8. Opitz A, Wagner J, Brütting W, Salzmann I, Koch N, Manara J, Pflaum J, Hinderhofer A, Schreiber F (2010) Charge separation at molecular donor–acceptor interfaces: correlation between morphology and solar cell performance. IEEE J Sel Top Quant 16:1707–1717. doi:10.1109/JSTQE.2010.2048096
9. Forrest SR (2011) The limits to organic photovoltaic cell efficiency. MRS Bull 30:28–32. doi:10.1557/mrs2005.5
10. Wilke A, Mizokuro T, Blum RP, Rabe JP, Koch N (2010) IEEE J Sel Top Quant 16:1732–1737. doi:10.1109/JSTQE.2010.2042035
11. Salzmann I, Duhm S, Heimel G, Oehzelt M, Kniprath R, Johnson RL, Rabe JP, Koch N (2008) Tuning the ionization energy of organic semiconductor films: the role of intramolecular polar bonds. J Am Chem Soc 130:12870–12871. doi:10.1021/ja804793a
12. Wagner J, Gruber M, Hinderhofer A, Wilke A, Bröker B, Frisch J, Amsalem P, Vollmer A, Opitz A, Koch N, Schreiber F, Brütting W (2010) High fill factor and open circuit voltage in organic photovoltaic cells with diindenoperylene as donor material. Adv Funct Mater 20:4295–4303. doi:10.1002/adfm.201001028
13. Pfützner S (2012) Studies on organic solar cells composed of fullerenes and zinc-phthalocyanines. Ph.D. thesis, TU Dresden. http://nbn-resolving.de/urn:nbn:de:bsz:14-qucosa-83486
14. Opitz A, Frisch J, Schlesinger R, Wilke A, Koch N (2013) Energy level alignment at interfaces in organic photovoltaic devices. J Electron Spectrosc Relat Phenom 190:12–24. doi:10.1016/j.elspec.2012.11.008
15. Lau KM, Tang JX, Sun HY, Lee CS, Lee ST, Yan D (2006) Interfacial electronic structure of copper phthalocyanine and copper hexadecafluorophthalocyanine studied by photoemission. Appl Phys Lett 88:173513. doi:10.1063/1.2198484
16. Krause S, Casu MB, Schöll A, Umbach E (2008) Determination of transport levels of organic semiconductors by UPS and IPS. New J Phys 10:085001. doi:10.1088/1367-2630/10/8/085001
17. Wilke A, Endres J, Hörmann U, Niederhausen J, Schlesinger R, Frisch J, Amsalem P, Wagner J, Gruber M, Opitz A, Vollmer A, Brütting W, Kahn A, Koch N (2012) Correlation between interface energetics and open circuit voltage in organic photovoltaic cells. Appl Phys Lett 101:233301. doi:10.1063/1.4769360
18. Han W, Yoshida H, Ueno N, Kera S (2013) Electron affinity of pentacene thin film studied by radiation-damage free inverse photoemission spectroscopy. Appl Phys Lett 103:123303. doi:10.1063/1.4821445
19. Brinkmann H, Kelting C, Makarov S, Tsaryova O, Schnurpfeil G, Wöhrle D, Schlettwein D (2008) Fluorinated phthalocyanines as molecular semiconductor thin films. Phys Stat Sol A 205:409–420. doi:10.1002/pssa.200723391
20. Djurovich PI, Mayo EI, Forrest SR, Thompson ME (2009) Measurement of the lowest unoccupied molecular orbital energies of molecular organic semiconductors. Org Electron 10:515–520. doi:10.1016/j.orgel.2008.12.011
21. Sze SM, Ng KK (2007) Physics of semiconductor devices, 3rd edn. Wiley, New York. doi:10.1002/0470068329

22. Rand B, Burk D, Forrest S (2007) Offset energies at organic semiconductor heterojunctions and their influence on the open-circuit voltage of thin-film solar cells. Phys Rev B 75:115327. doi:10.1103/PhysRevB.75.115327
23. Riede M, Mueller T, Tress W, Schueppel R, Leo K (2008) Small-molecule solar cells-status and perspectives. Nanotechnology 19:424001. doi:10.1088/0957-4484/19/42/424001
24. Stübinger T, Brütting W (2001) Exciton diffusion and optical interference in organic donor–acceptor photovoltaic cells. J Appl Phys 90:3632–3641. doi:10.1063/1.1394920
25. Grob S, Gruber M, Bartynski AN, Hörmann U, Linderl T, Thompson ME, Brütting W (2014) Amorphous vs crystalline exciton blocking layers at the anode interface in planar and planar-mixed heterojunction organic solar cells. Appl Phys Lett 104:213304. doi:10.1063/1.4879839
26. Tang CW (1986) Two-layer organic photovoltaic cell. Appl Phys Lett 48:183–185. doi:10.1063/1.96937
27. Yu S, Klimm C, Schäfer P, Rabe JP, Rech B, Koch N (2011) Organic photovoltaic cells with interdigitated structures based on pentacene nanocolumn arrays. Org Electron 12:2180. doi:10.1016/j.orgel.2011.09.021
28. Yu S, Opitz A, Grob S, Resel R, Oehzelt M, Brütting W, Salzmann I, Koch N (2014) Performance enhancement of diindenoperylene-based organic photovoltaic cells by nanocolumn-arrays. Org. Electron. 15:2210–2217. doi:10.1016/j.orgel.2014.06.023
29. Salzmann I, Duhm S, Opitz R, Johnson RL, Rabe JP, Koch N (2008) Structural and electronic properties of pentacene-fullerene heterojunctions. J Appl Phys 104:114518. doi:10.1063/1.3040003
30. Peumans P, Uchida S, Forrest SR (2003) Efficient bulk heterojunction photovoltaic cells using small-molecular-weight organic thin films. Nature 425:158–162. doi:10.1038/nature01949
31. Hinderhofer A, Schreiber F (2012) Organic-organic heterostructures: concepts and applications. ChemPhysChem 13:628–643. doi:10.1002/cphc.201100737
32. Opitz A, Ecker B, Wagner J, Hinderhofer A, Schreiber F, Manara J, Pflaum J, Brütting W (2009) Mixed crystalline films of co-evaporated hydrogen- and fluorine-terminated phthalocyanines and their application in photovoltaic devices. Org Electron 10:1259–1267. doi:10.1016/j.orgel.2009.07.004
33. Hinderhofer A, Frank C, Hosokai T, Resta A, Gerlach A, Schreiber F (2011) Structure and morphology of coevaporated pentacene-peruoropentacene thin films. J Chem Phys 134:104702. doi:10.1063/1.3557476
34. Reinhardt JP, Hinderhofer A, Broch K, Heinemeyer U, Kowarik S, Vorobiev A, Gerlach A, Schreiber F (2012) Structural and Optical Properties of Mixed Diindenoperylene–Perfluoropentacene Thin Films. J Phys Chem C 116:10917–10923. doi:10.1021/jp211947y
35. Opitz A, Wagner J, Brütting W, Hinderhofer A, Schreiber F (2009) Molecular semiconductor blends: Microstructure, charge carrier transport, and application in photovoltaic cells. Phys Stat Sol A 206:2683–2694. doi:10.1002/pssa.200925238
36. Gruber M, Rawolle M, Wagner J, Magerl D, Hörmann U, Perlich J, Roth SV, Opitz A, Schreiber F, Müller-Buschbaum P, Brütting W (2013) Correlating structure and morphology to device performance of molecular organic donor-acceptor photovoltaic cells based on diindenoperylene (DIP) and C 60. Adv Energy Mater 3:1075–1083. doi:10.1002/aenm.201201012
37. Ishii H, Sugiyama K, Ito E, Seki K (1999) Energy level alignment and interfacial electronic structures at organic/metal and organic/organic interfaces. Adv Mater 11:605–625. doi:10.1002/(SICI)1521-4095(199906)11:8<605::AID-ADMA605>3.0.CO;2-Q
38. Cahen D, Kahn A (2003) Electron energetics at surfaces and interfaces: concepts and experiments. Adv Mater 15:271–277. doi:10.1002/adma.200390065
39. Koch N (2007) Organic electronic devices and their functional interfaces. ChemPhysChem 8:1438–1455. doi:10.1002/cphc.200700177
40. Ueno N, Kera S (2008) Electron spectroscopy of functional organic thin films: deep insights into valence electronic structure in relation to charge transport property. Prog Surf Sci 83:490–557. doi:10.1016/j.progsurf.2008.10.002

41. Wu C, Hirose Y, Sirringhaus H, Kahn A (1997) Electron-hole interaction energy in the organic molecular semiconductor PTCDA. Chem Phys Lett 272:43–47. doi:10.1016/S0009-2614(97)00481-8
42. Oehzelt M, Koch N, Heimel G (2014) Organic semiconductor density of states controls the energy level alignment at electrode interfaces. Nat Commun 5:4174. doi:10.1038/ncomms5174
43. Wang H, Amsalem P, Heimel G, Salzmann I, Koch N, Oehzelt M (2014) Adv Mater 26:925–930. doi:10.1002/adma.201303467
44. Akaike K, Koch N, Oehzelt M (2014) Fermi level pinning induced electrostatic fields and band bending at organic heterojunctions. Appl Phys Lett 105:223303. doi:10.1063/1.4903360
45. Wilke A, Amsalem P, Frisch J, Bröker B, Vollmer A, Koch N (2011) Electric fields induced by energy level pinning at organic heterojunctions. Appl Phys Lett 98:123304. doi:10.1063/1.3571286
46. Wagner J, Gruber M, Wilke A, Tanaka Y, Topczak K, Steindamm A, Hörmann U, Opitz A, Nakayama Y, Ishii H, Pflaum J, Koch N, Brütting W (2012) Identification of different origins for s-shaped current voltage characteristics in planar heterojunction organic solar cells. J Appl Phys 111:054509. doi:10.1063/1.3692050
47. Amsalem P, Niederhausen J, Wilke A, Heimel G, Schlesinger R, Winkler S, Vollmer A, Rabe J, Koch N (2013) Role of charge transfer, dipole-dipole interactions, and electrostatics in Fermi-level pinning at a molecular heterojunction on a metal surface. Phys Rev B 87:035440. doi:10.1103/PhysRevB.87.035440
48. Peumans P, Forrest SR (2001) Very-high-efficiency double-heterostructure copper phthalocyanine/C[sub 60] photovoltaic cells. Appl Phys Lett 79:126–128. doi:10.1063/1.1384001
49. Steindamm A, Brendel M, Topczak AK, Pflaum J (2012) Thickness dependent effects of an intermediate molecular blocking layer on the optoelectronic characteristics of organic bilayer photovoltaic cells. Appl Phys Lett 101:143302. doi:10.1063/1.4757297
50. Heremans P, Cheyns D, Rand BP (2009) Strategies for increasing the efficiency of heterojunction organic solar cells: material selection and device architecture. Acc Chem Res 42:1740–1747. doi:10.1021/ar9000923
51. Heinemeyer U, Hinderhofer A, Alonso MI, Ossó JO, Garriga M, Kytka M, Gerlach A, Schreiber F (2008) Uniaxial anisotropy of organic thin films determined by ellipsometry. Phys Stat Sol A 205:927–930. doi:10.1002/pssa.200777765
52. Heinemeyer U, Scholz R, Gisslén L, Alonso MI, Ossó JO, Garriga M, Hinderhofer A, Kytka M, Kowarik S, Gerlach A, Schreiber F (2008) Excitonphonon coupling in diindenoperylene thin films. Phys Rev B 78:085210. doi:10.1103/PhysRevB.78.085210
53. Birkholz M (2006) Thin film analysis by X-ray scattering. Wiley-VCH, Weinheim
54. Tolan M (1999) X-ray scattering from soft-matter thin films: materials science and basic research. Springer tracts in modern physics. Springer, Berlin
55. Forrest SR, Kaplan ML, Schmidt PH (1984) Organic-on-inorganic semiconductor contact barrier diodes. II. Dependence on organic film and metal contact properties. J Appl Phys 56:543–551. doi:10.1063/1.333944
56. Dürr AC, Schreiber F, Kelsch M, Carstanjen HD, Dosch H (2002) Morphology and thermal stability of metal contacts on crystalline organic thin films. Adv Mater 14:961–963. doi:10.1002/1521-4095(20020705)14:13/14<961::AID-ADMA961>3.0.CO;2-X
57. Hinderhofer A, Gerlach A, Broch K, Hosokai T, Yonezawa K, Kato K, Kera S, Ueno N, Schreiber F (2013) Geometric and electronic structure of templated C_{60} on diindenoperylene thin films. J Phys Chem C 117:1053–1058. doi:10.1021/jp3106056
58. Hinderhofer A, Gerlach A, Kowarik S, Zontone F, Krug J, Schreiber F (2010) Smoothing and coherent structure formation in organic-organic heterostructure growth. Eur Phys Lett 91:56002. doi:10.1209/0295-5075/91/56002
59. Aufderheide A, Broch K, Novák J, Hinderhofer A, Nervo R, Gerlach A, Banerjee R, Schreiber F (2012) Mixing-induced anisotropic correlations in molecular crystalline systems. Phys Rev Lett 109:156102. doi:10.1103/PhysRevLett.109.156102
60. Kitaigorodsky A (1984) Mixed crystals. Springer series in solid-state sciences. Springer, Berlin

61. Broch K, Aufderheide A, Raimondo L, Sassella A, Gerlach A, Schreiber F (2013) Optical properties of blends: Influence of mixing-induced disorder in pentacene:diindenoperylene versus peruoropentacene:diindenoperylene. J Phys Chem C 117:13952–13960. doi:10.1021/jp4019487
62. Banerjee R, Novák J, Frank C, Lorch C, Hinderhofer A, Gerlach A, Schreiber F. (2013) Evidence for kinetically limited thickness dependent phase separation in organic thin film blends. Phys Rev Lett 110:185506. doi:10.1103/PhysRevLett.110.185506
63. Broch K, Gerlach A, Lorch C, Dieterle J, Novák J, Hinderhofer A, Schreiber F (2013) Structure formation in peruoropentacene:diindenoperylene blends and its impact on transient effects in the optical properties studied in realtime during growth. J Chem Phys 139:174709. doi:10.1063/1.4827868
64. Broch K, Bürker C, Dieterle J, Krause S, Gerlach A, Schreiber F (2013) Impact of molecular tilt angle on the absorption spectra of pentacene:peruoropentacene blends. Phys Stat Sol RRL 7:1084–1088. doi:10.1002/pssr.201308085
65. Topczak AK, Roller T, Engels B, Brütting W, Pflaum J (2014) Nonthermally activated exciton transport in crystalline organic semiconductor thin films. Phys Rev B 89:201203(R). doi:10.1103/PhysRevB.89.201203
66. Lunt RR, Benziger JB, Forrest SR (2010) Relationship between crystalline order and exciton diffusion length in molecular organic semiconductors. Adv Mater 22:1233–1236. doi:10.1002/adma.200902827
67. Peumans P, Yakimov A, Forrest SR (2003) Small molecular weight organic thin-film photodetectors and solar cells. J Appl Phys 93:3693–3723. doi:10.1063/1.1534621
68. Lunt RR, Giebrink NC, Belak AA, Benziger JB, Forrest SR (2009) Exciton diffusion lengths of organic semiconductor thin films measured by spectrally resolved photoluminescence quenching. J Appl Phys 105:1233–1236. doi:10.1063/1.3079797
69. Brendel M, Krause S, Steindamm A, Topczak AK, Sundarraj S, Erk P, Höhla S, Fruehauf N, Koch N, Pflaum J (2015) The effect of gradual fluorination on the properties of FnZnPc thin films and FnZnPc/C-60 bilayer photovoltaic cells. Adv Mater 25:1565–1573. doi:10.1002/adfm.201404434
70. Dürr AC, Schreiber F, Ritley KA, Kruppa V, Krug J, Dosch H, Struth B (2003) Rapid roughening in thin film growth of an organic semiconductor (diindenoperylene). Phys Rev Lett 90:016104. doi:10.1103/PhysRevLett.90.016104
71. Ghosh AK, Feng T (1978) Merocyanine organic solar cells. J Appl Phys 49:5982–5989. doi:10.1063/1.324566
72. Settels V, Liu WL, Pflaum J, Fink RF, Engels B (2012) Comparison of the electronic structure of different perylene-based dye-aggregates. J Comput Chem 33:1544–1553. doi:10.1002/jcc.22986
73. Settels V, Schubert A, Tafipolski M, Liu WL, Stehr V, Topczak AK, Pflaum J, Deibel C, Fink RF, Engel V, Engels B (2014) Identification of ultrafast relaxation processes as a major reason for inefficient exciton diffusion in perylene-based organic semiconductors. J Am Chem Soc 136:9327–9337. doi:10.1021/ja413115h
74. Gieseking B, Schmeiler T, Müller B, Deibel C, Engels B, Dyakonov V, Pflaum J (2014) Effects of characteristic length scales on the exciton dynamics in rubrene single crystals. Phys Rev B 90:205305. doi:10.1103/PhysRevB.90.205305
75. Schuenemann C, Petrich A, Schulze R, Wynands D, Meiss J, Hein MP, Jankowski J, Elschner C, Alex J, Hummert M, Eichhorn KJ, Leo K, Riede M (2013) Diindenoperylene derivatives: a model to investigate the path from molecular structure via morphology to solar cell performance. Org Electron 14:1704–1714. doi:10.1016/j.orgel.2013.04.006
76. Hansen NH, Wunderlich C, Topczak AK, Rohwer E, Schwoerer H, Pflaum J (2013) Exciton interaction with a spatially defined charge accumulation layer in the organic semiconductor diindenoperylene. Phys Rev B 87:241202(R). doi:10.1103/PhysRevB.87.241202
77. Due to the notation of the relative quenching per cent, the nonradiative recombination rate cited in Ref. [76] has to be divided by a factor of 100.

78. Hörmann U, Wagner J, Gruber M, Opitz A, Brütting W (2011) Approaching the ultimate open circuit voltage in thiophene based single junction solar cells by applying diindenoperylene as acceptor. Phys Stat Sol RRL 5:241–243. doi:10.1002/pssr.201105238
79. Hörmann U, Kraus J, Gruber M, Schuhmair C, Linderl T, Grob S, Kapfinger S, Klein K, Stutzman M, Krenner H, Brütting W (2013) Quantification of energy losses in organic solar cells from temperature-dependent device characteristics. Phys Rev B 88:235307. doi:10.1103/PhysRevB.88.235307
80. Hörmann U, Lorch C, Hinderhofer A, Gerlach A, Gruber M, Kraus J, Sykora B, Grob S, Linderl T, Wilke A, Opitz A, Hansson R, Anselmo AS, Ozawa Y, Nakayama Y, Ishii H, Koch N, Moons E, Schreiber F, Brütting W (2014) V_{OC} from a morphology point of view: the influence of molecular orientation on the open circuit voltage of organic planar heterojunction solar cells. J Phys Chem C 118:26462–26470. doi:10.1021/jp506180k
81. Horlet M, Kraus M, Brütting W, Opitz A (2011) Diindenoperylene as ambipolar semiconductor: influence of electrode materials and mobility asymmetry in organic field-effect transistors. Appl Phys Lett 98:233304. doi:10.1063/1.3598423
82. Gruber M, Wagner J, Klein K, Hörmann U, Opitz A, Stutzmann M, Brütting W (2012) Thermodynamic efficiency limit of molecular donor-acceptor solar cells and its application to diindenoperylene/C60-based planar heterojunction devices. Adv Energy Mater 2:1100–1108. doi:10.1002/aenm.201200077
83. Rau U (2007) Reciprocity relation between photovoltaic quantum efficiency and electroluminescent emission of solar cells. Phys Rev B 76:085303. doi:10.1103/PhysRevB.76.085303
84. Widmer J, Tietze M, Leo K, Riede M (2013) Open-circuit voltage and effective gap of organic solar cells. Adv Funct Mater 23:5814–5821. doi:10.1002/adfm.201301048
85. Vandewal K, Tvingstedt K, Gadisa A, Inganäs O, Manca JV (2009) On the origin of the open-circuit voltage of polymer-fullerene solar cells. Nat Mater 8:904–909. doi:10.1038/nmat2548
86. Tvingstedt K, Malinkiewicz O, Baumann A, Deibel C, Snaith HJ, Dyakonov V, Bolink HJ (2014) Radiative efficiency of lead iodide based perovskite solar cells. Sci Rep 4:6071. doi:10.1038/srep06071
87. King RR, Bhusari D, Boca A, Larrabee D, Liu XQ, Hong W, Fetzer CM, Law DC, Karam NH (2011) Band gap-voltage offset and energy production in next-generation multijunction solar cells. Prog Photovolt Res Appl 19:797–812. doi:10.1002/pip.1044

Optoelectronic Properties of PCPDTBT for Photovoltaics: Morphology Control and Molecular Doping

Elizabeth von Hauff, Enrico da Como, and Sabine Ludwigs

Contents

E. von Hauff (✉)
Department of Physics and Astronomy, Vrije Universiteit Amsterdam, De Boelelaan 1081, 1081 HV Amsterdam, The Netherlands
e-mail: e.l.von.hauff@vu.nl

E. da Como
Department of Physics, University of Bath, Claverton Down Road, BA2 7AY Bath, UK
e-mail: e.da.como@bath.ac.uk

S. Ludwigs
Institute of Polymer Chemistry, University of Stuttgart, Pfaffenwaldring 55, 70569 Stuttgart, Germany
e-mail: sabine.ludwigs@ipoc.uni-stuttgart.de

K. Leo (ed.), *Elementary Processes in Organic Photovoltaics*,
Advances in Polymer Science 272, DOI 10.1007/978-3-319-28338-8_5

Abstract Donor–acceptor copolymers have recently been recognized as excellent materials for organic photovoltaic applications. Because of complex film formation properties, however, direct correlations between morphology and optical and electrical properties have yet to be established. Within our collaborative project within the German Science Foundation–funded program "Elementary Processes in Photovoltaics," we focused on the morphology control and molecular doping of poly[2,6-(4,4-bis(2-ethylhexyl)-4H-cyclopenta[2,1-*b*;3,4-*b*′]dithiophene)-*alt*-4, 7(2,1,3-benzothiadiazole)] to control the optoelectronic properties. The main results are highlighted in this chapter.

Keywords Doping • Donor-acceptor • copolymers • Morphology

1 Introduction

Light absorption and charge transport are two of the dominant factors that strongly influence organic solar cell performance. In general, increased light absorption and charge carrier mobility in the photoactive layer lead to higher power conversion efficiencies (PCEs) in devices. Poly(3-hexylthiophene) (P3HT) was for a long time the most promising conjugated polymer used as an electron donor in organic solar cells [1]. The main drawback of P3HT is its relatively large bandgap (>2 eV), which leads to an absorption covering a small portion of the solar spectrum. Therefore, the demand arose for new synthetic strategies to further reduce the bandgap and increase the charge carrier mobility. With this in mind, donor–acceptor (D–A) copolymers were developed for photovoltaic applications [2]. In such polymers, the electron-rich donor units are coupled to the electron-poor acceptor moieties, resulting in a charge-transfer band in the absorption spectrum at low energies. The observed charge-transfer transition of the D–A copolymer is redshifted compared with the absorption spectra of the homopolymers, resulting in a decrease in the polymer bandgap [3].

Brabec and coworkers [4] were the first to report on the D–A copolymer poly[2,6-(4,4-bis(2-ethylhexyl)-4H-cyclopenta[2,1-*b*;3,4-*b*′]dithiophene)-*alt*-4,7(2,1,3-benzothiadiazole)] (PCPDTBT), which is based on alternating cyclopentadithiophene (donor) and benzothiadiazole (acceptor) units (for chemical structure, see Fig. 1a). To enable solution processing of the polymer, ethyl hexyl side chains are attached to the bridging carbon between the two thiophenes. The extended conjugation of the planar backbone and its D–A nature result in small optical bandgaps of about 1.4 eV (890 nm) for this polymer [4]. The highest occupied molecular orbital and lowest unoccupied molecular orbital energy values were determined via electrochemistry to be −4.8 and −3.1 eV, respectively. Heeger and coworkers [5] showed organic solar cell data using PCPDTBT as donor and [6,6]-phenyl-C_{61}-butyric acid methyl ester (PCBM) as acceptor material and obtained ultrafast charge separation as well as efficiencies of up to 3.6 %. At that time, PCEs up to 7 % were predicted for this novel copolymer.

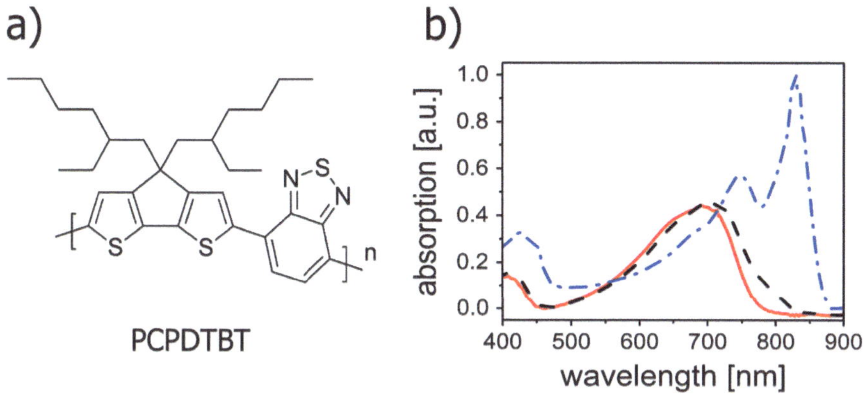

Fig. 1 (**a**) Chemical structure of PCPDTBT. (**b**) Absorption spectra of PCPDTBT dissolved in 2-methyltetrahydrofuran (0.25 mg/ml) at 67 °C (*red*), room temperature (*black*), and −163 °C (*blue*) (Adapted with permission from Wiley [17])

In 2007, Peet et al. [6, 7] showed that it was not possible to improve PCPDTBT/PCBM blend morphology using thermal annealing but that solvent additives with a high boiling point, such as 1,8-diiodooctane (DIO) or 1,8-octanedithiol, were necessary to achieve high solar cell efficiencies of up to 5.5 %. Temperature-dependent absorption experiments with DIO showed that a redshift in the absorption maximum to 800 nm is caused by aggregate formation. The origin of the 800-nm band was attributed to either planarization of the polymer backbone or an increased interchain coupling because of π stacking [7]. Efforts to increase the aggregation by thermal annealing did not further increase the aggregation but led to blue shifts in the absorption spectra.

Further information on the effect of solvent additives on PCPDTBT homopolymers and PCPDTBT:PCBM blends were gained by grazing incidence X-ray diffraction [8–11]. Although overall reflections were rather weak, the films showed increased reflection intensities in the diffraction pattern for films processed using solvent additives compared with those without solvent additives. The authors suggested that PCPDTBT shows long-range π stacking and packs with an edge-on orientation similar to P3HT.

Within our collaborative project we focused on morphology control and molecular doping of PCPDTBT and blends. We took a more fundamental approach to studying the PCPDTBT morphology and applied a range of techniques to the pristine polymer films such as solvent vapor annealing (SVA) [12]. A new crystalline dimer-like polymorph was discovered and correlations between crystal structures and optoelectronic properties were established [13]. Molecular doping is demonstrated as a versatile strategy to decrease recombination in PCPDTBT:PCBM blends and increase hole mobility in the polymer phase [14, 15]. In this chapter, we highlight some of our key results.

Section 2 summarizes advances in understanding and tuning the morphology of PCPDTBT thin films. Techniques such as ultraviolet–visible (UV-Vis) absorption spectroscopy, atomic force microscopy (AFM), and transmission electron microscopy (TEM)/electron diffraction (ED) analysis allowed insight into polymer packing and orientation. A new crystal structure was proposed for PCPDTBT and bilayer solar cells consisting of different polymer morphologies were tested.

In Sect. 3 molecular doping is discussed as a strategy to enhance the electrical properties of PCPDTBT. Electron microscopy, optical spectroscopy, and electrical characterization are applied to understand how molecular doping influences the optoelectronic response of the polymer. Proof-of-concept solar cells with doped PCPDTBT active layers are demonstrated. Finally, the polymer–doping interactions are carefully examined using infrared (IR) spectroscopy combined with theoretical modeling.

2 Tuning of Homopolymer Morphology and Correlation to Optical Properties

2.1 Optical Properties of PCPDTBT Solutions

Figure 1a shows the chemical structure of the D–A polymer under investigation: PCPDTBT. The polymer consists of an alternating cyclopentadithiophene donor unit and a benzothiadiazole acceptor moiety, which results in an optical bandgap of 1.6 eV in solution.

UV-Vis–near-IR (NIR)-absorption spectroscopy is routinely used to study the aggregation of conjugated polymers in solution. For example, P3HT was thoroughly investigated by Spano and Silva [16] and the observed redshift in the absorption spectrum together with a pronounced 610-nm absorption band was explained by planarizations of the polymer backbone and interchain couplings between polymer backbones. The absorption properties of PCPDTBT were investigated in dilute solutions of 2-methyltetrahydrofuran (MTHF) and are shown in Fig. 1b [17]. The dashed black line represents the spectrum measured at room temperature. It reveals a high-energy band attributed to a π–π^* transition at 420 nm as well as a low-energy band around 730 nm, which has been assigned to a charge-transfer transition. When the temperature was decreased from 67 °C down to −163 °C, the absorption spectrum became redshifted and the bands increased in intensity. When the polymer solution reached a temperature of 7 °C, an aggregation shoulder appeared at 830 nm. Upon further reduction of the solution temperature, the low-energy band split into two vibronic bands with maxima at 750 and 850 nm. Figure 1b also reveals the presence of an isosbestic point at 730 nm, indicating that the appearance of the aggregation band at 800 nm is caused by another polymer phase with increased conjugation length (redshift) and cannot be explained by the planarization of the polymer backbones during aggregation [17].

Such trends in the absorption spectra—a redshift, an increase in intensity, and an isosbestic point—have also been observed for other conjugated polymers such as P3HT [18, 19] and poly[2-methoxy-5-(2′-ethylhexyloxy)-p-phenylene vinylene [20]. They all seem to show similar aggregation behavior.

2.2 *Thin-Film Preparation by Spin Coating: The Role of Solvent Additives*

Very few studies focused on the absorption properties and morphology of pure PCPDTBT copolymer, and one of the most investigated solvents is chlorobenzene (CB). When we compare the solution and the thin-film spectra of PCPDTBT from CB, a broadening and redshift of the low-energy band from around 720 nm in solution to 770 nm in films is detected [7]. The absorption spectra of such films resemble the aggregated solution spectra presented in Fig. 1b [21]. In the thin film, two main maxima around 410 nm (high energy) and 770 nm (low energy) are found, whereby the 770-nm band is often accompanied by a shoulder at 725 nm. It has been noted that during aggregation in solution the low-energy band was shifted more than 100 nm up to 850 nm [17]. The small shift of the absorption of thin films processed from CB to 770 nm could therefore indicate a less aggregated phase of the polymer chains in the film compared to solution. Only for blends in the presence with PCBM or when additional solvent additives such as alkyl halides or alkyl thiols were added during processing were film spectra with maxima up to almost 800 nm reported [8–11]. These findings support the role of additives in enhanced polymer aggregation for films.

In our work we compared thin films of PCPDTBT that were spin coated from various solvents, and both the absorption and morphological properties were probed using AFM and UV–Vis absorption spectroscopy [21]. In Fig. 2, the topography images and absorption spectra are shown for four samples prepared using different processing conditions: spin coated from $CHCl_3$, 1-chloronaphthalene (1-CN), CB containing 2 % DIO, as well as spin coated from $CHCl_3$ followed by exposure to defined CS_2 vapor atmospheres. A quick glance at the AFM height images immediately convinces the reader that various film morphologies for the PCPDTBT polymer can be obtained. In Figs. 2a and b, the polymer was deposited using $CHCl_3$ and 1-CN, respectively. In the case of the film deposited using the low-boiling-point $CHCl_3$, a disordered structure of amorphous nature was formed, whereas for the film made using the high-boiling-point 1-CN, a fiber-like morphology with feature sizes of 40 ± 3 nm was obtained. Since the boiling point of 1-CN is significantly higher than that of $CHCl_3$, we attribute the increased order to longer drying times and therefore more time to organize.

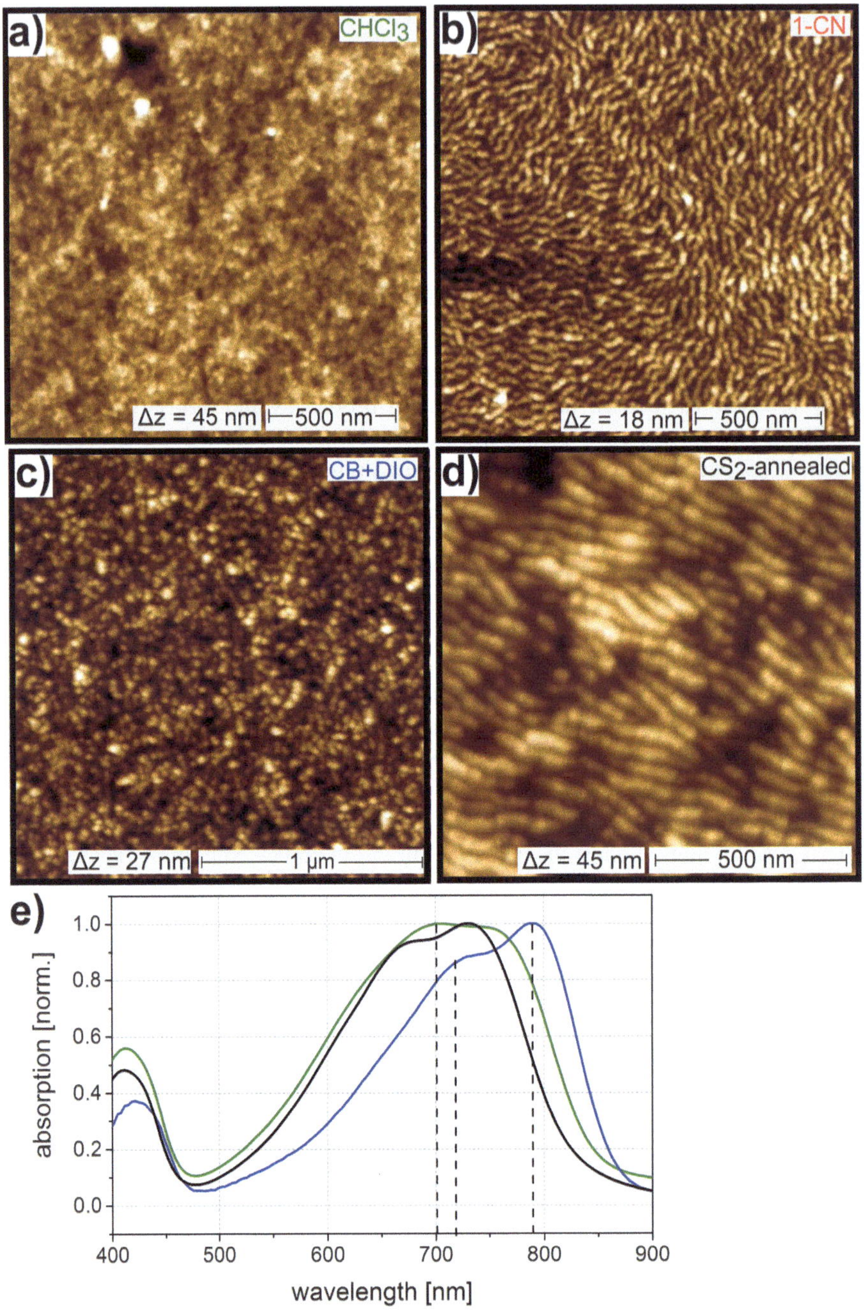

Fig. 2 Atomic force microscopy height images of thin PCPDTBT films prepared by spin coating from (**a**) chloroform (3 mg/ml), (**b**) 1-chloronaphthalene (3 mg/ml), (**c**) chlorobenzene/1,8-diiodooctane (DIO;3 mg/ml + 2 w% DIO), and (**d**) by solvent vapor annealing in CS_2; (**e**) corresponding absorption spectra of the thin films shown in (**a**), (**c**), and (**d**). (Adapted with permissions from Wiley and ACS [12, 21])

In accordance with literature protocols, we prepared films by spin coating from CB in the presence of 2 wt% DIO as additive, which is reported to have an edge-on orientation and long-range π stacking [8–11]. Figure 2c shows randomly oriented aggregates approximately 40 nm in size; no fiber formation was observed. The height image shown in Fig. 2d was taken on a sample that was first spin coated from chloroform solution and then exposed to CS_2 vapor (more details about SVA are given in the next part of this section). The image reveals a highly ordered fiber-like structure, with structures 40 nm in size.

The corresponding absorption spectra of the thin films shown in Figs. 2a and c are also depicted in Fig. 2e. The film spin coated from $CHCl_3$ is broad and featureless with a maximum absorption at 700 nm and can be considered amorphous for the most part. In contrast, the film made from CB/DIO shows a large redshift to 790 nm with a distinct shoulder at 725 nm. Hence, the presence of DIO seems to increase the aggregation tendency of PCPDTBT without causing the formation of fiber-like structures.

2.3 *Solvent Vapor Annealing as a Method to Crystallize PCPDTBT*

SVA has been used by us to thin polymer films to gain better control over the crystallization process [22, 23]. This solution-based technique is a good alternative to conventional thermal annealing because the conditions are milder and thermal degradation can be avoided. The thin film is first dissolved to create a highly concentrated solution, after which crystallization can be induced by reducing the vapor pressure in a stepwise manner. SVA allows for the tuning of both nucleation density and film drying time, yielding ordered and defined structures.

$CHCl_3$–spin-coated PCPDTBT films were used for the SVA experiments because of their amorphous nature, as evidenced by the broad absorption profile and the featureless, homogeneous AFM topography images (Figs. 2a and e). Forty-nanometer-thick films were prepared by spin coating from a 3-mg/ml $CHCl_3$ solution on different substrates and solvent vapor annealed in either CS_2 or CB [12]. The height image shown in Fig. 2d was taken on a sample that was prepared as described above and then exposed to CS_2 vapor. The image reveals highly ordered fiber-like structures. Unexpectedly, the absorption profile of the CS_2-annealed film (Fig. 2e) is not redshifted as in the case of the CB/DIO-processed film, the charge-transfer band becomes narrower, the absorption maximum is shifted to 740 nm, and a distinct shoulder emerges at 650 nm. Despite the presence of ordered structures in both AFM images shown in Figs. 2c and d, the measured absorption spectra differ significantly.

Although both CS_2 and CB were investigated as solvents in the annealing process, the CB-annealed samples are described in further detail in this section. Since

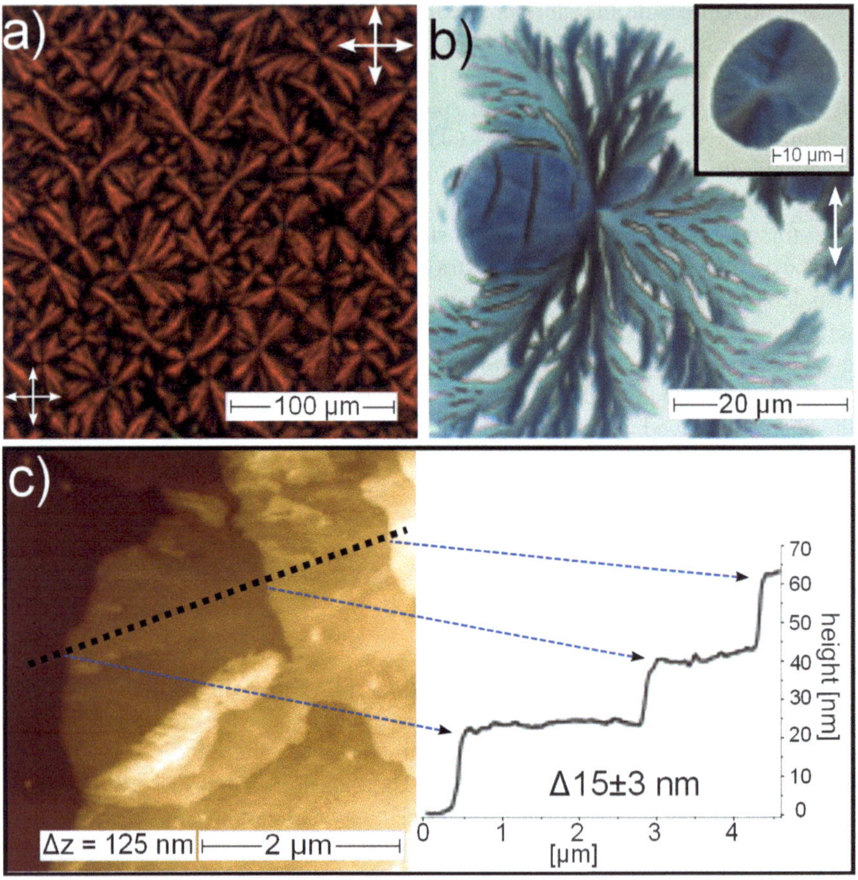

Fig. 3 Morphology of PCPDTBT thin films prepared by chlorobenzene (CB) solvent vapor annealing. The spin-coated films (from $CHCl_3$) were swollen around 80–90 % CB vapor pressure and recrystallized at vapor pressures between 70 and 73 %. (**a**) Polarized microscopy (POM) image of a thin film consisting of spherulites. (**b**) POM image of a single spherulite next to a terrace-like structure; inset: magnification of the terrace-like structure. (**c**) Atomic force microscopy height image and height profile illustrating distinct height steps with respect to the substrate. (Adapted and reproduced with permission from Wiley [12])

CB has a rather high boiling point (132 °C), the annealing protocol was modified and the samples were exposed to CB vapor at 50 °C [12]. The obtained CB-annealed films were then characterized using several microscopy-based techniques, such as polarized microscopy (POM), AFM, and scanning electron microscopy (SEM), on the micrometer to nanometer length scale. Additionally, the films underwent a structural characterization using X-ray and electron diffraction. Figure 3 shows the POM images of the CB-annealed films, revealing a strong birefringence [12] and spherulitic structures that are reminiscent of P3HT spherulites we published earlier [22].

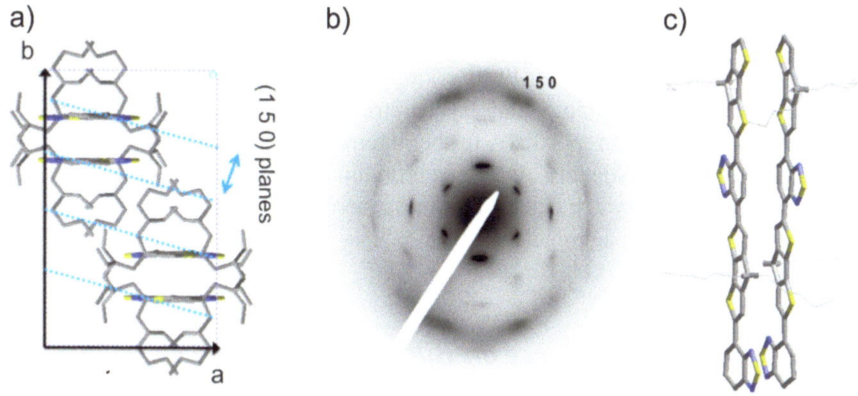

Fig. 4 (**a**) Proposed dimer-like structure. (**b**) Experimental electron diffraction patterns of the [001] zone from the model shown in (**a**). (**c**) Segregated stacking of the CPDT and BT units of PCPDTBT in the dimers. (Adapted with permission from the American Chemical Society [13])

The POM images in Fig. 3b revealed a second, deep blue circular structure in addition to the spherulites in films of low nucleation densities. The AFM investigations of such circular structures evidenced the formation of terrace-like structures with distinct vertical steps (15 ± 3 nm) (image shown in Fig. 3c). We assigned these structures to standing polymer chains. Using TEM/ED, we found distinct ED patterns for the CB-annealed samples, showing reflections up to the sixth order and high symmetry (Fig. 4). Assuming standing chains for the terrace-like structure, the ED patterns were indexed corresponding to a [001] zone and the unit cell parameters were determined to be $a = 12.4$ Å, $b = 19.3$ Å, and $c = 23.6$ Å [12, 13].

By comparing our ED pattern with other reports in the literature, we assigned the crystal structure to the Pccn space group. Several other conjugated polymers such as poly(9,9′-di-*n*-octyl- 2,7-fluorene) have already been seen to pack in a dimer-like fashion [24]. The six intense, inner reflections of the ED pattern form a quasi-hexagonal pattern and were indexed as $\pm 2\ 0\ 0$ and $\pm 1\ \pm 1\ 0$, which corresponds to an orthorhombic unit cell with distances in the range of 9–11 Å [13]. The two intense outer reflections were indexed as $\pm 1\ 5\ 0$ for PCPDTBT and correlate to a short π-stacking distance of 3.6–3.8 Å between two polymer chains comprising a dimer. We emphasize that no long-range π stacking was observed in this as-yet-unreported crystal structure. Based on the information of the CB-annealed films, the CS_2-annealed films could be assigned to the same crystal structure, the only difference being that in CS_2-annealed films the chains are oriented in-plane with respect to the substrate [12].

To briefly summarize the morphological findings, PCPDTBT spin coated from $CHCl_3$ yields a rather amorphous thin film. In the literature it is reported that both neat PCPDTBT films and PCPDTBT/fullerene blends deposited using CB/DIO undergo aggregation and possess an edge-on orientation with long-range π stacking. Furthermore, when the polymer is exposed to either CS_2 or CB solvent vapor, a new crystal structure with dimer-like packing of PCPDTBT chains is obtained. The differences in the morphology and crystal structure are accompanied by distinct changes in the absorption spectra. A further interpretation of the optical phenomena is currently under investigation. One recent finding, for example, is that long-range π stacking is necessary for enhanced photogeneration of polaron pairs via direct excitation of interchain aggregates [25].

2.4 Morphology Influence on Bilayer Solar Cell Performance

The enhancement of organic photovoltaic device performance achieved over the last 10 years has in many cases been associated with the optimization of the active layer morphology. Therefore, since we can prepare PCPDTBT films consisting of three different morphologies—amorphous (spin coated from $CHCl_3$), dimer-like (CS_2-annealed), and long-range π-stacked edge-on orientation (CB/DIO-film)—we have investigated the effect of polymer morphology on bilayer solar cell performance. The bilayer device structure was chosen to ensure that the fullerene acceptor will not disturb the polymer chain packing. The following device structure was used: indium tin oxide (ITO) | poly(3,4-ethylenedioxythiophene) polystyrene sulfonate | PCPDTBT (40 nm) | evaporated C60 (40 nm) | bathocuproine | Al (see inset of Fig. 5a). The solar cell devices were characterized under standard AM 1.5G conditions; *J–V* curves as well as external quantum efficiencies (EQEs) were measured and are shown in Fig. 5 [12].

We found that despite the three different polymer morphologies, all *J–V* curves are quite comparable. Short-circuit current densities (J_{sc}) between 4.4 and

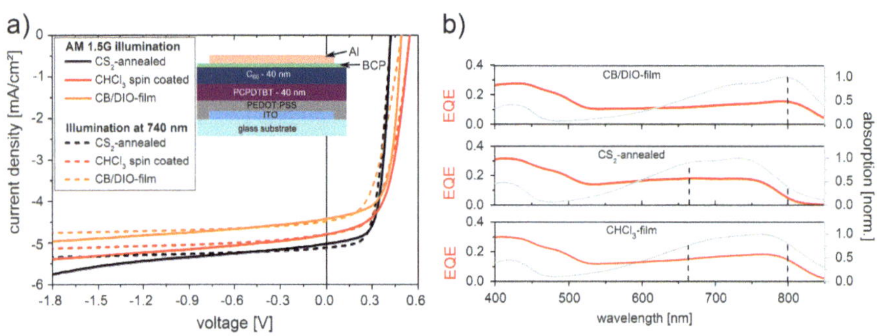

Fig. 5 Influence of polymer morphology on solar cell performance. (**a**) *J–V* curves and (**b**) external quantum efficiency spectra (*red*) for bilayer PCPDTBT/C60 cells are compared to the absorption spectra of the pure polymer layers (*gray*) without C60 (*right scale*). (Adapted with permission from Wiley [12])

5.0 mA/cm^2 were measured and high fill factors (FFs) between 57 and 66 % were obtained. The open-circuit potentials (V_{oc}) varied from 0.55 to 0.42 to 0.50 V for the $CHCl_3$–spin-coated, CS_2–annealed, and CB/DIO–spin-coated polymer films, respectively. Good charge transport in PV devices is confirmed by the high FFs and low field dependence of the photocurrent at reverse bias. Overall PCEs between 1.4 and 1.5 % were obtained regardless of the polymer film morphology.

The EQE spectra shown in Fig. 5b confirm that the three different morphologies are present in the three bilayer devices. Interestingly, all three devices display higher photon-to-electron conversion ratios at high energies (400–500 nm) than those in the low-energy region (700–850 nm) despite the relative lower absorption intensity in the respective spectral regions. For all three devices, a maximum EQE of approximately 30 % was obtained at 420 nm. Regardless of polymer morphology, splitting of the charge-transfer excitons (CTEs) at the polymer/fullerene heterojunctions seems to be not limited by an energy barrier.

3 Molecular Doping to Tune the Optoelectronic Response of the Polymer

In addition to the macroscopic control of polymer ordering and conformation, recent efforts have focused on applying versatile strategies for tuning the electrical properties. For example, molecular doping [26, 27], that is, the use of a strong electron acceptor to p-dope the polymer via ground-state charge transfer, is a promising strategy for enhancing carrier transport. To date, molecular doping has been applied to tune the electrical properties of interfaces between organic semiconductors and electrode materials to create transport layers [28]. This strategy has not yet been adequately investigated because of the need to elucidate the complex interplay between dopant-induced charge and morphological variations induced by the charged dopant molecules. Both effects can significantly influence the electrical properties of the active layer.

In the following, the influence of doping pristine PCPDTBT films and PCPDTBT:PCBM blends with the well-known electron acceptor F4-TCNQ is discussed. The goal is to develop molecular doping as a strategy to enhance the photocurrent in PCPDTBT:PCBM solar cells.

3.1 The Influence of Molecular Doping on Polymer Morphology: Correlating Dopant Distribution with Polymer Conductivity

In the literature, saturation or even a reduction in carrier conductivity with doping concentration has been reported [29]. The ionized dopant molecule is expected to

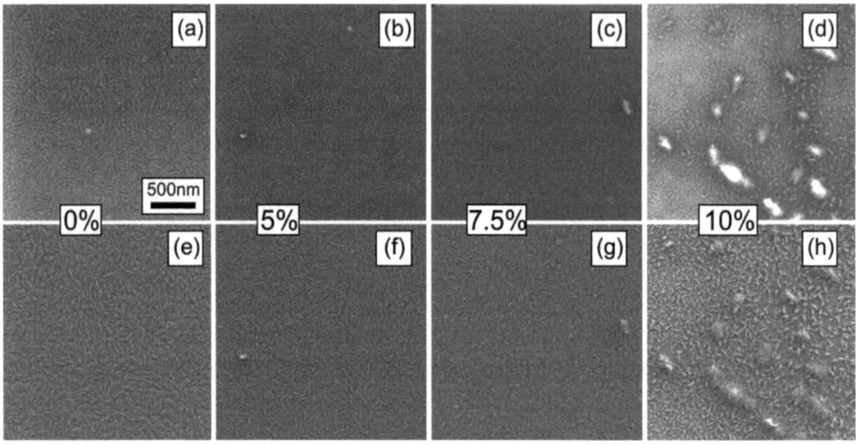

Fig. 6 Scanning electron microscopy images of PCPDTBT films doped with F4-TCNQ at w/w% of (**a**, **e**) 0 %, (**b**, **f**) 5 %, (**c**, **g**) 7.5 %, and (**d**, **h**) 10 %. (**a**–**d**) Images recorded with InLens and (**e**–**h**) SE2 detector. Scale bar is 500 nm for all images. (Reproduced with permission from [31])

influence polymer conformation and interactions; however, very few studies [15, 30, 31] report on the morphology of doped polymer films. Here we report on the correlation between the F4-TCNQ distribution and the hole conductivity in PCPDTBT films [31]. To demonstrate the generality of our findings, these results are compared to doped P3HT films.

SEM was applied to monitor the morphology of the polymer film as a function of the F4-TCNQ concentration. To discriminate between morphological features on the film surface and in the bulk, we applied two detectors. The SEM data are presented in two rows: At the *top* we recorded images with the InLens detector, whereas at the *bottom* we have used a detector for secondary electrons (SE2) placed close to the films. The SE2 detector gives a good contrast for the surface, while the InLens detector can look at material properties (atomic number contrast) and charging [32, 33].

Figure 6 shows the SEM images of PCPDTBT films doped at a weight percent (w/w%) of 0 % (a, d), 5 % (b, e), 7.5 % (c, f), and 10 % (d, g) with F4-TCNQ. The top row (a–d) includes micrographs recorded using the InLens detector, while the bottom row (e–g) includes micrographs recorded with the SE2 detector. The pristine PCPDTBT films demonstrate fiber-like structures, consistent with the semicrystalline behavior described earlier in the chapter. The number and size of the features increase with the doping concentration, and we attribute this to F4-TCNQ clustering in the polymer film.

Figure 7 shows SEM images of doped P3HT. The morphology of the undoped polymer is less structured than that of PCPDTBT. The image contrast additionally increases with doping concentration. For P3HT we observe the development of contrast at lower F4-TCNQ concentrations (5 %) than for PCPDTBT. At 10 %,

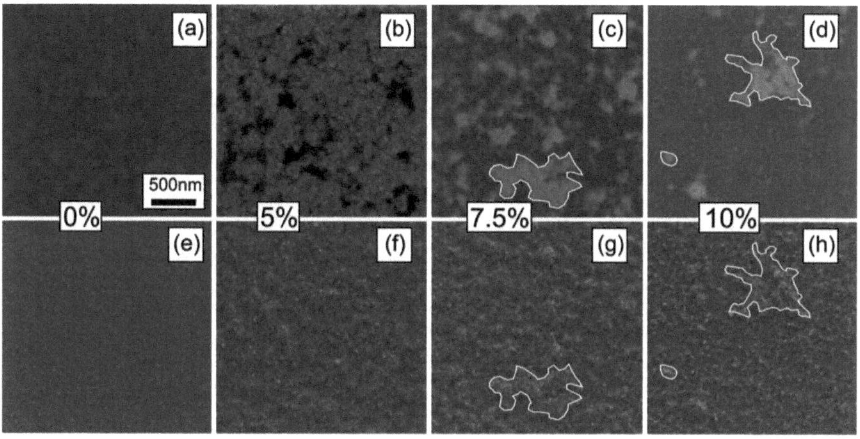

Fig. 7 Scanning electron microscopy images of P3HT films doped with F4-TCNQ at w/w% of (**a**, **e**) 0 %, (**b**, **f**) 5 %, (**c**, **g**) 7.5 %, and (**d**, **h**) 10 %. (**a**–**d**) Images recorded with InLens and (**e**–**h**) SE2 detector. In panels (**c**, **d**, **g**, **h**) the regions with bright contrast with the InLens detector have been marked to be clearly distinguished also in the SE2 images. Scale bar is 500 nm for all images. (Reproduced with permission from [31])

clusters are as large as 500 nm (Fig. 7d). The grayscale contrast of InLens images differs and is material dependent.

The SEM analysis is sensitive to features restricted to a 15-nm depth from the surface. To probe deeper into the bulk of the film, we applied TEM (Fig. 8). In the transmission detection configuration, darker areas are ascribed to regions with a high electron scattering probability; that is, these regions have a high concentration of F4-TCNQ [31]. At 5 % doping, P3HT exhibits dopant-rich areas, consistent with the SEM images (panel f). For PCPDTBT (panel b) we notice small clusters (smaller than 100 nm), which are not observable in SEM.

Confocal photoluminescence (PL) was applied to complement the electron microscopy and identify dopant-induced variations in the PL map of the polymers [34, 35] (Fig. 9). Confocal PL is particularly suited to probe dopant distributions because of localized exciton-polaron quenching. Dark areas in the map correspond to a high concentration of dopants [36, 37]. Confocal PL microscopy has a resolution similar to the diffraction limit of the excitation light (~400 nm). Thus, the morphological features in the SEM for the pristine PCPDTBT cannot be discerned in PL microscopy. Notably, upon doping the overall PL intensity in the images is quenched homogeneously as the insets show. This behavior is expected as dopant clusters in the polymer films smaller than 200 nm cannot be resolved. The PL images indicate that excitons, which have a short diffusion length (10 nm), are quenched with a high probability by dopant clusters smaller than 200 nm. This results in a uniform PL quenching pattern. In contrast, the doped P3HT films show changes in the PL microscopy maps on a length scale of a micron. This is in accordance with

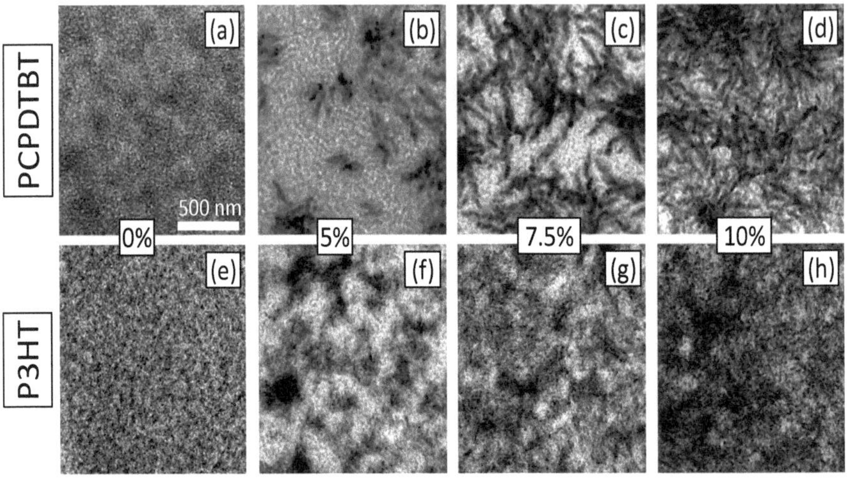

Fig. 8 Transmission electron microscopy images for PCPDTBT (**a**–**d**) and P3HT (**e**–**h**) films doped weight % of (**a**, **e**) 0 %, (**b**, **f**) 5 %, (**c**, **g**) 7.5 %, and (**d**, **h**) 10 % with F4-TCNQ as indicated in the figure. Acceleration voltage of 100 kV. Scale bar is 500 nm for all images. (Reproduced with permission from [31])

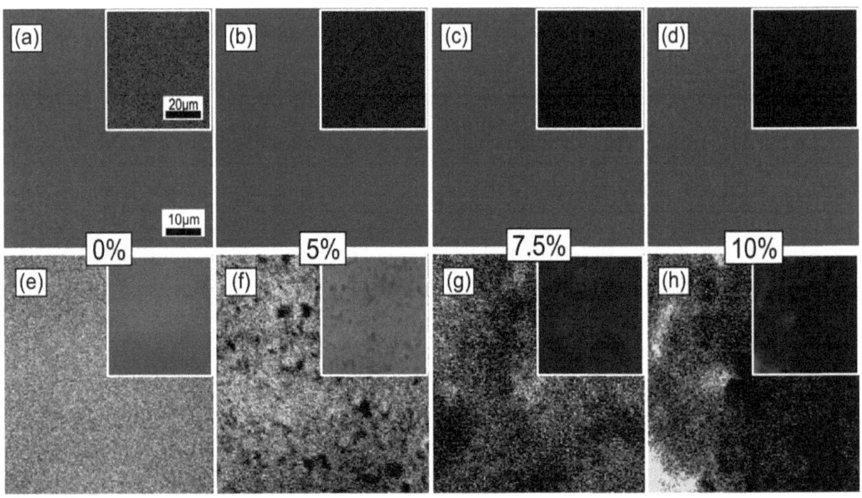

Fig. 9 Photoluminescence images for PCPDTBT (**a**–**d**) and P3HT (**e**–**h**) thin films doped with F4-TCNQ at different w/w% ratios as indicated in the figure. Images were postprocessed for optimal contrast; the insets show the unprocessed images on a common photoluminescence intensity scale. Excitation wavelength is 532 nm and scale bar 10 μm for all images. (Reproduced with permission from [31])

the SEM results, which show bright areas of similar sizes because of the F4-TCNQ–rich clusters.

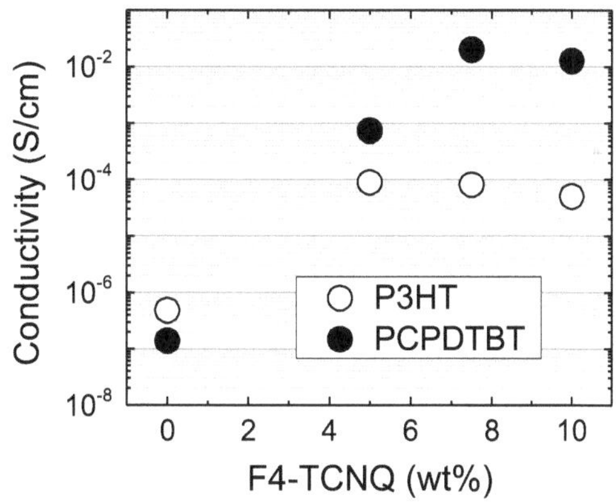

Fig. 10 Electrical conductivity of P3HT (*closed symbols*) and PCPDTBT (*open symbols*) as a function of F4-TCNQ doping expressed as w/w% ratio with respect to the polymer. Conductivity extracted from four probe experiments as detailed in text. (Adapted from [31])

To correlate the morphology of the doped polymers with their electrical properties, we studied the conductivity as a function of doping on films that were blade coated on a prepatterned ITO interdigitated planar structure. Figure 10 shows the change in conductivity of PCPDTBT (open symbols) and P3HT (closed symbols) thin films (100- ± 10-nm thickness) vs. F4-TCNQ concentration. In both polymers the conductivity increases with the dopant concentration by two orders of magnitude before saturation. In P3HT, the conductivity saturates at 5 % doping and decreases at higher concentrations (>7.5 %). In PCPDTBT, saturation occurs at 7.5 %. The saturation in conductivity is consistent with the dopant clustering at these F4-TCNQ concentrations that has been observed using electron microscopy and PL mapping.

Electrostatic forces between dopant and semiconductor should not be disregarded in the film formation process and thus in the final morphological structure [38, 39]. The structure of PCPDTBT, which has an intrinsic dipole [40, 41] arising along the polymer backbone because of the alternation of donor and acceptor moieties, can electrostatically screen these interaction forces and thus result in a different morphology with respect to what is observed for a homopolymer like P3HT. We consider the role of molecular structure on semiconductor–dopant interactions later in the chapter.

3.2 *The Influence of Molecular Doping on Hole Transport in Pristine PCPDTBT Films and PCPDTBT:PCBM Blends*

Low concentrations of F4-TCNQ increase the conductivity of the polymer without causing substantial excitonic quenching [31]. This can be understood from the perspective that low concentrations of dopant promote trap filling in the polymer [14, 15, 39, 42]. This suggests that there is an optimal dopant concentration at which trap filling occurs, but the charge density is low enough to minimize the effects of exciton-polaron annihilation.

This indicates that molecular doping is a versatile way to microscopically tune the electrical properties of the polymer phase. It allows us to increase the photocurrent without reducing the extended bulk heterojunction interface required for efficient charge separation. This is a limitation of strategies based on macroscopic morphology control, as described earlier in the chapter. However, studies on the molecular doping of polymers in the literature have led to mixed results about the influence on carrier mobility [15, 43–46], and there have been few reports of molecular doping applied to bulk heterojunction devices [15, 44, 47, 48], as the complex influence of the dopant on morphology and energetics can prevent controlled doping without additional detrimental effects, such as recombination and trapping [43, 45, 46] and morphological variations [30].

Field-effect measurements are a reliable method for discriminating between electron and hole transport in blends of organic semiconductors [49]. Thin-film field-effect transistors (FETs) are three-terminal devices. The channel current is made up of majority carriers, and the sign of the majority carriers is determined by the polarity of the bias applied to the gate electrode relative to the grounded source electrode [50]. The gate-source voltage (V_{gs}) results in the accumulation of charge in the channel, and the drain-source voltage (V_{ds}) results in current flow in the channel. FET measurements have been applied in the literature to study variations in electron and hole transport as a function of blend processing [15, 51].

Figure 11 shows (a) the hole channel current and (b) the hole field-effect mobility taken from FETs prepared with pristine PCPDTBT films (closed symbols) and PCPDTBT:PC_{60}BM blends (open symbols) as a function of the F4-TCNQ doping concentration (0–0.3 %). The ON/OFF behavior of the device is determined by the low conductivity of the organic semiconductor [50]; at higher F4-TCNQ concentrations, reliable FETs were not possible [15].

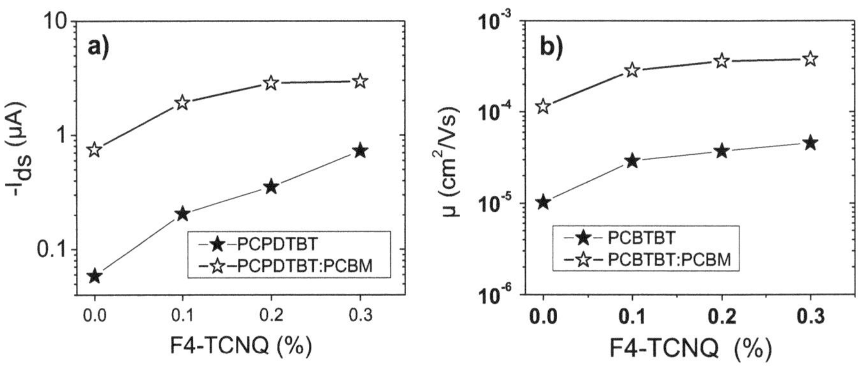

Fig. 11 (**a**) Source-drain current I_{ds} at source-drain voltage, $V_{ds} = -10$ V, and gate voltage $V_{gs} = -60$ V as a function of F4-TCNQ concentration for PCPDTBT (*closed stars*) and PCPDTBT:PCBM (*open stars*) thin films. (**b**) Field-effect mobility values, μ, extracted from the linear regime as a function of F4-TCNQ for the same samples in panel (**a**). (Reproduced with permission from [15])

Both the channel current and the field-effect mobility increase when the polymer is blended with the fullerene. This effect has been observed in amorphous polymers, such as derivatives of poly(p-phenylene vinylene) (PPV), and was attributed to effects such as electronic interactions [52, 53] and/or hole transport by the fullerene [54, 55], as well as changes in polymer conformation induced by the fullerene [56, 57]. Studies that compared the electronic properties and Raman spectra of PPV blended with silica nanoparticles separately and blended with fullerene [58] demonstrated that the effect is mainly morphological and is consistent with increased planarity of the polymer [57]. Blending PCPDTBT (undoped) with the fullerene results in a nearly tenfold increase in both channel current and field-effect mobility of the sample. The addition of F4-TCNQ to the samples results in an additional increase in channel current and field-effect mobility. The effect is slightly more pronounced in the pristine PCPDTBT film (1×10^{-5} to 5×10^{-5} cm^2/Vs) than in the blend (1×10^{-4} to 4×10^{-4} cm^2/Vs). In these samples, the enhancement in the electrical properties induced by the F4-TCNQ is less than that induced by the fullerene. This is consistent with "electrical tuning," that is, trap filling, of the polymer phase by adding small concentrations of the F4-TCNQ molecule rather than macroscopic morphological manipulation [31].

Increasing the hole mobility should positively impact the photocurrent in solar cells. In addition, the initial geminate recombination of bound pairs at the polymer/fullerene interface can be reduced by tuning the electronic properties of the donor and acceptor domains [59, 60]. The separation of CTEs into free carriers is strongly dependent on the local mobility within the polymer and fullerene domains, and CTE recombination has been anticorrelated with the photocurrent in solar cells [61].

3.3 *The Influence of Molecular Doping on Charge Separation and Recombination in PCPDTBT:PCBM Blends*

CTEs have been studied with PL spectroscopy [61, 62]. The CTE has been shown to dissociate if the electrostatic energy induced by polarizing the surrounding environment becomes comparable to the binding energy of the CTE [60]. Charge delocalization is known to facilitate charge separation [59, 60, 62].

Figure 12 shows PL spectra of PC$_{60}$BM, PCPDTBT, and a blend of these two materials in a 1:1 weight ratio (left panel). The pristine materials demonstrate the typical PL signatures [63]. In the spectra from the blend film, however, a broad unstructured emission is observed that peaks at 1.17 eV. PL signatures from the PCBM and PCPDTBT, expected to appear at energies above 1.4 eV, are not detectable in the blend. This indicates efficient exciton diffusion and charge separation at the D–A interface. The PL signature at 1.17 eV is ascribed to radiative recombination of the CTE [62, 63].

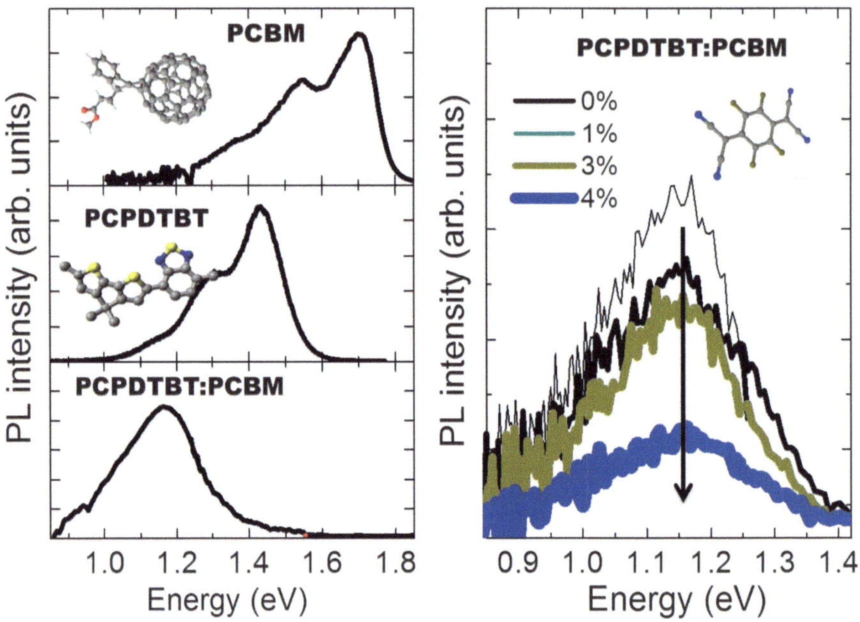

Fig. 12 *Left*: Photoluminescence (PL) spectra and chemical structures of PCBM, PCPDTBT, and PCPDTBT/PCBM (1:1 weight ratio) blend. *Right*: PL spectra of the charge-transfer exciton emission for blends doped with F4-TCNQ at 0 % (*thin black*), 1 % (*black*), 3 % (*dark yellow*), and 4 % (*blue*). The chemical structure of F4-TCNQ is shown in the inset. Excitation was performed at 3.1 eV, and PL spectra intensity in right panel was corrected considering the number of absorbed photons. (Reproduced with permission from [14])

Doping the blend results in the quenching of the CTE emission. Figure 12 (right panel) shows how the PL intensity from CTE emission is quenched in the presence of increasing F4-TCNQ concentration. Figure 13 shows the normalized PL decay transients of the CTE emission at 1.17 eV after exciting the blends with a 150-fs laser pulse for PCPDTBT:PCBM blends doped at 0, 4, and 5 % with F4-TCNQ. No significant differences were observed in the PL lifetimes of samples doped between 0 and 4 %. Only at 5 % does the decay rate decrease, from $\tau = 300$ ps to $\tau = 150$ ps. This indicates that above 5 % other deactivation processes such as exciton-polaron quenching start to dominate.

The influence of doping on the CTE emission can be explained by the fact that the increased carrier density induced by the dopant influences the rate of recombination via nonradiative transitions, resulting in a reduced CTE PL emission. Additionally, dopant-induced charge increases the probability of CTE dissociation through a competitive channel, which favors charge separation.

Time-resolved photoinduced absorption spectroscopy (TAS) is a sensitive pump-probe technique for studying the nature and dynamics of excited species such as

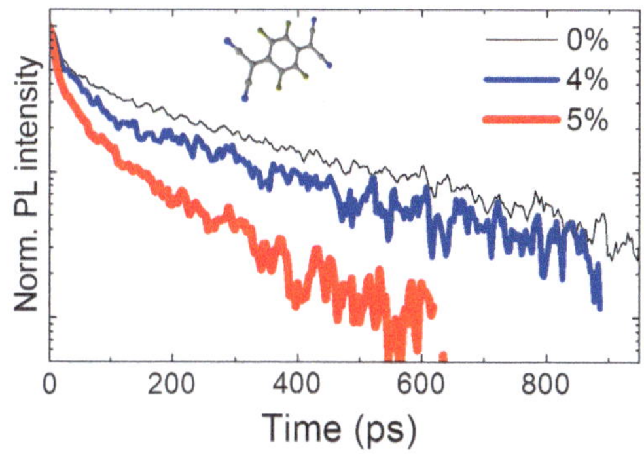

Fig. 13 Normalized transient decays of photoluminescence recorded at 1.17 eV for PCPDTBT/PCBM blends doped with 0 % (*black*), 4 % (*blue*), and 5 % (*red*) concentrations of F4-TCNQ. Excitation was performed at 1.65 eV with 150-fs pulses and detection was achieved with a near-infrared streak camera with a resolution of 5 ps. (Reproduced with permission from [14])

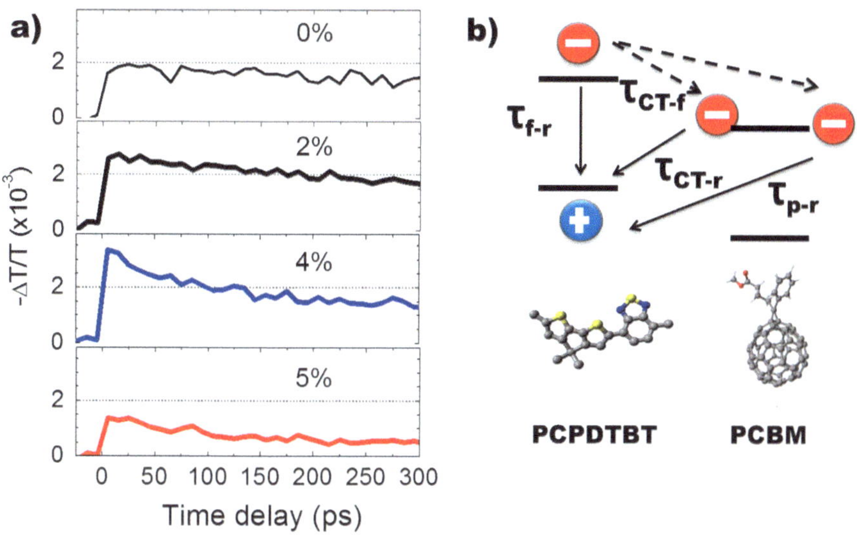

Fig. 14 (**a**) Photoinduced differential transmission transients of PCPDTBT:PCBM blends doped at concentrations between 0 and 5 % with F4-TCNQ. The same laser pump (3.1 eV) and probe (0.4 eV) configuration was maintained for all the measurements. (**b**) Energy-level scheme of PCPDTBT and PCBM with the respective highest occupied molecular orbital and lowest unoccupied molecular orbital together with the most significant electronic transitions depicted. (Adapted from [14])

excitons and polarons in organic semiconductors. Polarons in organic semiconductors typically give rise to a TAS signal in the near- to middle-IR region of the spectrum [64]. Two prominent polaron transitions have been reported in the literature for PCPDTBT:PCBM blends at 0.4 and 0.9 eV [65]. Figure 14a shows the TAS data performed on the same samples as for the PL studies using a pump excitation of 3.1 eV and a probe laser beam centered at 0.4 eV. The 0.9-eV polaron signal overlaps with excitonic features in the NIR and can complicate the analysis of the polaron dynamics [65]. The signal amplitude, expressed as differential transmission of the probe pulse $\Delta T/T$, is proportional to the density of polarons resulting from photoexcitation. The data clearly exhibit a difference in the time-zero amplitudes, $\Delta T/T$ ($t = 0$). Within the first 50 ps, the blends doped at 2 and 4 % (black and blue traces) reveal larger signal amplitudes than the sample doped at 0 % (thin black line) by a factor of 1.4 and 1.7, respectively. This means that, within the time resolution of the measurement (~150 fs), there is a larger density of polarons upon photoexcitation in the doped blends. For higher doping concentrations (5 %), the amplitude of the trace is lower than that of the undoped blend. At times longer than 50 ps, the blends doped at 4 and 5 % both show an increased decay rate of the polaron signal, indicating increased recombination phenomena.

These results reveal that blends that are doped up to 2 % with F4-TCNQ demonstrate an increased polaron density at short timescales, which does not impact recombination dynamics (compared to the undoped blend). The additional reduction in CTE emission suggests that low doping concentrations can be beneficial for reducing geminate recombination without inducing further unwanted and detrimental effects.

The PL and TAS transients can be quantified by modeling the time constants of the transitions with coupled rate equations. The model was based on the schematic in Fig. 14b, which illustrates the possible electronic transitions after photoexcitation. The dashed arrows indicate the fate of Frenkel excitations formed on the PCPDTBT, which can form either CTE states or free polarons with probabilities dependent on the rates $1/\tau_{\text{CTE-}f}$ and $1/\tau_{p\text{-}f}$, respectively. The features in the PL spectra can be related to recombination of the Frenkel excitons ($\tau_{f\text{-}r}$) and CTEs ($\tau_{\text{CT-}r}$). The polaronic features observed in the TAS transients are described by the rate constants depicted with the straight arrows, which correspond to the time constants of polaron generation from Frenkel excitons ($\tau_{p\text{-}f}$) and CTEs ($\tau_{\text{CT-}f}$). A model that considered polaron formation solely from CTE states was not able to fit the PL and TAS data, indicating this is not the only transition leading to free charge after photoexcitation. A value of 200 fs [66] was used for $\tau_{p\text{-}f}$ and $\tau_{\text{CT-}f}$ in the undoped samples. The recombination time constant of the CTE ($\tau_{\text{CT-}r}$) was fixed at 300 ps and the recombination constant of Frenkel excitons ($\tau_{f\text{-}r}$) at 125 ps for blends doped at 4 % and below. At higher doping concentrations, higher recombination rates were used to account for the faster decay rates observed [14]. The model resulted in a good fit of the experimental transients, as observed by the solid lines in Figs. 13 and 14.

One of the most important observations involves the differences in the signal amplitudes of PL and TAS. These can be quantitatively described by varying the initial probability of CTE formation. At a 2 % doping concentration, this corresponds to 1/500 fs^{-1} and decreases to 1/950 fs^{-1} at a doping concentration of 4–5 %. Variations in $\tau_{p\text{-}f}$ were not considered, as no increase in the rise time of photogenerated polarons was observed in the doped samples, and the increase in polaron density could be related to faster CTE dissociation. The increase in the TAS decays in samples doped at 5 % were modeled by increasing the recombination rates of Frenkel excitons and polarons, suggesting an increasing significance of Auger recombination and exciton quenching by free polarons at higher doping concentrations. These results suggest that photoexcitations in doped PCPDTBT:PCBM blends are less likely to form CTEs. This favors the formation of polarons and reduces geminate recombination for blends that are doped up to 4 % with F4-TCNQ.

To understand these effects, we consider that increased carrier densities are expected to screen the Coulombic interaction between bound electron–hole pairs, analogous to effects in inorganic semiconductors [67]. Additionally, increased charge density can polarize the surrounding neutral molecular environment [60], effectively weakening the binding energy of initial photoexcitations in PCPDTBT as well as CTEs formed at the D–A interface. From the point of view of electrical transport, holes generated in the ground state by the dopants are expected to fill trap states in the electronic density of states and in particular the low-energy tail, which in many cases is responsible for exciton localization in organic semiconductors [68].

3.4 *Molecular Doping to Enhance Photocurrents in PCPDTBT:PCBM Blends*

Figure 15 shows the current density–voltage (*J–V*) characteristics of PCPDTBT: PCBM solar cells (a) under illumination and (b) in the dark. The active layers of the solar cells were prepared under the same conditions as the organic layers used in the FET structures discussed earlier in the chapter. An increase in the short-circuit current density (J_{sc}) from 9.42 to 10.31 mA/cm^2 in the solar cell with the active

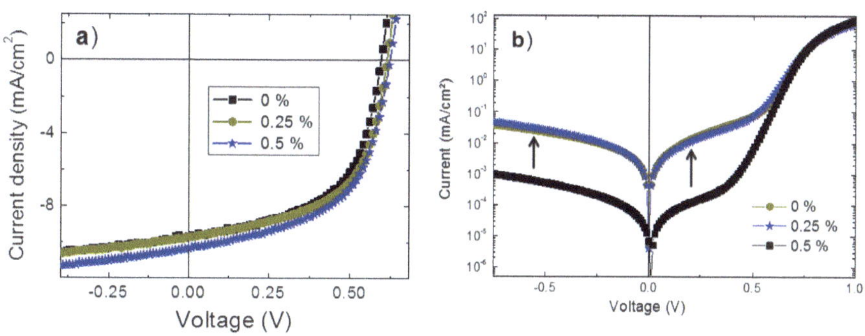

Fig. 15 (**a**) Current density as a function of voltage, *J–V*, curves recorded under illumination with an AM 1.5G solar simulator for PCPDTBT:PCBM solar cells doped with different amounts of F4-TCNQ in w/w%: 0 (*black squares*), 0.25 (*green dots*), and 0.5 (*blue stars*). (**b**) *J–V* characteristics of the devices in the dark in a log-linear scale. See text for extracted solar cell efficiencies and other parameters. (Reproduced with permission from [15])

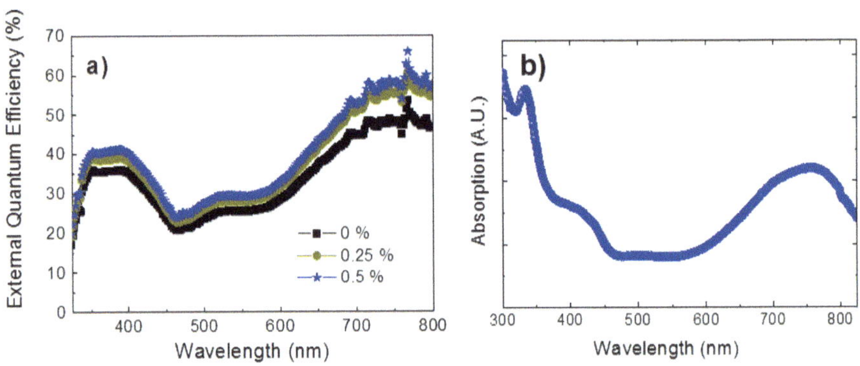

Fig. 16 (**a**) External quantum efficiency spectra for PCPDTBT:PCBM solar cells doped with different amounts of F4-TCNQ in w/w%: 0 (*black squares*), 0.25 (*green dots*), and 0.5 (*blue stars*). (**b**) Light absorption spectra for a 1:1 PCPDTBT:PCBM blend prepared with the same methods as for the solar cells. (Reproduced with permission from [15])

layer doped at 0.5 % with F4-TCNQ is apparent. The FF drops slightly from 56 % in the cell with the undoped active layer (0 %) to 55 % in the cell with the 0.5 % doped active layer. Overall, however, the improvement in the J_{sc} (9 %) and in the V_{oc} (from 0.61 to 0.63 V) results in an increase in the efficiency. The dark *J–V* curves reveal that the current density increases upon doping. These results agree with the FET measurements, which demonstrated saturation in the conductivity and mobility values at doping concentrations around 0.3 %. Interestingly, in the case of the illuminated *J–V*, the current continues to increase even at higher doping concentrations, demonstrating the positive impact of doping not just on electrical transport, but also on charge separation.

Figure 16a shows the EQE spectra of the solar cells. The NIR band (730 nm) can be assigned to photocurrent originating from the PCPDTBT absorption, while the UV band is likely from a superposition of PCBM and PCPDTBT contributions. Figure 16b shows the absorption spectrum of the undoped blend for reference. The NIR EQE contribution due to PCPDTBT undergoes a visible enhancement for doping concentrations above 0.1 %. This indicates that the collection of photogenerated charges is more efficient in a spectral region where PCPDTBT is the main absorber of light in the blend system. On the other hand, we did not observe a clear trend for the shorter-wavelength absorption. The generally low FF and EQE values from our PCPDTBT:PCBM solar cells are consistent with the values reported in the literature and have been attributed to poor charge transport and collection, which lead to enhanced recombination [4, 69].

The relative increase in efficiency (20 %) is a result of the increase in the photocurrent upon doping. The decrease in V_{oc} is related to an increase in the reverse saturation current (J_{sat}) [70], which can increase with doping [71]. Additionally, an imbalance in the density of electrons and holes can lead to a decrease in the built-in field and thus to a deviation from the ideal diode behavior of the cell, which will impact the device's FF [72].

3.5 Does Molecular Disorder Influence the Doping Efficiency of Low-Bandgap Copolymers?

The experiments presented here as well as recent studies by other groups indicate that doping in D–A copolymers is less effective than in homopolymers. Zhang and Blom [42] reported that D–A copolymers require concentrations of several w/w% in order to improve electrical behavior. Addressing these questions requires a detailed understanding of the mechanism of doping at the molecular level, ideally with a combined experimental and theoretical approach. We used IR spectroscopy, an experimental technique that can detect with high sensitivity charge carriers in organic molecules [73, 74]. Further, we combined the experiments with theoretical modeling of the vibrational spectra by ab initio density functional theory quantum chemical calculations. We chose to focus our attention on the polymer vibrations because of their high sensitivity to the presence of charges on conjugated molecules through the strong enhancement of the absorption intensity [73].

Figure 17a shows the Fourier transform IR spectra of PCPDTBT as pure compound and in a series of samples at different doping concentrations (% molar ratios) with F4-TCNQ. The molar ratio refers to the moles of F4-TCNQ per moles of polymeric repeat unit. This is a more useful definition of concentration to analyze the following results on microscopic interactions compared to the w/w% we used in previous sections. The doped samples with molar ratio above 4 % have additional optically active vibrations in the region 1000–1300 cm^{-1}, which are practically absent in pure PCPDTBT. In particular, from the inset one can notice the increase in the absorption intensity of modes at 1055, 1080, 1120, and 1295 cm^{-1} as the doping concentration reaches 6.7 %. We assign the enhancement of molecular vibrations between 1000 and 1300 cm^{-1} to the presence of hole polarons delocalized on the backbone. Figure 17b shows the IR signals up to 6000 cm^{-1} for the same PCPDTBT thin films. Starting from 6.7 %, one can observe a prominent polaronic band, consistent with the charge-induced vibrational modes shown in Fig. 17a and with results in Sect. 3.2, where free polarons were speculated to be present only above 4 % w/w, which corresponds to a molar ratio of ~7 %.

An interesting strategy to probe the effect of D–A moieties on the polymeric chain is to look at the spectra for the pristine and doped PCPDT. This is a model homopolymer consisting only of bithiophene units. Figure 17c shows IR spectra of PCPDT with the same F4-TCNQ molar concentrations. The intensity enhancement of vibrational modes in the region from 1000 to 1300 cm^{-1} can be discerned at lower concentrations than those for PCPDTBT. Vibrational bands are at 1083, 1108, and 1182 cm^{-1}. The remarkable aspect is that here the effect of doping is already observed at 1 % F4-TCNQ, with a weak enhancement of the unstructured band

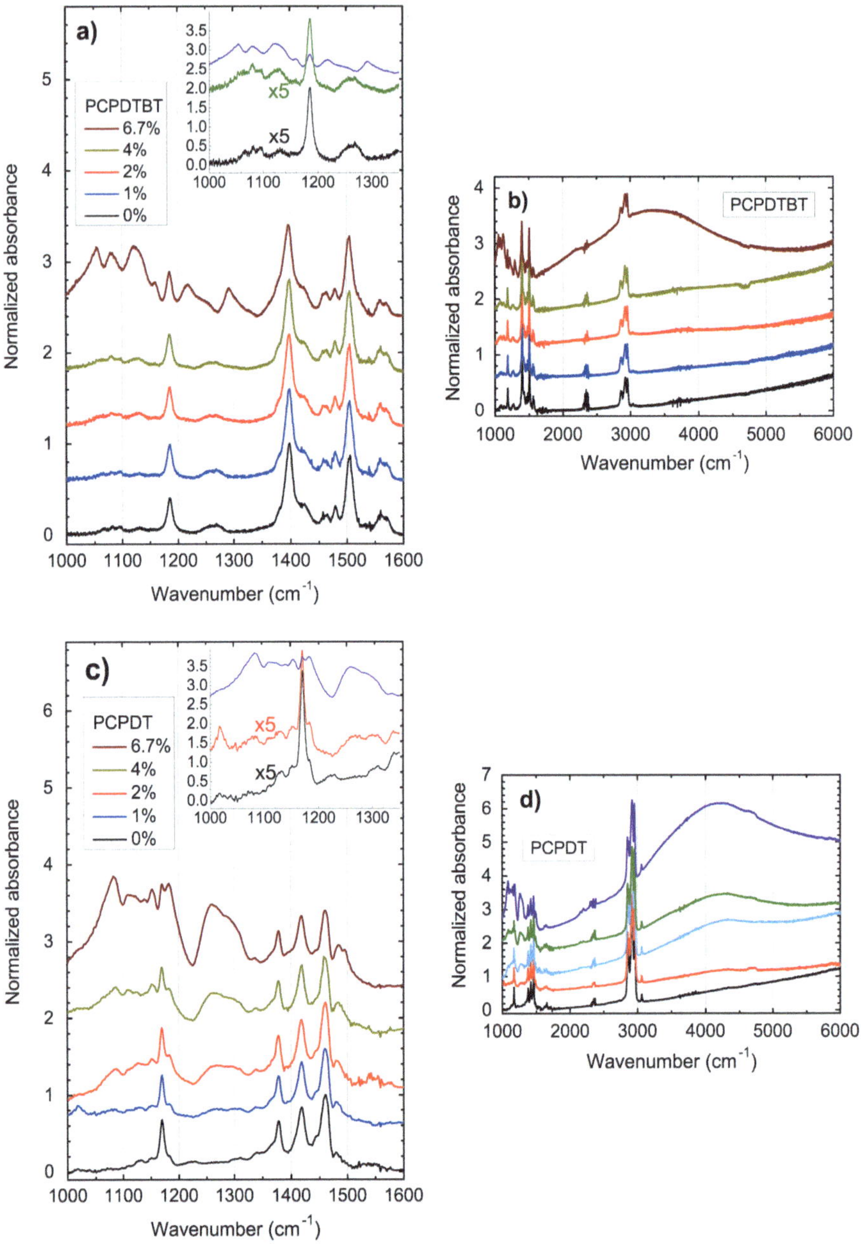

Fig. 17 Fourier transform infrared spectra in the interval of backbone vibrations for (**a**) PCPDTBT and (**c**) PCPDT thin films, pure and doped with F4-TCNQ (see legends). The molar percentage is calculated using the number of moles of dopant divided by the moles of polymeric repeat unit. The insets in (**a**, **c**) show the low wavenumber part, with the intensity of neutral and doped samples' spectra (4 and 1 % for PCPDT-BT and PCPDT, respectively). (**b**, **d**) show larger spectral intervals for PCPDTBT and PCPDT, respectively. (Adapted from [77])

between 1225 and 1335 cm^{-1} (inset Fig. 17c). Only at 2 % does the region between 1000 and 1300 cm^{-1} show well-developed vibrational modes. Pristine PCPDT is instead dominated by three modes discernible at 1378, 1418, and 1478 cm^{-1}.

Panel (d) of the same figure shows the enlarged spectral range with the polaronic transition peaking at ~4100 cm^{-1}. The polaron band of PCPDT is seen above 2 %, while signatures of dopant-enhanced vibrational modes appear below 2 %. From Figs. 17a and c we note that F4-TCNQ, in both neutral and charged states, does not significantly contribute to the IR absorption in the 1000–1600 cm^{-1} spectral region or above, because of the low doping concentrations of our study. This is different from to the studies by Koch and coworkers [75].

Figure 18a shows the calculated spectrum (lower black curve) for a PCPDTBT polymer segment made of four repeat units and one donating CPDT moiety, including one side chain on the central CPDT moiety. The computed spectrum is in very good agreement with the experiments reported in Fig. 17a for pure PCPDTBT. Generally, the computed vibrational modes are blue-shifted compared to the experimental ones. Following standard procedures we have rescaled the computed spectrum by 0.93 [76], to match with the experimental peak at 1398 cm^{-1} of the pristine PCPDTBT film. The bands for the modes at 1398 cm^{-1} are reproduced as the most intense, followed by the peaks around 1500 cm^{-1} and at 1560 cm^{-1}. Next we show the spectrum of a complex in which the dopant F4-TCNQ is cofacially aligned with one of the CPDT moieties of the PCPDTBT strand (indicated as "Complex 1," blue curve). The computed IR spectrum changes dramatically and clearly shows very intense modes at 1000–1200 cm^{-1}. We point out that this spectrum has been scaled down in intensity by a factor of more than 20 compared to that of the neutral polymer. This highlights the strong electron–molecular vibration coupling and charge-induced enhancement for this intermolecular geometry. In this figure we show two additional spectra, which correspond to the same polymer strand without F4-TCNQ but with a positive charge on the chain (red curve), that is, a hole polaron (cation), and the spectrum of Complex 2 where F4-TCNQ is in a different intermolecular configuration, that is, in proximity to the A moiety. The IR spectrum calculated for the hole polaron exhibits similar features to the complex with the F4-TCNQ on the CPDT unit presented above, with similar characteristics in terms of intensity of the modes. The spectrum of the complex with the dopant on the BT is, instead, more similar to that of the pristine neutral polymer strand. The similarities between the hole polaron and the Complex 1 computed spectra indicate that for F4-TCNQ on the homopolymer D moiety, charge transfer occurs and a positive polaron is formed. When the F4-TCNQ is instead on the BT moiety, charge transfer and polaron formation do not occur.

For PCPDTBT we calculated the IR spectra for the homopolymer strands and in complexes with F4-TCNQ. In Fig. 18b we report spectra for a PCPDT polymer strand in its neutral form (black curve at the bottom), a complex of the strand with F4-TCNQ (blue curve in the middle), and the same polymer strand with a net positive charge (red at the top). Also, for this polymer the spectra of the hole

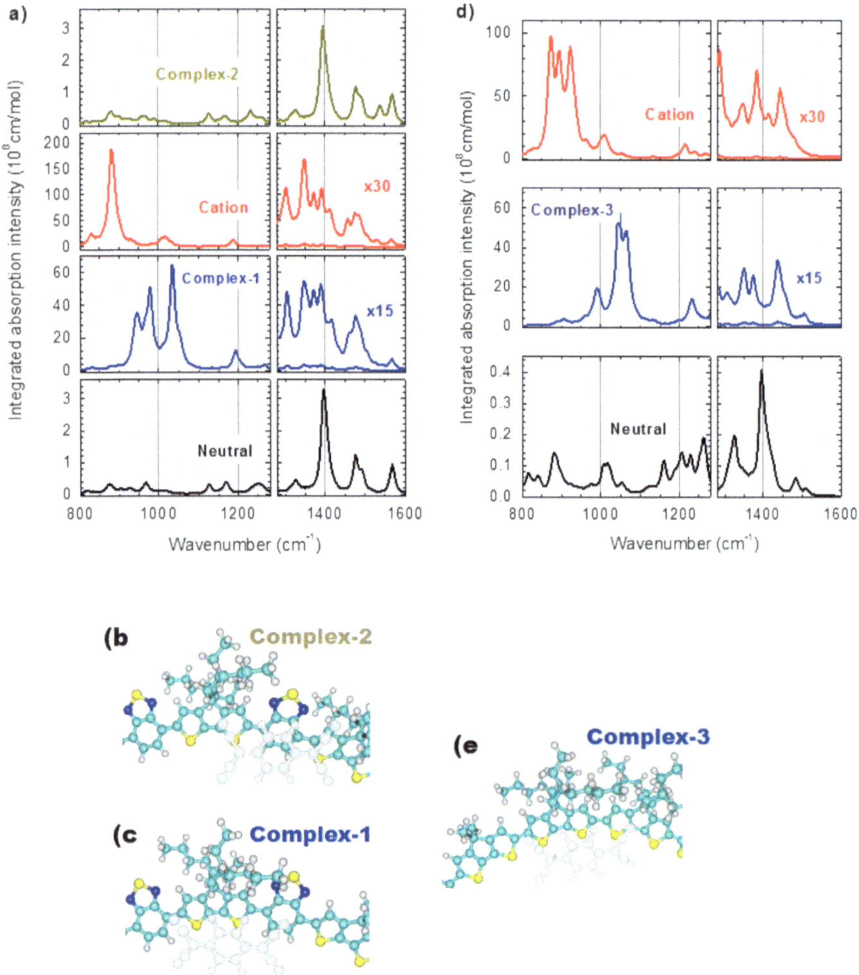

Fig. 18 (**a**) Calculated vibrational absorption spectra of PCPDTBT in different configurations: single polymer strand without F4-TCNQ (Neutral), with an F4-TCNQ molecule localized close to the CPDT moiety (Complex 1), without F4-TCNQ molecule and one electron less in the highest occupied molecular orbital (HOMO) (Cation), and with an F4-TCNQ molecule localized close to the BT moiety (Complex 2). (**b**, **c**) Image zooms of the structures of Complex 2 and Complex 1, respectively. F4-TCNQ is colored in gray. (**d**) Calculated vibrational absorption spectra of PCPDT in different configurations: single polymer strand without F4-TCNQ (Neutral), with F4-TCNQ (Complex 3), and without F4-TCNQ and one electron less on the HOMO (Cation). (**e**) Image zoom of the structure used to compute Complex 3 spectrum. The distances between the atoms of F4-TCNQ and the closest atoms on the polymer backbone are on the order of 0.40, 0.38, and 0.42 nm for Complex 1, 2, and 3 respectively. (Adapted from [77])

polaron and the complex are similar and demonstrate the transfer of one electron from the polymer to the dopant.

The most important outcome of our work is that doping of conjugated polymers depends on the molecular arrangement and disorder. By "disorder," we refer to the position of the dopant molecule with respect to the D and A moieties in the polymer backbone. This picture agrees with the findings on the homopolymer, where charge transfer occurs with similar amounts of charge transferred almost independently with respect to the F4-TCNQ position along the polymer backbone. From our ab initio quantum chemical methods we have calculated the total Gibbs energy between complexes 1 and 2, which is 0.2 eV more stable for Complex 1.

The experimental spectra for PCPDTBT (Fig. 17a) indicate that for all the molar ratios that we have considered, there is an intense contribution from the molecular vibrations characteristic of the pristine polymer. Even at the highest molar concentration (brown curve), the vibrational bands between 1350 and 1550 cm^{-1} remain the most intense. This concentration would correspond to one dopant molecule per 10 monomer units, which is comparable to a situation close to our theoretical simulations. From Fig. 18a, our calculations on the doped chain suggest that those characteristic vibrations should be much weaker in intensity with respect to the charge-enhanced vibrational bands from 1000 to 1300 cm^{-1}. A large enhancement of vibrational modes in other conjugated polymers in the presence of charges has recently been reported, and we are confident that the effect observed in our calculated spectra reflects the experimental situation of a completely doped copolymer film, that is, where each F4-TCNQ molecule generates a hole on the PCPDTBT. Thus, there seems to be a discrepancy that might be generated by the fact that not all F4-TCNQ molecules are effectively undergoing integer charge transfer. Thus, the experimental spectrum is simply a sum of the spectra we present in Fig. 18a considering charged and neutral polymer strands.

The dopant could segregate in the film in a pure phase separated from the polymer, as we observed in the sections above. Dopant molecules not in contact with PCPDTBT are inactive and do not contribute to doping. With the low concentrations we explore here, this is very improbable. However, our calculations suggest a different explanation: Even if the vast majority of F4-TCNQ molecules are dispersed, there are structural intermolecular configurations that do not lead to charge transfer, namely, doping. It is important that polymers in devices are prepared with molecularly dispersed dopants; this can be obtained by low doping concentrations [29] and proper solvents. However, this work shows that specifically in D–A polymer, F4-TCNQ close to the acceptor moiety forms a very weak charge-transfer complex that is only partially active in terms of charge transfer.

Acknowledgments We thank many coworkers who gave a crucial contribution to the results presented in this chapter. These colleagues are Sybille Allard, Manfred Bayer, Martin Brinkmann, Andras Deak, Felix Deschler, Antonietta De Sio, Daniele Di Nuzzo, Bernhard Ecker, Jochen Feldmann, Florian Fischer, Claudio Fontanesi, Tillmann Godde, Alamgir Karim, Anna Köhler, Rebecca Jones, Dieter Neher, Jurgen Parisi, Uli Scherf, Gisela Schulz, Stefan Schumacher, and Ali V. Tunc. The work was supported by the German Science Foundation (DFG) by the priority program SPP1355, "Fundamental Processes in Organic Photovoltaics" (HA 4447, CO751, LU1445).

References

1. Tremel K, Ludwigs S (2014) P3HT revisited: from molecular scale to solar cell devices. Springer, Heidelberg
2. Svensson M, Zhang F, Veenstra SC, Verhees WJH, Hummelen JC, Kroon JM, Inganäs O, Andersson MR (2003) Adv Mater 15:988
3. Reynolds JR (2010) Acc Chem Res 43:1396
4. Mühlbacher D, Scharber M, Morana M, Zhu Z, Waller D, Gaudiana R, Brabec C (2006) Adv Mater 18:2884
5. Hwang I-W, Soci C, Moses D, Zhu Z, Waller D, Gaudiana R, Brabec CJ, Heeger AJ (2007) Adv Mater 19:2307
6. Peet J, Kim JY, Coates NE, Ma WL, Moses D, Heeger AJ, Bazan GC (2007) Nat Mater 6:497
7. Peet J, Cho NS, Lee SK, Bazan GC (2008) Macromolecules 41:8655
8. Agostinelli T, Ferenczi TAM, Pires E, Foster S, Maurano A, Müller C, Ballantyne A, Hampton M, Lilliu S, Campoy-Quiles M, Azimi H, Morana M, Bradley DDC, Durrant J, Macdonald JE, Stingelin N, Nelson J (2011) J Polym Sci Part B Polym Phys 49:717
9. Gu Y, Wang C, Russell TP (2012) Adv Energy Mater 2:683
10. Rogers JT, Schmidt K, Toney MF, Kramer EJ, Bazan GC (2011) Adv Mater 23:2284
11. Rogers JT, Schmidt K, Toney MF, Bazan GC, Kramer EJ (2012) J Am Chem Soc 134:2884
12. Fischer FSU, Trefz D, Back J, Kayunkid N, Tornow B, Albrecht S, Yager KG, Singh G, Karim A, Neher D, Brinkmann M, Ludwigs S (2014) Adv Mater 27:1223
13. Fischer FSU, Kayunkid N, Trefz D, Ludwigs S, Brinkmann M (2015) Macromolecules 48:3974
14. Deschler F, Da Como E, Limmer T, Tautz R, Godde T, Bayer M, Von Hauff E, Yilmaz S, Allard S, Scherf U, Feldmann J (2011) Phys Rev Lett 107:127402
15. Veysel Tunc A, De Sio A, Riedel D, Deschler F, Da Como E, Parisi J, von Hauff E (2012) Org Electron 13:290
16. Spano FC, Silva C (2014) Annu Rev Phys Chem 65:477
17. Scharsich C, Fischer FSU, Wilma K, Hildner R, Ludwigs S, Köhler A (2015) J Polym Sci Part B Polym Phys 53:1416
18. Clark J, Silva C, Friend RH, Spano FC (2007) Phys Rev Lett 98:206406
19. Scharsich C, Lohwasser RH, Sommer M, Asawapirom U, Scherf U, Thelakkat M, Neher D, Köhler A (2012) J Polym Sci Part B Polym Phys 50:442
20. Köhler A, Hoffmann ST, Bässler H (2012) J Am Chem Soc 134:11594
21. Fischer FSU, Tremel K, Saur A-K, Link S, Kayunkid N, Brinkmann M, Herrero-Carvajal D, Navarrete JTL, Delgado MCR, Ludwigs S (2013) Macromolecules 46:4924
22. Crossland EJW, Rahimi K, Reiter G, Steiner U, Ludwigs S (2011) Adv Funct Mater 21:518
23. Crossland EJW, Tremel K, Fischer F, Rahimi K, Reiter G, Steiner U, Ludwigs S (2012) Adv Mater 24:839
24. Brinkmann M (2007) Macromolecules 40:7532
25. Di Nuzzo D, Viola D, Fischer FSU, Cerullo G, Ludwigs S, Da Como E (2015) J Phys Chem Lett 6:1196
26. Falkenberg C, Uhrich C, Maennig B, Riede MK, Leo K (2008) J Appl Phys 6999:S9990
27. Chan CK, Zhao W, Kahn A, Hill IG (2009) Appl Phys Lett 94:24
28. Walzer BMK, Pfeiffer M, Leo K (2007) Chem Rev 107:1233
29. Duong DT, Wang C, Antono E, Toney MF, Salleo A (2013) Org Electron 14:1330
30. Duong DT, Phan H, Hanifi D, Jo PS, Nguyen TQ, Salleo A (2014) Adv Mater 26:6069

31. Deschler F, Riedel D, Deák A, Ecker B, Von Hauff E, Da Como E (2015) Synth Met 199:381
32. Reimer L (1998) Scanning electron microscopy. Springer, Berlin
33. Egerton RF (2005) Physical principles of electron microscopy: an introduction to TEM, SEM, and AEM. Springer, New York
34. Da Como E, Loi MA, Murgia M, Zamboni R, Muccini M (2006) J Am Chem Soc 128:4277
35. Loi MA, da Como E, Dinelli F, Murgia M, Zamboni R, Biscarini F, Muccini M (2004) Nat Mater 4:81
36. Bolinger JC, Traub MC, Adachi T, Barbara PF (2011) Science 331:565
37. Reineke S, Walzer K, Leo K (2007) Phys Rev B 75:125328
38. Zhang Y, de Boer B, Blom PWM (2010) Phys Rev B 81:085201
39. Zhang Y, de Boer B, Blom PWM (2009) Adv Funct Mater 19:1901
40. Tautz R, Da Como E, Limmer T, Feldmann J, Egelhaaf H-J, von Hauff E, Lemaur V, Beljonne D, Yilmaz S, Dumsch I, Allard S, Scherf U (2012) Nat Commun 3:970
41. Carsten B, Szarko JM, Son HJ, Wang W, Lu L, He F, Rolczynski BS, Lou SJ, Chen LX, Yu L (2011) J Am Chem Soc 133:20468
42. Zhang Y, Blom PWM (2010) Appl Phys Lett 97:083303
43. Arkhipov VI, Heremans P, Emelianova EV, Adriaenssens GJ, Bässler H (2003) Appl Phys Lett 82:3245
44. Zhang Y, Zhou H, Seifter J, Ying L, Mikhailovsky A, Heeger AJ, Bazan GC, Nguyen T-Q (2013) Adv Mater 25:7038
45. Pingel P, Schwarzl R, Neher D (2012) Appl Phys Lett 100:143303
46. Yoo S-J, Kim J-J (2015) Macromol Rapid Commun 36:984
47. Li A, Miao X, Deng X (2013) Synth Met 168:43
48. Stelzl FF, Würfel U (2012) Phys Rev B 86:075315
49. von Hauff E, Dyakonov V, Parisi J (2005) Sol Energy Mater Sol Cells 87:149
50. Horowitz G (1998) Adv Mater 10:365
51. Von Hauff E, Parisi J, Dyakonov V (2006) J Appl Phys 100:043702
52. Wise AJ, Precit MR, Papp AM, Grey JK (2011) ACS Appl Mater Interfaces 3:3011
53. Bruevich VV, Makhmutov TS, Elizarov SG, Nechvolodova EM, Paraschuk DY (2007) J Chem Phys 127:104905
54. Tuladhar SM, Poplavskyy D, Choulis SA, Durrant JR, Bradley DDC, Nelson J (2005) Adv Funct Mater 15:1171
55. Anthopoulos TD, Tanase C, Setayesh S, Meijer EJ, Hummelen JC, Blom PWM, de Leeuw DM (2004) Adv Mater 16:2174
56. Melzer C, Koop EJ, Mihailetchi VD, Blom PWM (2004) Adv Funct Mater 14:865
57. McGehee MD, Cates NC, Gysel R, Dahl JEP, Sellinger A (2010) Chem Mater 22:3543
58. Tunc AV, Giordano AN, Ecker B, Da Como E, Lear BJ, Von Hauff E (2013) J Phys Chem C 117:22613
59. Veldman D, Ipek O, Meskers SCJ, Sweelssen J, Koetse MM, Veenstra SC, Kroon JM, van Bavel SS, Loos J, Janssen RAJ (2008) J Am Chem Soc 130:7721
60. Gelinas S, Rao A, Kumar A, Smith SL, Chin AW, Clark J, van der Poll TS, Bazan GC, Friend RH (2013) Science 343:512
61. Hallermann M, Da Como E, Feldmann J, Izquierdo M, Filippone S, Martín N, Jüchter S, von Hauff E (2010) Appl Phys Lett 97:023301
62. Hallermann M, Kriegel I, Da Como E, Berger JM, Von Hauff E, Feldmann J (2009) Adv Funct Mater 19:3662
63. Morana M, Azimi H, Dennler G, Egelhaaf HJ, Scharber M, Forberich K, Hauch J, Gaudiana R, Waller D, Zhu ZH, Hingerl K, van Bavel SS, Loos J, Brabec CJ (2010) Adv Funct Mater 20:1180
64. Österbacka R, Jiang X, An C, Horovitz B, Vardeny Z (2002) Phys Rev Lett 88:226401
65. Di Nuzzo D, Aguirre A, Shahid M, Gevaerts VS, Meskers SCJ, Janssen RAJ (2010) Adv Mater 22:4321
66. Österbacka R (2000) Science 287:839
67. Cingolani R, Ploog K (1991) Adv Phys 40:535

68. Kersting R, Lemmer U, Mahrt RF, Leo K, Kurz H, Bässler H, Göbel EO (1993) Phys Rev Lett 70:3820
69. Morana M, Wegscheider M, Bonanni A, Kopidakis N, Shaheen S, Scharber M, Zhu Z, Waller D, Gaudiana R, Brabec C (2008) Adv Funct Mater 18:1757
70. Perez MD, Borek C, Forrest SR, Thompson ME (2009) J Am Chem Soc 131:9281
71. Yim K-H, Whiting GL, Murphy CE, Halls JJM, Burroughes JH, Friend RH, Kim J-S (2008) Adv Mater 20:3319
72. Ecker B, Egelhaaf HJ, Steim R, Parisi J, Von Hauff E (2012) J Phys Chem C 116:16333
73. Zamadar M, Asaoka S, Grills DC, Miller JR (2013) Nat Commun 4:2818
74. Neugebauer H (2004) J Electroanal Chem 563:153
75. Pingel P, Zhu LY, Park KS, Vogel JO, Janietz S, Kim EG, Rabe JP, Bredas JL, Koch N (2010) J Phys Chem Lett 1:2037
76. Andersson MP, Uvdal P (2005) J Phys Chem A 109:2937
77. Di Nuzzo D, Fontanesi C, Jones R, Allard S, Dumsch I, Scherf U, von Hauff E, Schumacher S, Da Como E (2015) Nat Commun 6:6460

Interplay Between Microscopic Structure and Intermolecular Charge-Transfer Processes in Polymer–Fullerene Bulk Heterojunctions

Vladimir Dyakonov, Hannes Kraus, Andreas Sperlich, Robert Magerle, Mario Zerson, and Martin Dehnert

Contents

Abstract We provide an overview of the development and application of nonconventional techniques that allowed us to probe the molecular excited states in conjugated polymers with respect to their molecular orientation and the nanomechanical properties of the topmost surface layer. To probe triplet and charge-transfer excited states and their fate, we applied the optically

V. Dyakonov (✉) • H. Kraus • A. Sperlich
Fakultät für Physik und Astronomie, Julius-Maximilian Universität Würzburg, Am Hubland, 97074 Würzburg, Germany
e-mail: dyakonov@physik.uni-wuerzburg.de; kraus@physik.uni-wuerzburg.de; sperlich@physik.uni-wuerzburg.de

R. Magerle • M. Zerson • M. Dehnert
Fakultät für Naturwissenschaften, Technische Universität Chemnitz, Reichenhainer Straße 70, 09126 Chemnitz, Germany
e-mail: robert.magerle@physik.tu-chemnitz.de; mario.zerson@physik.tu-chemnitz.de; martin.dehnert@physik.tu-chemnitz.de

K. Leo (ed.), *Elementary Processes in Organic Photovoltaics*, Advances in Polymer Science 272, DOI 10.1007/978-3-319-28338-8_6

detected magnetic resonance technique extended for angular resolution. The surface morphology and three-dimensional depth profiles of the topmost surface layer were obtained with multi-set point intermittent contact mode atomic force microscopy. These studies were performed on model systems such as poly(3-hexylthiophene) as well as on the novel, low-bandgap copolymer poly [(4, 8-bis-(2-ethylhexyloxy)-benzo(1,2-*b*:4,5-*b*9)dithiophene)-2,6-diyl-*alt*-(4-(2-ethylhexyl)-3-fluorothieno[3,4-*b*]thiophene-)-2-carboxylate-2-6-diyl)].

Keywords AFM • Atomic force microscopy • Bulk heterojunction • Charge-transfer states • CTS • Depth profiling • EPR • Nanomechanical properties • ODMR • OPV • Organic photovoltaics • Polymer surfaces • Triplet excitons • Volume imaging

1 Introduction

In organic photovoltaics, the ongoing search for new materials and the steadily improving understanding of the fundamental processes taking place in polymer–fullerene solar cells have led to power conversion efficiencies exceeding 10 % [1]. In addition to the promising reports of record efficiency values, there is an ongoing discussion about a crucial step between photoexcitation and photocurrent extraction, namely, the dissociation of the singlet exciton at the donor–acceptor (D–A) interface. In recent years, numerous experiments have shown that this process is strongly influenced by charge-transfer states (CTSs) situated at the heterojunction [2–5]. Although the impact of these CTSs for charge carrier photogeneration is generally accepted, they are assigned controversial roles in the various photophysical models presented in the literature. For example, in the model proposed in [6] and in [7], the CTSs can be populated either from the ground state or from polaron pairs formed after the prompt separation of singlet excitons at the interface. In another model that is commonly accepted to date and supported by, for example, Bredas et al. [8] and Clarke et al. [5], after the diffusion of the photo-induced singlet exciton to the polymer–fullerene interface, an electron transfer from donor to acceptor can occur. This transfer may result in the formation of an initially "hot" CTS, that is, a CTS with excess thermal energy. In this picture, free charge carriers can be created by dissociation of the "hot" CTS or, if their Coulomb binding energy is overcome by an additional driving force, of the relaxed CTS. However, the former process is experimentally hardly distinguishable from a direct dissociation of singlet excitons to free charge carriers at the interface [9, 10]. Furthermore, all these microscopic processes are strongly dependent on the microstructure of the bulk heterojunctions.

The formation of a bulk heterojunction, that is, a bicontinuous, three-dimensional (3D) network of electron- and hole-conducting materials, is a key requirement for efficient organic photovoltaic (OPV) devices. Blends of the semiconducting semicrystalline polymer poly(3-hexylthiophene) (P3HT) and the fullerene derivate

[6,6]-phenyl-C61-butyric acid methyl ester ($PC_{60}BM$) are among the most studied combinations of materials for polymer-based OPV devices. The fundamental structure formation processes in P3HT:$PC_{60}BM$ blends are the miscibility of the components and the crystallization of both the semicrystalline polymer and the fullerene derivate. P3HT forms lamellar ~10-nm-wide crystals that are embedded in an amorphous matrix. The detailed microstructure and crystallinity depend on the stereo-regularity of the polymer backbone, the structure of the side chains, the average molecular weight, and the polydispersity of the polymer, as well as the details of the film processing conditions [11–13]. In general, a large degree of order within the crystalline regions is considered to be essential for a high charge-carrier mobility [14, 15]. Processing and annealing conditions can either improve or deteriorate the electrical properties of P3HT. In addition, when cast from solution, crystalline P3HT fibers can be formed and deposited on the surface [16, 17].

The nanoscale morphology of blends is known to play a crucial role in the device performance. In thin films, preferential aggregation of one component at the film surface (surface segregation) and the film–substrate interface is a strong driving force for structure formation, which can lead to a layering of the blend components known as vertical phase segregation. The film morphology is frequently inferred from atomic force microscopy (AFM) images of the surface of thin films and related to changes in the material's electronic properties, which may affect the performance of the OPV device. The detailed surface structure is of particular importance for the charge-transfer process between the polymer and electrodes. So far, little attention has been paid to the distribution of the amorphous components at the surface of P3HT films, whereas most studies have addressed the morphology of crystalline regions. Thus, the correlation between the complex hierarchical microstructure of semicrystalline polymers, their blends with fullerene derivates, and the electronic properties is yet to be established.

During the last decade great progress in understanding of the role of CT states at D–A interfaces for the performance of OPV devices has been achieved. The same is true for the understanding of the complex microstructure of the model conjugated polymers and their blends. On the other hand, a gap still exists in understanding the correlations between the formation of clearly assigned CTSs and/or excitons and the microstructure of the photovoltaic D–A blends. The detailed three-dimensional structure of the polymer–fullerene film is of extreme importance but also experimentally extremely challenging to characterise. TEM tomography images give a credible impression of how the 3D structure of a bulk heterojunction may look [18]. The desired study of the same sample by two or more relevant techniques, one of which is sensitive to the three-dimensional microstructure, is not easily conceivable. There is a clear demand on simple but effective tools for imaging the 3D structure of absorber layers of real solar cells. In recent years a number of different approaches have been developed for high-resolution 3D imaging based on AFM [19] and other scanning probe techniques [20]. Among these, multi-set point intermittent contact (MUSIC) mode AFM enables 3D depth profiling of the near-surface region of polymer films and other soft materials [21].

2 Optically Detected Magnetic Resonance: ODMR

2.1 Spin Hamiltonian Formalism

The common ground for all magnetic resonance methods is the spin. A free electron, by itself, possesses a spin of $S = 1/2$. In a classical picture, the spin will precess in an external magnetic field and can herein occupy $2S + 1$ unique positions. In a quantum mechanical picture, the $S = 1/2$ has two possible Zeeman energy states and an $S = 1$ triplet has three energetic states. The precession mode of the spins can be switched by an electromagnetic field, which is in case of electrons in the μeV microwave range for laboratory magnetic field strengths. The involved resonance energies of all possible electron paramagnetic resonance (EPR) transitions are fully described by the spin Hamiltonian formulated by Abragam and Pryce [22]. The important contributions to the measurements presented in this chapter are the Zeeman splitting and the anisotropic dipolar zero-field interaction (ZFI), as shown in Fig. 1:

$$\mathcal{H} = \boldsymbol{g}\mu_B \boldsymbol{S}\vec{B} + \boldsymbol{S}^T\boldsymbol{D}\boldsymbol{S},$$

where $\boldsymbol{S}$ is the spin operator vector, $\vec{B}$ the magnetic field vector, $\boldsymbol{g}$ the g-factor tensor, $\boldsymbol{D}$ the ZFI tensor, and μ_B the Bohr magneton. Both $\boldsymbol{g}$ and $\boldsymbol{D}$ have only main axis components. Tensors and both vector and scalar operators are denoted in bold.

When the energy of microwave photons and the Zeeman splitting are equal, transitions between spin states can be induced. Such transitions lead to a detectable absorption of microwaves. Without an external magnetic field, the magnetic sublevels of a spin state with an arbitrary multiplicity should be degenerate. In real molecules, this is usually not the case because of the ZFI. Two effects can give rise to a ZFI: spin-orbit coupling of ground and excited states, and anisotropic

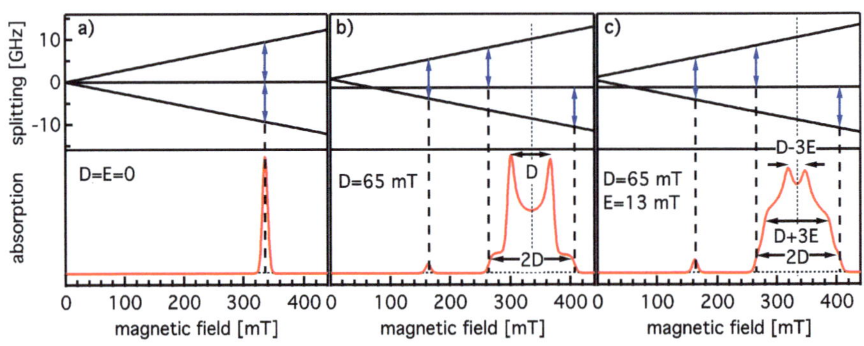

Fig. 1 Zeeman and zero-field splitting for an exemplary triplet spin $S = 1$ system (*top*). The zero-field interaction tensor can be (**a**) isotropic ($D = E = 0$), (**b**) axial ($D > 0$, $E = 0$), or (**c**) rhombic ($D \neq E > 0$). This symmetry is reflected in the corresponding absorption powder spectra (*bottom*). The first-order forbidden $\Delta m_S = \pm 2$ half-field transition is only detectable for $D \gg 0$ in (**b**) and (**c**)

dipolar interaction of the participating spins. The ZFI tensor $\boldsymbol{D} = (D/3 - E, D/3 + E, -2D/3)$ is set traceless and symmetric and can be described by the parameters D and E. In a triplet system, the D parameter describes an axial anisotropy of the ZFI, which lifts the degeneracy between $|0>$ and the $|\pm 1>$ manifold. The electron distribution is flattened along the z-axis in the sample coordinate system. The E parameter now introduces a rhombic anisotropy, additionally elongating the electron distribution along the x-axis [23]. The orientation dependence of the ZFI parameters D and E on the Euler angles (θ, ϕ) is as follows:

$$D = D_0 \left(\cos^2\theta - \tfrac{1}{3}\right),$$
$$E = E_0 \sin^2\theta \ \cos(2\phi).$$

Under the influence of an external magnetic field, the Zeeman splitting is added to the ZFI. The influence of the rotation of B against the sample's molecular or crystal axis (also called the c-axis) is depicted in Fig. 3. In highly oriented samples, the spectrum shown at a certain angle can be measured directly. In anisotropic, unoriented samples, all transitions from all possible Euler angles are summed and lead to the also shown powder spectrum (Fig. 1, bottom row). For samples that are neither highly crystalline nor completely unoriented, a compromise between the summed powder and oriented spectra has to be found. Especially when ordering of paramagnetic centers in a plane is probable, as expected for very thin samples, the following approach is viable. The simulated spectra are computed by weighting the transition probability for a certain orientation with a factor that depends on how much this orientation is aligned with the paramagnetic center c-axis. This is expressed by an ordering parameter λ, which influences the calculation of the transition probability P as follows:

$$P(\theta) = e^{\frac{\lambda}{2}\left(3\cos^2\theta - 1\right)}.$$

At negative ordering parameters $\lambda < 0$, the paramagnetic centers are rather ordered perpendicular to the external magnetic field, while $\lambda > 0$ means an inclination of the centers to the field. At $\lambda = 0$, all orientations are equal, yielding a powder spectrum. The higher the value, the sharper the weighting peaks become, describing a higher degree of ordering.

The half-field transition is the first-order forbidden dipole transition with $\Delta m_S = \pm 2$, which is spectrally situated at half the magnetic field or double the g-factor ($g = 4$) of the principal $\Delta m_S = \pm 1$ transitions ($g = 2$). Because of its forbidden nature, the amplitude of this transition is very weak and usually does not allow for spectral deconvolution. However, its very presence is a clear and undisputable qualitative argument for a high-spin system.

2.2 ODMR Method

Until now, the EPR conditions were shown in a generalized way, without catering to the exact mechanism of signal generation. In classic EPR, the observable is the microwave absorption, or rather, the deviation of the microwave absorption under EPR conditions. However, in optically detected magnetic resonance (ODMR), the observable is the variation of the intensity of an optical transition caused by EPR-induced transitions in the spin manifold. Under EPR conditions, populations of the participating Zeeman substates are equalized. If we assume different population or depopulation rates for the Zeeman sublevels, these EPR transitions will influence the rate for the optical transition, leading to a variation in the photoluminescence (PL) intensity as depicted in Fig. 2.

ODMR samples were prepared from solution by spin coating on 3-mm-wide herasil substrates, which were then placed in EPR tubes. Sample handling occurred inside a nitrogen-filled glovebox to avoid degradation. The EPR tubes were sealed under a 20-mbar helium atmosphere. ODMR measurements were performed using a modified X-Band EPR spectrometer (Bruker E300). The EPR tube was placed in an optical resonant cavity (Bruker ER4104OR) equipped with a goniometer for exact sample orientation and a continuous-flow helium cryostat (Oxford ESR 900). A 70-mW, 532-nm laser was used for excitation. PL from the sample was collected with a Si photodiode, excluding the excitation wavelengths by a 550-nm longpass filter. The preamplified (Femto DLPCA-200) variation of the photodiode current, a result of resonant microwave irradiation (Anritsu synthesizer), was recorded by a lock-in amplifier, referenced by TTL-modulating the 2W microwave power in the kHz range. For better comparability, the ODMR signal is always normalized to the corresponding PL intensity, leading to the relative change in photoluminescence signal *d*PL/PL.

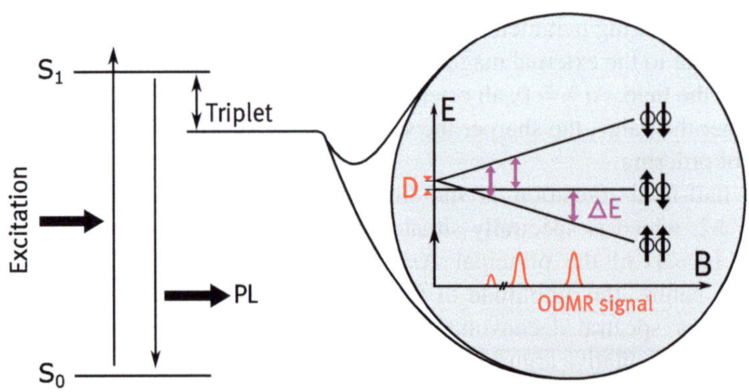

Fig. 2 For the optically detected magnetic resonance technique, the variation of an optical transition (here the sample photoluminescence), caused by electron paramagnetic resonance–induced transitions in the spin manifold, is measured

Simulations were performed with MATLAB using the program pepper from the EasySpin package [24]. For non-Boltzmann equilibrium spin populations, pepper correctly calculates microwave absorptive and emissive curves [25]. However, in ODMR we do not observe the microwave sign but rather observe the influence of population changes on the PL. To account for this, the microwave emissive transition was inverted for ODMR simulations.

2.3 *ODMR Transitions with Regard to Sample Orientation*

Most conjugated polymers have a preferential stacking orientation toward a substrate surface. They may stack with the molecular backbone either lying flat on the substrate (face-on) or standing up (edge-on), as illustrated in Fig. 3. An important

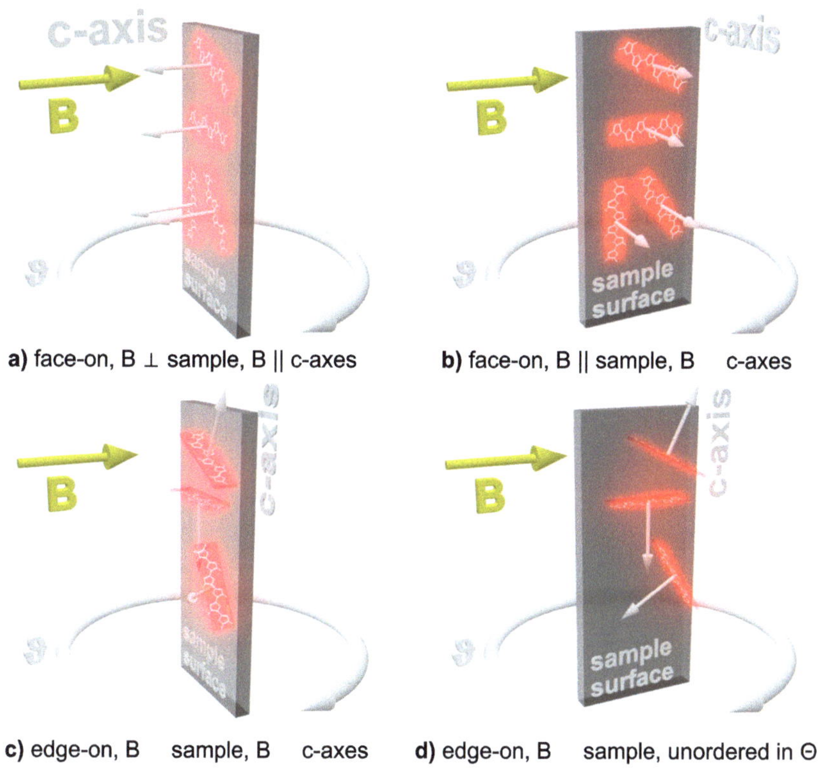

Fig. 3 Illustration of sample surface orientation toward magnetic field and molecular *c*-axis for molecules that orient (**a**), (**b**) face-on or (**c**), (**d**) edge-on with respect to the sample surface. This partial molecular ordering can be reconstructed from the optically detected magnetic resonance spectra

difference concerning the ODMR method is that in a face-on configuration, the molecular plane is parallel to the sample surface, and the *c*-axis is orthogonal. This means that when the magnetic field is parallel to the sample surface, it is perpendicular to all molecular *c*-axes regardless of their long-axis orientation on the sample. Inversely, when the magnetic field is perpendicular to the sample surface, again all *c*-axes are parallel to the *B*-field. In an edge-on configuration, the *c*-axes are parallel to the sample surface. However, in this plane, they are completely anisotropic. This means that only at $B \perp$ sample surface, all *c*-axes are perpendicular to *B*. In all other orientations, a fraction is parallel, and another perpendicular. This means that in edge-on configurations, we can expect a singularity spectrum at $B \perp$ sample surface, and pseudo-powder spectra for other orientations, and in face-on configurations, we can expect defined spectra for $B \perp$ sample and for $B \parallel$ sample.

2.4 ODMR on P3HT

Figure 4 depicts the P3HT ODMR spectrum, which consists of a symmetrical "wing"-like signature centered around an intense peak. At first glance, it has strong similarities to the example spectrum in Fig. 1c. Even the half-field $\Delta m_S = \pm 2$

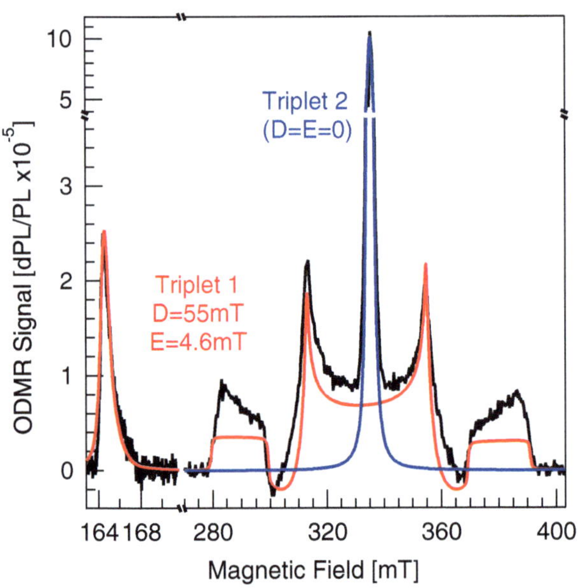

Fig. 4 Optically detected magnetic resonance spectrum of a poly(3-hexylthiophene) thin film at 5 K. B $\perp$ sample surface (0°); compare to Fig. 3d. The spectrum has been simulated with two triplet spin systems, with triplet 1 (*red*) having $D = 55$ mT, $E = 4.6$ mT and an ordering parameter in the ϕ-direction of $\lambda = 1.3$. Triplet 2 (*blue*) has $D = E = 0$. The line widths are considered unimportant, as they cannot be ascertained in a broad pseudo-powder spectrum

transition is observed. However, because of the partly crystalline nature of this polymer, molecular ordering needs to be included to reproduce the ODMR spectrum in simulation. Directly ascertaining D and E from stationary points in the spectrum leads to false values, such as those reported by Swanson et al. in [26], where the ODMR of P3HT was first reported. A basic approach is to assign the central peak at $g = 2$ (335 mT) to a loosely bound triplet species with almost no ZFI ($D = E = 0$) and the broad "wing"-like signature to an axial triplet, and determine the outermost turning points to obtain 2D. This yields $D \approx 55$ mT but no proper value for E [27].

In thin P3HT films, the individual monomers orient edge-on to the substrate [28], which we also confirmed for this sample by X-ray diffractometry (not shown). The simulations were done with ZFI parameters for two triplets, with triplet 2 having $D = E = 0$ and triplet 1 having $D = 55$ mT, $E = 4.6$ mT and an ordering parameter in the ϕ -direction of $\lambda = 1.3$, which is consistent with the preferential orientation depicted in Fig. 3c and d. The negative sign of the triplet 1 signature at 305 and 365 mT is caused by a non-Boltzmann population of the spin sublevels, with the population order from lowest to highest Zeeman sublevel ($m_S = -1$, 0, 1). A nonequilibrium population is to be expected when assuming triplet–triplet annihilation as the signal generation path. The simulations are not a perfect match, but the stationary points and the signal sign can accurately be reproduced. To further improve upon this, both θ and ϕ ordering would have to be accounted for simultaneously, which was not possible with EasySpin at the time of publication.

We can estimate the extent of the triplet $S = 1$ wavefunction from the D parameter [29], which amounts to 0.37 nm. This value, together with the fact that the ZFI parameters were found to be identical for different samples, and hence independent of morphological variations, corroborates the long-standing working model of the triplet exciton having a high degree of localization on one thiophene ring. For triplet 2 we can estimate the distance between loosely interacting charges in a triplet state to be at least 2–3 nm, to fit with the observed narrow linewidth and vanishing ZFI. Interesting supplementary measurements would be on a sample with P3HT on a nickel oxide layer, as it changes the P3HT configuration to preferentially face-on [30].

2.5 ODMR on PTB7

The thienothiophene derivative poly[(4,8-bis-(2-ethylhexyloxy)-benzo(1,2-*b*:4,5-*b*9)dithiophene)-2,6-diyl-*alt*-(4-(2-ethylhexyl)-3-fluorothieno[3,4-*b*]thiophene-)-2-carboxylate-2-6-diyl)] (PTB7) is a third-generation D–A block copolymer with a low bandgap of 1.8 eV [31]. In comparison with second-generation donor polymers like P3HT, its low-energy gap improves PTB7's absorption at longer wavelengths up to 800 nm. In conjunction with the fullerene $PC_{70}BM$ as acceptor, PTB7 yields reproducible power conversion efficiencies above 9 % [32]. Another advantage concerning charge carrier transport in solar cells is the expected face-on stacking configuration of PTB7 [33], facilitating vertical charge transport to the contacts. The

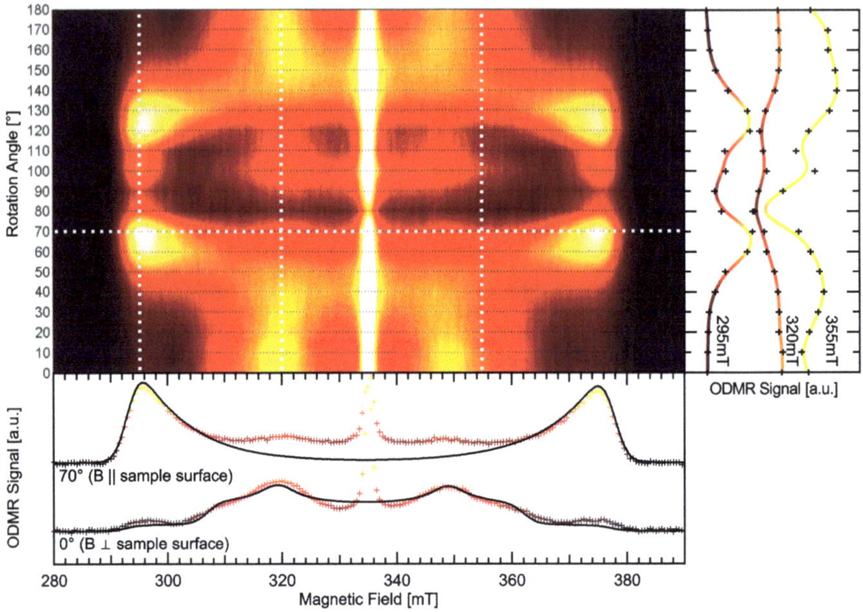

Fig. 5 Optically detected magnetic resonance contour plot for a PTB7 film sample with differing sample orientations at 5 K. Cuts at different magnetic field positions (*right inset*) and spectra for 0° and 70° (*inset bottom*) are marked by the *dotted white lines*

face-on configuration is also reflected in ODMR spectra. A series of spectra were recorded and combined in the contour plot shown in Fig. 5. Exemplary spectra for 0° and 70° orientations are shown plotted in the figure; cuts at different magnetic field values are plotted to the right. The individual spectra are similar to P3HT, with a broad and a central narrow signature. Here the strong angular dependence is additionally highlighted.

The ZFI parameters for the broad triplet signature in PTB7 were taken from [34] and slightly optimized, yielding $D = 42.3$ mT and $E = 4.3$ mT. As expected, the change in ODMR spectral shape of the face-on PTB7 can be, unlike P3HT, explained considering only an ordering in respect to the θ angle and a Boltzmann equilibrium distribution of the Zeeman states. The spectrum for 0° is simulated with an ordering parameter of $\lambda(0°) = -0.4$, which is consistent with an ordering of the *c*-axis perpendicular to the magnetic field. Conversely, the spectrum for 70° is simulated with $\lambda(70°) = 2$, where the *c*-axis ordering should be along the magnetic field. Summarily, the deconvolution of the ODMR of PTB7 corroborates the assumption of the materials face-on configuration.

3 Microscopic Structure Studied with AFM Imaging

3.1 MUSIC Mode AFM

Amplitude modulation AFM, also known as tapping- or intermittent contact (IC) mode AFM, is widely used for imaging surfaces of polymers and other soft materials. It allows one to measure the shape of the surface (the height image) simultaneously with the phase image from which local mechanical properties of the specimen such as elastic modulus and viscoelastic response can be inferred. On soft polymeric materials, however, the tip can indent several tens of nanometers into the surface, which can cause artifacts in both height and phase images [35, 36].

The position of the unperturbed surface and the tip indentation can be determined from point-wise measurements of the amplitude and the phase as a function of the tip–sample distance [36]. The tip indentation can be used as a depth coordinate for reconstructing 3D depth profiles (volume images) of the tip–sample interaction [21, 37]. For example, the amplitude–phase distance (APD) curves can be deconvoluted to discriminate between conservative and dissipative contributions to the tip–sample interaction [38–40]. MUSIC mode AFM is a variant of this approach [41]. It allows us to use one data set of APD curves to reconstruct height and phase images that correspond to different amplitude set points. The resulting height and phase images are free of feedback-loop artifacts, further enhancing the image quality. This is advantageous for imaging polymer surfaces [21, 37] as well as individual objects on the nanometer scale [41, 42].

The principle of the measurement is illustrated in Fig. 6. The AFM tip is oscillating at or close to its resonance frequency (~300 kHz) with a free amplitude A_0 of typically 40 nm. The amplitude A and the phase φ of the oscillating cantilever are recorded as a function of the tip–sample distance d on a typically 50×50 large array of positions on the surface. From this set of APD curves, the position of the unperturbed surface, z_0, where the tip starts to interact with the surface, can be determined. Technically, we determine z_0 as the position where the phase starts to deviate from the phase of the free oscillation. Upon approaching the surface further, the cantilever oscillation is damped as the tip interacts with the surface at the inflection point of the oscillation. On a compliant surface, the tip starts to indent into the surface [35, 36]. For a given amplitude $A < A_0$, the tip indentation is the additional distance Δd that the tip must approach the surface to reach the same damped amplitude A as on a stiff reference surface that does not allow for any indentation [43]. The value of the tip indentation depends on the definition of the unperturbed surface (Fig. 6b). The position of z_0 is often defined by the first appearance of an attractive interaction as the tip approaches the surface [21, 36, 43, 44]. In case of strong attractive tip–sample interactions (van der Waals forces, capillary forces), this choice can lead to astonishingly large values for the tip indentation $\tilde{z}_0$. In these cases, an alternative definition of the unperturbed sample surface is the position z_B, where a balance of attractive and repulsive forces is

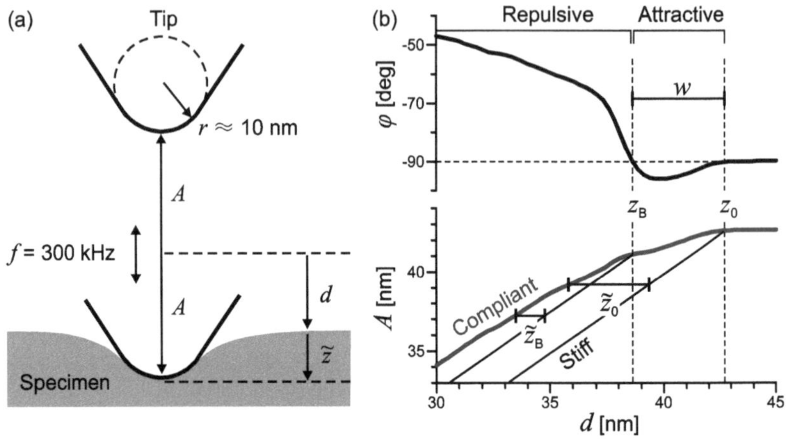

Fig. 6 (**a**) Sketch of an oscillating atomic force microscopy tip indenting into a compliant specimen (after [47]). (**b**) The phase φ and amplitude A are measured as a function of the tip–sample distance d (approach curves). The *thin solid lines* indicate the amplitude on a stiff reference surface. (Adapted from [42])

observed in the phase–distance curves [42]. At this point, the transition from the attractive to the repulsive regime of the tip–sample interaction occurs.

The tip indentation can be used as a depth coordinate relative to the unperturbed surface z_0 for plotting a depth profile of the phase φ or other quantities that can be determined from the measured amplitude and phase, for instance, the conservative contribution to the tip–sample interaction k_{TS} [40] and E_{dis}, the energy dissipated between the tip and the sample during one tip-oscillation cycle, which is a measure of the dissipative part of the tip–sample interaction [38].

3.2 *Surface Structure of P3HT*

Measurements of the tip indentation into the surface of semicrystalline polypropylenes with a low degree of crystallinity have shown that the edges of crystalline lamellae are covered by an approximately 10-nm-thick amorphous layer [21, 44, 45]. This motivated us to investigate surfaces of semiconducting semicrystalline polymers. For the first example, we studied a highly regioregular type of P3HT with a molecular weight $M_n = 10.2$ kg/mol and a polydispersity index of 1.19 [37].

Figure 7 shows MUSIC mode AFM images of the unperturbed surface z_0, the maximum dissipated energy E_{dis}^{max}, and the conservative contribution to the tip–sample interaction, k_{TS}. The unperturbed surface z_0 is smooth, and no signs of a contrast between crystalline lamellae and amorphous regions are visible. The interaction between the tip and the stiff crystalline lamellae starts several

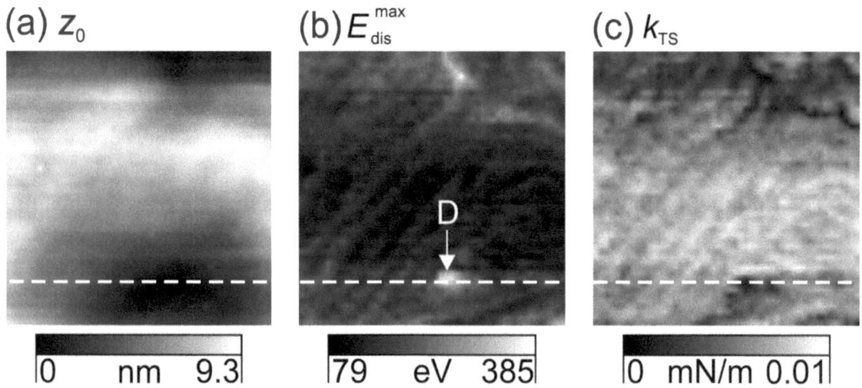

Fig. 7 Multi-set point intermittent contact mode atomic force microscopy images of a 100-nm-thick P3HT film annealed at 230 °C for 20 min. Within the (250 × 250)-nm^2-large area, amplitude–phase distance curves were measured at 50 × 50 points. (**a**) Unperturbed surface z_0, reconstructed for an amplitude set point $A/A_0 = 1$. (**b**) Maximum dissipated energy E_{dis}^{max}. (**c**) k_{TS} reconstructed for $A/A_0 = 0.90$. The *dashed line* marks the position of the depth profiles of k_{TS} and E_{dis} shown in Fig. 9d and e, respectively. In (**b**), a ~ 30-nm-large domain with high values of E_{dis}^{max} is marked by the letter *D*. (Adapted from [37])

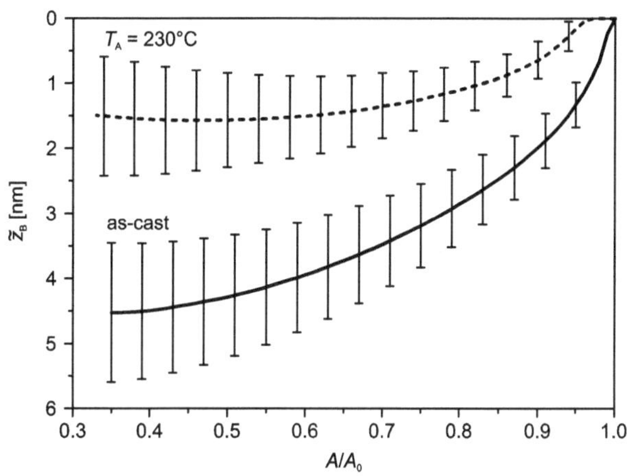

Fig. 8 Laterally averaged tip indentation $\tilde{z}_B$ as a function of the amplitude set point A/A_0 for the as-cast P3HT film and the P3HT film annealed at 230 °C for 20 min. The *error bars* indicate the standard deviation of the data, which reflects the lateral variations of the tip indentation within the imaged area. (Data from [37])

nanometers below the surface of the film. The crystalline lamellae are visible as dark and bright stripes in the E_{dis} and k_{TS} maps, respectively.

Figure 8 shows the laterally averaged tip indentation $\tilde{z}_B$ as a function of the amplitude set point A/A_0 for the as-cast and the annealed P3HT film. In both cases, the tip indentation $\tilde{z}_B$ increases with decreasing amplitude set point A/A_0. On the as-cast film, the laterally averaged tip indentation $\tilde{z}_B$ is 4.5 nm (for $A/A_0 = 0.40$) and after 20 min annealing at 230 °C, it decreases to 1.5 nm (for $A/A_0 = 0.40$). The maximal tip indentation $\tilde{z}_B$ can be identified with the thickness of a compliant, amorphous layer that covers the entire surface including the edges of crystalline lamellae. This is in accordance with earlier reports of an amorphous surface layer on semicrystalline polypropylenes [21, 44–46].

The tip indentation can be used as a depth coordinate that starts at the position of the unperturbed surface z_0. From this reference point, 3D depth profiles of the parameters characterizing the tip–sample interaction can be reconstructed. In Fig. 9, depth profiles of k_{TS} and E_{dis} are shown for the as-cast and the annealed P3HT film along with parts of conventional intermittent contact mode AFM phase images (Fig. 9c and f) where the positions of the profiles are indicated. One profile corresponds to one horizontal line in the array of APD data. For the annealed P3HT film, this line is also indicated in the maps shown in Fig. 7. The k_{TS} depth profiles shown in Fig. 9 are plotted with a color code that allows us to distinguish between attractive and repulsive regimes of the tip–sample interaction. The width of the attractive regime, w, is caused by attractive van der Waals and capillary forces between the tip and the specimen [42]. In the k_{TS} depth profiles shown in Fig. 9a and b, a white stripe marks the position of z_B where $k_{TS} = 0$ mN/m. It is located ~2 nm below the position of the unperturbed surface z_0. The repulsive regime of the tip–sample interaction is shown as a red band. It can be identified with the compliant, amorphous layer covering the surface of the P3HT film. On the as-cast film, the amorphous layer is on average 4.5 nm thick, and after annealing at 230 °C

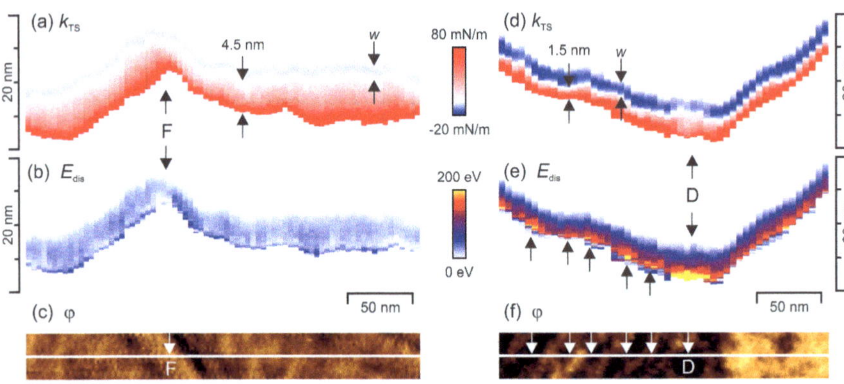

Fig. 9 (**a**, **b**) Depth profiles of k_{TS} and E_{dis} for an as-cast P3HT film at the position indicated in the intermittent contact mode atomic force microscopy (AFM) phase image shown in (**c**). (**d**, **e**) Depth profiles of k_{TS} and E_{dis} for the P3HT film, annealed at 230 °C for 20 min, measured at the position indicated in the intermittent contact mode AFM phase image shown in (**f**) and in Fig. 7. The arrows mark the local maxima of E_{dis}, which correspond to the amorphous regions between the crystalline lamellae. The width of the attractive regime, w, is marked in the k_{TS} depth profiles shown in (**a**) and (**d**). It corresponds to indentation depths where k_{TS} is negative. This range is shown in *blue colors,* whereas the repulsive regime is shown in *red colors.* (Adapted from [37])

its thickness decreases to (on average) 1.5 nm. The depth profiles in Fig. 9 show the local variations of the amorphous surface layer that covers the crystalline lamellae as well as the P3HT fibril that was deposited on the surface of the as-cast film (marked with F in Fig. 9a and c).

The presence of an amorphous surface layer can have important consequences for OPV devices. For instance, it can influence the charge-transfer process between P3HT and electron acceptor materials as well as the charge extraction at the electrodes. It might also influence the alignment of electronic levels at these interfaces. Because annealing causes changes within the amorphous surface layer, possible contributions of amorphous surface layers toward electronic properties and/or device performance should be considered.

4 Conclusions

Studies of correlations between the microscopic structure and the formation of CTSs and triplet excitons in conjugated polymers by nonconventional magnetic resonance techniques and scanning probe microscopy are reported in this chapter. We developed and applied the spin-sensitive technique (ODMR), and it allowed us to probe the molecular excited states in the pure polymers with respect to their origin and molecular orientation. In particular, we investigated model systems such as P3HT as well as the novel, low-bandgap copolymer PTB7. With the ODMR technique we were able to monitor the formation and parameters of CTSs and triplet excitons, formed upon photoexcitation, and to determine the orientation of monomers. For P3HT and PTB7, the monomer orientation relative to the substrate was found to be oriented edge-on and face-on, respectively. The surface morphology and 3D depth profiles of the nanomechanical properties of the topmost surface layer were obtained with MUSIC mode AFM. The data show that the entire surface of P3HT films is covered with a thin amorphous surface layer. This can have important consequences for charge-transfer processes between P3HT and electron acceptor materials, as well as the charge extraction at the electrodes in OPV devices. We note that although the length scales to which those two methods are sensitive are quite different, it was possible to distinguish between the crystalline and amorphous phases within the polymer. To obtain a complete picture of correlations between the blend morphology and structure–electronic properties, an additional structural characterization using X-ray scattering methods on the same samples is needed.

Acknowledgments The reported work was completed in close collaboration with S. Väth from Julius-Maximilians Universität Würzburg, E.-C. Spitzner and C. Riesch from Technische Universität Chemnitz, and R. Lohwasser and M. Thelakkat from Universität Bayreuth. We gratefully acknowledge their contributions. This work was supported by the Deutsche Forschungsgemeinschaft within the SPP 1355 (DY18/6-1,2,3 and MA1744/3-1,2,3). R.M. acknowledges further funding by the Volkswagen Foundation and the Deutsche Forschungsgemeinschaft.

References

1. Green MA, Emery K, Hishikawa Y, Warta W, Dunlop ED (2015) Prog Photovolt Res Appl 23:1
2. Ohkita H, Cook S, Astuti Y, Duffy W, Tierney S, Zhang W, Heeney M, McCulloch I, Nelson J, Bradley DDC, Durrant JR (2008) J Am Chem Soc 130:3030
3. Drori T, Sheng C-X, Ndobe A, Singh S, Holt J, Vardeny ZV (2008) Phys Rev Lett 101:037401
4. Vandewal K, Gadisa A, Oosterbaan WD, Bertho S, Banishoeib F, van Severen I, Lutsen L, Cleij TJ, Vanderzande D, Manca JV (2008) Adv Funct Mater 18:2064
5. Clarke TM, Durrant JR (2010) Chem Rev 110:6736
6. Jarzab D, Cordella F, Gao J, Scharber M, Egelhaaf H-J, Loi MA (2011) Adv Energy Mat 1:604
7. Provencher F, Sakowicz M, Brosseau C, Latini G, Beaupré S, Leclerc M, Reynolds LX, Haque SA, Leonelli R, Silva C (2012) J Polym Sci B Polym Phys 50:1395
8. Bredas J-L, Norton JE, Cornil J, Coropceanu V (2009) Acc Chem Res 42:1691
9. Hwang I-W, Soci C, Moses D, Zhu Z, Waller D, Gaudiana R, Brabec CJ, Heeger AJ (2007) Adv Mater 19:2307
10. Howard IA, Laquai F (2010) Macromol Chem Phys 211:2063
11. Hugger S, Thomann R, Heinzel T, Thurn-Albrecht T (2004) Colloid Polym Sci 282:932
12. Gang L, Vishal S, Yan Y, Yang Y (2005) J Appl Phys 98:043704
13. Wu Z, Petzold A, Henze T, Thurn-Albrecht T, Lohwasser RH, Sommer M, Thelakkat M (2010) Macromolecules 43:4646
14. Sirringhaus H, Brown PJ, Friend RH, Nielsen MM, Bechgaard K, Langeveld-Voss BMW, Spiering AJH, Janssen RAJ, Meijer EW, Herwig P, de Leeuw DM (1999) Nature 401:685
15. Kline RJ, McGehee MD, Kadnikova EN, Liu J, Fréchet JMJ, Toney MF (2005) Macromolecules 38:3312
16. Oosterbaan WD, Vrindts V, Berson S, Guillerez S, Douheret O, Ruttens B, D'Haen J, Adriaensens P, Manca J, Lutsen L, Vanderzande D (2009) J Mater Chem 19:5424
17. Ihn KJ, Moulton J, Smith P (1993) J Polym Sci B Polym Phys 31:735
18. van Bavel SS, Sourty E, de With G, Loos J (2009) Nano Lett 9:507
19. Magerle R (2000) Phys Rev Lett 85:2749
20. Alekseev A, Efimov A, Loos J, Matsko N, Syurik J (2014) Eur Polym J 52:154
21. Spitzner E-C, Riesch C, Magerle R (2011) ACS Nano 5:315
22. Abragam A, Pryce MHL (1951) Proc Royal Soc A Math Phys Eng Sci 205:135
23. El-Sayed MA (1970) Pure Appl Chem 24:475
24. Stoll S, Schweiger A (2006) J Magn Reson 178:42
25. Gonen O, Levanon H (1984) J Phys Chem 88:4223
26. Swanson LS, Shinar J, Yoshino K (1990) Phys Rev Lett 65:1140
27. Swanson LS, Lane P, Shinar J, Wudl F (1991) Phys Rev B 44:10617
28. Chu CW, Yang H, Hou WJ, Huang J, Li G, Yang Y (2008) Appl Phys Lett 92:103306
29. Jeschke G (2002) Macromol Rapid Commun 23:227
30. Marks TJ (2012) Inorganic and organic/plastic photovoltaics. In: NSF/ONR workshop on key scientific and technological issues for development of next-generation organic solar cells, pp 1–30
31. Liang Y, Xu Z, Xia J, Tsai ST, Wu Y, Li G, Ray C, Yu L (2010) Adv Mater 22, E135
32. He Z, Zhong C, Su S, Xu M, Wu H, Cao Y (2012) Nat Photonics 6:593
33. Szarko JM, Guo J, Liang Y, Lee B, Rolczynski BS, Strzalka J, Xu T, Loser S, Marks TJ, Yu L (2010) Adv Mater 22:5468
34. Niklas J, Beaupré S, Leclerc M, Xu T, Yu L, Sperlich A, Dyakonov V, Poluektov OG (2015) J Phys Chem B 119:7407
35. Bar G, Thomann Y, Brandsch R, Cantow H-J (1997) Langmuir 13:3807
36. Knoll A, Magerle R, Krausch G (2001) Macromolecules 34:4159
37. Zerson M, Spitzner E-C, Riesch C, Lohwasser R, Thelakkat M, Magerle R (2011) Macromolecules 44:5874

38. Cleveland JP, Anczykowski B, Schmid AE, Elings VB (1998) Appl Phys Lett 72:1613
39. Garcia R, Gómez CJ, Martinez NF, Patil S, Dietz C, Magerle R (2006) Phys Rev Lett 97:016103
40. Schröter K, Petzold A, Henze T, Thurn-Albrecht T (2009) Macromolecules 42:1114
41. Spitzner E-C, Riesch C, Szilluweit R, Tian L, Frauenrath H, Magerle R (2012) ACS Macro Lett 1:380
42. Spitzner E-C, Röper S, Zerson M, Bernstein A, Magerle R (2015) ACS Nano 9:5683
43. Höper R, Gesang T, Possart W, Hennemann OD, Boseck S (1995) Ultramicroscopy 60:17
44. Dietz C, Zerson M, Riesch C, Franke M, Magerle R (2008) Macromolecules 41:9259
45. Dietz C, Zerson M, Riesch C, Gigler AM, Stark RW, Rehse N, Magerle R (2008) Appl Phys Lett 92:143107
46. Sakai A, Tanaka K, Fujii Y, Nagamura T, Kajiyama T (2005) Polymer 46:429
47. San Paulo A, Garcia R (2001) Surf Sci 471:71

Nanoscale Morphology from Donor–Acceptor Block Copolymers: Formation and Functions

David Heinrich, Martin Hufnagel, Chetan Raj Singh, Matthias Fischer, Shahidul Alam, Harald Hoppe, Thomas Thurn-Albrecht, and Mukundan Thelakkat

Contents

D. Heinrich • M. Hufnagel • C.R. Singh • M. Thelakkat
Applied Functional Polymers, Macromolecular Chemistry I, University of Bayreuth, Universitätsstrasse 30, 95440 Bayreuth, Germany
e-mail: mukundan.thelakkat@uni-bayreuth.de

M. Fischer • T. Thurn-Albrecht (✉)
Institut für Physik, Martin-Luther-Universität, von-Danckelmann Platz 3, 06099 Halle, Germany
e-mail: thomas.thurn-albrecht@physik.uni-halle.de

S. Alam • H. Hoppe
Center for Energy and Environmental Chemistry Jena (CEEC), Friedrich Schiller University Jena, Philosophenweg 7, 07743 Jena, Germany
e-mail: harald.hoppe@uni-jena.de

K. Leo (ed.), *Elementary Processes in Organic Photovoltaics*,
Advances in Polymer Science 272, DOI 10.1007/978-3-319-28338-8_7

Abstract General design principles of donor–acceptor block copolymers are reviewed and specific results arising from block copolymers consisting of semicrystalline poly(3-hexylthiophene-2,5-diyl) (P3HT) blocks of appreciably high molecular weight and acceptor blocks carrying pendant perylene bisimides or fullerene derivatives are summarized. The chapter is structured according to the building blocks P3HT, poly(perylene bisimide acrylate), and a polystyrene copolymer grafted with phenyl-C61-butyric acid methyl ester used for the synthesis of the corresponding block copolymers, and in each part the synthetic challenges, structure formation, and consequences for charge transport, and in some cases photovoltaic properties, are addressed.

Keywords Crystallization • Donor-acceptor block copolymers • Microphase separation • Poly(3-hexylthiophene-2,5-diyl) • Scattering techniques • Self-assembly

1 Introduction

Conjugated polymer–based organic solar cells have been successfully developed over the past 20 years [1–7], reaching current power conversion efficiencies (PCEs) over 10 % [8–15]. One major requirement, which arises from the fact that organic semiconductors generally exhibit strong exciton binding energies one order of magnitude above the thermal energy (k_BT) and only small exciton diffusion lengths (a few nanometers), constitutes the existence of a so-called donor–acceptor (D–A) interface to facilitate exciton dissociation and thus charge generation at this heterojunction interface. Thus, only an intimate blend of donor and acceptor phases enables efficient charge generation throughout the bulk of the photoactive layer. Here the acceptor can be distinguished from the donor simply by having a larger electron affinity, thus forming a type II heterojunction. In other words, both the lowest unoccupied molecular orbital (LUMO) as well as the highest occupied molecular orbital (HOMO) of the acceptor need to have lower energy than the donor. In addition to charge generation, their efficient extraction at the two opposing selective electrodes displays a second requirement for successful photovoltaic operation. Hence, donor and acceptor phases need to form a bicontinuous interpenetrating network of a certain coarse-grained lateral extension, providing percolation paths for both types of charge carriers (i.e., holes and electrons). To minimize charge recombination events, these phases need to be well separated at a length scale roughly corresponding to the exciton diffusion length [16], and both the charge generation and extraction benefit from the existence of semicrystalline phases. This three-dimensional mixing of donor and acceptor phases on the nanoscale is called the bulk heterojunction morphology [17, 18].

Commonly, the formation of such bulk heterojunctions is practically obtained from blending the two molecular or polymeric components in a common organic solvent and casting a film from it, resulting in a morphology governed by a kinetically locked nonequilibrium distribution of those two phases. The disadvantage of this approach lies in the fact that the scale of phase separation can be subject to coarsening processes with time, resulting in an increase of domain sizes as a result of phase separation [19, 20] or unfavorable phase segregation toward an electrode [21–23], reducing the amount of charge generation and extraction over time. These processes are generally termed *morphological degradation* and need to be prevented to ensure long-term efficient device operation. Hence several approaches have been developed to lock the scale of phase separation within the bulk heterojunction, among which (1) cross linking of single phases by, for example, epoxy-based agents [19, 24], (2) increasing the glass transition temperature of the blend [22, 25–28], (3) compatibilization of the donor and acceptor phases via bifunctional surfactants [29, 30], and finally, (4) the use of microphase-separated D–A diblock copolymers as a single component with microstructures at thermodynamic equilibrium [31, 32] have been pursued. In the last approach, in addition to the viability of stable morphology, the scale of phase separation can be finely adjusted via precise control of the individual donor and acceptor blocks lengths to meet the restrictions of the exciton diffusion lengths. Furthermore, by definition of the respective volume fraction via the individual block lengths, certain phase distributions can be readily obtained, as predicted by the phase diagram of block copolymers. As an illustration, Fig. 1 summarizes various resulting phase distributions dependent on the relative volume fractions of phases A and B.

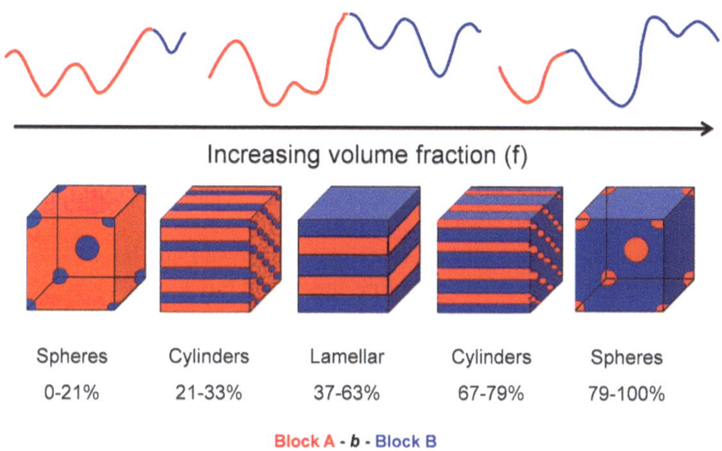

Fig. 1 Schematic representation of the most commonly observed morphologies of coil–coil block copolymers dictated by relative volume fractions. The typical range of percent volume fractions for the morphologies is also shown. *Left* to *right*: spherical, cylindrical, lamellar, inverse-cylindrical, inverse-spherical phases. The *red phase* represents polymer block A and the *blue phase* represents polymer block B. (Reproduced with permission from [33])

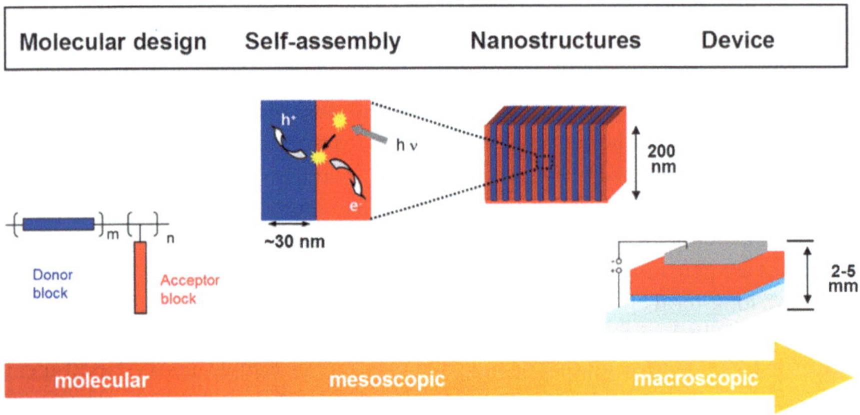

Fig. 2 The schematic shows the goal pursued: via control of molecular structure by design leading to morphological control through self-assembly of nanostructures, yielding functional control in devices spanning all length scales from molecular (nanoscopic) over mesoscopic nanostructures to macroscopic device levels

In this chapter, this approach toward creating microstructures in D–A block copolymers is reviewed. So far, polymer physics has described mostly the phase behavior of coil–coil (amorphous or liquid-like) block copolymers, that is, where both blocks exhibit a rather small persistence length (or Kuhn segment length), making them rather flexible. However, as already pointed out above, semicrystalline phases inhere the advantage of improved charge-transport and -generation capabilities, which triggered many authors to choose the so-called semicrystalline block copolymers (often referred to as rod–coil or rod–rod) to aim for well-defined phase separation in conjunction with crystallinity within the individual hole- and/or electron-transporting domains. With that approach, the goal of the majority of the studies was to obtain a controlled self-assembly of the donor and acceptor phases via molecular design of the functional blocks, yielding nanostructures suitable for photovoltaics and enabling efficient charge generation and transport over the whole lateral area of photovoltaic devices (compare with Fig. 2).

Several attempts for reaching this goal were followed and have been described in the literature so far, from realizing first functional block copolymers with a dual function of electron- and hole-transporting phases [34–43], to using oligomers with a rather small scale of phase separation [44], up to the current rather successful approaches based on diblock copolymers exhibiting a poly(3-hexylthiophene-2,5-diyl) (P3HT)–based donor block combined with a fullerene-based acceptor block [45, 46]. Because of the large diversity of chemical structures and the huge number of publications in this field, we limit our discussions to P3HT systems in which the acceptor block carries either perylene bisimides (PBIs) or fullerene derivatives (C_{60}, C_{70}) as pendant groups (see Fig. 3). For reviews of different kinds of semiconductor block copolymers, we draw attention to published work by Mori et al. [43], Horowitz et al. [34], Thelakkat et al. [39, 40], and Scherf et al. [47].

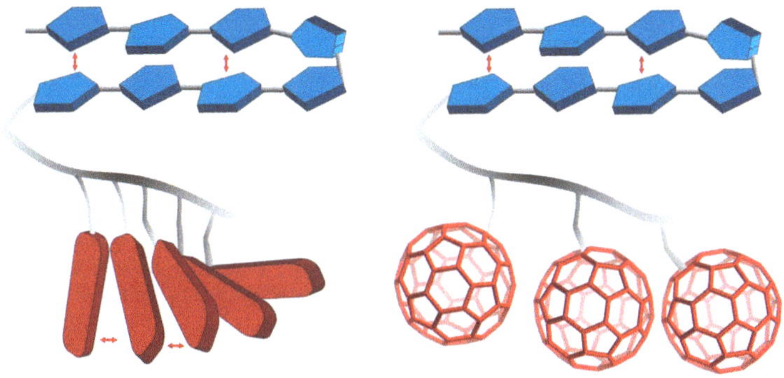

Fig. 3 Simplified schematics of P3HT containing block copolymers carrying either perylene bisimides or fullerene derivatives as pendant groups in the acceptor block. The *double-headed red arrows* indicate the π–π interaction of the thiophene units (resp., perylene bisimide units), leading to crystallinity and improved charge transport in these systems

The chemical structures of some of the reported D–A block architectures are given in Fig. 4. Hashimoto et al. [48] developed the most successful P3HT-b-poly(C_{60}) systems in which the fullerenes were grafted to a P3HT copolymer backbone, delivering PCEs of 2.46 % in single-layer devices using polymer 4 (Fig. 4). The best-performing device so far was indeed obtained using a non–fullerene-based acceptor block using a poly(9,9-dioctylfluorene)-co-(4,7-di-2-thienyl-2,1,3-benzothiadiazole) copolymer approaching 3 % PCE [51]. This success was largely the result of a considerably high open-circuit voltage of more than 1.2 V.

In this chapter, studies are reviewed which focus on an improved understanding of the self-assembling process by bringing the block copolymers first into the melt, accompanied by the formation of a microphase separation, and followed by a controlled crystallization upon slow cooling. A major challenge that still remains to be solved consists of obtaining the desired vertical phase orientation within the photoactive layer.

To understand and control the D–A nanostructure in block copolymer systems as introduced above, it is important to identify the driving forces responsible for structure formation. The classical microphase separation refers to systems with two amorphous blocks and is driven by their incompatibility, quantified by the product χN (N is the degree of polymerization, and χ is theFlory–Huggins interaction parameter). In addition, the phase diagram depends on the volume fraction; cf. Fig. 1. The situation becomes more complex if additional ordering processes take place within the individual blocks, which can also lead to a separation of the components. Crystallization is such an ordering process, but the formation of liquid-crystalline phases can also influence microphase separation [53]. The complexity of microphase separation in crystalline–amorphous systems as well as in double-crystalline block copolymers has, been studied in detail by, for instance, Register et al. using polyethylene as a model block [54, 55]. In general, it is useful to first study the individual components of such a complex block

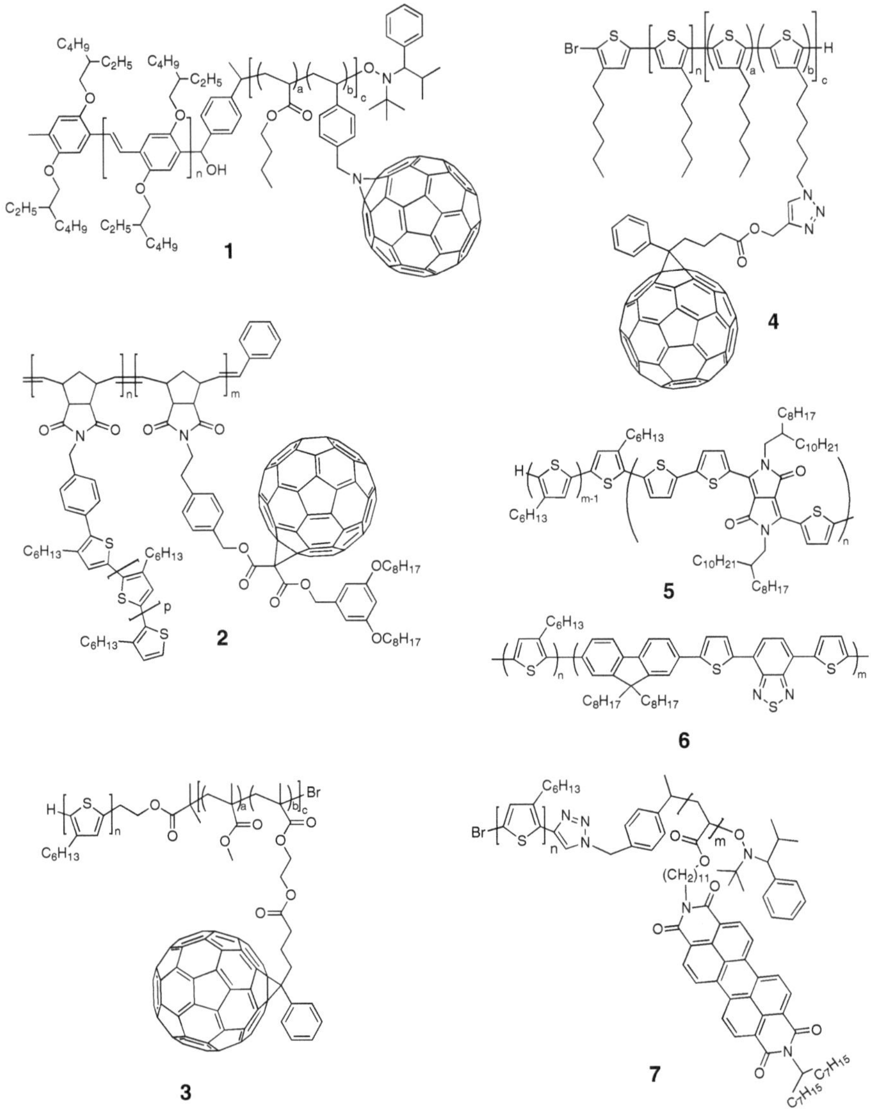

Fig. 4 Chemical structures of some of the selected P3HT-*b*-Poly(C_{60}) systems (polymers 1–4) [29, 48–50], all conjugated polymers (polymers 5, 6) [51, 52], and a P3HT-b-PPerAcr system 7 [40], which were tested in polymer solar cells

copolymer system by themselves before studying structure formation in the fully functionalized block copolymers. As P3HT can crystallize and as most D–A block copolymers synthesized until now contain P3HT as the donor block, the case of crystallization is most important. Schematically, this situation is illustrated in Fig. 5.

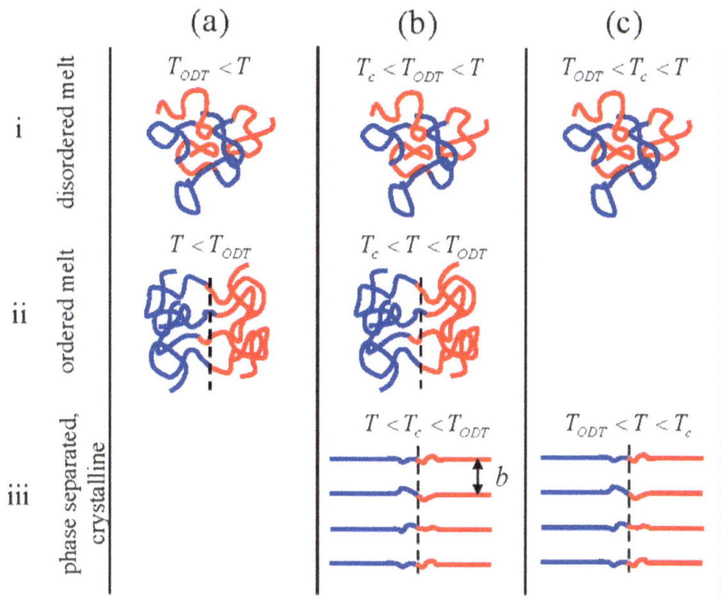

Fig. 5 Schematic illustration of possible ordering scenarios for block copolymers upon cooling (*top* to *bottom*). (**a**) A block copolymer with two noncrystallizable blocks forms a disordered melt at temperatures above the order–disorder temperature T_{ODT} (**i**), while for $T < T_{\mathrm{ODT}}$ (**ii**) in the ordered melt the well-known microphase-separated morphologies develop. For block copolymers with one or two crystallizable blocks, the crystallization of the individual blocks occurs either from an ordered melt (**b**) or directly from the disordered melt (**c**) depending on the relative locations of T_{ODT} and T_{c}. Here, for simplicity, only one melting/crystallization temperature T_{c} was assumed. For case (**b**) very often crystallization overrides the existing microphase structure, leading to a nanostructure consisting of lamellar crystals as in case (**c**). The case where for (**b**) the microphase structure stays intact upon crystallization is called confined crystallization. (Reproduced with permission from [56])

An additional aspect to be taken into account when considering structure formation in D–A block copolymers is the dependence on processing pathways. Drying from solution and cooling from the melt might not necessarily lead to the same nanostructure. Nevertheless it makes sense to first attemp to determine the equilibrium structure and to study deviations from it in a second step. To get information about phase diagrams, temperature-dependent in situ techniques are necessary, such as scattering techniques, complemented by imaging techniques used mostly at room temperature. To understand the structure formation in D–A block copolymers, in the following we first detail the controlled synthesis leading to structure formation in individual blocks.

2 Donor Building Block: P3HT

2.1 Controlled Synthesis as a Key for Structure Formation

Polythiophene, one of the most commonly used and most studied conjugated polymers, was earlier synthesized as an unsubstituted and insoluble derivative [57, 58]. Later irregular alkyl-substituted polythiophene [59] and in recent years regioregular poly(3-hexylthiophene-2,5-diyl), P3HT was prepared by Ni-catalyzed syntheses [60, 61]. A further development was the synthesis of the active Grignard monomer species through a Grignard metathesis reaction [62]. This route, which starts from 2,5-dibromo-3-hexylthiophene, will be referred to as the McCullough route, whereas the Yokozawa route uses 2-bromo-3-hexyl-5-iodothiophene as the starting component (Fig. 6). Yokozawa et al. [64, 66] and McCullough et al. [65, 67] were able to show that the Ni-catalyzed polymerization of **2a** follows a chain growth mechanism and gives increased control over the properties of the targeted polymer. With this method it was possible to synthesize regioregular P3HT with narrow distributions and predictable molecular weights, which can be seen from the size exclusion chromatography (SEC) curves of a series of P3HTs with different molecular weights (Fig. 7a). The method is generally termed Kumada catalyst-transfer polymerization (KCTP).

Upon addition of the nickel catalyst, the active species **2a** forms the nickel inserted dimer **4;** species **2b** does not take part in the polymerization because of

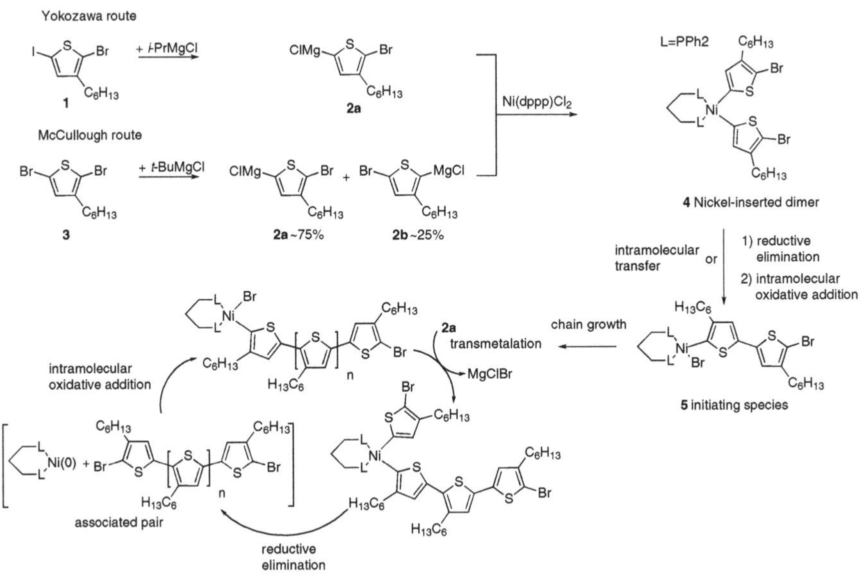

Fig. 6 Formation of the active Grignard monomer **2a** by Yokozawa and McCullough's route, and chain growth mechanism of the Kumada catalyst-transfer polymerization. (Reproduced with permission from [63])

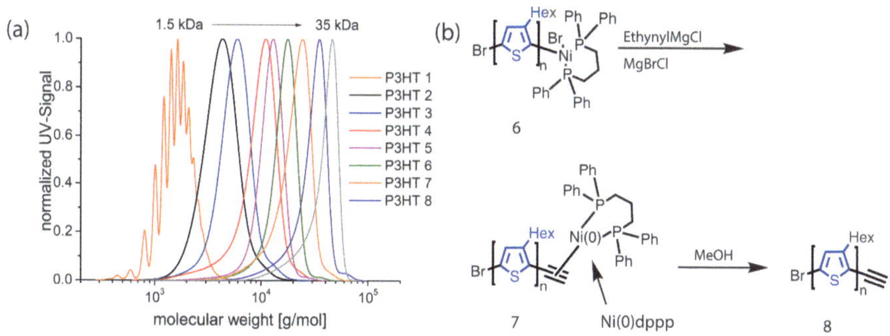

Fig. 7 (**a**) Size exclusion chromatography (SEC) curves of a series of P3HTs with different molecular weights and narrow distributions. The SEC was calibrated against polystyrene standards. (**b**) Formation of P3HT-alkyne via endcapping with ethynyl magnesium chloride and quenching in methanol

the sterical hindrance of the hexyl chain. One regio defect is always generated at the initial step from the inserted dimer **4** to the initiating species **5**. In the additional chain growth step, only head-to-tail couplings occur. After the monomer is consumed, the living chain remains active until the reaction is quenched with a suitable reagent. If P3HT is intended to be used in block copolymer systems or other, more complex architectures, it is important to control the end groups of the obtained polymers. Lohwasser et al. [63] showed that a detailed understanding of the mechanism involved makes it possible to perfectly control the end groups of P3HT. It has been shown that the quenching agent has a great effect on the final product [68]. Methanol as quenching agent was shown to lead to chain–chain coupling via disproportionation, which has a detrimental effect that cannot be observed after quenching the polymerization in dilute HCl. It has also been shown that a complete Grignard monomer formation was crucial to achieve almost 100 % H/Br end groups. When LiCl was used as an additive, the complete consumption of t-BuMgCl and the complete formation of the active species could be assured [69]. LiCl accelerates the active monomer formation and increases the molecular weight of the final polymer by also incorporating the second sterically disfavored monomer species **2b**. Wu et al. [70] found that the detrimental effect of the incorporation of the second monomer on the regioregularity is minimal. This is because of the lower reactivity of the sterically hindered monomer **2b,** which is only incorporated into the polymer chain once most of the majority species **2a** is consumed. A simple way to obtain functional end groups is highly advantageous when P3HT is intended to be part of a block copolymer or more complex polymer architectures. Jeffries et al. [71] reported a straightforward method to obtain a series of end groups simply by adding a functional Grignard reagent in order to end cap the polymer. While this method proved to be highly versatile and efficient, it did lead only to monocapped products for a variety of end-capping agents. Only end groups like vinyl or alkyne, which form stable π-complexes with the nickel catalyst, did not lead to dicapped products. The formation of dicapped products in cases where the catalyst was not bound to the end group can be explained by the effect of a random catalyst walking along the polymer chain, which was observed by Tkachov et al. [72]. The authors were able to show that the catalyst is not bound to one chain end but can move along the chain and initiate the polymerization at the other end of the polymer. This process, aside from having possibly negative effects on end capping, also leads to a change of position of regiodefects, which will not stay at one chain end but may rather be in the middle of the chain at the end of the polymerization. Especially alkyne-functionalized P3HT can be an interesting starting material for the synthesis of block copolymers containing P3HT. To obtain such polymers in high yield, it is again crucial to quench the polymerization in the appropriate media. Lohwasser et al. [73] showed that, in the case of alkyne functionalization, dilute HCl leads to a hydration and hydrohalogenation of the end group. Methanol, on the other hand, appeared to be a good choice in this particular case. The nickel is now in the Ni(0) state and no disproportionation reaction with the methanol is therefore possible (see Fig. 7b).

2.2 *Structure Elucidation in Bulk and Thin Films of P3HT*

2.2.1 Temperature-Dependent Phase Diagram

The structure of P3HT has been studied in detail by many authors, for instance, [74–76]. Depending on the conditions of sample preparation such as solvent, temperature, and molecular weight, it can exist in an amorphous phase with coiled chain conformations or in an aggregate phase containing planarized chains with a tendency to form semicrystalline domains. It has also now been well established that P3HT can adopt different morphologies [76]. Using selected samples of P3HT with well-defined molecular weight and high regioregularity from the series of materials mentioned above, Wu et al. [77] performed temperature-dependent in situ small- and wide-angle X-ray scattering experiments complemented by differential scanning calorimetry (DSC) and atomic force microscopy (AFM) experiments.

Figure 8 shows the results of DSC measurements on a series of P3HTs with varying molecular weights. Excluding the sample with the lowest molecular weight, all samples melt in a similar temperature range around 220 °C. P3HT 3 shows an additional peak around 60 °C, whose structural origin becomes clear when considering the temperature-dependent wide-angle X-ray scattering (WAXS) patterns shown in Fig. 9 together with a sketch of the microstructure of P3HT. As described in the figure caption, the reflections indicating order in the crystal across the layers separated by side chains disappear around 60 °C, in line with the observed transition in the DSC. The samples with higher molecular weight do not show this transition, most likely because the corresponding ordering process at low temperatures is kinetically suppressed. This interpretation of the phase transition was later confirmed by nuclear magnetic resonance (NMR) experiments in which a

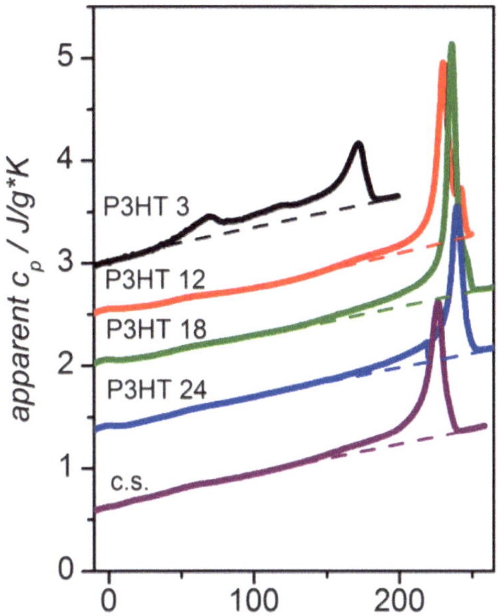

Fig. 8 Differential scanning calorimetry thermograms of samples of P3HT with a well-defined molecular weight. The sample names indicate the molecular weight M_n as obtained by matrix assisted laser desorption/ionization time-of-flight mass spectrometry; polydispersities are about 1.15. C.S. is a commercial sample with a broader molecular weight distribution. All data were obtained during a second heating run with a rate of 20 K/min. Other than with P3HT 24, all curves are shifted vertically for clarity. *Straight lines* below the melting peaks are used as a background signal subtracted for integrations to determine the melting enthalpy ΔH_m. Note the transition around 60 °C for P3HT 3 caused by side chain melting. (Reproduced with permission from [78])

structural disordering process of the alkyl side chains, not affecting the rigidity and conformation of the backbones, could also be observed [79].

2.2.2 Semicrystalline Morphology of P3HT

That P3HT is semicrystalline is confirmed by the observation of a diffuse scattering signal in the X-ray diffraction pattern appearing underneath the Bragg reflections reflecting the crystalline fraction of the sample (Fig. 9). The corresponding morphology can be visualized by AFM. In the lower row of Fig. 10, a well-developed semicrystalline morphology, consisting of lamellar or fibrous crystals separated by amorphous domains on a scale of some 10 nm, is visible. These samples were crystallized by cooling from the melt. Directly after spin coating, crystallization is largely suppressed, as the images in the upper row show. A more quantitative analysis of the semicrystalline morphology is possible with SAXS, which allows a measurement of the long period of the semicrystalline structure. The result is shown in Fig. 11 together with the contour length of the corresponding molecular

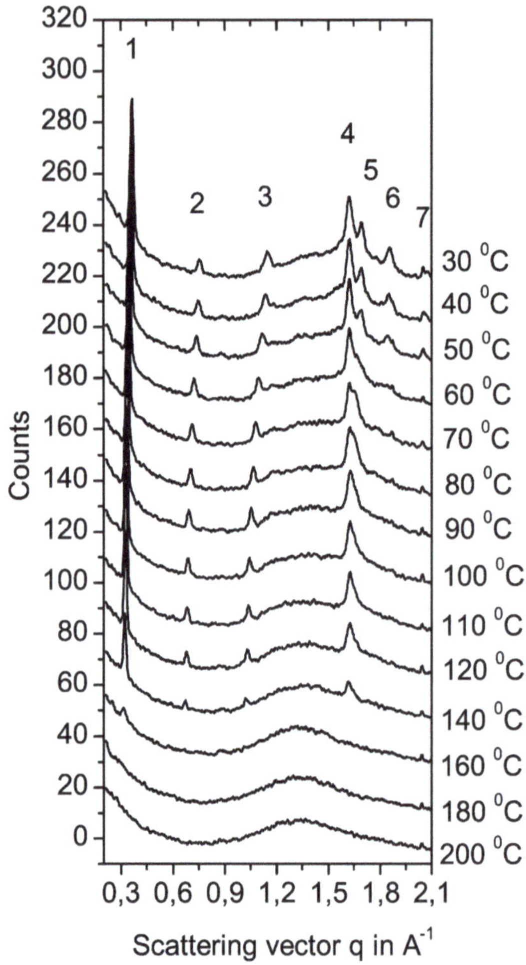

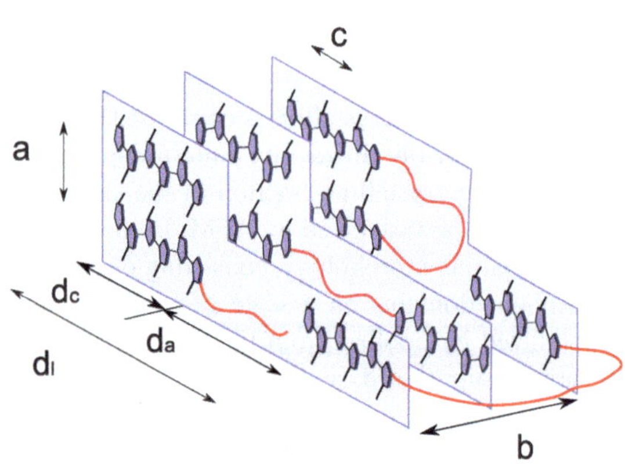

Fig. 9 *Top*: temperature-dependent wide-angle X-ray powder scattering diagrams of P3HT 3 reflecting the crystal structure. All data were taken during heating. Bragg reflections are numbered from low to high scattering vector. Peaks 1–3 are caused by alternating layers of main and side chains in the a-direction, while peak 4 reflects the π stacking. Peaks 5 and 6 carry mixed indices (hk0) and can only exist if crystal packing in different layers of main chains is in register. This kind of order is lost at the phase transition at 60 °C, which can therefore be attributed to side chain disordering. Peak 7 contains a contribution from the substrate and remains therefore visible up to the highest temperatures. *Bottom:* scheme of the typical microstructure of regioregular P3HT. a, b, c, crystal lattice parameters; d_c, thickness of lamellar crystals; d_a, thickness of amorphous layers, d_l, long period (scheme not to scale: the long period is about one order of magnitude larger than the lattice parameters a, b, c). (Reproduced with permission from [77])

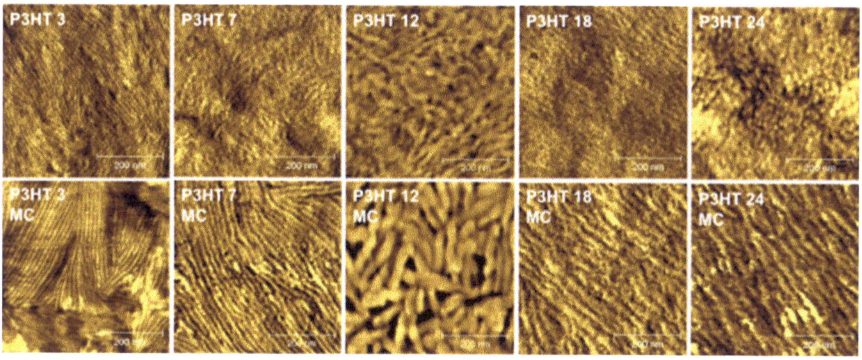

Fig. 10 Atomic force microscopy phase images of spin-cast (*top*) and melt-crystallized (*bottom*) films of P3HT with different molecular weights. The scan area for each sample is $500 \times 500\ nm^2$. (Reproduced with permission from [80])

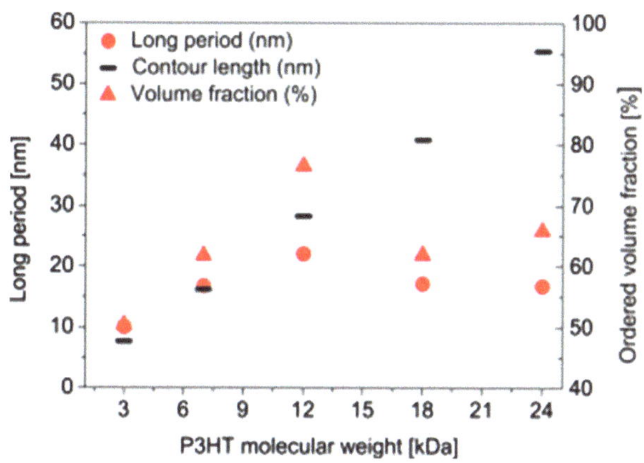

Fig. 11 *Left axis*: long period (*circles*) as measured by small-angle X-ray scattering in bulk samples and calculated contour lengths (*black bars*) vs. molecular weight of P3HT (matrix assisted laser desorption/ionization time-of-flight mass spectrometry). *Right axis*: volume fraction (*triangle*) of more ordered P3HT domains in melt-crystallized films determined by analysis of ellipsometric data. (Reproduced with permission from [80])

weight. The results show that for low molecular weight the polymer crystallizes as extended chain crystals but starts to form folded chain crystals around a molecular weight of 12 kg/mol [matrix assisted laser desorption/ionization time-of-flight mass spectrometry (MALDI-TOF MS)].

2.2.3 Quantitative Determination of Crystallinity

A common method to determine the crystallinity of a semicrystalline polymer is based on comparing the melting enthalpy as determined by DSC to the melting

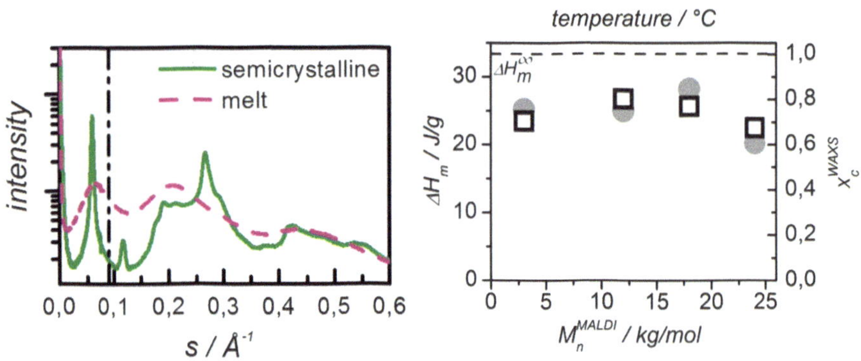

Fig. 12 *Left*: X-ray diffraction pattern of P3HT 18 in the semicrystalline state at 40 °C (*solid line*) after cooling from the melt and in the melt (*dashed line*). The *vertical dot-dashed line* indicates the range in which the intensities were compared to determine the crystallinity. *Right*: crystallinities determined by wide-angle X-ray scattering (*squares*) and differential scanning calorimetry melting enthalpies (*gray circles*) of P3HT showing the same trend with molecular weight. The *broken horizontal line* indicates the extrapolated melting enthalpy of a 100 % crystalline sample. (Reproduced with permission from [78])

enthalpy of a 100 % crystalline sample. The latter value has to be determined independently by an absolute method such as X-ray diffraction of NMR. We therefore attempted a quantitative determination of the crystallinity based on temperature-dependent SAXS/WAXS measurements on the above-mentioned series of P3HT with different molecular weights. The analysis is based on the evaluation of the scattered intensity from the amorphous regions as shown in Fig. 12, providing an easy and fast method to determine the crystallinity in the class of side chain–substituted polymers. The resulting values for the crystallinity of our P3HTs are in the range of 68–80 % at room temperature depending on the molecular weight. Based on these values, an extrapolated reference melting enthalpy of a 100 % crystalline material was determined ($\Delta H_{m,\infty} = 33 \pm 3$ J/g) for use in DSC measurements. This value is consistent with a previous estimation based on NMR measurements [79] and includes a substantial correction to previously used values [81]. For higher molecular weights a decrease of the crystallinity was observed that can be explained by the onset of chain folding as deduced from the analysis of the SAXS patterns. An in-depth analysis of the scattering patterns indicated that the crystalline regions of P3HT exhibit a large amount of internal disorder, considerably larger than typically found in other synthetic polymers. The results are described in detail in [78].

2.3 Structural Correlation with Charge Transport in Bulk and Thin Films

Based on the results described above, the effect of crystallinity and molecular weight on the hole mobility in P3HT was investigated in single-carrier devices using the space charge limited current (SCLC) method [80]. To vary the crystallinity in otherwise identical materials, we prepared two samples; one remained in the disordered state after spin coating, while the other was crystallized from the melt. As Fig. 10 shows and as confirmed by X-ray scattering, crystallization is largely suppressed after spin coating because of the fast drying process. The results of the charge-transport investigation in Fig. 13 show that there is a strong increase of up to one order of magnitude in hole mobility upon crystallization. Even stronger was the effect of molecular weight, displaying a strong rise over nearly three orders of magnitude from the lowest to intermediate molecular weights. Beyond 12 kg/mol, the hole mobility decreased again and leveled off for highest molecular weights. This reduction thus occurred at a molecular weight, where chain folding sets in, resulting in a decrease in the overall crystallinity and an increase in the amorphous volume fraction. This result is in good agreement with the general model of chain organization, leading to predicted changes in certain material properties according to Vikar et al. [82]. In this case, however, the dependency is exponential, whereas the crystallinity itself displays the same behavior on a linear scale (compare with Fig. 11).

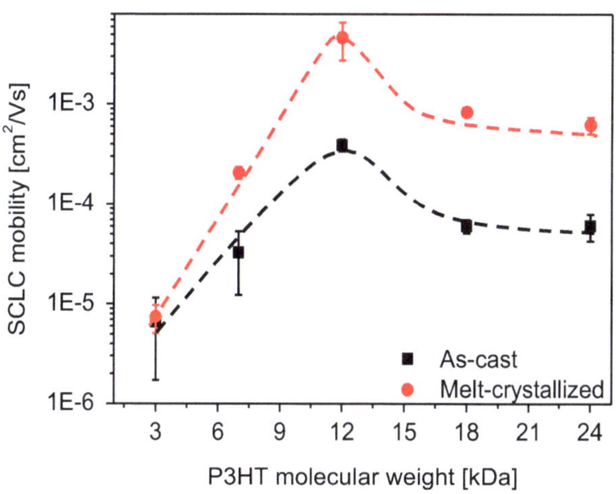

Fig. 13 Space charge limited current charge carrier mobility of holes in as-cast (*square*) and melt-crystallized (*circle*) P3HT films of different molecular weights (matrix assisted laser desorption/ionization time-of-flight mass spectrometry). Each data point represents the average of three to four sets of devices produced for varying film thicknesses in the range of 250–350 nm. *Dashed lines* are shown as a visual guide for the observed trend of charge carrier mobility. (Reproduced with permission from [80])

It should be noted that the charge carrier mobility obtained for the melt-crystallized P3HT sample with $M = 12$ kg/mol is one of the highest reported values for bulk charge transport through a film, thus reflecting the extraordinary quality of the material synthesized as described above. Based on these charge carrier mobility studies, the intermediate molecular weight sample could be identified as an ideal building block for the donor within block copolymers.

3 Acceptor Building Block: Poly(Perylene Bisimide Acrylate)/Poly(PBI)

3.1 Controlled Synthesis

Polymers with pendant perylene bisimide side chains (PPBI) were first reported by Lindner et al. [83] in 2004. The authors synthesized an asymmetrical PBI monomer with a solubilizing swallow-tail substituent and an acrylate group for polymerization. This acrylate monomer was polymerized with nitroxide-mediated radical polymerization (NMRP) [84]. It was possible to obtain moderate molecular weights by this method, and even block copolymers were obtained via sequential polymerization [40]. The homo-polymerization of this high-molecular-weight PBI monomer is not optimal. Narrow distributions, which are characteristic for controlled radical polymerizations, could not be obtained. In addition, the polymerization has to be performed at high monomer concentrations and the resulting polymer shows a limited solubility. This makes it difficult to achieve high molecular weights and also limits the choice of side chains that can be introduced. These synthetic restrictions were overcome by introducing the Copper-catalyzed azide-alkyne cycloaddition (CuAAC) "click" chemistry concept [85] and combining it with controlled radical polymerization. The first attempt to obtain PPBI polymer using this concept was reported by Tao et al. [86] in 2009.

Lang et al. [87] investigated this concept in detail and compared it to the conventional synthesis approach. The polymer backbone is synthesized independently and afterward decorated with the PBI moieties. The authors synthesized poly(propargyloxystyrene) via NMRP with alkyne functionality at each monomer unit and also PBIs with azide functionality. Alkynes can undergo a very efficient reaction at room temperature with azides in the presence of a Cu(I) catalyst and form stable 1,4-substituted triazoles [86]. It was reported that the polymer backbone can be almost quantitatively decorated with the PBI moieties as proven by Fourier transform infrared and proton NMR spectroscopy. These polymers had very narrow distributions as good as 1.16 and molecular weights of up to 15.000 g/mol. The "clicked" polymers were compared with poly(PBI acrylates) (synthesized directly via NMRP) with different alkyl spacers. The comparison of the phase behavior showed a strong dependence on spacer lengths. In a further study, Lang et al. [88] investigated the effect of the spacer length and different phase behavior of PPBI

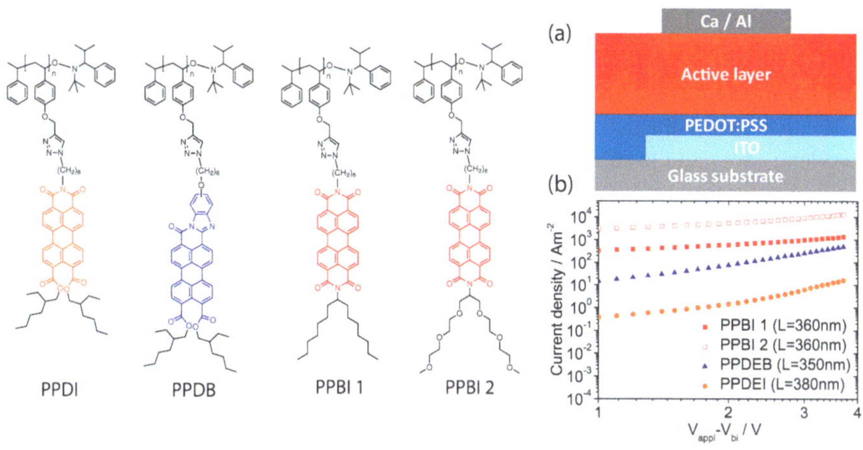

Fig. 14 *Left*: pendant perylene side chains synthesized via "click chemistry" and *right*: (**a**) schematic of an electron-only device and (**b**) *J–V* characteristics for electron-only devices of compounds PPBI 1, PPBI 2, PPDB, and PPDI with active layer thickness *L*. (Reproduced with permission from [89])

polymers with hydrophilic side chains. Two sets of polymers were synthesized as shown in Fig. 14: **PPBI 1**—polymers with PPBIs with hydrophobic alky swallow tails; and **PPBI 2**—a set with hydrophilic oligo ethylene glycol swallow tails. In both cases three polymers were synthesized, using three different spacers [$(CH_2)_6$, $(CH_2)_8$, and $(CH_2)_{11}$]. Systematically investigating the influence of the PBI substitutions could be done because the approach of a polymer-analogous introduction of different pendant groups to a single precursor polymer made it possible to obtain a set of highly comparable polymers. All polymers had molecular weights up to 60.000 g/mol (SEC) and narrow distributions below 1.09. This also showed that this modular synthesis method is reliable even for higher molecular weights. A marked difference between hydrophilic PPBIs with OEG- and hydrophobic PPBIs with alkyl swallow tails was reported. The hydrophilic PPBIs were all found to be amorphous with the spacer length influencing the glass transition temperature, T_g, of the polymers. An increase in spacer length reduced T_g. A similar trend was observed for the T_g of the hydrophobic polymers with generally higher transition temperatures. If the spacer length did not exceed $(CH_2)_8$, the polymers were not amorphous but liquid crystalline. The introduction of the new hydrophilic side chains claimed to increase the χ-parameter through a polar–apolar driving force and could be interesting for a supposed block copolymer implementation. Two other new perylene derivatives were introduced utilizing this reliable concept [89]. In this case, the electronically active perylene core was modified itself, tuning the absorption properties of the polymers. A poly(perylenediesterbenzimidazole) (**PPDB**) and poly(perylenediesterimide) (**PPDI**) with a blue-—respectively, redshifted—absorption with respect to PBIs were successfully synthesized (see Fig. 14).

3.2 Charge Carrier Transport in Polymeric PBIs

As the same parent scaffold polymer was used for "clicking" the different perylene derivatives, the influence on charge-transport properties by modifying the pendant perylene core and the substituents on PBI could be well compared [89]. In organic field-effect transistor (OFET) devices, poor performance with high threshold voltages, hysteresis, and low on/off ratios were reported for each material. Electron mobility values of the OFET measurements are summarized in Table 1. Since the charge transport in OFET geometry is determined by a thin channel of charge at the gate–dielectric interface, the results can be heavily influenced by the interface effects, wetting/dewetting issues, and unfavorable alignment of the polymers within the channel. Thus, the SCLC method is better suited to compare the bulk charge-transport properties of these polymers. The SCLC electron mobilities are usually determined by fitting measured *J*–*V* characteristics using the Mott–Gurney equation in electron-only SCLC devices [90].

The mobility values obtained from the SCLC devices are also mentioned in Table 1. The typical *J*–*V* curves of electron-only devices for the four polymers mentioned above are depicted in Fig. 14 together with a schematic of an electron-only device. A comparison of the mobilities shows that except for PPDI, all other polymers are good electron-transport materials. The electron mobility in PPDI, 5×10^{-6} cm^2/Vs, was two orders of magnitude lower than that of in PPDB, 6×10^{-4} cm^2/Vs. Nevertheless, in comparison to PPDB, better electron transport was reported for both PPBI polymers, that is, PPBI 1 and PPBI 2. A direct comparison of PPBI 1 and PPBI 2 showed that not only the core of the π-conjugation system, but also the substituent have an impact on the charge-transport properties of the material. PPBIs with hydrophilic OEG, PPBI 2, showed a major increase of one order of magnitude in electron mobility over PPBI 1 with hydrophobic alky tails. The reported electron mobility in PPBI 2 was 1×10^{-2} cm^2/Vs, which is among the highest bulk electron mobility values ever reported for polymers [91, 92]. However, the X-ray diffraction data suggested an liquid crystalline SmC structure for PPBI 1, whereas it suggested an amorphous phase for PPBI 2 [88, 89]. Thus, the less-ordered PPBI 2 was surprisingly superior in terms of electron mobility.

Table 1 Organic field-effect transistor (OFET) and space charge limited current (SCLC) (obtained from single-carrier devices with an active layer thickness of ca. 350 nm) mobility values for PDEB, PDEI, PPBI 1, and PPBI 2 [89, 90]

Polymer	OFET electron mobility (μ_e) (cm^2/Vs)	SCLC electron mobility (μ_e) (cm^2/Vs)
PPDB	–	6×10^{-4}
PPDI	–	5×10^{-6}
PPBI 1	2×10^{-6}	1×10^{-3}
PPBI 2	1×10^{-6}	1×10^{-2}

3.3 Effect of Polymer Architecture on the Structure of PBIs

From the different types of acceptor polymers introduced above, one, namely an acrylate with PPBIs with hydrophobic alkyl swallow tails, was selected for detailed structural investigations. This type of polymer was used later in the D–A block copolymers. To study the effect of the polymer architecture on structure, we included a low-molecular-weight model compound, PBI, as a reference material (cf. Fig. 15) [93].

The results of polarized light microscopy had already suggested that there are structural differences between the low-molecular-weight model compound PBI and the corresponding polymer poly(perylene bisimide acrylate) (PPerAcr) (cf. Fig. 15, right). Through a combination of DSC, optical microscopy, and temperature-dependent SAXS/WAXS, it was shown that both compounds display a lamello-columnar packing. While the PBI crystallizes, the PPerAcr suppresses order, leading to only a 2D lamello-columnar liquid-crystalline phase as schematically shown in Fig. 15. Most likely the reduced order in the polymeric compound is because of the quenched chemical disorder in the atactic polymer. In thin films, the a-axis is oriented perpendicularly to the substrate. Similarly as for P3HT, the ordering is suppressed directly after spin coating and a higher order is beneficial for charge transport, here electron transport.

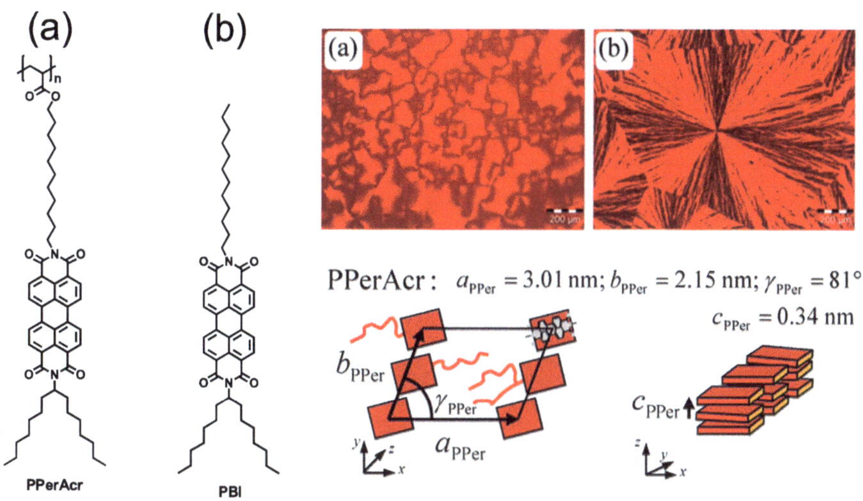

Fig. 15 *Left*: chemical structures of the investigated materials: (**a**) perylene bisimide side chain polymer with polyacrylate backbone (PPerAcr); (**b**) asymmetrically substituted low-molecular-weight perylene bisimide (PBI); *right top*: polarized light optical microscopy: PPerAcr (**a**) and PBI (**b**) at room temperature after cooling from the melt. Scale bar represents 200 μm in both cases; *right bottom*: proposed lamella-columnar liquid-crystalline structure for the polymer PPerAcr at $T = 20\,°C$. Packing in the a–b-plane is not correlated with stacking in the c-direction. (Reproduced with permission from [93])

4 Acceptor Building Block: Poly(fullerenes)/PPCBM

4.1 Controlled Synthesis

Basically, the preparation of pendant fullerene polymers can be achieved either by polymer-analogous modification of a preformed polymer [46, 94–98] or directly by polymerization of fullerene-derivatized monomers using organometallic catalysis [100–103]. While many polymer-analogous approaches involve C_{60} as reagent, which often leads to polymer cross linking and multiaddition, we applied a well-controlled fullerene-grafting method using Steglich esterification [50]. A series of well-soluble fullerene-grafted copolymer PPCBMs with high contents of pendant phenyl-C_{61}-butyric acid methyl ester (PCBM) between 30–64 wt% were synthesized (Fig. 16) [104]. The tailor-made precursor copolymers poly(4-methoxystyrene-*stat*-4-*tert*-butoxystyrene) obtained here by reversible addition fragmentation chain transfer (RAFT) polymerization were functionalized via an efficient polymer-analogous esterification. Both the grafting density and the PCBM content could easily be tuned by the monomer ratio in the precursor copolymers. The resulting PCBM-grafted copolymers exhibit low-molecular-weight dispersity and

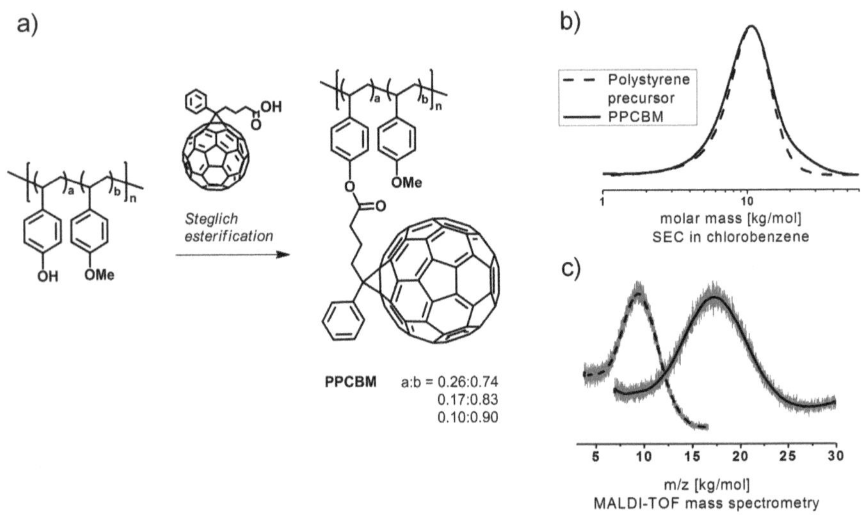

Fig. 16 (**a**) Synthesis route toward PPCBM acceptor polymers starting from a poly(4-*tert*-butoxystyrene-*stat*-4-methoxystyrene) copolymer with subsequent *tert*-butylether deprotection. The hydroxyl-functionalized precursor copolymer is grafted with phenyl-C_{61}-butyric acid ($PC_{61}BA$) using an optimized Steglich esterification protocol. (**b**) Characteristic size-exclusion chromatography traces in chlorobenzene at 50 °C of a PPCBM polymer (*solid line*) with a monomer ratio of $a{:}b$ = 0.23:0.77 and 46 wt% PCBM in comparison to the corresponding precursor copolymer (*dashed line*). (**c**) Matrix assisted laser desorption/ionization time-of-flight mass spectrometry ALDI-TOF mass spectra confirming the molar mass growth after successful grafting of the precursor (*dashed line*) with $PC_{61}BA$ moieties yielding PPCBM (*solid line*)

no cross linking as a result of the controlled monofunctional grafting reaction,that is, Steglich esterification [50] of phenyl-C_{61}-butyric acid with the hydroxyl moieties of the precursor copolymer. The synthesized acceptor copolymers retain the optical and electrochemical properties of the incorporated PCBM independent of their fullerene weight fraction.

4.2 Structure Formation and Charge Transport

Earlier studies on side chain polymers carrying pendant C_{60} have shown a correlation of increasing electron mobility with increasing C_{60} content [105]. In these systems, for a C_{60} content of 23–60 wt%, a rather low electron mobility of 10^{-9} to 10^{-7}cm^2/Vs was determined by the SCLC method [14, 106]. Further, these fullerene polymers exhibit a C_{60} aggregation starting at a threshold of 12–13 vol% of incorporated C_{60}. The improved charge transport reported by Fang et al. [103] of pendant C_{60} polynorbornenes in OFET was attributed to the confined organization of fullerenes along the polymer chain. The potential for the application of pendant fullerene polymers as a suitable acceptor material in polymer solar cells was successfully demonstrated by Eo et al. [102].

Our structural studies using AFM, transmission electron microscopy, and X-ray diffraction reveal a homogeneous and amorphous morphology of the PPCBMs both in thin films and in bulk phase. In contrast to pristine PCBM or blends of polystyrene and PCBM, the strong tendency for nanocrystal formation of PCBM is fully suppressed in the PCBM-grafted copolymers. Additionally, the absence of nanocrystal formation in PPCBM was maintained even after prolonged thermal annealing (Fig. 17).

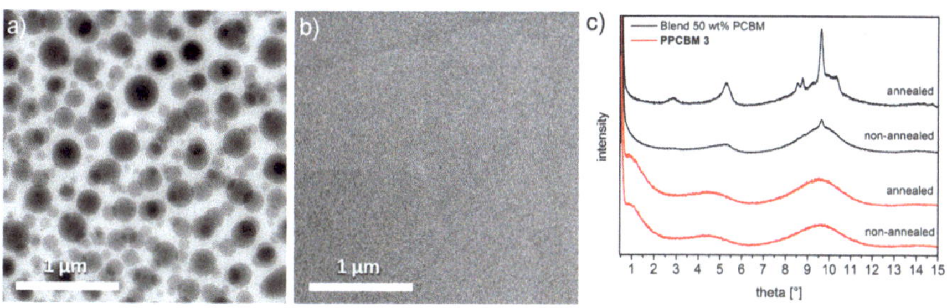

Fig. 17 Transmission electron microscopy images representing thin films made of (**a**) a blend consisting of 50 wt% PCBM content and 50 wt% polystyrene precursor and (**b**) a PPCBM acceptor polymer with 51 wt% of incorporated $PC_{61}BM$. By the covalent attachment of the $PC_{61}BM$ moieties to the polymer backbone, the demixing of the fullerenes and polymer is fully restrained, leading to a homogeneous morphology. (**c**) Powder X-ray diffractograms of the blend sample and PPCBM before and after annealing at 140 °C for 3 h, indicating a suppression of nanocrystal formation in the PPCBM acceptor polymer. (Reproduced with permission from [104])

The electron-transport properties were studied using the SCLC method. The maximum electron mobility μ_e of 1×10^{-4} cm^2/Vs was achieved for 37 wt% of incorporated phenyl-C_{61}-butyric methyl ester ($PC_{61}BM$). Despite the dilution of the electronically active fullerene moieties with the insulating polymer backbone, the $PC_{61}BM$-grafted copolymers exhibited exceptionally high charge transport compared to blend systems between polystyrene copolymers and $PC_{61}BM$. For 30 wt% $PC_{61}BM$ content, the grafted copolymer exhibited three orders of magnitude better mobility than the corresponding blend. Thus, an efficient charge carrier percolation is facilitated by the homogeneous distribution of $PC_{61}BM$ in the copolymer. Charge transport in the blends relies on nanocrystal formation and is improved by increasing the $PC_{61}BM$ content. On the other hand, the $PC_{61}BM$-grafted copolymers exhibit excellent mobility and no nanocrystal formation even after thermal annealing. This can be of great advantage if issues of long-term stability in devices have to be addressed using fullerene materials. The modular synthetic approach presented here can also be transferred to other fullerene derivatives.

5 Donor–Acceptor Block Copolymers: P3HT-*block*-PPerAcr

5.1 Synthesis of P3HT-block-PPerAcr

P3HT-b-PPerAcr copolymers constitute one of the fully functionalized block copolymers carrying both donor and acceptor moieties obtained by controlled synthesis of the individual blocks, maintaining appreciable solubility in common solvents [37, 86, 107, 108]. The synthetic strategy of this system can be taken as a general criterion for the design of D–A block copolymer systems. P3HT-b-PPerAcr polymers were first synthesized by Sommer et al. [107] from a P3HT macroinitiator. For this, P3HT with a molecular weight of 8900 g/mol (SEC) and a narrow PDI was first synthesized by a modified version of the aforementioned Yokozawa route and in situ functionalized to obtain a macroinitiator suitable for NMRP. This macroinitiator was then used to polymerize a PBI acrylate. A set of block copolymers with small polydispersities and different PBI contents was obtained. The low degree of polymerization of these block copolymers prevented them from phase separating into well-defined and ordered structures in the melt. By a modified synthetic strategy, Lohwasser et al. [56] obtained higher molecular weights, which led to a sufficient increase of the χN-parameter. P3HT with a high molecular weight (19.7 kg/mol in SEC, equivalent to a MALDI-TOF MS molecular weight of 12.4 kg/mol) and an alkyne end group was synthesized and converted into a macroinitiator suitable for NMRP via CuACC "click" chemistry. This macroinitiator was then used to polymerize the PBI acrylate monomer. This new method also prevented the formation of triblocks, which can occur when the macroinitiator is synthesized in situ as reported before. In this way, the synthesis of a series of D–A block copolymers P3HT-b-PPerAcr with sufficient purity,

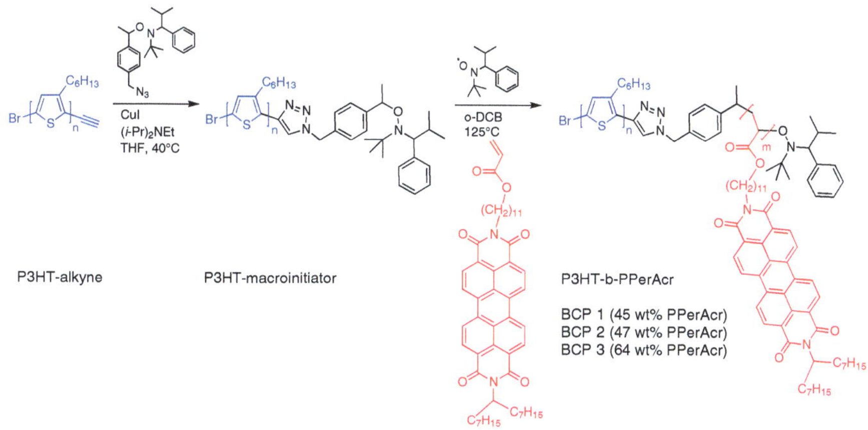

Fig. 18 Synthetic scheme for P3HT-b-PPerAcr block copolymers starting from P3HT-alkyne and its conversion to a macroinitiator suitable for nitroxide-mediated radical polymerization of PerAcr monomer. The listed samples BCP 1–3 are used for further investigations described ahead

a narrow distribution (PDI < 1.20), and a high molecular weights was realized (see Fig. 18). The structure formation and microphase separation in two selected block copolymers with 47 and 64 wt% of PPerAcr are discussed in detail in the following.

5.2 Structural Elucidation of P3HT-b-PPerAcr

Transmission electron microscopy images of two exemplary block copolymers of the type introduced above are shown in Fig. 19. The patterns are very similar to classical microphase structures as they had not been observed before on D–A block copolymers. Combined SAXS/WAXS measurements showed that indeed these samples microphase separate at elevated temperatures, where both components are in the molten state.

The data in Fig. 20 demonstrate that subsequent cooling leads to crystallization but leaves the microphase structure intact, as it becomes obvious from the unchanged position of the peak in the SAXS pattern reflecting the nanostructure. These results show that to achieve well-defined classical microphase structures, the key is high molecular weight leading to a large enough incompatibility χN, as it induces microphase separation at high temperatures with subsequent confined crystallization. The observed microstructures fitted the respective volume fractions well, and the crystalline packing within the individual blocks was analogous to those in the respective homopolymers. For the first time, typical lamellar or cylindrical phase-separated structures as known for amorphous coil–coil systems were realized for a crystalline–liquid crystalline, D–A block copolymer [56]. A

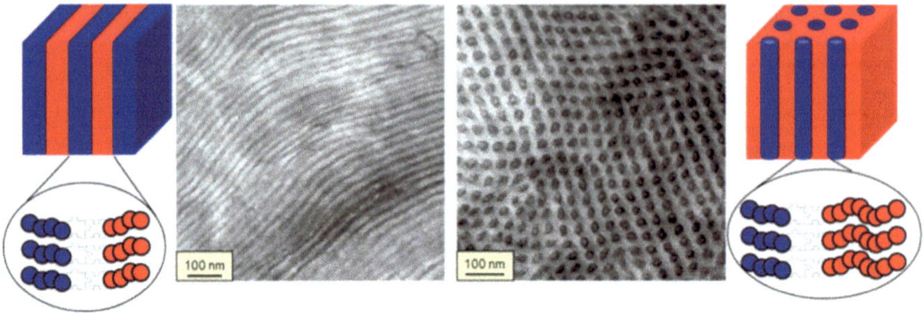

Fig. 19 Illustration of the microphase separation in donor–acceptor block copolymers, P3HT-b-PPerAcr with different volume fractions of PPerAcr, and transmission electron micrographs of the two exemplary block copolymers exhibiting lamellar (BCP 2) and cylindrical (BCP 3) microphase structures

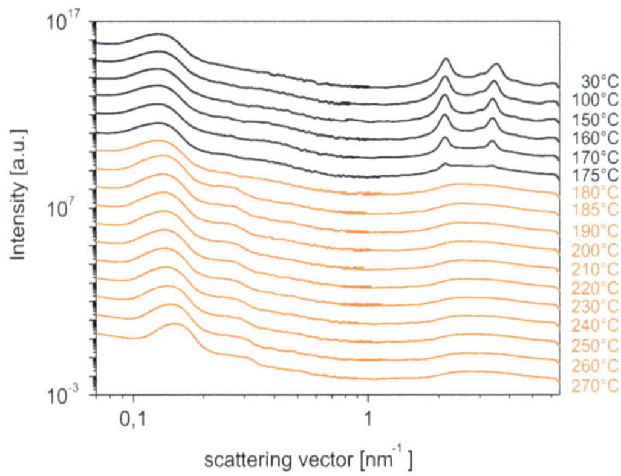

Fig. 20 Combined small-angle and wide-angle X-ray diffraction pattern of the asymmetric P3HT-b-PPerAcr (BCP 3) (cylindrical microstructures) recorded during cooling from the melt. Temperatures at which the block copolymer is in the molten state are marked in orange. Note that the position of the small-angle peak reflecting the nanostructure remains unchanged upon crystallization (Bragg peaks at high scattering vector), indicating confined crystallization. (Reproduced with permission from [56])

similar block copolymer synthesized with the above-mentioned method exhibited only crystallization-induced microphase separation (cf. Fig. 5).

Next, structural and electronic properties of thin films prepared from these block copolymers using a combination of X-ray scattering, AFM, and vertical charge-transport measurements in diode devices were studied [109]. Exemplary results are shown in Fig. 21. Generally, the well-defined microphase structures found in bulk could also be prepared in thin films, but block copolymer self-assembly and crystallization of the individual components are interrelated, and well-developed microdomains form only after an annealing step above the melting temperature of both components. In addition, alignment parallel to the substrate induced by interfacial interactions was observed. The copolymers sustain ambipolar charge

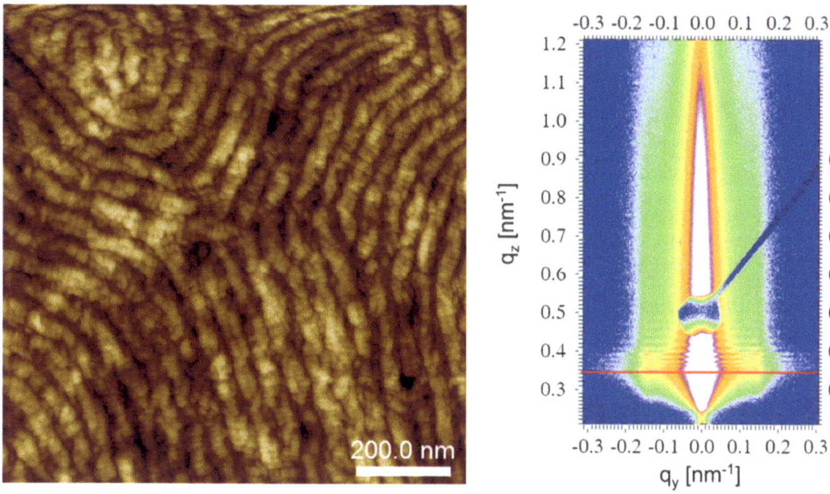

Fig. 21 (**a**) Atomic force microscopy height image of an annealed thin film of P3HT-b-PPerAcr with cylindrical microdomains (BCP3) measured with a Bruker MultiMode 8 using PeakForce Tapping™ mode (nanoscope 5 controller) and showing cylindrical microstructures lying flat on the substrate. We attribute the small-scale structure within the cylinders to P3HT crystals. (**b**) Two-dimensional grazing-incidence small-angle X-ray scattering image of a similarly prepared film. The specular reflection is masked by the beam stop. The *red line* indicates Yoneda's position. The two peaks around $q_y = 0.14\ \text{nm}^{-1}$ reflect the lateral arrangement of cylindrical microdomains. (Reproduced with permission from [109])

transport, but experiments on samples prepared with different thermal treatment show that the exact values of electron and hole mobilities depend strongly on orientation and connectivity of the microdomains as well as the molecular order within the domains. Generally, the measured mobilities were lower than in pure homopolymers, similarly as in blend morphologies, as they are used for organic solar cells. Apart from optimization of domain sizes, an important further step toward usage of the investigated materials in solar cell devices will be to achieve vertical alignment of the microphase structure.

5.3 *Solar Cell Devices Based on P3HT-b-PPerAcr*

In general, the photovoltaic performance of the P3HT-b-PPerAcr systems investigated up to now was weak for a variety of reasons, as discussed below. First, photovoltaic devices based on the P3HT-b-PPerAcr diblock copolymers [110] were realized with a PPerAcr weight fraction of 55 wt% and two different block lengths of the P3HT block of 8.9 and 17.0 kg/mol. Here a strong dependence on the block length was found: Whereas the smaller-molecular-weight block copolymer yielded only about 3 % maximum external quantum efficiency (EQE), the larger

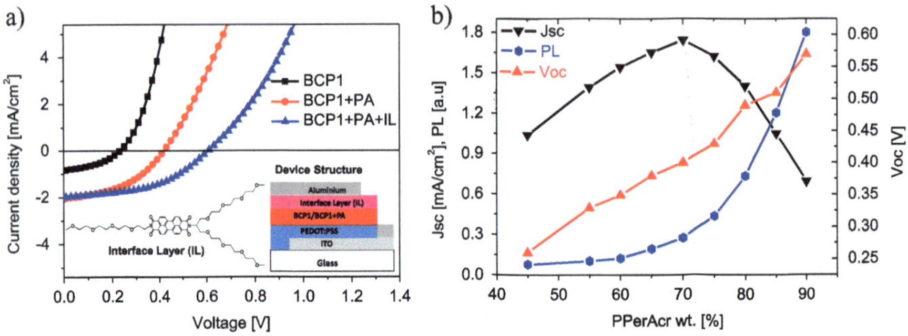

Fig. 22 (**a**) Current–voltage characteristics of photovoltaic devices based on pristine P3HT-b-PPerAcr (BCP1) (*black squares*), those with additional PPerAcr (PA) homopolymers (*red circles*), and finally the device with an additional perylene bisimide interface layer. (**b**) Dependence of the short-circuit photocurrent, open-circuit voltage, and photoluminescence intensity of P3HT-b-PPerAcr:PPerAcr block copolymer:homopolymer blends. (Reproduced with permission from [111])

resulted in a maximum of 31 % EQE and thus produced one order of magnitude larger photocurrent. However, overall device efficiencies remained low at 0.007 and 0.17 % PCE as a result of the small open-circuit voltages and small fill factors. Nevertheless, an important lesson has been learned: For efficient charge transport, a certain scale of coarse graining in the donor and acceptor domains is required.

The next step of improvement was reported about 2 years later with the same kind of block copolymer, exhibiting a PPerAcr weight fraction of 45 % and a higher total molecular weight of about 27 kg/mol [111]. The corresponding results are shown in Fig. 22. While the pristine block copolymer delivered a PCE of 0.1 % only, the addition of PPerAcr homopolymer yielded a threefold improvement, yielding an overall PCE of about 0.34 %. This improvement was largely based on doubling the photocurrent to 2 mA/cm^2 and a relative increase of the open-circuit voltage by another 60 %. Using steady-state photoluminescence measurements, this increase could be partially assigned to a statistically increased PPerAcr domain size, as the strong rise in the corresponding luminescence signal shows.

A surprising result was the steady increase of the open-circuit voltage with increasing PPerAcr fraction, reaching a maximum of 575 mV for the largest weight fraction (90 wt%) investigated. The key for understanding the rise in the open-circuit voltage was provided by X-ray photoelectron spectroscopy measurements on identically processed films, by which it was found that the PPerAcr surface fraction was increasing over-stoichiometrically compared to the bulk fraction. Since the electron-extracting electrode of the photovoltaic devices was situated at this interface with the photoactive layer, an enrichment of the electron-transporting PPerAcr phase could be related to the formation of a hole-blocking layer, thus preventing charge recombination at the electron-extracting electrode and improving the open-circuit voltage by a higher maintained quasi-Fermi level of the electrons. To prove this hypothesis, an alcohol-soluble perylene derivative, PBI, was deposited

Table 2 Summary of the photovoltaic parameters obtained for the P3HT-b-PPerAcr system (BCP 1) upon domain coarsening via homopolymer blending and blocking layer insertion using the perylene bisimide interlayer [111]

Active layer	J_{sc} (mA/cm^2)	V_{oc} (V)	Fill factor (%)	η (%)	R_S (Ω)	R_{SH} (Ω)
BCP 1	0.84	0.23	41	0.08	12.1	1277
BCP 1+PPerAcr	2.00	0.41	45	0.34	18.5	1851
BCP 1+PPerAcr+IL	1.96	0.61	47	0.56	26.7	3002

as an extra layer to function as a hole-blocking electron-transporting layer on top of the photoactive layer. At the optimum blend ratio yielding the maximum photocurrent, the application of this PBI interlayer resulted in an increase of the open-circuit voltage yielding reasonable 610 mV and a PCE of 0.56 % [111]. Table 2 collects the device parameters from these experiments.

The block copolymers BCP 2 (lamellar morphology, 47 wt% PPerAcr) and BCP 3 (cylindrical morphology, 64 wt% PPerAcr) were used to undertake a first study of the interrelation between morphology and solar cell performance. Different thermal treatments were applied to vary the morphology: In the film obtained after spin coating, microphase separation as well as crystallization of the components are largely suppressed by the fast drying process. One film was annealed above the melting temperature of PPerAcr but below the melting temperature of P3HT. In this film both components crystallize, but there is no time for the development of a well-defined and ordered microphase structure because of the fast crystallization of the P3HT component. A third film was first annealed at a high temperature where both components are molten, leading to the development of a well-defined microphase structure with subsequent crystallization during cooling. Of course, this latter treatment also leads to an orientation of the microdomains parallel to the substrate, preventing a good charge transport to the electrodes; a high-performance photovoltaic device is not yet produced in this way. As an example, the results of the photovoltaic characterization of BCP 3, for which the structural analysis is shown above, are shown in Fig. 23. Clearly, the increase in crystallinity induced by the thermal treatment at the intermediate temperature led to a considerable increase in the photocurrent, whereas the unfavorable orientation of the microdomains induced by annealing in the melt state again suppressed charge transport, leading to a very low short-circuit current and an even worse photovoltaic performance than in the "as-cast" case. These results demonstrate the combined influence of crystallinity and block copolymer morphology upon photovoltaic parameters.

As an outlook for further studies, therefore either a more favorable vertical microphase orientation or a bicontinuous phase like the gyroid structure may be most beneficial for obtaining high-performance organic photovoltaic devices based on self-assembling single-component block copolymers.

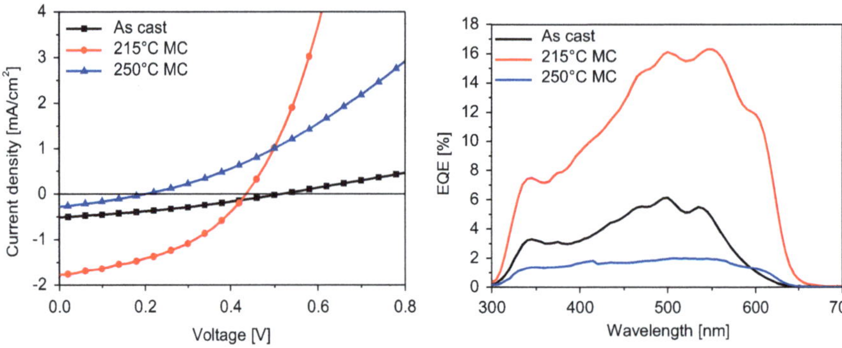

Fig. 23 Current–voltage characteristics (*J–V*, *left*) and external quantum efficiency (EQE, *right*) obtained from BCP 3 prepared in the following way: (i) as-cast (*black squares*), (ii) annealing at an intermediate temperature (215 °C) leading to crystallisation without well-ordered microphase separation (*red circles*), and (iii) annealing at high temperatures (250 °C) leading to the formation of well-defined microdomains oriented parallel to the substrate (*blue triangles*). Clearly, the unfavorable orientation of the lying cylinders results in the worst photovoltaic performance of all cases. (Reproduced with permission from [33])

6 Donor–Acceptor Block Copolymer: P3HT-*block*-PPCBM

6.1 Controlled Synthesis Without Cross Linking

The very first reports on block copolymers carrying a poly(p-phenylene vinylene) conjugated block and a fullerene pendant block are related to the work of Hadziioannou et al. [94]. Then the development of controlled P3HT polymerization and the capability of specific end functionalization of P3HT opened new perspectives for the synthesis of D–A block copolymers comprising fullerene-grafted acceptor blocks. A variety of C_{60}-decorated block copolymers was reported [31, 35, 38, 97, 112]. However, an elegant method for a controlled fullerene attachment without multiaddition or cross linking was first shown by Russel et al. using Steglich esterification [50] and Hashimoto et al. using alkyne-azide click chemistry [48].

We present a straightforward method for the preparation of novel D–A block copolymers based on an acceptor block with pendant $PC_{61}BM$ or its C_{70} analogue [6,6]-phenyl-C_{71}-butyric acid methyl ester ($PC_{71}BM$) and a regioregular P3HT as donor [41]. Our strategy is based on a modular combination of azide-terminated polystyrene copolymers with alkyne-terminated P3HT using CuAAc to form the block copolymer. The polystyrene precursor poly(4-*tert*-butoxystyrene-*stat*-4-methoxystyrene) can be tailored precisely using controlled radical polymerization methods such as RAFT polymerization or NMRP. Subsequent acidic hydrolysis was used to deprotect the *tert*-butylether group, yielding a hydroxyl-functionalized copolymer. Polymer coupling with ethynyl-terminated regioregular P3HT [73] using CuAAC proceeds equally well for both RAFT and NMRP precursor copolymers. Because of the modularity of the P3HT-b-PS_{OH} block copolymer synthesis, a wide range of individual polymer designs can be realized in terms of the block

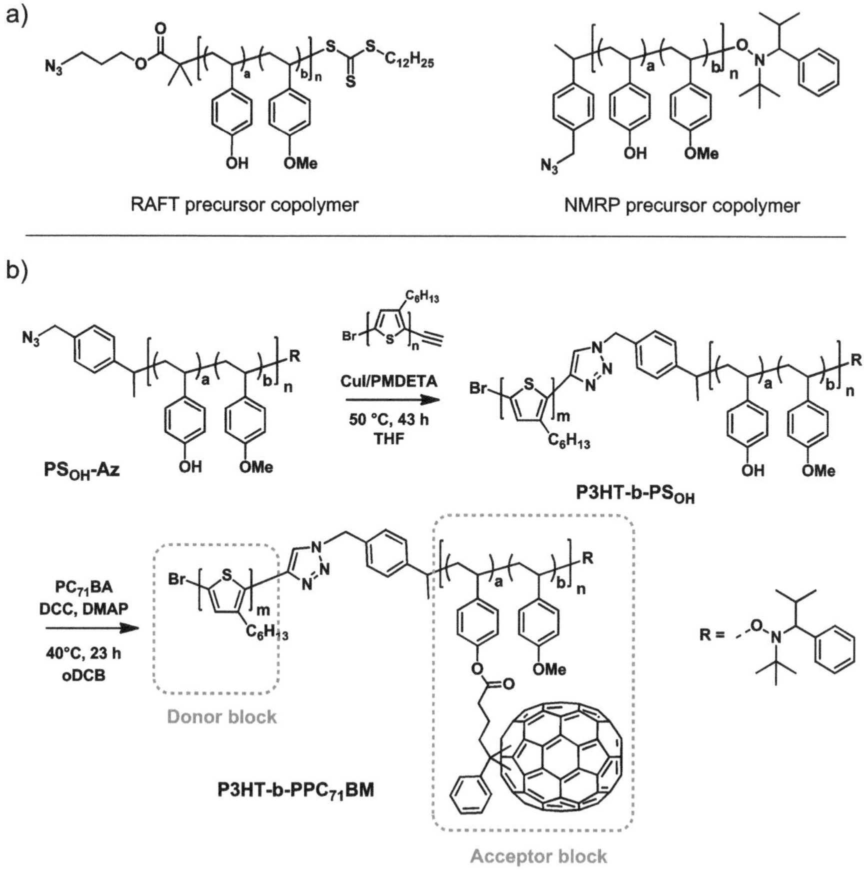

Fig. 24 (**a**) Azide-terminated poly(4-hydroxystyrene-*stat*-4-methoxystyrene) copolymers obtained either by reversible addition fragmentation chain transfer polymerization or nitroxide-mediated radical polymerization of 4-methoxystyrene and 4-*tert*-butoxystyrene with subsequent polymer-analogous deprotection of the *tert*-butyl ether group. (**b**) Synthesis route of the donor–acceptor block copolymer P3HT-b-PPC$_{71}$BM via copper(I)-catalyzed alkyne-azide cycloaddition and Steglich esterification with PC$_{71}$BA

lengths, grafting density with fullerene moieties, and the final D–A composition after PCBM grafting. Lastly, the fullerene acceptors are covalently attached to the preformed block copolymer to give the P3HT-b-PPCBM D–A block copolymers.

To enhance the optical properties of the block copolymers, PC$_{71}$BM as the state-of-the-art acceptor in organic photovoltaics was introduced to these polymer systems. The esterification reaction of phenyl-C$_{71}$-butyric acid (PC$_{71}$BA) to the hydroxyl groups of the polystyrene precursor was first optimized in homopolymers and yielded near-quantitative conversion. An analogous grafting reaction with P3HT-b-PS$_{OH}$ yielded the fully functionalized D–A block copolymer P3HT-b-PPC$_{71}$BM (Fig. 24). As a result of incorporating C$_{70}$, the D–A block copolymer exhibits enhanced absorption in the whole visible range of 300–600 nm.

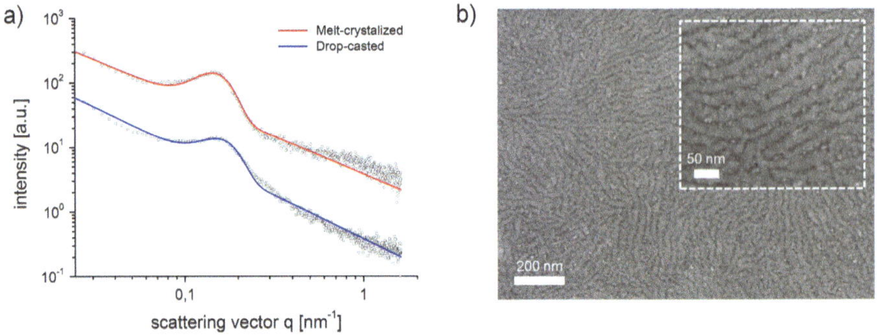

Fig. 25 (**a**) Grazing-incidence small-angle X-ray scattering data of P3HT-b-PPC$_{71}$BM showing horizontal intensity profiles around Yoneda's position at room temperature of films prepared by drop-casting and spin coating with subsequent annealing in the melt at 240 °C. The peaks indicate a periodic nanostructure that is consistent with bulk data from small-angle X-ray scattering measurements. (**b**) Scanning electron microscope image of P3HT-b-PPC$_{71}$BM prepared by drop-casting of a 2 wt% dichlorobenzene solution. The inset shows a magnified section of the film. (Reproduced with permission from [41])

6.2 Structure Formation

Even though different synthetic approaches to realize fullerene-containing block copolymers have been reported, no microphase separation with ordered D–A domains was observed for C_{60}-grafted D–A block copolymers up to now. Structural investigations on these block copolymers are rare and show either a loss of nanoscale structure because of fullerene aggregation [49] or a disordered phase-separated morphology [48, 50].

In contrast, our structural analysis of the P3HT-b-PPC$_{71}$BM block copolymer based on SAXS in transmission and in grazing-incidence geometry (GISAXS) as well as scanning electron microscopy (SEM) gave clear evidence for the formation of a periodic nanostructure of 37 nm in bulk and in thin films (Fig. 25). Temperature-dependent SAXS measurements both in the melt at 240 °C and in solid state at room temperature show the same periodic nanostructure. The fact that the SAXS peak is almost unchanged after cooling to room temperature, that is, after crystallization of the P3HT component, suggests that the nanostructure is already caused by a liquid–liquid phase separation in the melt [56]. Compared to the block copolymers described above, the high T_g in the acceptor block here seems to be a hurdle for the formation of equally well-developed microdomains on cooling from melt. Remarkably, the observed nanoscale morphology is nearly independent of the processing method, that is, from solution or by melt crystallization above the melting point of P3HT at 240 °C. SAXS and GISAXS structural analysis correlates with the domain size observed in SEM.

7 Conclusion and Outlook

The biggest challenges in designing and synthesizing block copolymers carrying semicrystalline P3HT segments of appreciably high molecular weight and acceptor blocks carrying PPBIs or fullerene derivatives have been successfully accomplished. Some of the initial problems such as end-group fidelity of the first block are no longer issues, and the problems of subsequent polymerization using P3HT macroinitiator have been resolved by making use of the modular approach provided by click chemistry. This has led to well-defined P3HT-b-PPerAcr block polymers with required volume fractions and low polydispersity. One of the biggest tasks in the synthesis of pendant fullerene polymers is the attainment of sufficient fullerene content without compromising on solubility. Additionally, the multivalent attachment of fullerene to polymer backbone leading to cross linking and insoluble materials reported earlier could be avoided by adopting a polymer-analogous esterification. Thus, a combination of all these synthetic techniques involving NMRP, KCTP, and click chemistry has facilitated the tailored synthesis of such complex D–A structures exhibiting microphase separation and thus suitable for structure–property correlation studies. Note that using two P3HT-b-PPerAcr block copolymers of sufficient molecular weight and either 47 or 64 wt% PPerAcr in the acceptor block allows one to obtain microphase separation and subsequent confined crystallization via cooling from the melt, yielding lamellar in the former case and cylindrical microdomains in the latter case. In the synthesis, further efforts are needed to obtain controlled lengths of novel low-bandgap donor segments with low polydispersity and to integrate them into block copolymers.

In terms of the microscopic structure, the basic result of our investigations is that the crystalline–liquid crystalline D–A block copolymers, P3HT-b-PPerAcr, under study are in fact very complex systems, but the driving forces and principles underlying the structure formation are very similar to more conventional materials. There is generally a competition between liquid–liquid microphase separation and the formation of order on a molecular scale via crystallization or the formation of a liquid-crystalline phase. P3HT shows the typical features of semicrystalline polymers, and PPerAcr forms a side chain liquid crystal. The morphology of block copolymers from such materials depends on the relative values of the different transition temperatures and on the mobility of the materials involved. In addition, there is not necessarily one well-defined equilibrium structure; the structure may depend on the processing conditions (thermal history, melt, or solvent processing). Also, the effect of the thin-film geometry on self-assembly is rather similar to simpler systems studied in the past. Because of interfacial interactions, microdomains tend to align parallel to the plane of the film. Moreover, crystallization under the constraints set by the microphase structure can lead to orientation of the crystals. It is important to note that for the systems studied, the crystallization of the individual components did not destroy the microdomains; that is, the crystallization was confined.

With respect to the application of diblock copolymers in photovoltaic devices, substantial progress in understanding the underlying principles has been made. At first, a certain scale of phase separation appeared to be required in order to allow efficient charge extraction. This has been shown by comparing D–A block copolymers bearing different molecular weights and via blending in a PPerAcr homopolymer, where the original block composition included only 45 wt% of PPerAcr before and 70 wt% after homopolymer addition. With microphase-separated block copolymers, the preferential alignment of the microdomains parallel to the electrode surface unfortunately constitutes a major hindrance for efficient charge extraction. However, as desired, the block copolymers exhibit ambipolar charge transport, but as the transport and structure are coupled, further optimization is necessary and probably also possible. An important further step toward the materials investigated here being used in solar cell devices will be to achieve vertical alignment of the microphase structure. This task is not trivial, as the two most established methods to orient thin-film block copolymers cannot be applied without difficulty. Electric field-induced alignment is not efficient most likely because of the residual conductivity of the materials, and solvent annealing is difficult in crystalline systems, as it requires compatibilization of the components as well as dissolution of the crystals. Nevertheless, at this point it has already become clear that the block copolymer architecture allows the preparation of well-defined and stable D–A nanostructures with the shape and size of the domains determined by the architecture and molecular weight of the polymer.

Acknowledgments We acknowledge financial support from the German Research Foundation (DFG) within the priority program SPP1355, projects HO 3981/6, Th 807/3, and TH 1281/1. Part of the results was obtained from experiments on beamlines ID10 and Dubble at the European Synchrotron Radiation Facility (ESRF), Grenoble, France.

References

1. Sariciftci NS, Smilowitz L, Heeger AJ, Wudl F (1992) Science 258:1476
2. Sariciftci NS, Braun D, Zhang C, Srdanov VI, Heeger AJ, Stucky G, Wudl F (1993) Appl Phys Lett 62:585
3. Yu G, Gao J, Hummelen JC, Wudl F, Heeger AJ (1995) Science 270:789
4. Yu G, Heeger AJ (1995) J Appl Phys 78:4510
5. Halls JJM, Walsh CA, Greenham NC, Marseglia EA, Friend RH, Moratti SC, Holmes AB (1995) Nature 376:498
6. Hoppe H, Sariciftci NS (2004) J Mate Res 19:1924
7. Hoppe H, Sariciftci NS (2008) Photoresponsive Polymers II 214:1
8. Chen JD, Cui CH, Li YQ, Zhou L, Ou QD, Li C, Li YF, Tang JX (2015) Adv Mater 27:1035
9. Liu C, Yi C, Wang K, Yang YL, Bhatta RS, Tsige M, Xiao SY, Gong X (2015) ACS Appl Mater Interfaces 7:4928
10. Huang F (2015) Sci China Chem 58:190
11. Liu YH, Zhao JB, Li ZK, Mu C, Ma W, Hu HW, Jiang K, Lin HR, Ade H, Yan H (2014) Nat Commun 5
12. Zhang SQ, Ye L, Zhao WC, Yang B, Wang Q, Hou JH (2015) Sci China Chem 58:248
13. Wu HB (2015) Sci China Chem 58:189
14. Li YF (2015) Sci China Chem 58:188

15. Etxebarria I, Ajuria J, Pacios R (2015) Org Electron 19:34
16. Kimber RGE, Walker AB, Schröder-Turk GE, Cleaver DJ (2010) Phys Chem Chem Phys 12:844
17. Hoppe H, Sariciftci NS (2006) J Mater Chem 16:45
18. Hoppe H, Glatzel T, Niggemann M, Schwinger W, Schaeffler F, Hinsch A, Lux-Steiner MC, Sariciftci NS (2006) Thin Solid Films 511:587
19. Drees M, Hoppe H, Winder C, Neugebauer H, Sariciftci NS, Schwinger W, Schaffler F, Topf C, Scharber MC, Zhu ZG, Gaudiana R (2005) J Mater Chem 15:5158
20. Nguyen LH, Hoppe H, Erb T, Gunes S, Gobsch G, Sariciftci NS (2007) Adv Funct Mater 17:1071
21. van Bavel SS, Barenklau M, de With G, Hoppe H, Loos J (2010) Adv Funct Mater 20:1458
22. Synooka O, Eberhardt KR, Singh CR, Hermann F, Ecke G, Ecker B, von Hauff E, Gobsch G, Hoppe H (2014) Adv Energy Mater 4:10
23. Sachs-Quintana IT, Heumuller T, Mateker WR, Orozco DE, Cheacharoen R, Sweetnam S, Brabec CJ, McGehee MD (2014) Adv Funct Mater 24:3978
24. Wantz G, Derue L, Dautel O, Rivaton A, Hudhomme P, Dagron-Lartigau C (2014) Polym Int 63:1346
25. Vandenbergh J, Conings B, Bertho S, Kesters J, Spoltore D, Esiner S, Zhao J, Van Assche G, Wienk MM, Maes W, Lutsen L, Van Mele B, Janssen RAJ, Manca J, Vanderzande DJM (2011) Macromolecules 44:8470
26. Cardinaletti I, Kesters J, Bertho S, Conings B, Piersimoni F, D'Haen J, Lutsen L, Nesladek M, Van Mele B, Van Assche G, Vandewal K, Salleo A, Vanderzande D, Maes W, Manca JV (2014) J Photonics Energy 4:1–12. Article Number 040997
27. Derue L, Dautel O, Tournebize A, Drees M, Pan HL, Berthumeyrie S, Pavageau B, Cloutet E, Chambon S, Hirsch L, Rivaton A, Hudhomme P, Facchetti A, Wantz G (2014) Adv Mater 26:5831
28. Khiev S, Derue L, Ayenew G, Medlej H, Brown R, Rubatat L, Hiorns RC, Wantz G, Dagron-Lartigau C (2013) Polym Chem 4:4145
29. Sivula K, Ball ZT, Watanabe N, Frechet JMJ (2006) Adv Mater 18:206
30. Bicciocchi E, Haeussler M, Rizzardo E, Scully AD, Ghiggino KP (2015) J Polym Sci, Polym Chem Ed 53:888
31. Heuken M, Komber H, Erdmann T, Senkovskyy V, Kiriy A, Voit B (2012) Macromolecules 45:4101
32. Johnson K, Huang YS, Huettner S, Sommer M, Brinkmann M, Mulherin R, Niedzialek D, Beljonne D, Clark J, Huck WTS, Friend RH (2013) J Am Chem Soc 135:5074
33. Singh CR (2013) Dissertation, TU Ilmenau, Germany
34. Yassar A, Miozzo L, Gironda R, Horowitz G (2013) Prog Polym Sci 38:791
35. de Boer B, Stalmach U, van Hutten PF, Melzer C, Krasnikov VV, Hadziioannou G (2001) Polymer 42:9097
36. de Boer B, Stalmach U, Melzer C, Hadziioannou G (2001) Synth Met 121:1541
37. Zhang QL, Cirpan A, Russell TP, Emrick T (2009) Macromolecules 42:1079
38. van der Veen MH, de Boer B, Stalmach U, van de wetering KI, Hadziioannou G (2004) Macromolecules 37:3673
39. Sommer M, Huettner S, Thelakkat M (2010) Complex Macromol Syst II 228:123
40. Sommer M, Huettner S, Thelakkat M (2010) J Mater Chem 20:10788
41. Hufnagel M, Fischer M, Thurn-Albrecht T, Thelakkat M (2015) Polym Chem 6:813
42. Darling SB (2009) Energy Environ Sci 2:1266
43. Nakabayashi K, Mori H (2014) Materials 7:3274
44. Bu LJ, Guo XY, Yu B, Qu Y, Xie ZY, Yan DH, Geng YH, Wang FS (2009) J Am Chem Soc 131:13242
45. Sary N, Richard F, Brochon C, Leclerc N, Leveque P, Audinot JN, Berson S, Heiser T, Hadziioannou G, Mezzenga R (2010) Adv Mater 22:763
46. Miyanishi S, Zhang Y, Tajima K, Hashimoto K (2010) Chem Commun 46:6723
47. Scherf U, Gutacker A, Koenen N (2008) Acc Chem Res 41(9):1086

48. Miyanishi S, Zhang Y, Hashimoto K, Tajima K (2012) Macromolecules 45:6424
49. Barrau S, Heiser T, Richard F, Brochon C, Ngov C, van de Wetering K, Hadziioannou G, Anokhin DV, Ivanov DA (2008) Macromolecules 41:2701
50. Lee JU, Cirpan A, Emrick T, Russell P, Ho W, Russell TP, Jo WH (2009) J Mater Chem 19:1483
51. Guo CH, Lin YH, Witman MD, Smith KA, Wang C, Hexemer A, Strzalka J, Gomez ED, Verduzco R (2013) Nano Lett 13:2957
52. Ku S-Y, Brady MA, Treat ND, Cochran JE, Robb MJ, Kramer EJ, Chabinyc ML, Hawker CJ (2012) J Am Chem Soc 134:16040
53. Olsen BD, Segalman RA (2008) Mater Sci Eng R Rep 62:37
54. Loo YL, Register RA, Ryan AJ (2002) Macromolecules 35:2365
55. Li S, Myers SB, Register RA (2011) Macromolecules 44:8835
56. Lohwasser RH, Gupta G, Kohn P, Sommer M, Lang AS, Thurn-Albrecht T, Thelakkat M (2013) Macromolecules 46:4403
57. Yamamoto T, Sanechika K, Yamamoto AJ (1980) Polym Sci Polym Lett Ed 18:9
58. Lin JWP, Dudek LP (1980) J Polym Sci, Polym Chem Ed 18:2869
59. Jen K-Y, Miller GG, Elsenbaumer RL (1986) J Chem Soc, Chem Commun 17:1346
60. McCullough RD, Lowe RD (1992) J Chem Soc, Chem Commun 1:70
61. Chen T-A, Rieke RD (1992) J Am Chem Soc 114:10087
62. Loewe RS, Khersonsky SM, McCullough RD (1999) Adv Mater 11:250
63. Lohwasser RH, Thelakkat M (2011) Macromolecules 44:3388
64. Yokoyama A, Miyakoshi R, Yokozawa T (2004) Macromolecules 37:1169
65. Sheina EE, Liu J, Iovu MC, Laird DW, McCullough RD (2004) Macromolecules 37:3526
66. Miyakoshi R, Yokoyama A, Yokozawa T (2005) J Am Chem Soc 127:17542
67. Iovu MC, Sheina EE, Gil RR, McCullough RD (2005) Macromolecules 38:8649
68. Miyakoshi R, Yokoyama A, Yokozawa T (2004) Macromol Rapid Commun 25:1663
69. Krasovskiy A, Straub BF, Knochel P (2006) Angew Chem Int Ed 45:159
70. Wu S, Huang L, Tian H, Geng Y, Wang F (2011) Macromolecules 44:7558
71. Jeffries-El M, Sauve G, McCullough RD (2005) Macromolecules 38:10346
72. Tkachov R, Senkovskyy V, Komber H, Sommer J-U, Kiriy A (2010) J Am Chem Soc 132:7803
73. Lohwasser RH, Thelakkat M (2012) Macromolecules 45:3070
74. Prosa TJ, Winokur MJ, McCullough RD (1996) Macromolecules 29:3654
75. Kline RJ, DeLongchamp DM, Fischer DA, Lin EK, Richter LJ, Chabinyc ML, Toney MF, Heeney M, McCulloch I (2007) Macromolecules 40:7960
76. Brinkmann M (2011) J Polym Sci B: Polym Phys 49:1218
77. Wu ZY, Petzold A, Henze T, Thurn-Albrecht T, Lohwasser RH, Sommer M, Thelakkat M (2010) Macromolecules 43:4646
78. Balko J, Lohwasser RH, Sommer M, Thelakkat M, Thurn-Albrecht T (2013) Macromolecules 46:9642
79. Pascui OF, Lohwasser RH, Sommer M, Thelakkat M, Thurn-Albrecht T, Saalwachter K (2010) Macromolecules 43:9401
80. Singh CR, Gupta G, Lohwasser RH, Engmann S, Balko J, Thelakkat M, Thurn-Albrecht T, Hoppe H (2013) J Polym Sci B Polym Phys 51:943
81. Malik S, Nandi AK (2002) J Polym Sci B 40:2073
82. Virkar AA, Mannsfeld S, Bao ZA, Stingelin N (2010) Adv Mater 22:3857
83. Lindner SM, Thelakkat M (2004) Macromolecules 37:8832
84. Hawker CJ, Bosman AW, Harth E (2001) Chem Rev 101:3661
85. Kolb HC, Finn MG, Sharpless KB (2001) Angew Chem Int Ed 40:2004
86. Tao Y, McCulloch B, Kim S, Segalman RA (2009) Soft Matter 5:4219
87. Lang AS, Neubig A, Sommer M, Thelakkat M (2010) Macromolecules 43:7001
88. Lang AS, Thelakkat M (2011) Polym Chem 2:2213
89. Lang AS, Muth M-A, Heinrich CD, Carassco-Orozco M, Thelakkat M (2013) J Polym Sci B Polym Phys 51:1480

90. Mott NF, Gurney RW (1940) Electronic processes in ionic crystals. The Clarendon Press, Oxford
91. Steyrleuthner R, Schubert M, Jaiser F, Blakesley JC, Chen Z, Facchetti A, Neher D (2010) Adv Mater 22:2799
92. Mueller CJ, Singh CR, Fried M, Huettner S, Thelakkat M (2015) Adv Funct Mater 25:2725–2736
93. Kohn P, Ghazaryan L, Gupta G, Sommer M, Wicklein A, Thelakkat M, Thurn-Albrecht T (2012) Macromolecules 45:5676
94. Stalmach U, de Boer B, Videlot C, van Hutten PF, Hadziioannou GJ (2000) Am Chem Soc 122:5464–5472
95. Hawker CJ (1994) Macromolecules 27:4836–4837
96. Liu B, Bunker CE, Sun Y-P (1996) Chem Commun 1241
97. Yang C, Lee JK, Heeger AJ, Wudl FJ (2009) Mater Chem 19:5416–5423
98. Heuken M, Komber H, Voit B (2012) Macromol Chem Phys 213:97–107
99. Zhang N, Schricker S, Wudl F, Prato M (1995) Chem Mater 7:441–442
100. Kim J, Yun MH, Lee J, Kim JY, Wudl F, Yang C (2011) Chem Commun 47:3078–3080
101. Eo M, Lee S, Park MH, Lee MH, Yoo S, Do Y (2012) Macromol Rapid Commun 33:1119–1125
102. Fang L, Liu P, Sveinbjornsson BR, Atahan-Evrenk S, Vandewal K, Osuna S, Jiménez-Osés G, Shrestha S, Giri G, Wei P, Salleo A, Aspuru-Guzik A, Grubbs RH, Houk KN, Bao Z (2013) J Mater Chem C 1:5747
103. Hufnagel M, Muth M-A, Brendel JC, Thelakkat M (2014) Macromolecules 47:2324
104. Adamopoulos G, Heiser T, Giovanella U, Ouldsaad S, Vandewetering K, Brochon C, Zorba T, Paraskevopoulos K, Hadziioannou G (2006) Thin Solid Films 511–512:371–376
105. Perrin L, Nourdine A, Planes E, Carrot C, Alberola N, Flandin L (2013) J Polym Sci B Polym Phys 51:291–302
106. Sommer M, Lang AS, Thelakkat M (2008) Angw Chem Int Ed 47:7901
107. Rajaram S, Armstrong PB, Kim BJ, Fréchet JMJ (2009) Chem Mater 21:1775
108. Gupta G, Singh CR, Lohwasser RH, Himmerlich M, Krischok S, Müller-Buschbaum P, Thelakkat M, Hoppe H, Thurn-Albrecht T (2015) ACS Appl Mater Interfaces. doi:10.1021/am5049948
109. Sommer M, Hüttner S, Steiner U, Thelakkat M (2009) Appl Phys Lett 95:183308(1–3)
110. Singh CR, Sommer M, Himmerlich M, Wicklein A, Krischok S, Thelakkat M, Hoppe H (2011) Phys Status Solidi RRL 5:247
111. Gholamkhass B, Holdcroft S (2010) Chem Mater 22:5371–5376

Donor–Acceptor Dyes for Organic Photovoltaics

Alhama Arjona-Esteban, Martin Robert Lenze, Klaus Meerholz, and Frank Würthner

Contents

Abstract Small-molecule π-systems bearing donor (D) and acceptor (A) groups constitute an interesting class of dyes because of their tunable strong absorption, which covers the visible and near-infrared range. The dipolarity associated with D–A structures directs antiparallel stacking arrangements in the solid state, thus reducing the dipolar disorder at the supramolecular level. Their straightforward synthesis and purification make them good candidates for photovoltaic application with power conversion efficiencies >6 %. This review summarizes the application of D–A dyes, and in particular merocyanines, in organic photovoltaics in recent years.

A. Arjona-Esteban • F. Würthner (✉)
Institut für Organische Chemie and Center for Nanosystems Chemistry, Universität Würzburg, Am Hubland, 97074 Würzburg, Germany
e-mail: wuerthner@chemie.uni-wuerzburg.de

M.R. Lenze • K. Meerholz (✉)
Department of Chemistry, University of Cologne, Luxemburgerstr. 116, 50939 Cologne, Germany
e-mail: klaus.meerholz@uni-koeln.de

K. Leo (ed.), *Elementary Processes in Organic Photovoltaics*,
Advances in Polymer Science 272, DOI 10.1007/978-3-319-28338-8_8

Keywords Bulk heterojunctions • D-A dyes • Dipole moment • Merocyanines • Organic photovoltaics • Tandem solar cell

1 Introduction

Silicon-based photovoltaics are now a well-established technology that offers power conversion efficiencies (PCEs) up to 30 % for tandem devices (http://solaerotech.com/products/terrestrial-solar-cells/ (22 May 2015)). However, the manufacturing costs for these kinds of cells remain too high compared with other energy resources, which points out the need to search for alternatives. With the entrance of organic materials into the world of photovoltaics, a wide range of possibilities appeared. Organic compounds can be prepared rather inexpensively by following established synthetic strategies, and their absorption and electrochemical properties can be tuned through suitable molecular design. Although organic solar cells (OSCs) are a relatively novel technology, they hold the promise of a substantial price reduction and faster return on investment in comparison with silicon-based counterparts. They can be processed in vacuum or from solution, thus allowing for simple and inexpensive roll-to-roll processes; when using flexible substrates, lightweight solar cells of various dynamic shapes can be obtained.

The field of OSCs was first inaugurated with small molecules [1–3], whose efficiencies were rapidly overcome by functional polymers [4], resulting in great improvements in PCEs from about 1 % to over 8 % in the last decade. Although semiconducting polymers show good charge carrier mobilities, they usually present low absorption coefficients and challenging purification. On the contrary, small molecules have the advantages of easy synthesis and purification, monodispersity, and a well-defined structure, and in some cases they allow for vacuum deposition (VAC) techniques, which offer better control over thin-film morphologies than solution processing (SOL). In this respect huge efforts have been made in recent years to improve the performance of bulk heterojunction (BHJ) solar cells based on either polymers or small molecules, and efficiencies approaching 10 % have been reported [5–11]. While these efficiencies demonstrate a great scientific advancement, they usually refer to very small solar cells of about 1 cm^2, which do not meet the demands for successful commercialization. With respect to the acceptor materials used so far in OSCs, fullerene derivatives such as [6,6]-phenyl C_{61}-butyric acid methyl ester ($PC_{60}BM$), [6,6]-phenyl C_{71}-butyric acid methyl ester ($PC_{70}BM$), C_{60}, and C_{70} are the widest used acceptor materials for OSCs and therefore allow for a comparison of the solar cell performance. Although other electron-poor organic semiconductor materials have been successfully applied recently to organic photovoltaics (OPVs) [12, 13], in this review they are not taken into account.

Among the huge variety of molecules used so far as donor materials in active layers of OSCs, an increasing number of molecules with alternating donor (D) and acceptor (A) subunits have been recently reported [14–16]. The use of alternating electron-donating and electron-withdrawing functional groups allows the optical bandgap to be lowered as a result of intramolecular charge transfer, therefore

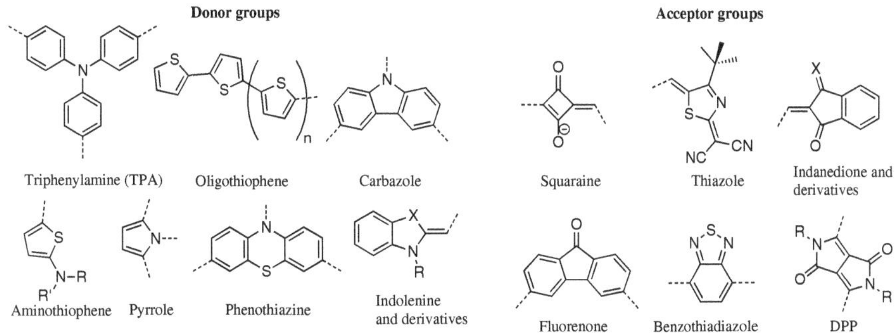

Fig. 1 Examples of donor (*left*) and acceptor (*right*) moieties applied in dipolar dyes for bulk heterojunction solar cells

offering the possibility of tuning the optical absorption and the highest occupied molecular orbital (HOMO)/lowest unoccupied molecular orbital (LUMO) levels of the molecules through the careful design of their structure. While the majority of research groups focused their efforts on either alternating (D–A)$_n$ polymers and D–A–D or A–D–A small molecules, our groups decided in 2006 to focus our attention on dipolar D–A dyes. Notably, the unsymmetrical D–A structure introduces a dipole moment in these molecules, which is considered to be detrimental for charge carrier transport in amorphous organic materials [17]. On the other hand, earlier work from our groups demonstrated that highly dipolar molecules preferentially assemble in centrosymmetric dimers, causing the dipolar character to vanish at the supramolecular level [18, 19]. Furthermore, our earlier studies demonstrated photorefractive effects in single-component glassy materials of D–A dyes, which is only possible if charge carriers are generated and exhibit a decent mobility [20]. Major advantages of D–A dyes are that the combination of available donor and acceptor moieties offers an almost infinite library of compounds with well-adjustable molecular properties. The most common donor and acceptor groups are listed in Fig. 1.

2 Merocyanine Dyes

Merocyanines are the archetypical class of D–A dyes, where the push (D) and pull (A) character of the end groups exert a pronounced impact on the connecting polymethine chain. Thus, the electronic properties of merocyanine dyes depend on the length of the polymethine chain and the strength of the donor and acceptor moieties, which determine the dipole moment and the absorption properties (Fig. 2). In general, their electronic structure can be described by two resonance structures, a neutral one and a zwitterionic one [21, 22].

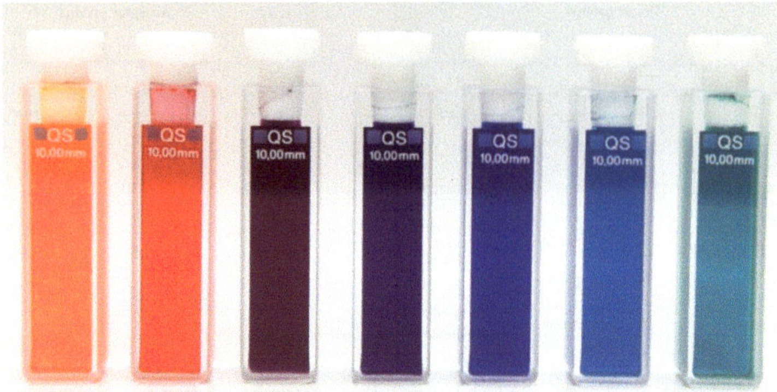

Fig. 2 The optical properties of merocyanine dyes are easily modified by carefully adjusting the strengths of the electron-rich and electron-poor end groups

In the case of weakly electron-donating and electron-withdrawing end groups, the corresponding merocyanine dye is usually polyene-like, displaying alternating double C=C and single C–C bonds within the methine bridge and small dipole moments in the ground state. Strengthening of the end groups increases the transfer of electron density from the donor unit to the acceptor group and leads to a leveling of the bond lengths and concomitantly a large ground-state dipole moment. At the so-called *cyanine limit*, the chromophore displays no bond length alternation at all [21, 22]. For molecules in the *cyanine limit*, the geometric and electron density changes upon optical excitation are minimal, resulting in narrow and most intense absorption bands with extinction coefficients of about 10^5 $M^{-1}cm^{-1}$ [23]. Further polarization of the polymethine chain by particularly strong donor and acceptor groups may lead to dyes whose ground state is even dominated by the zwitterionic structure. This is, however, only possible for short polymethine chains and heterocyclic donor and acceptor end groups that take advantage of aromatization in the zwitterionic structure [24–26].

Interestingly, probably because of the outstanding absorption properties and lack of knowledge about charge-transport mechanisms [27], a few studies on merocyanine dyes can be found in the early literature of OPVs. Thus, in 1978 Morel et al. [28] reported on a Schottky diode consisting of a layer of dye **1** sandwiched between aluminum and silver electrodes with a surprisingly high efficiency of 0.70 % for such a simple device configuration.

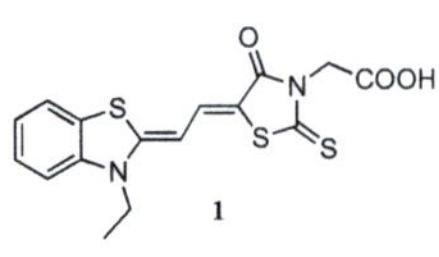

1

A small number of other groups also reported studies on merocyanine-based solar cells with similar PCEs [29–33]. Furthermore, vacuum-processed organic thin-film transistors (OTFTs) based on merocyanine dyes were reported by Kudo et al.

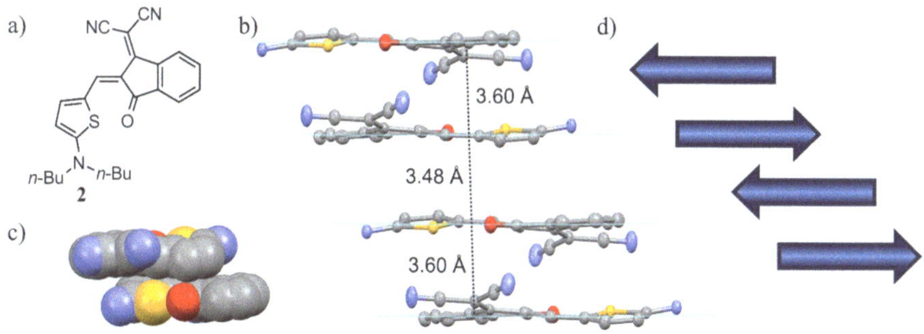

Fig. 3 Example of the antiparallel arrangement of a merocyanine dye (**2**) in the solid state [36]. (**a**) Chemical structure of dye **2**. (**b**) π-stack of **2** with antiparallel packing motif. (**c**) Space-filling view of the close dimer of **2** [alkyl chains and protons in (**b**) and (**c**) are omitted]. (**d**) Schematic representation of **2** stack showing the antiparallel orientation of the dipole moments (*arrows*). Adapted with permission from [36]. © 2011 Wiley-VCH Verlag GmbH & Co. KGaA, Weinheim

[34] in 1984. The modest hole mobilities of $\sim 10^{-5}$ cm^2/Vs were rationalized by the formation of deep trap states near the SiO_2–dye interface. Besides these early publications on D–π–A chromophores, in the following years merocyanine dyes disappeared from organic electronics and photovoltaics but found application as spectral sensitizers in photography and in dye-sensitized solar cells, as markers in biology and medicine, and in the field of photorefractive materials [21, 22] and nonlinear optics [35].

Unsymmetrical push–pull systems such as merocyanine dyes were thereafter presumably not applied in BHJ OSCs because of their high dipole moments. According to Bässler's model [17], completely randomly arranged highly dipolar molecules are supposed to limit the transport of charge carriers in amorphous solid-state materials because of a broadening of the density of states. However, the study of merocyanines in several groups (including ours) involved in research on nonlinear optical materials have demonstrated that the majority of these compounds arrange in the solid state as centrosymmetric dimers with small transversal or longitudinal displacement (see Fig. 3), resulting in the annihilation of the dipole moment on the supramolecular level and leading to an entirely new situation concerning charge transport.

3 Solution-Processed BHJ Solar Cells

Following work on A–D–A and D–A–D structures by other groups [14–16, 37], the first application of merocyanine D–π–A dyes was reported by our groups in 2008 in solution-processed BHJ solar cells [38]. In this first study, two series of chromophores bearing 2-aminothiophene or indolenine donor units combined with electron-withdrawing groups of different strengths were blended with $PC_{60}BM$.

While the increase of merocyanine content enhanced photon absorption, a larger amount of $PC_{60}BM$ resulted in more favorable charge-transport properties. For this series of chromophores a correlation between the open-circuit voltage V_{OC} and the HOMO level of the donor material could be found. The most efficient solar cells were fabricated with thiazole-containing chromophore **3**, showing film absorption at the longest wavelength of 650 nm of the series. This enables a good overlap of the absorption spectrum with the solar irradiance and originates in the best short-circuit current J_{SC} of 6.3 mA/cm^2 among the investigated D–A dyes. In combination with 70 wt% $PC_{60}BM$, a PCE of 1.7 % was reached, which at that time was among the best results in the field of solution-processed small-molecule solar cells. After this first promising result, the solar cells based on **3** were optimized by varying the top electrode. Ba/Ag provided the best results, mainly because of an increase in J_{SC}. The efficiency of these cells, 2.1 %, was even higher under reduced illumination conditions, where a PCE of 2.7 % was achieved [39].

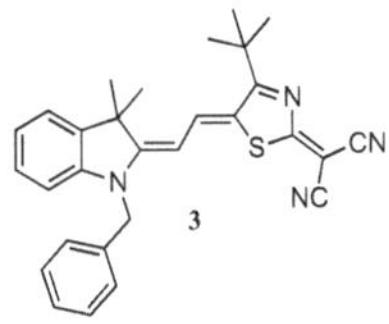

3

Further studies on solution-processed solar cells based on merocyanine dyes established some general rules that helped to define lead structures among the vast available library of possible D–A molecules. In order to obtain homogeneous films of sufficient thickness by spin coating, a solubility of at least 20 mg/ml in chlorobenzene is desired. From an energy viewpoint, frontier molecular orbital (FMO) levels of about −3.7 eV (LUMO) and −5.7 eV (HOMO) are preferred when using $PC_{60}BM$ as the acceptor material in order to ensure efficient charge carrier generation and minimize energy losses. Thus, the optical bandgap should be around 2.0 eV (560–690 nm). Finally, a high tinctorial strength (with $\mu_{eg}{}^2M^{-1} \geq 0.20$ D^2 mol/g) will allow high absorption (optical density about 0.4–0.5) while still having a relatively thin active layer of about 40–60 nm.

Several acceptor moieties were investigated in combination with the 2-aminothiophene donor unit, thus revealing the tunability of molecular properties by structural design. In Fig. 4a the studied acceptor groups are listed in order of their electron-withdrawing strength. From the ultraviolet–visible (UV-Vis) absorption spectra depicted in Fig. 4c, it becomes clear that increasing the acceptor strength decreases the LUMO level while maintaining an almost constant HOMO level (Fig. 4b), thus shifting the absorption toward the near-infrared (NIR) region. For several dyes very high extinction coefficients $\varepsilon > 10^5$ M^{-1}cm^{-1} and dipole moments μ_g in the range 5.7–13.1 D are observed in solution (Fig. 4c), as matches their cyanine-type character. Solution-processed BHJ cells were built in combination with $PC_{60}BM$ using the device structure indium tin oxide (ITO)/poly(3, 4-ethylenedioxythiophene) polystyrene sulfonate (PEDOT:PSS) (40 nm)/dye:

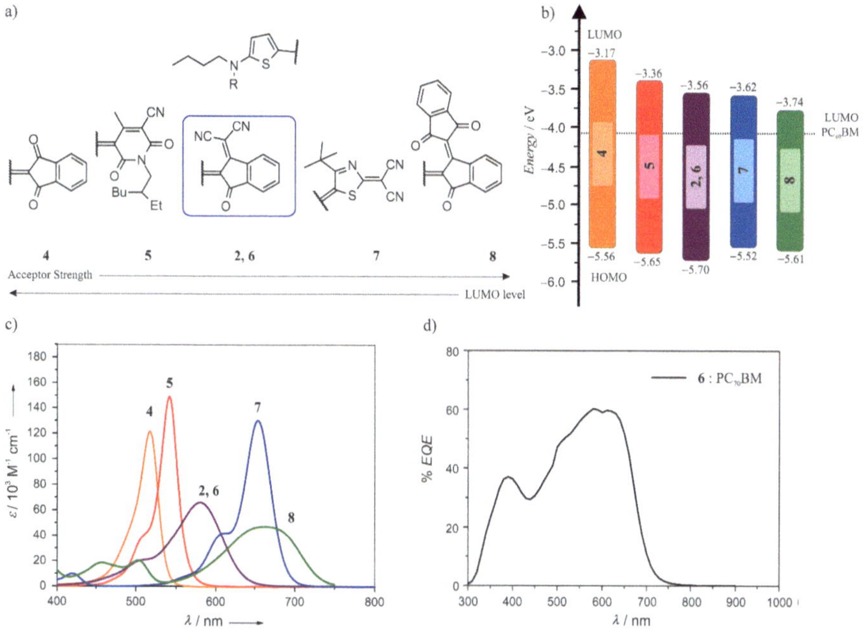

Fig. 4 (**a**) Molecular structures of the investigated dyes. R=n-Bu for all dyes except **6**, where R=Et. The energy levels depend on the acceptor moiety. The best-performing solar cells from solution could be obtained for dyes **2** and **6** [36]. (**b**) Schematic representation of frontier molecular orbital levels of molecules **2** and **4**–**8** with respect to the lowest unoccupied molecular orbital level of $PC_{60}BM$. (**c**) Ultraviolet–visible spectra of **2** and **4**–**8** in dichloromethane (10^{-5} M, 298 K). (**d**) External quantum efficiency of the optimized solution-processed BHJ device ITO/MoO_3/**6**:$PC_{70}BM$/Ba/Ag reaching a power conversion efficiency of 4.5 %. Adapted with permission from [36]. © 2011 Wiley-VCH Verlag GmbH & Co. KGaA, Weinheim

$PC_{60}BM$/aluminum (120 nm) with active layer thicknesses of about 50–60 nm. An initial comparison of the molecules showed that derivatives containing indandione-malononitrile acceptors displayed the highest V_{OC} and J_{SC}, reaching a PCE of 3.0 % for dye **6**. Moreover, solar cell fabrication was optimized by replacing $PC_{60}BM$ with $PC_{70}BM$, which enhanced the cell absorption. Further substituting PEDOT:PSS by molybdenum trioxide (MoO_3) and using a Ba/Ag metal electrode provided the best efficiency for the cell based on **6**, which reached a PCE up to 4.5 % and an external quantum efficiency (EQE) up to 60 % without posttreatment (Fig. 4d) [36].

Continuing with an indandione-malononitrile acceptor unit, variations were made on the indolenine donor by bridging the alkyl chain, thus creating a more rigid structure with planar geometry. Both ethylene (**10**) and propylene (**11**) bridging groups were used in an attempt to improve solid-state packing and solar cell morphology. The synthesized dyes showed similar energy levels and absorption properties. Interestingly, dye **11** had a better performance for relatively high contents of dye vs. $PC_{60}BM$, thus displaying the highest J_{SC} of 8.24 mA/cm^2 for this series.

Together with a V_{OC} of 0.94 V, **11** gave the best solar cells, with a PCE of 2.6 % [40].

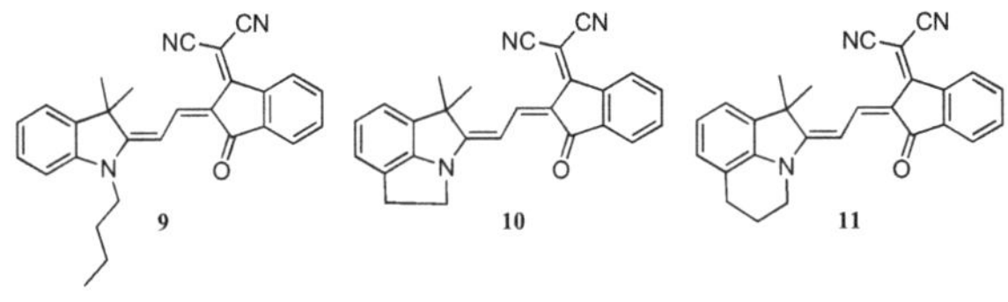

9 10 11

In addition, chromophores featuring NIR absorption have been developed and tested in solution-processed BHJ solar cells. The indolenine-containing dye **12** presents a π-extended system with an absorption maximum at 736 nm in dichloromethane solution. The dicyanopyrrolidinone acceptor group provides suitable FMO levels for application in solar cells using $PC_{60}BM$ as acceptor material. Despite the low bandgap of these dyes, V_{OC} up to 0.74 V could be measured and the PCE reached 1.0 % [41]. π-extension of the methylene bridge of merocyanine dyes is a straightforward way to obtain NIR-absorbing chromophores [42, 43]. Thus, thiophene-based dyes **13–16** were synthesized, and their performance in solution-processed solar cells was evaluated with respect to their shorter methylene bridge derivatives. Chromophores **13–16** displayed absorption maxima in the range 656–764 nm in solution and extended the photovoltaic response up to 900 nm in BHJ solar cells (Fig. 5b). Despite the favorable absorption properties, these dyes showed lower J_{SC} than their shorter counterparts, which was attributed to a less advantageous packing in the solid state. Nevertheless, devices based on **16**, which showed the highest efficiency, were optimized by replacing the PEDOT:PSS hole-collecting layer by MoO_3 and using $PC_{70}BM$ as acceptor material to afford reasonably good V_{OC} (0.74 V), J_{SC} (6.65 mA/cm^2), and fill factor (FF) (47 %) values, thus reaching a PCE of 2.3 % [44].

Two reasons underlie the current interest in merocyanine-based organic semiconductor research on the impact of molecular packing on the functional properties absorption (λ_{max}, ε_{max}), hole carrier mobility (μ_h), and photovoltaic performance (PCE): First, dipolar disorder is known to slow down charge carrier transport in amorphous materials as given in BHJ blends. Accordingly, the question prevails whether these D–A molecules indeed pack in a perfect and complete antiparallel arrangement or what distribution of mutual packing arrangements is given in BHJ materials composed of D–A dyes. Second, the high transition dipole moments $\mu^2{}_{eg}$ of merocyanine dyes should produce pronounced excitonic coupling in closely π-stacked arrangements according to Kasha theory [45], which might be useful for color tuning and enhancement of the exciton diffusion length. To address these issues, our groups have explored the effect of intramolecular folding-driven aggregation of bis-merocyanine dyes on the absorption properties and solar cell performance. Bis-merocyanine **17** underwent folding upon decreasing the solvent polarity (from tetrahydrofuran to methylcyclohexane—see Fig. 6a—or from

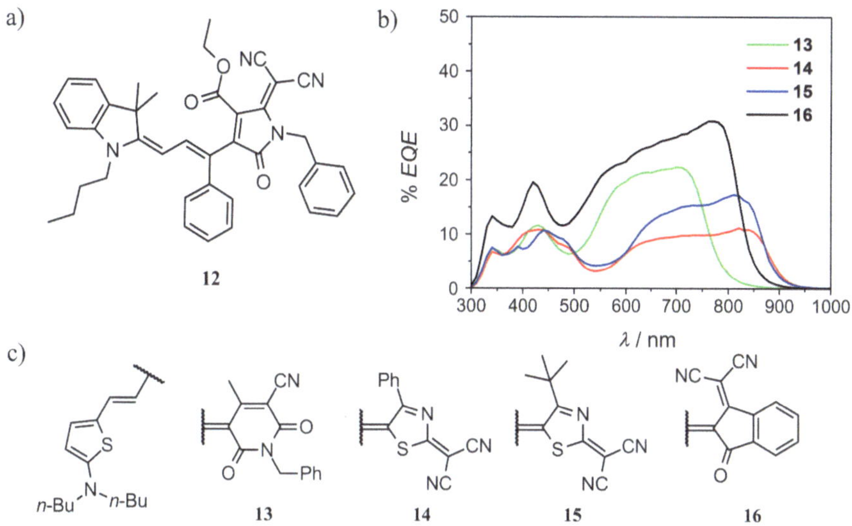

Fig. 5 (**a**) Molecular structure of near-infrared absorbing dye **12**. (**b**) External quantum efficiency of bulk heterojunction solar cells with π-extended MCs **13–16** in the device setup ITO/PEDOT:PSS/dye:$PC_{60}BM$/Ca/Ag. Note the redshift of the spectral responsivity upon π-extension from dye **6** (Fig. 4d) to dye **16** (*black line*) [44]. (**c**) Molecular structure of dyes **13–16**. Adapted with permission from [44]. © 2014 American Chemical Society

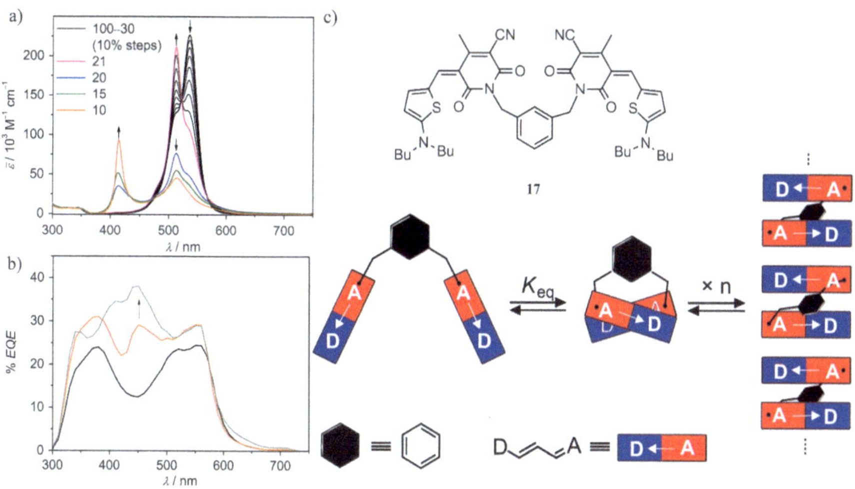

Fig. 6 (**a**) Solvent-dependent ultraviolet–visible spectra of **17** in tetrahydrofuran (THF)/methylcyclohexane (MCH) mixtures starting in pure THF (*black*). (**b**) External quantum efficiency spectra of **17**:$PC_{60}BM$ (30:70) blends depending on the casting solvent mixture of chloroform ($CHCl_3$) (*black*), $CHCl_3$/toluene (*orange*), and $CHCl_3$/*o*-xylene (*gray*). (**c**) Chemical structure of **17** and schematic representation of the supramolecular aggregation pathway for dipolar merocyanine dyes composed of π-conjugated electron donor (D)–electron acceptor (A) subunits into cofacially stacked H-aggregates [46]. Adapted with permission from [46]. © 2013 The Royal Society of Chemistry

chloroform to *o*-xylene; see Fig. 6b), thus forming a sandwich dimer of two stacked dyes with a hypsochromically displaced absorption maximum. Furthermore, adding an unpolar solvent favored the formation of supramolecular π stacks, which could be supported by UV-Vis spectroscopy via detecting an H-band at ~400 nm (Fig. 6). The impact of supramolecular polymerization on the solar cell performance was subsequently evaluated by processing the BHJ under aggregation-favoring conditions. As a consequence, a nearly twofold increase in photocurrent generation was observed as a result of harvesting high-energy photons in the 400–500 nm region [46]. While the choice of the chromophore and sandwich-type packing motif that reduced the absorption strength and led to a hypsochromic shift away from the highest solar flux was clearly not favorable for producing high-performance solar cells, this work illustrates the impact of dye–dye contacts on relevant parameters for solar light harvesting, photocurrent generation, and solar cell performance.

4 Vacuum-Deposited Solar Cells Based on Merocyanine Dyes

4.1 BHJ Active Layers

While solution-processing (SOL) by spin coating is a cheap and well-established procedure that is widely used for polymer solar cells, small molecules enable the use of VAC vacuum-processing (VAC), which allows more controlled growth of high-quality thin films from highly purified materials with reproducible layer thickness as well as complex multilayer architectures. Direct comparisons of both methods for the same molecules are not frequent. Therefore, in order to clearly evaluate the possibilities offered by vacuum techniques in comparison with SOL, we manufactured solar cells for a series of merocyanine dyes, which were chosen to display enough solubility as well as thermal stability. All devices had the same structure, ITO/PEDOT:PSS/merocyanine:acceptor/BPhen/Ag [47]. As acceptor material, $PC_{60}BM$ and C_{60} were used for SOL and VAC, respectively. Thus, lower V_{OC} values by ca. 100 mV were found for VAC devices because of the higher electron affinity of C_{60} compared to that of $PC_{60}BM$. On the other hand, most of the VAC devices featured a higher J_{SC} (Fig. 7a) and FF, thus showing a higher overall performance compared to their SOL counterparts. Since the hole mobility and absorption did not change between solution- and vacuum-processed devices, the better performance of the latter may be explained through much more efficient charge carrier generation or reduced recombination of the charges, which is reflected in the higher EQE of vacuum-processed cells (Fig. 7b). Furthermore, for dye **11** a much smaller phase separation within the active layer was found in vacuum-deposited devices, which would infer the presence of a larger interface for exciton dissociation. Subsequently, the high PCE obtained for vacuum-processed devices based on **11** was improved by replacing PEDOT:PSS with MoO_3. Further optimization of the active layer thickness and composition enhanced the J_{SC}, which

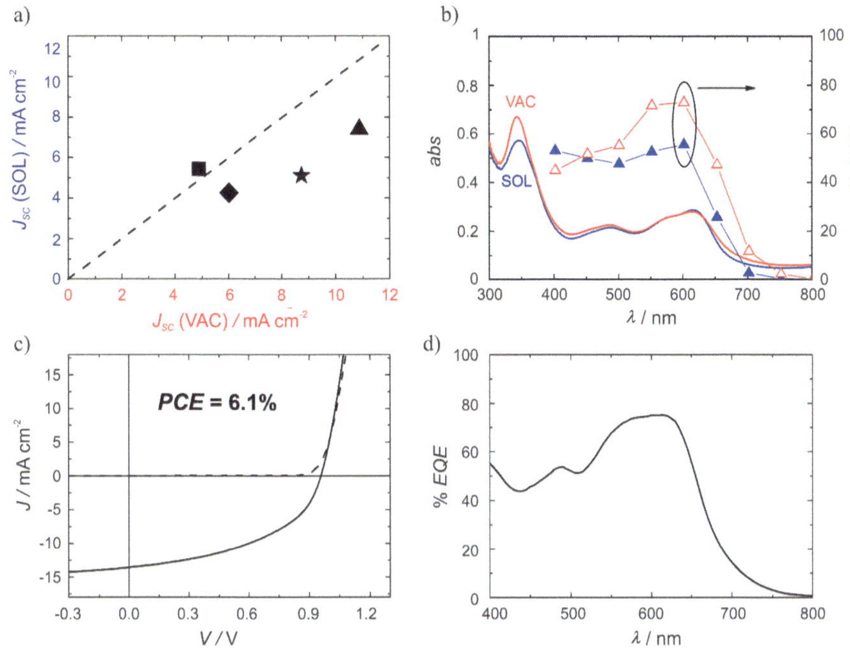

Fig. 7 (**a**) Comparison of the short-circuit current J_{SC} of merocyanine-based bulk heterojunction solar cells fabricated by solution (SOL) or vacuum (VAC) processing for dyes **HB094** (*square*), **HB226** (**19**, *diamond*), **9** (*star*), and **11** (*triangle*) [47]. Adapted with permission from [47]. © 2010 Wiley-VCH Verlag GmbH & Co. KGaA, Weinheim. (**b**) Thin-film absorption spectra (*solid lines*) as well as external quantum efficiency (EQE) values (*symbols*) of the SOL (*filled triangles*) and VAC (*open triangles*) active layers with dye **11**. (**c**) *J–V* response in the *dark* (*dashed*) and under simulated AM1.5 illumination (*solid*) of the optimized vacuum-processed solar cell based on dye **11** in the setup ITO/MoO_3/dye:C_{60}(55 nm)/BPhen/Ag. (**d**) EQE of the optimized device based on dye **11** under short-circuit conditions [48]. Adapted with permission from [48]. © 2011 Wiley-VCH Verlag GmbH & Co. KGaA, Weinheim

was highest at 55 % C_{60} content, reaching EQE values up to 75 %. For the best device, a PCE of 6.1 % could be measured [48] (Fig. 7c, d).

4.2 *Tandem Devices and Ternary Blends*

The main idea of a tandem solar cell is to combine different semiconductors with complementary absorption bands to maximize the light harvesting. In Fig. 8 a typical structure for a tandem solar cell is shown. To avoid thermalization losses, the material with the larger optical bandgap should be placed in the first subcell. As a consequence, only light of a longer wavelength will reach the second subcell, where the material with a narrower bandgap is placed. A recombination layer between both subcells should ensure that charge carriers meet at the same energy level so that

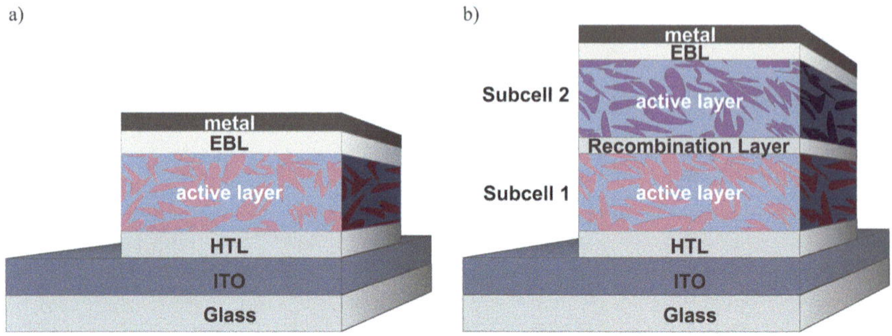

Fig. 8 Schematic representation of (**a**) a single bulk heterojunction solar cell and (**b**) a tandem solar cell structure composed of two subcells. In the first subcell the material with a larger bandgap absorbs the high-energy light. Subsequently, the unabsorbed long-wavelength light reaches the second subcell, where the narrow-bandgap material is placed. A recombination contact between the subcells ensures efficient recombination of the charge carriers. *EBL*, exciton blocking layer; *HTL*, hole-transporting layer

recombination can take place without energy losses. The generated photocurrent of a tandem solar cell is limited by the smallest photocurrent of the subcells, while the V_{OC} of the tandem cell is the sum of the V_{OC} values of the two subcells.

In 2013, we reported [49] a tandem solar cell bearing the well-established ZnPc chromophore **18** as long-wavelength absorber combined with merocyanine **19**, which absorbs in the shorter 400–600 nm spectral region. Optimization of the single cells showed efficiencies of up to 2.6 % and 3.1 % for **18** and **19**, respectively. The tandem solar cell was next fabricated taking the parameters of the best single cells and Al/MoO_3 as the recombination layer (Fig. 9a). The V_{OC} of the tandem solar cell (1.60 V) almost equals the sum of the single cells' V_{OC} (0.73 V and 0.93 V for **18** and **19**, respectively), showing a nearly loss-free contact between both subcells (Fig. 9b). Furthermore, the J_{SC} of 5.7 mA/cm^2 measured for the tandem solar cell is in excellent agreement with the results from optical simulations. In this case, the cell based on ZnPc **18** is the limiting cell with a J_{SC} of 6.8 mA/cm^2 in the corresponding single-junction setup. The FF of 50 % remains between the values of both subcells. Overall, the tandem cell showed an improved performance, with a PCE of 4.5 %.

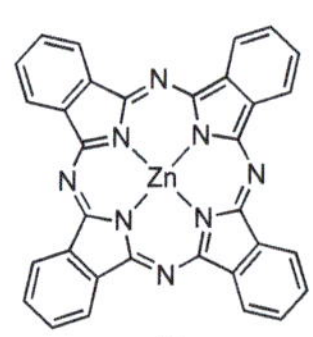

18

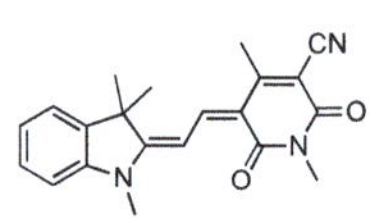

19

Tandem solar cells may also be fabricated from two subcells bearing the same absorber material. This was performed with dye **9** in a combined structure of

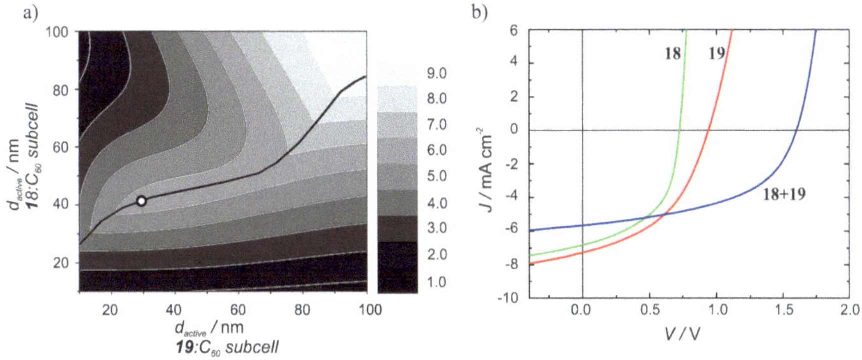

Fig. 9 (**a**) Two-dimensional contour plot of the total calculated short-circuit current density $J_{SC,cal}$ of the tandem solar cell as a function of the active layer thicknesses in the subcells assuming an internal quantum efficiency of 100 %. The *black line marks* those subcell thickness combinations with identical $J_{SC,cal}$. The highlighted data point (*white*) indicates the subcell thickness combination used in the experimental tandem device. (**b**) Experimental J–V response under illumination of the optimized single-junction devices based on **18** and **19** as well as that of the fabricated tandem solar cell (**18+19**) [49]. Adapted with permission from [49]. © 2013 Elsevier B.V.

bulk and planar heterojunction (PHJ), where the active layer of each subcell was sandwiched between donor- and acceptor-only layers [50]. The advantages of such a cell structure have been discussed in the literature [51]. Indeed, because of the extra donor and acceptor layers, direct contact between the active layer and the electrodes is avoided, thus improving the charge collection at the electrodes and leading to a higher J_{SC}. A slight decrease in the FF was observed in comparison to pure BHJ devices. Overall, a final PCE of 4.7 %, higher than the single-cell performance (4.3 %), could be obtained as a result of a very high V_{OC} value of 2.1 V.

An alternative to the tandem cell structure is the ternary concept, where the complementary absorbers are mixed together in the same layer. Dyes **20** and **21** were mixed in different ratios in combination with C_{60} as acceptor material in vacuum-deposited BHJ [52]. The dependence of the V_{OC} on the dye ratio was found to be almost linear and increased with a higher ratio of dye **20**, that is, the dye with the lower HOMO level (Fig. 10a). This observation, accompanied by the fact that the best performance was obtained for a 50 % mixture of the dyes, suggested the presence of three different phases—**20**-rich, **21**-rich, and C_{60}-rich phases—thus providing parallel BHJ devices within the blend (Fig. 10b). The phase separation could be attributed to the very different values of the molecular dipole moments, which are 4.0 D and 13.0 D for **20** and **21**, respectively. Because the Coulomb attraction depends on the products of the molecular dipole moments, the formation of homoaggregates is preferred, thus leading to separated phases in the bulk through a self-sorting process [53].

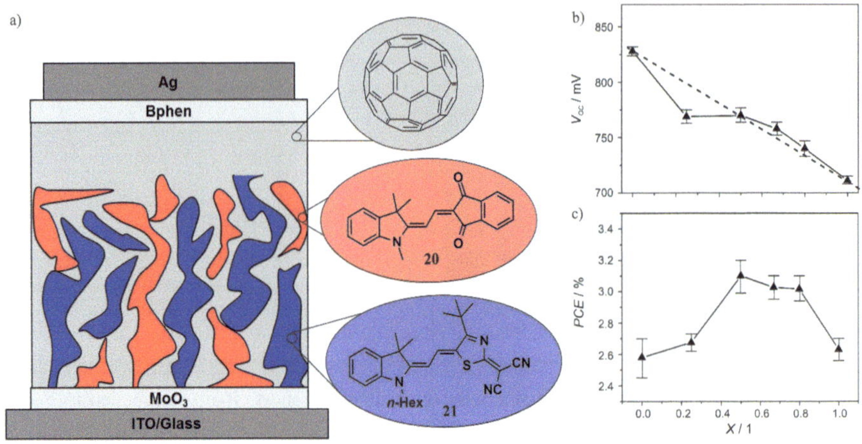

Fig. 10 (**a**) Schematic representation of the ternary blend containing phase-separated domains of **20** and **21** as well as C_{60}. The molecular structures of **20** and **21** are depicted. The (**b**) V_{OC} and (**c**) the power conversion efficiency depend on the dye **21** content in the ternary BHJ blend. *X* denotes the weight fraction of dye **21** with respect to dye **20**. The C_{60} content was 60 wt% and the active layer thickness 39 nm in all cases. The *dashed line* in (**b**) is a linear fit to the data [52]. Adapted with permission from [52]. © 2010 Wiley-VCH Verlag GmbH & Co. KGaA, Weinheim

4.3 PHJ Active Layers

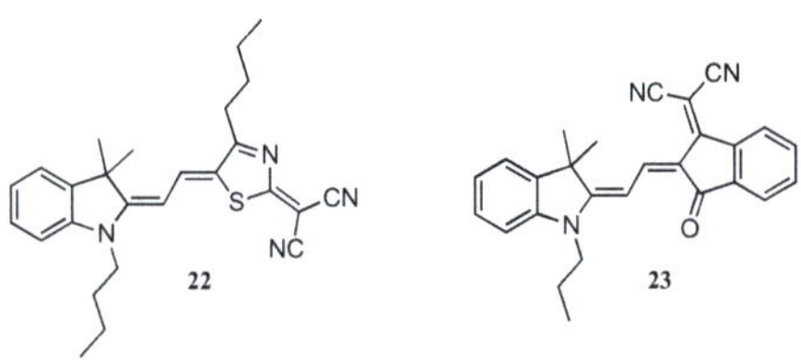

Until now much effort has been devoted to the development of new semiconductor materials with improved absorption and conducting properties. Nevertheless, cell structure optimization is equally important. Studies about solar cell morphology have demonstrated that dye orientation in the bulk plays an important role in solar cell efficiency. The use of transition metal oxide buffer layers between the ITO electrode and the dye film, for example, modifies its crystallinity. Indeed, using vanadium oxide (V_2O_5)–modified ITO substrates led to highly crystalline films of molecule **22**, which was not the case for other buffer layers such as tungsten trioxide (WO_3) or MoO_3. Application of these buffer layers in PHJ solar cell fabrication improved PCEs because of an increase in V_{OC} from 0.69 to 0.74 V while maintaining a very high FF of 68 % [54]. Furthermore, deposition of the chromophores on a heated substrate or postannealing of the latter above the glass transition temperature turns to be a suitable tool to reorient the molecules toward faster exciton dissociation and to increase the crystallinity through a centrosymmetric arrangement of the dye pairs, reducing the dipolar disorder. In the case of PHJ solar cell based on **23**, whose glass transition temperature is 77 °C, a twofold increase in FF was observed after annealing the cell at 80 °C for 5 min [55]. This was attributed to a 45° tilting of the molecules upon annealing, while in as-deposited films they tend to stand with their long axis perpendicular to the surface. Such arrangements have also been observed for highly crystalline merocyanine layers in OTFT studies, where the transition from amorphous to crystalline films improved the mobility by four to five orders of magnitude [56–58].

5 Other D–A Dyes

Several other research groups have contributed to the development of OSCs based on D–A dyes as well, as we will now briefly discuss. Triphenylamine (TPA) donor groups have been widely applied to linear and star-shaped systems in organic semiconductors. In particular, Roncali and coworkers [37] investigated combinations with various acceptor groups that led to symmetrical D–A–D chromophores. Unsymmetrical chromophores bearing triaryl groups have also recently been studied by several groups. In 2008, a series of D–A compounds consisting of oligothiophene derivatives end-capped with triarylamine groups on one side and (di-, tri-) cyanovinyl electron-withdrawing units on the other side were described by M.S. Wong et al. [59]. The conjugation length of the central oligothiophene core was varied and its influence on solar cell performance studied. For this purpose, PHJ solar cells were prepared by vacuum deposition of C_{60} on top of the donor material. It was found that increasing the conjugation length from two to four central thiophene units increased the absorption coefficient while decreasing the optical bandgap. While the V_{OC} remained very similar for all molecules, devices based on **24c** showed a much higher FF of 48 %, thus reaching the maximal PCE of this series of 2.7 % after thermal annealing. After their early work on symmetric TPA dyes [37], in 2011 Roncali's group [60] published a series of triphenylamine-based chromophores with different acceptor end-capped arms. Interestingly, the dye containing two dicyanovinyl groups showed an almost 100 nm redshift and a broader absorption than that containing only one acceptor group although the extinction coefficient was reduced from 104,700 to 57,500 $M^{-1}cm^{-1}$. The best performance was obtained for compound **25** blended with $PC_{60}BM$, where a very high V_{OC} of 1.07 V and a PCE of 2.0 % were obtained.

Later, a series of D–π–A molecules bearing electron-rich bis-fluorenylaniline combined with different bridging units such as thiophene or indenothiophene and end-capped with several acceptor moieties was reported by Ko and coworkers [61]. The dicyanovinyl-containing chromophores had the greatest intramolecular charge transfer, which appeared as a redshifted absorption as well as a higher extinction coefficient ($\varepsilon = 40{,}200\ M^{-1}cm^{-1}$). Consequently, a much higher J_{SC} was found for

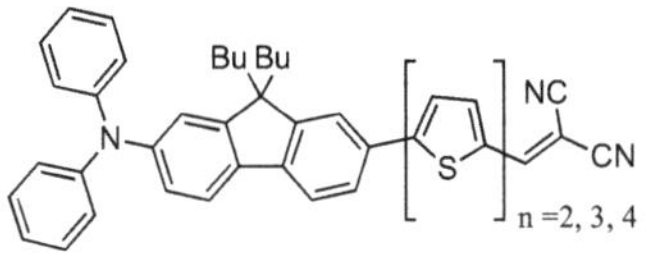

24a-c (n = 2, 3, 4)

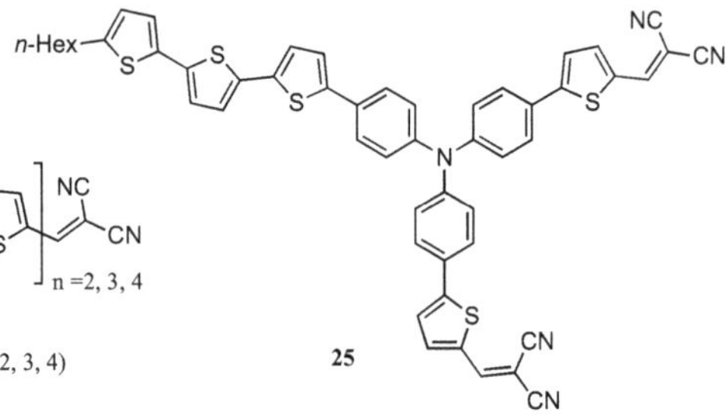

25

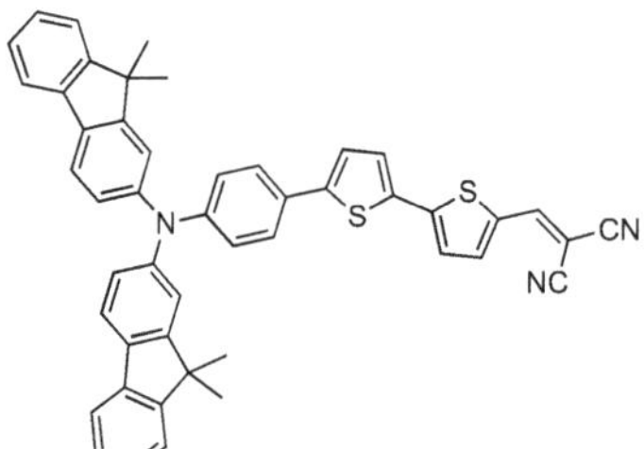

26

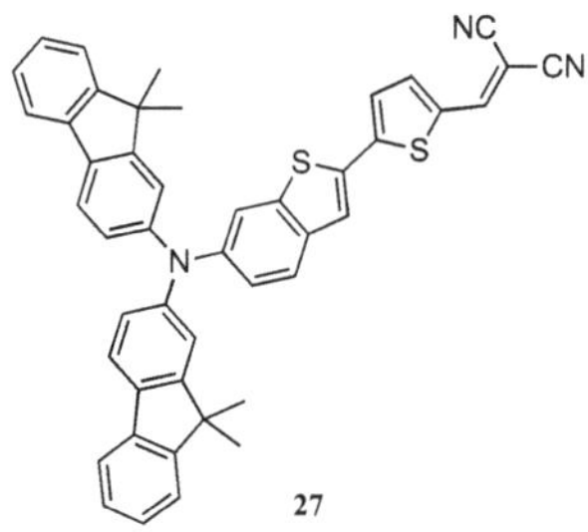

27

26 than for its homologues, reaching an efficiency of 3.7 % in solution-processed BHJ solar cells blended with $PC_{70}BM$ upon the addition of 1-chloronaphthalene [62]. Continuing with the same donor moiety, a more rigid benzothiophene bridging unit was chosen (**27**), which slightly increased light harvesting and improved the PCE up to 4.0 % in the blend with $PC_{70}BM$ using a titanium oxide functional layer as the optical spacer [63]. Additional studies about the effect of bridging units have demonstrated that using a naphthalene diimide moiety between the donor and acceptor subunits provides a redshift of the absorption up to 630 nm and a twofold improvement in the PCE (up to 3.0 %) compared to the reference compound [64].

K.-T. Wong's group [65] entered the field of BHJ solar cells in 2011 and reported the highest PCE values to date for D–A dye-based OSCs. They published a series of structures based on a triarylamine electron-donating group and acceptor groups such as benzothiadiazole [65], pyrimidine [51], and dicyanovinylene [66]. All compounds could be vacuum deposited and were tested in planar mixed heterojunction (PMHJ) solar cells using C_{70} as the acceptor material. Among the reported compounds, the pyrimidine-containing dye **28** [51] obtained an impressive PCE of 6.4 %, with a very high EQE of almost 80 % at 400 nm and over 50 % from 300–700 nm, together with the highest J_{SC} of 16.35 mA/cm^2 among this series. Furthermore, PCEs up to 6.8 % have been reported for dye **29** [67]. Although the higher J_{SC} was observed for devices based on the homologue compound containing a thiophene bridging unit [65], **29** allows for a better balance between the J_{SC} and V_{OC}, leading to the best FF of this series, 54 %. Furthermore, the same compound demonstrated its suitability for application in simple BHJ solar cells. After layer thickness optimization, PCEs of about 8 % were reported, which is among the best efficiencies reported to date for vacuum-processed devices based on small molecules [68]. Recently, compound **29** was combined with C_{70} in solution-processed BHJ solar cells. Although the FF was lower than for vacuum-deposited devices, the V_{OC} and J_{SC} remained nearly the same, providing devices with PCEs up to 5.9 % after solvent vapor annealing [69].

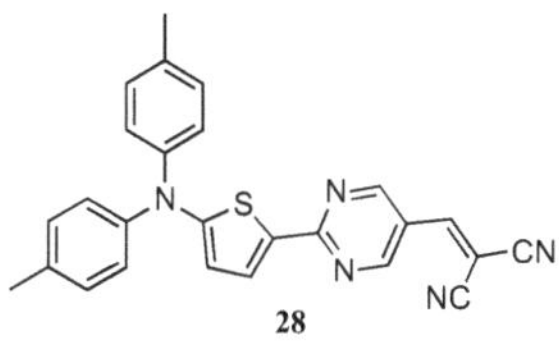

28

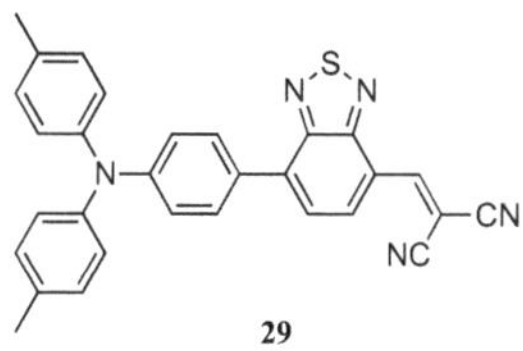

29

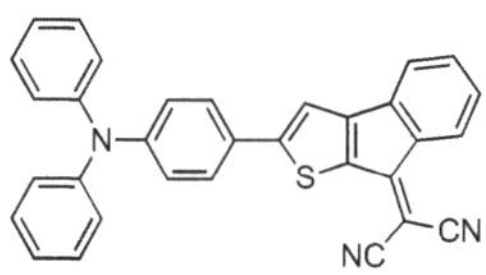

30

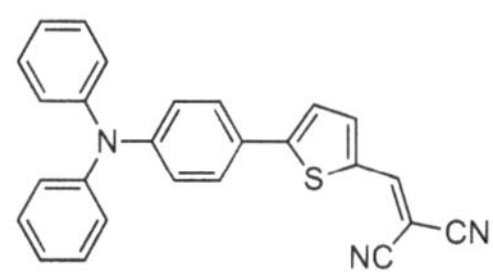

31

Reducing molecular size to the basic donor and acceptor units is a recent trend that has gained increasing interest [70]. Playing with rigidity has shown that covalent cyclization of the acceptor unit represents an efficient way of modifying intramolecular charge transfer, thus shifting the overall broader absorption bathochromically and improving the FF. Blanchard-Desce and Roncali [71] obtained efficiencies up to 3.0 % for compound **30** in thermally annealed PHJ devices with C_{60} as acceptor material. The nonbridged derivative **31** also showed promising results and was therefore optimized for vacuum-deposited PMHJ solar cells, where a 4,4′-bis[*N*-(1-naphtyl)-*N*-phenylamino]biphenyl buffer layer was added in order to prevent an otherwise S-shaped *J*–*V* curve. Finally, the active layer was tuned to show a composition gradient that favored exciton dissociation and charge transport, yielding an overall efficiency of 4 % [72].

In addition, Wang et al. [73] tested in OPVs some chromophores originally developed for dye-sensitized solar cells. They first used three indoline-based carboxy dyes as donor materials in solution-processed PHJ. For the best performing dye an esterification was performed in order to render the molecule thermally stable thus enabling its application in vacuum-deposited BHJ. The optimized cell based on **32** showed PCEs up to 5.1 %. Palomares' group [74] has also recently published a series of dipolar molecules bearing an indoline donor unit and a dicyanovinyl acceptor linked by different bridging groups, including thiophene, bithiophene, cyclopentadithiophene, and benzothiadiazole moieties. They built BHJ solar cells by spin coating the molecules with $PC_{70}BM$ as the acceptor material; solvent annealing was performed before deposition of the cathode. Devices based on dye **33** showed the slowest recombination dynamics, thus providing a high V_{OC} of 0.91 V and the best performance among this series, with a PCE of 3.4 %.

A study reported by Yin [75] introduced a D–A dye **35** with an embedded diketopyrrolopyrrole unit which displayed a PCE of 5.9 % after blending with $PC_{70}BM$. The success of such kind of structures has been emphasized by Wong et al. [76]. Recently, they studied six molecules in vacuum-deposited PMHJs where the donor unit diarylaminothiophene, present in dye **28**, was replaced by a coplanar dithienopyrrol donor moiety and combined with several acceptor subunits. Chromophore **34** containing a pyrimidine–dicyanovinylene acceptor showed the best performance, with a PCE of 5.6 % and a high FF of 52 %

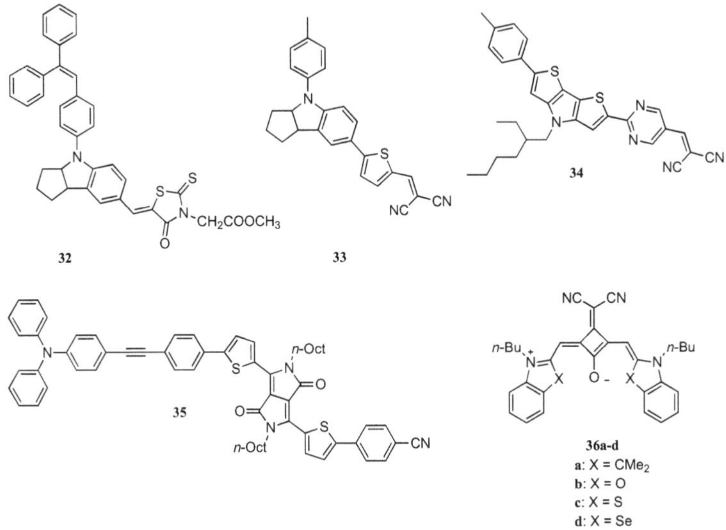

Our groups investigated dipolar cisoid squaraine chromophores as NIR absorbing donor materials for solution-processed BHJ solar cells. The synthesized dyes featured identical donor moieties and a central squaric acid ring bearing a dicyanovinyl substituent, which generated a cisoid arrangement of the two donor units and concomitantly a dipole moment on the molecule in the range of 4.5–6.4 D. The synthesized molecules displayed redshifted absorption bands at about 700 nm in solution as well as low-lying FMO levels. Solar cells fabricated with **36c**:$PC_{60}BM$ (60:40) featured a high J_{SC} of 12.6 mA/cm^2 but a PCE of only 1.79 % after thermal annealing. Interestingly, the tight packing of the molecules in the crystal reveals strong intermolecular interactions, which translate into a redshift and broadening of the optical absorption in the solid state [77]. It is noteworthy that for the selenium derivative **36d,** highly ordered thin films were obtained that showed charge carrier mobilities > 1 cm^2/Vs in organic thin-film transistors [78].

6 Conclusion

Donor–acceptor dyes and in particular merocyanine dyes have been shown to be useful chromophores for photovoltaic applications. Their straightforward synthesis allows us to combine an infinite number of different electron-rich and electron-poor moieties, thus generating an enormous library of compounds whose optical properties can be easily tuned. Indeed, dyes absorbing from the UV region up to the NIR region have been synthesized in simple one- or two-step procedures. Furthermore, merocyanines display very high extinction coefficients, thus featuring good absorption of light even in very thin films with low dye content. For many years, the dipole moments that characterize this kind of push-pull chromophores have been the reason not to use them as organic semiconductors, arguing that they should act as traps for charge carriers. However, our studies have demonstrated that these dyes self-assemble in dimer units with no overall dipole moment, reducing the dipolar disorder in particular in crystalline materials as well as in amorphous materials. Their suitability as semiconductors has also recently been demonstrated through application in transistor devices, where hole mobilities up to 0.18 cm^2/Vs could be achieved [56–58].

Another advantage of merocyanine dyes is their relatively small size, which allows for both SOL and VAC. SOL has been shown to be an inexpensive and useful tool to test the suitability of merocyanine dyes as donor materials for OSCs. Our best solution-processed devices yielded a PCE up to 4.5 % [36]. Furthermore, the outstanding molecules could later be vacuum deposited, which led to even better performances because of the higher purity of the layers and the better control of the blend morphology. Indeed, efficiencies up to 6.1 % were measured for dye **11** [48].

Unfortunately, merocyanine dyes have some drawbacks. Although they present very high absorption in solution, this is not always reflected in the solid state because of the disfavored ordering of the molecules, which tend to stand perpendicular to the substrate [55–58, 79]. As a consequence, relatively low J_{SC} values are obtained for the fabricated solar cells. Furthermore, most merocyanines show low charge carrier mobility compared to the acceptor material. Also, a field-dependent generation of charge carriers might exist because of a nonoptimized active layer morphology/crystallinity in merocyanine-based devices [80]. This may be a reason for the low FFs that are typically found for merocyanine-based BHJ solar cells. Nevertheless, the overall efficiency of a solar cell is a device property influenced by several factors. Therefore, further research may focus less on developing new donor materials but more on optimizing the active layer morphology, cell structure, and interface architecture.

Acknowledgment This work was financially supported by the DFG Priority Program SPP1355 (projects ME 1246/19 and WU 317/10). We thank all our co-workers who were involved in the research reviewed here as well as our collaboration partners from academia and industry.

References

1. Tang CW (1986) Appl Phys Lett 48:183
2. Chamberlain GA (1983) Solar Cells 8:47
3. Wöhrle D, Meissner D (1991) Adv Mater 3:129
4. Gunes S, Neugebauer H, Sariciftci NS (2007) Chem Rev 107:1324

5. Zhou J, Zuo Y, Wan X, Long G, Zhang Q, Ni W, Liu Y, Li Z, He G, Li C, Kan B, Li M, Chen Y (2013) J Am Chem Soc 135:8484
6. Zhou J, Wan X, Liu Y, Zuo Y, Li Z, He G, Long G, Ni W, Li C, Su X, Chen Y (2012) J Am Chem Soc 134:16345
7. Liu Y, Chen CC, Hong Z, Gao J, Yang YM, Zhou H, Dou L, Li G, Yang Y (2013) Sci Rep 3:3356
8. van der Poll TS, Love JA, Nguyen T-Q, Bazan GC (2012) Adv Mater 24:3646
9. Chao YH, Jheng JF, Wu JS, Wu KY, Peng HH, Tsai MC, Wang CL, Hsiao YN, Wang CL, Lin CY, Hsu CS (2014) Adv Mater 26:5205
10. He ZC, Zhong CM, Su SJ, Xu M, Wu HB, Cao Y (2012) Nat Photonics 6:591
11. Kyaw AK, Wang DH, Wynands D, Zhang J, Nguyen TQ, Bazan GC, Heeger AJ (2013) Nano Lett 13:3796
12. Lin Y, Zhan X (2014) Mater Horiz 1:470
13. Anthony JE (2011) Chem Mater 23:583
14. Mishra A, Bäuerle P (2012) Angew Chem Int Ed 51:2020
15. Lin Y, Li Y, Zhan X (2012) Chem Soc Rev 41:4245
16. Li YW, Guo Q, Li ZF, Pei JN, Tian WJ (2010) Energy Environ Sci 3:1427
17. Dieckmann A, Bässler H, Borsenberger PM (1993) J Chem Phys 99:8136
18. Würthner F, Yao S (2000) Angew Chem Int Ed 39:1978
19. Würthner F, Yao S, Debaerdemaeker T, Wortmann R (2002) J Am Chem Soc 124:9431
20. Würthner F, Yao S, Schilling J, Wortmann R, Redi-Abshiro M, Mecher E, Gallego-Gomez F, Meerholz K (2001) J Am Chem Soc 123:2810
21. Marder SR, Kippelen B, Jen AKY, Peyghambarian N (1997) Nature 388:845
22. Würthner F, Wortmann R, Meerholz K (2002) ChemPhysChem 3:17
23. Würthner F, Archetti G, Schmidt R, Kuball H-G (2008) Angew Chem Int Ed 47:4529
24. Marder SR, Perry JW (1994) Science 263:1706
25. Würthner F, Thalacker C, Matschiner R, Lukaszuk K, Wortmann R (1998) Chem Commun 1739
26. Beverina L, Pagani GA (2014) Acc Chem Res 47:319
27. Bässler H (1993) Phys Status Solidi 175:15
28. Morel DL, Ghosh AK, Feng T, Stogryn EL, Purwin PE, Shaw RF, Fishman C (1978) Appl Phys Lett 32:495
29. Ghosh AK, Feng T (1978) J Appl Phys 49:5982
30. Chamberlain GA, Cooney PJ, Dennison S (1981) Nature 289:45
31. Chamberlain GA, Malpas RE (1980) Faraday Discuss Chem Soc 70:299
32. Moriizumi T, Kudo K (1981) Appl Phys Lett 38:85
33. Skotheim T, Yang JM, Otvos J, Klein MP (1982) J Chem Phys 77:6151
34. Kudo K, Yamashina M, Moriizumi T (1984) Jpn J Appl Phys 23:130
35. Shirinian V, Shimkin A (2008) In: Strekowski L (ed), vol 14. Springer, Heidelberg, p 75
36. Bürckstümmer H, Tulyakova EV, Deppisch M, Lenze MR, Kronenberg NM, Gsänger M, Stolte M, Meerholz K, Würthner F (2011) Angew Chem Int Ed 50:11628
37. Roncali J (2009) Acc Chem Res 42:1719
38. Kronenberg NM, Deppisch M, Würthner F, Lademann HW, Deing K, Meerholz K (2008) Chem Commun 6489
39. Kronenberg NM, Bürckstümmer H, Deppisch M, Würthner F, Meerholz K (2011) J Photonics Energy 1:011101
40. Bürckstümmer H, Kronenberg NM, Gsänger M, Stolte M, Meerholz K, Würthner F (2010) J Mater Chem 20:240
41. Bürckstümmer H, Kronenberg NM, Meerholz K, Würthner F (2010) Org Lett 12:3666
42. Blanchard-Desce M, Alain V, Bedworth PV, Marder SR, Fort A, Runser C, Barzoukas M, Lebus S, Wortmann R (1997) Chem Eur J 3:1091
43. Alain V, Thouin L, Blanchard-Desce M, Gubler U, Bosshard C, Günter P, Muller J, Fort A, Barzoukas M (1999) Adv Mater 11:1210

44. Zitzler-Kunkel A, Lenze MR, Kronenberg NM, Krause A-M, Stolte M, Meerholz K, Würthner F (2014) Chem Mater 26:4856
45. Kasha M, Rawls HR, El-Bayoumi MA (1965) Pure Appl Chem 11:371
46. Zitzler-Kunkel A, Lenze MR, Meerholz K, Würthner F (2013) Chem Sci 4:2071
47. Kronenberg NM, Steinmann V, Bürckstümmer H, Hwang J, Hertel D, Würthner F, Meerholz K (2010) Adv Mater 22:4193
48. Steinmann V, Kronenberg NM, Lenze MR, Graf SM, Hertel D, Meerholz K, Bürckstümmer H, Tulyakova EV, Würthner F (2011) Adv Energy Mater 1:888
49. Steinmann V, Umbach TE, Schadel M, Krumrain J, Lenze MR, Bürckstümmer H, Würthner F, Meerholz K (2013) Org Electron 14:2029
50. Steinmann V, Kronenberg NM, Lenze MR, Graf SM, Hertel D, Bürckstümmer H, Würthner F, Meerholz K (2011) Appl Phys Lett 99:193306
51. Chiu SW, Lin LY, Lin HW, Chen YH, Huang ZY, Lin YT, Lin F, Liu YH, Wong KT (2012) Chem Commun 48:1857
52. Ojala A, Bürckstümmer H, Stolte M, Sens R, Reichelt H, Erk P, Hwang J, Hertel D, Meerholz K, Würthner F (2011) Adv Mater 23:5398
53. Safont-Sempere MM, Fernández G, Würthner F (2011) Chem Rev 111:5784
54. Ojala A, Bürckstümmer H, Hwang J, Graf K, von Vacano B, Meerholz K, Erk P, Würthner F (2012) J Mater Chem 22:4473
55. Ojala A, Petersen A, Fuchs A, Lovrincic R, Pölking C, Trollmann J, Hwang J, Lennartz C, Reichelt H, Höffken HW, Pucci A, Erk P, Kirchartz T, Würthner F (2012) Adv Funct Mater 22:86
56. Liess A, Huang L, Arjona-Esteban A, Lv A, Gsänger M, Stepanenko V, Stolte M, Würthner F (2015) Adv Funct Mater 25:44
57. Huang L, Stolte M, Bürckstümmer H, Würthner F (2012) Adv Mater 24:5750
58. Lv A, Stolte M, Würthner F (2015) Angew Chem Int Ed 54:10512
59. Xia PF, Feng XJ, Lu J, Tsang S-W, Movileanu R, Tao Y, Wong MS (2008) Adv Mater 20:4810
60. Ripaud E, Rousseau T, Leriche P, Roncali J (2011) Adv Energy Mater 1:540
61. Lee JK, Kim J, Choi H, Lim K, Song K, Ko J (2014) Tetrahedron 70:6235
62. Ko HM, Choi H, Paek S, Kim K, Song K, Lee JK, Ko J (2011) J Mater Chem 21:7248
63. Kim J, Ko HM, Cho N, Paek S, Lee JK, Ko J (2012) RSC Adv 2:2692
64. Bobe SR, Gupta A, Rananaware A, Bilic A, Bhosale SV, Bhosale SV (2015) RSC Adv 5:4411
65. Lin LY, Chen YH, Huang ZY, Lin HW, Chou SH, Lin F, Chen CW, Liu YH, Wong KT (2011) J Am Chem Soc 133:15822
66. Lin HW, Lin LY, Chen YH, Chen CW, Lin YT, Chiu SW, Wong KT (2011) Chem Commun 47:7872
67. Chen YH, Lin LY, Lu CW, Lin F, Huang ZY, Lin HW, Wang PH, Liu YH, Wong KT, Wen J, Miller DJ, Darling SB (2012) J Am Chem Soc 134:13616
68. Zou Y, Holst J, Zhang Y, Holmes RJ (2014) J Mater Chem A 2:12397
69. Lin H-W, Chang J-H, Huang W-C, Lin Y-T, Lin L-Y, Lin F, Wong K-T, Wang H-F, Ho R-M, Meng H-F (2014) J Mater Chem A 2:3709
70. Jeux V, Demeter D, Leriche P, Roncali J (2013) RSC Adv 3:5811
71. Leliege A, Regent C-HL, Allain M, Blanchard P, Roncali J (2012) Chem Commun 48:8907
72. Choi JW, Kim C-H, Pison J, Oyedele A, Tondelier D, Leliege A, Kirchner E, Blanchard P, Roncali J, Geffroy B (2014) RSC Adv 4:5236
73. Wang Z, Wang X-F, Yokoyama D, Sasabe H, Kido J, Liu Z, Tian W, Kitao O, Ikeuchi T, Sasaki S-i (2014) J Phys Chem C 118:14785
74. Montcada NF, Cabau L, Kumar CV, Cambarau W, Palomares E (2015) Org Electron 20:15
75. Gao H, Li Y, Wang L, Ji C, Wang Y, Tian W, Yang X, Yin L (2014) Chem Commun 50:10251
76. Lu H-I, Lu C-W, Lee Y-C, Lin H-W, Lin L-Y, Lin F, Chang J-H, Wu C-I, Wong K-T (2014) Chem Mater 26:4361
77. Mayerhöffer U, Deing K, Gruß K, Braunschweig H, Meerholz K, Würthner F (2009) Angew Chem Int Ed 48:8776

78. Gsänger M, Kirchner E, Stolte M, Burschka C, Stepanenko V, Pflaum J, Würthner F (2014) J Am Chem Soc 136:2351
79. Winkler S, Frisch J, Amsalem P, Krause S, Timpel M, Stolte M, Würthner F, Koch N (2014) J Phys Chem C 118:11731
80. Proctor CM, Albrecht S, Kuik M, Neher D, Nguyen T-Q (2014) Adv Energy Mater 4:1400230

Controlled Morphologies by Molecular Design and Nano-Imprint Lithography

Thomas Pfadler, Claudia M. Palumbiny, Wojciech Pisula, Holger C. Hesse, Xinliang Feng, Klaus Müllen, Peter Müller-Buschbaum, and Lukas Schmidt-Mende

Contents

T. Pfadler (✉) • L. Schmidt-Mende
Department of Physics, University of Konstanz, M680, 78457 Konstanz, Germany
e-mail: Thomas.Pfadler@uni-konstanz.de; lukas.schmidt-mende@uni-konstanz.de

C.M. Palumbiny • P. Müller-Buschbaum
Physik-Department, Lehrstuhl für Funktionelle Materialien, Technische Universität München, James-Franck-Straße 1, 85747 Garching, Germany
e-mail: claudia.palumbiny@ph.tum.de; muellerb@ph.tum.de

H.C. Hesse
Physik-Department, Lehrstuhl für Funktionelle Materialien, Technische Universität München, James-Franck-Straße 1, 85747 Garching, Germany

Department of Electrical and Computer Engineering, Lehrstuhl für Elektrische Energiespeichertechnik - EES, Arcisstraße 21, 80333 München, Germany
e-mail: holger.hesse@tum.de

W. Pisula • K. Müllen
Department of Synthetic Chemistry, Max Planck Institute for Polymer Research, Ackermannweg 10, 55128 Mainz, Germany
e-mail: pisula@mpip-mainz.mpg.de; muellen@mpip-mainz.mpg.de

X. Feng
Professur für Makromolekulare Funktionsmaterialien, Technische Universität Dresden, Mommsenstr. 10, 01062 Dresden, Germany
e-mail: xinliang.feng@tu-dresden.de

K. Leo (ed.), *Elementary Processes in Organic Photovoltaics*,
Advances in Polymer Science 272, DOI 10.1007/978-3-319-28338-8_9

Abstract In this review we report on the recent advances concerning molecular design and nano-imprint techniques to improve the morphology of organic solar cells. Advanced scattering techniques allow us to resolve issues pertaining to the alignment and crystallization of the photoactive materials of nanostructured solar cells. Interfacial design at a donor–acceptor heterojunction is one of the key issues for improved device performance.

Keywords Advanced scattering techniques • Discotic molecules • Molecular-assembly • Nanoimprint lithography

1 Introduction

To date the efficiencies of organic solar cells have not been able to compete with those of conventional semiconductor photovoltaics. One major challenge is the control of the interfacial area between the donor (D) and acceptor (A) materials. In this work we report on recent advances in achieving higher control over morphology through the design of materials and nano-imprint lithography. Embossing techniques— also feasible in principle for large-scale roll-to-roll production—can be used to fabricate photovoltaic devices with a controlled and large interface between the donor and acceptor materials and with high charge carrier mobility. Advanced scattering techniques give full details about the alignment and crystallization of the organic materials inside the nanostructures from the molecular to the mesoscopic structural levels. This full control and knowledge about the structure permit a systematic investigation of the device performance's dependence on the thin-film morphology. Ideal heterojunction solar cells have a large interfacial area between the acceptor and donor materials, which is designed in a way that exciton separation is promoted and charge recombination is suppressed.

2 Self-Organization of Discotic Liquid Crystals

Great attention in materials science is given to the spontaneous self-organization of small molecular building blocks to form functional materials. The well-defined supramolecular assemblies are built as a result of intermolecular noncovalent forces such as hydrogen bonds, π stacking, or dipole–dipole interactions [1–3]. Conjugated molecules such as organic semiconductors assemble into complex superstructures, enabling the transport of charge carriers through precisely established pathways. This approach opens the door for organic electronics. Conjugated liquid crystals (LCs) represent an attractive alternative to surpass the disadvantages of traditional organic semiconductors [4–6]. The thermotropic properties of LCs lead to mesophases that significantly differ in molecular order and dynamics to the isotropic melt and the crystal state[7]. Because of the increased molecular dynamics, structural defects such as grain boundaries, which act as charge carrier traps and limit the performance of the semiconductor in the device, can be self-healed in a straightforward way [8]. Thus, under appropriate (post)processing, it is possible to achieve macroscopic defect-free monodomains in the device.

LCs are classified according to their molecular shape, which dictates the type of phase and superstructure. Discotic LCs typically consist of a flat, rigid core substituted by flexible aliphatic chains [9, 10]. In most cases the core is based on a large polycyclic aromatic hydrocarbon unit (PAH) that is built up from a two-dimensional (2D) flat hexagonal carbon network. PAHs are especially interesting for electronic applications because of their extended π-orbital overlap. For this reason, larger aromatic cores enhance the superstructure stability and order favoring the charge carrier migration through the semiconductor. The π-stacking interactions between the PAH cores and local phase separation between the rigid aromatic part and the flexible disordered side chains initiate a self-assembly into columnar stacks that serve as pathways for the charge carriers [11]. The conductivity along the columns is several orders of magnitude higher than that in the perpendicular direction [12, 13], and local charge carrier mobilities up to 1.1 cm^2/Vs have been found for discotic hexa-*peri*-hexabenzocoronene (HBC; Fig. 2, **1a**) [14].

An understanding of the structure–property relationship and control over the phase behavior of discotic PAHs is of fundamental importance in optoelectronic applications. Current research has focused on the design and synthesis of new discotic systems to systematically evaluate the influence of specific structural features on the self-assembly and properties. The thermotropic properties can be adjusted by using alkyl side chains to tune the attractive intermolecular forces within the stack. The alkyl substituents are also necessary to ensure the solubility of the compounds and their solution processing into the device since unsubstituted PAHs are not soluble in common organic solvents [15]. Discotic LCs assemble in two main superstructures. In the nematic phase the disc-shaped molecules organize orientationally without a long-range translational order [16]. In columnar structures the building blocks are packed on top of each other, surrounded by aliphatic side chains [17]. The intermolecular core–core distance in the column amounts for many discotics to 3.5 Å. The close intermolecular distance ensures the mentioned extensive π-orbital overlap and a charge carrier migration [18]. Typically, the LC state is a mesophase appearing between a perfectly regular crystal and an amorphous isotropic melt. Some discotics show solely an LC phase over the entire temperature range, without the existence of the other two main states. Each of the three phases exhibits a specific supramolecular organization of the molecules (Fig. 1). In the crystalline phase the molecules possess the highest order, while the side chains are fully or partially crystallized. To optimize the orbital interactions between molecules, the aromatic cores are tilted toward the columnar axis (Fig. 1) [19]. This tilting corresponds to a lateral shift between cores initialed by the π–σ interactions according to the Hunter and Sanders model [20]. The molecular dynamics significantly increase in the LC phase in which the discs undergo longitudinal and rotary motions around their axis (Fig. 1). In the columnar structures, the molecules pack on top of each other without any tilting, as in the crystalline phase. The columnar stacks are typically arranged in a hexagonal 2D

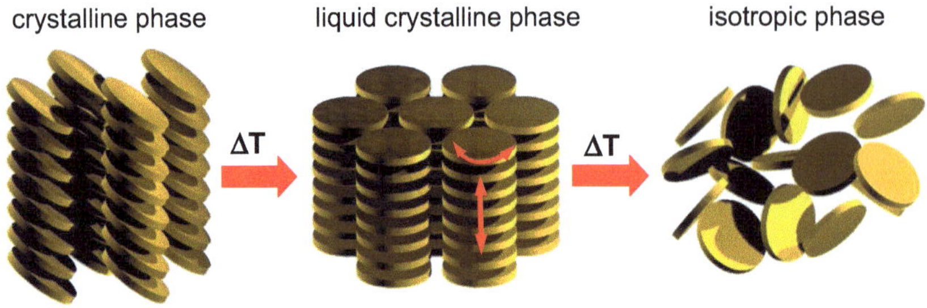

Fig. 1 Schematic illustration of the main phases of discotic columnar liquid crystals: crystalline, liquid crystalline, and isotropic

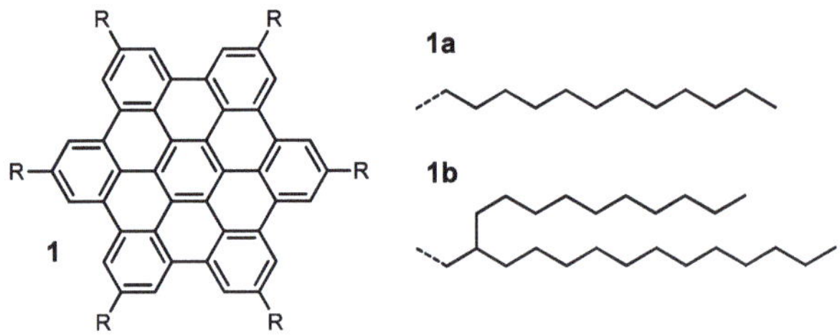

Fig. 2 Hexa-alkyl–substituted hexa-*peri*-hexabenzocoronenes with different side chains

lattice (Fig. 1), whereby the aliphatic side chains between these stacks adapt a fluid-like state at elevated temperatures [21]. As mentioned above, this mesophase can be exploited to self-heal structural defects due to the low viscosity of the material. The self-healing can occur during annealing at higher temperatures, leading to monodomains over large macroscopic areas.

The phase transition temperatures between those states are dictated by the core size and side chain length [22, 23]. As a result of more intensive π-stacking interactions, molecules with a larger aromatic core show higher phase transition temperatures and eventually isotropic phases already above the decomposition temperature. The intermolecular forces and thus the phase transition temperatures can be reduced by using alkyl substituents that possess higher steric demand and increase the overall disorder. The steric character of the flexible side chains can be modified because its geometry also affects the intracolumnar order. A close correlation for different discotic systems has been reported between the number of carbon atoms in the linear alkyl substituent and the phase transition temperature [24–26]. Hexa-dodecyl–substituted HBC (Fig. 2, **1a**) shows phase transition temperatures to the LC phase at 107 °C and to the isotropic melt around 420 °C (Fig. 2) [19]. Because of high π interactions, larger PAHs do not show any clearing point [27]. Engineering of the thermotropic behavior is more effective using branched instead of linear aliphatic substituents. The steric demand is significantly increased for branched side chains because of their more space-filling nature and pronounced rotational freedom [28–30]. The steric demand is especially high when the branching point is located close to the aromatic core, lowering the clearing point of HBC even to only 46 °C (Fig. 2, **1b**) [31]. At the same time, the steric hindrance of the bulky side chains considerably decreases the intermolecular interactions and thus the intracolumnar order [32, 33]. As a consequence of the disordered columnar phase, the charge carrier transport along the columnar pathway is greatly hampered. For this reason, a compromise is necessary among intermolecular interactions, the intracolumnar order, and the accessibility of the mesophase in designing new discotics.

Functional groups capable of further noncovalent interactions significantly extend the molecular toolbox to control the thermotropic properties and self-assembly of discotics [34]. The use of such specific units permits a more defined control over the supramolecular organization [35]. For instance, hydrogen bonds can be located in the side chains as noncovalent interactions to control the molecular assembly [36, 37]. Depending on the position of the function groups, the attractive forces can act intra- and intercolumnarly [38, 39]. As a result, the LC phase transition becomes irreversible and an intracolumnar nontilted packing motif is established. Dipole functionalization is an alternative to hydrogen bonds to tune the molecular dynamics and intracolumnar assembly. One possibility is to introduce dipoles within the aromatic core, inducing an antiparallel stacking of the discs [40, 41]. Dipole moments can also be formed at the columnar periphery either within the alkyl substituent or close to the aromatic core. In the latter case, intercolumnar dipole interactions stabilize the LC phase and improve the molecular order by generating a helical superstructure [42].

The twist angle between stacked discotic PAHs also influences the π-orbital overlap between neighboring molecules. A close relationship between the transfer integral and the intermolecular rotation angle has been found for PAHs of various symmetries and sizes (Fig. 3) [43]. These findings indicate a crucial rotation angle for optimized charge carrier hopping between discs along the columnar pathways. The substituents can be used not only to control the thermotropic behavior of discotics, but also to establish a helical organization with a specific molecular twisting angle. Attaching bulky and rigid substituents at the aromatic core forces a lateral molecular rotation to overcome the peripheral high steric hindrance [44]. The rotation angle strongly depends on the bulkiness of the attachment. Hexa-alkylphenyl–substituted HBC reveals a propeller configuration with phenyl groups rotated out-of-plane toward the aromatic core (Fig. 4) [45]. In the columnar structures the molecules are rotated laterally by 20° to each other, forming a helical assembly with a pitch of 1 nm as indicated by quantum-chemical calculations (Fig. 4). To gain additional control over the helical packing of discotics, the hydrophilic and hydrophobic nature of the flexible substituents can be adjusted [46–48]. Following this approach, the rotation angle of the C_3 symmetric semitriangular-shaped **3a** is changed from 60° for the alkyl-substituted derivative to 40° for an alkyl/triethyleneglycol alternatingly decorated aromatic core **3b** (Fig. 5) [42]. The variation in twist angle is attributed to hydrophilic–hydrophobic interactions between the alkyl and triethyleneglycol substituents in the core periphery. As predicated by calculations based on Marcus theory, a twist angle

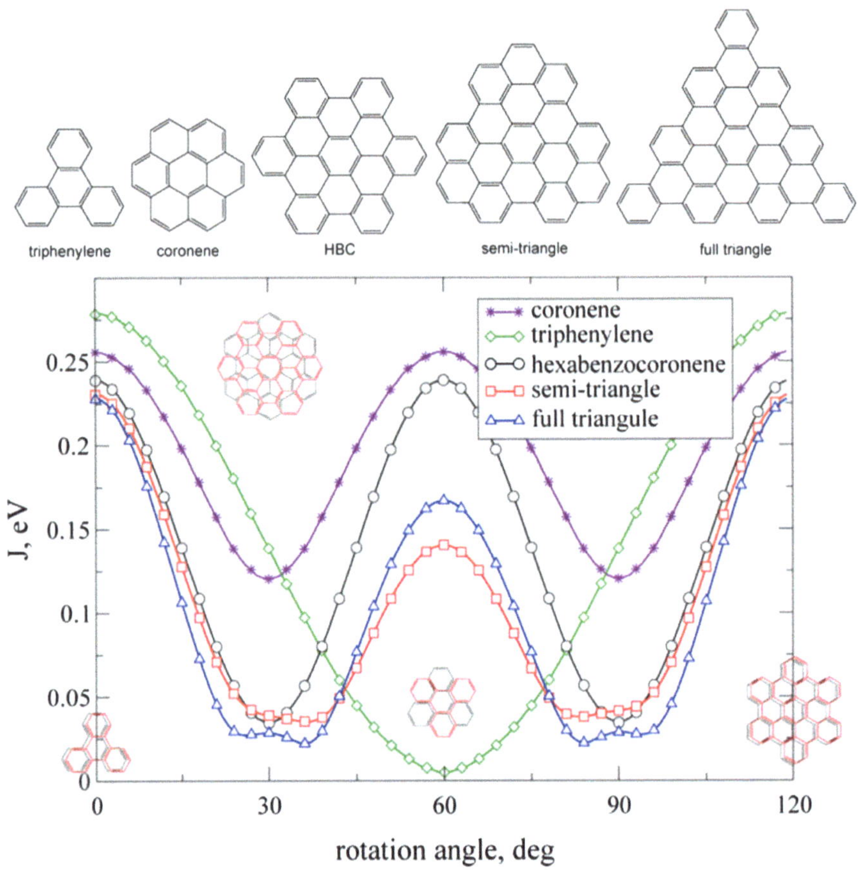

Fig. 3 Absolute value of the transfer integral J as a function of the azimuthal rotation angle for a range of symmetric polyaromatic hydrocarbon cores. Interdisc distance of 0.36 nm fixed for all cases. (Reprinted with permission from [43]. Copyright 2009, Nature Publishing Group)

of 40° leads to a high orbital integration and thus to higher local charge carrier mobility within the stack.

Besides the discussed molecular packing and order, the microscopic organization of discotics on the surface is crucial for the charge carrier transport in electronics. The device performance of discotic semiconductors is influenced by the microscopic order and by the orientation of the molecules toward the charge-transport direction. In general, transport over meso- and macroscopic scales is highly affected by the defect density in the solution-deposited film. Since discotics show two characteristic orientations on surfaces, their arrangement can be adopted accordingly to the type of electronic device. In field-effect transistors the charge carrier transport takes place parallel to the surface. For this reason, an edge-on ordering is necessary in which the molecular core is oriented perpendicular to the surface (Fig. 6a).

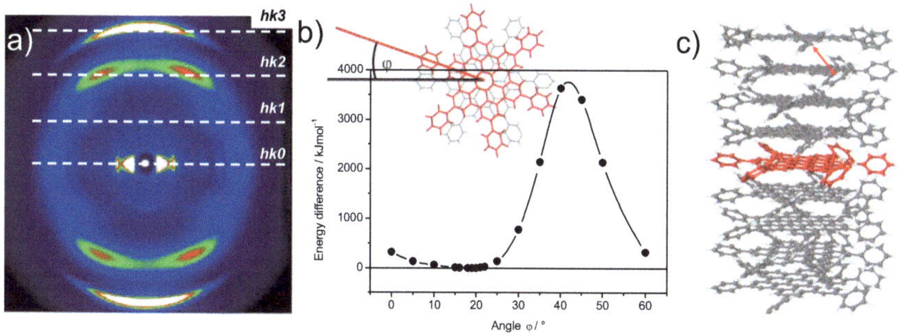

Fig. 4 (**a**) Fiber two-dimensional wide-angle X-ray scattering pattern of 2 with indicated layer lines. (**b**) Definition of the twisting angle in the columnar stack for 2 and dependence of the energy of a hexa-*peri*-hexabenzocoronene dimer on the twisting angle. (**c**) Columnar structure of 2 (shown here for the specific case of a nonamer) with a twisting angle of 20° as suggested by quantum-chemical calculations. (Reproduced with permission from [45]. Copyright 2007, American Chemical Society)

Ideally, the columnar structures as charge pathways are uniaxially oriented in the direction of the applied electric field between the source and the drain. This long-range orientation can be achieved by using suitable processing methods such as Langmuir–Blodgett [49], zone casting [27], solution shearing [50], or dip coating [51]. Contrary to this, upon cooling from the isotropic melt discotics can also arrange spontaneously in a face-on fashion where the aromatic cores lie flat on the surface. After nucleation at the surface and the organization of a mono-/multilayer, a homeotropic phase propagates into the bulk film, with columns aligned perpendicular to the substrate (Fig. 6b). This stack orientation favors a vertical charge transport in, for example, diode-type devices with top and bottom electrodes.

Typically, homeotropic phases occur in films that are confined between two surfaces, which results in large, nonbirefringent monodomains [52], where the surface energy and roughness do not seem to play an essential role in the alignment [53]. Nevertheless, successful alignment has been reported in open films for a few discotics, allowing their implementation in photovoltaics [54]. The open film alignment has even been achieved in bilayers consisting of a donor and an acceptor discotic layer [55]. For phase separation in the bilayer, both compounds show a low degree of miscibility, where the film was spin coated sequentially from two different solvents. Alternatively, to achieve the necessary confinement by two surfaces, it is possible to deposit a sacrificial polymer layer on top of a discotic thin film [56]. After alignment, the additional polymer layer is removed with selective solvent. In terms of the alignment mechanism, it was assumed that heteroatoms located at the core periphery can improve interactions between the molecule and the surface, favoring the homeotropic orientation [32]. This hypothesis has been rebutted as PAH-based discotics without heteroatoms are also orientable, indicating another type of mechanism [57]. It was found that instead a liquid-crystalline phase leads to the homeotropic phase, while compounds without a mesophase could not be oriented in the specific manner [58].

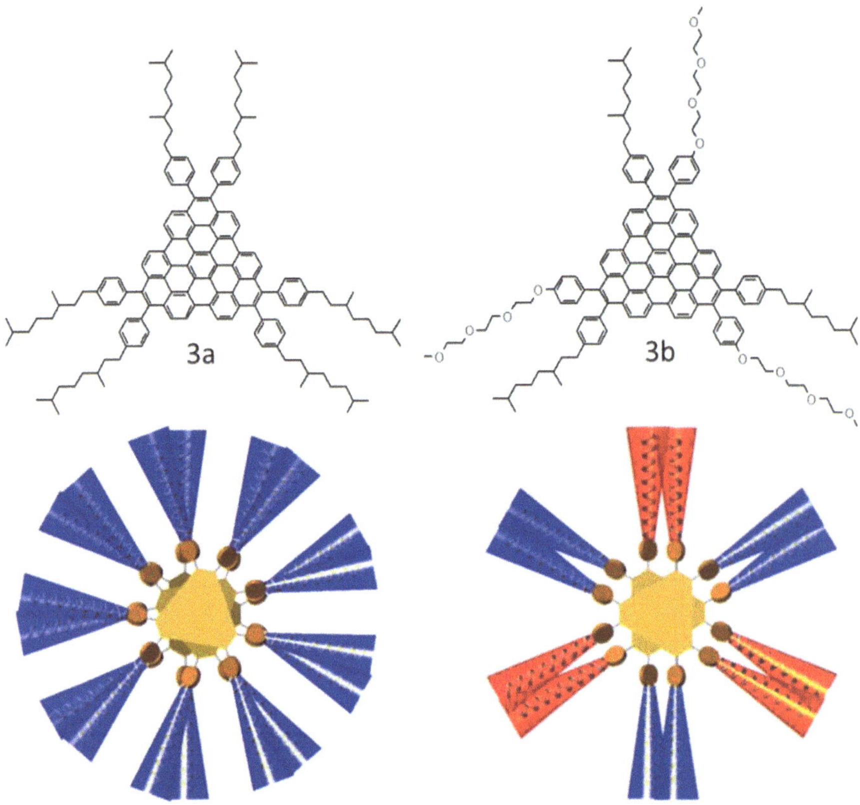

Fig. 5 Semitriangle-shaped polycyclic aromatic hydrocarbons 3 and their corresponding packing in *top view. Blue* and *red* substituents indicate alkyl and triethyleneglycol chains. (Reprinted with permission from [43]. Copyright 2009, Nature Publishing Group)

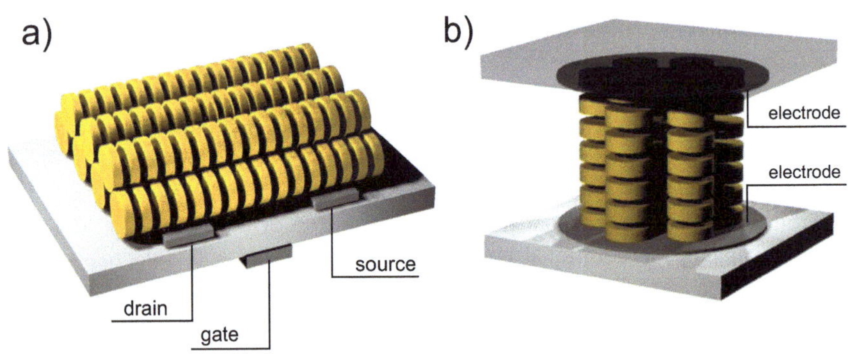

Fig. 6 Schematic illustration of (**a**) edge-on orientation of the molecules desired in field-effect transistors and (**b**) face-on arrangement allowing a homeotropic alignment considered to favor the performance of photovoltaic cells. (Reprinted with permission from [5]. Copyright 2009, Wiley-VCH Verlag GmbH)

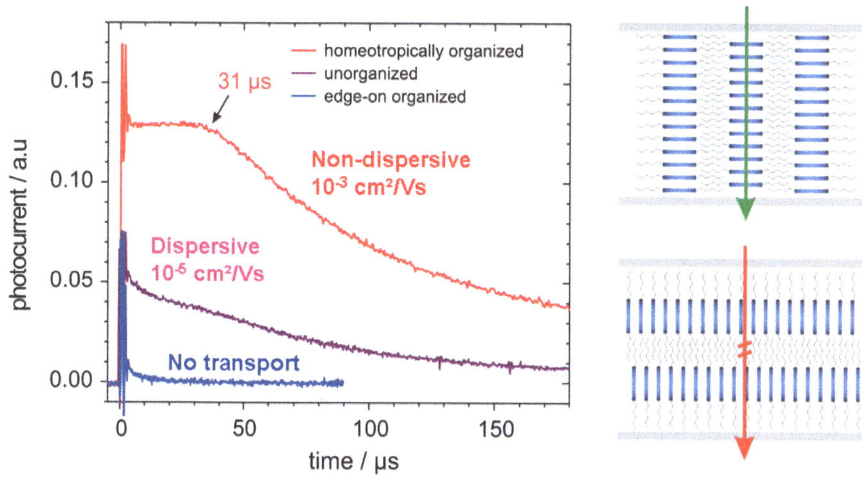

Fig. 7 (*Left*) Examples of time-of-flight hole current transients of discotic hexa-*peri*-hexabenzocoronenes organized differently on the surface: face-on, weakly organized, and edge-on arranged toward the surface as illustrated schematically (*right*). (Reprinted with permission from [60]. Copyright 2006, Wiley-VCH Verlag GmbH)

The exciton diffusion length of discotic columnar LCs is not affected by the homeotropic alignment of the film, whereas the short-circuit photocurrent and thus the charge carrier transport are significantly enhanced in oriented layers vs. in nonoriented layers [59]. Time-of-flight (TOF) is a valuable method to gain information about the impact of order and alignment on homeotropic monodomains on the charge carrier transport through the active layer [60]. For edge-on discotic HBCs, no TOF signal can be detected since the columnar stacks are arranged perpendicular to the main transport direction (Fig. 7). In contrast, homeotropic alignment results in the appearance of a reasonable signal where the degree of order influences the shape of the transient plot. A dispersive photocurrent indicates poor order and a low degree of alignment that traps charges at columnar defects and grain boundaries (Fig. 7). On the other hand, a nondispersive signal with a pronounced plateau is recorded for highly ordered and well-oriented films with a low trap density.

3 Structural Analysis Based on Advanced Scattering Experiments

Understanding the self-assembly and spontaneous self-organization of conjugated molecules into complex superstructures requires dedicated analytical techniques that provide information from a molecular to a mesoscopic scale. In particular, it is challenging to detect the complex morphologies that are crucial for high-efficiency organic solar cells. Structures need to be detected from the crystalline arrangement of the molecules to mesoscopic domain sizes, because conductivity is driven by the crystalline parts and the morphology is strongly linked to the device performance. Moreover, the active layer of an organic solar cell is sandwiched between multiple other layers, all of which are necessary for proper device function. Advanced scattering experiments using X-rays or neutrons have proven to be well suited for this task [61–63]. Since the devices make use of a thin-film geometry, which limits the scattering volume, the use of a reflection geometry turned out to be beneficial for such investigations. The successfully applied X-ray scattering techniques are grazing-incidence wide-angle X-ray scattering (GIWAXS) [61, 62, 64], grazing-incidence small-angle X-ray scattering (GISAXS) [61, 62, 64–67], and X-ray reflectivity (XRR) [61, 63]. Software packages developed in the framework of distorted-wave Born approximation allow the modeling of 2D GISAXS distributions [68–70]. The analogous neutron-scattering techniques are grazing-incidence wide-angle neutron scattering (GIWANS), grazing-incidence small-angle neutron scattering [71, 72], and neutron reflectivity. With the GIWAXS/GIWANS techniques, the crystalline parts of the films are addressed. The orientation of molecules and polymer chains in the crystalline domains with respect to the interfaces given by the electrodes is accessible. In addition, information about the size of the crystalline domains can be obtained. As conductivity typically is anisotropic in polymer crystals, it will strongly depend on the orientation of the crystalline domains and is thus of fundamental importance for understanding the current flow in the device. Last but not least, scattering investigations are not performed using only hard X-rays but also using polarized soft X-rays scattering (P-SoXS) [73, 74] with energies typically below 2 keV and in the range of electronic transitions in atoms. Thus, the technique offers chemical sensitivity and tunability to specific bond types. P-SoXS provides contrast matching through the choice of energy as well as lateral structure information and molecular orientation at interfaces.

In grazing-incidence scattering techniques, the vertical and lateral structures of a thin film are measured in reflection geometry; the incident beam k_i impinges onto the sample, from which it is then reflected and diffracted. To probe the vertical structure of a thin film, XRR probes the specular scattering q_z as a function of the incident angle α_i, where the incident angle equals the exit angle. Using this technique, one can determine the film thickness as well as the vertical film composition [75–77]. The importance of the vertical film composition on the performance of organic photovoltaics (OPVs) was shown to depend on processing parameters such as the choice of solvent or temperature [76, 78]. Ruderer et al. showed for the model system poly(3-hexylthiophene):phenyl-C_{61}-butyrate methyl ester (P3HT:PCBM) that the vertical composition of the material mixture, the material composition profile, and the corresponding photovoltaic device performances change with the solvent [76]. This holds not only for polymeric systems but also for pure small-molecule devices [75] or nanoporous titania films [79].

One of the limiting parameters in the transformation from light to electrical current by photovoltaic devices is the exciton diffusion length. Within a certain time

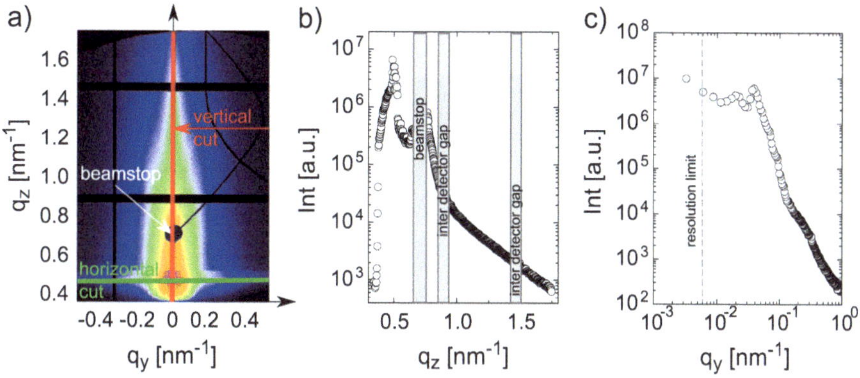

Fig. 8 (**a**) Sample two-dimensional grazing-incidence small-angle X-ray scattering data showing the relevant positions for typical line cuts: (**b**) vertical line cut [depicted in (**a**) in *red*] and (**c**) horizontal line cut [depicted in (**a**) in *green*]

the exciton has to reach a D–A interface for exciton separation into charge carriers for the final electrical current. Hence, for the application of polymeric or small-molecule thin films in photovoltaic devices, not only the vertical but also the lateral structures are of great importance [80]. Grazing-incidence X-ray scattering (GIXS) is a nondestructive measurement technique and provides exactly this information from the film volume with good statistics [61, 62, 66, 67]. In contrast to XRR, in GIXS the diffuse scattering is analyzed instead of the specular reflection, which is blocked by a beamstop in the case of GIXS. Depending on the distance between the sample and the detector, different angles are detected and with these angles different length scales of structures in the film are probed.

The film morphology is important for charge separation in the active layer of OPVs. To probe the relevant length scales for OPVs, GISAXS with typical sample-detector distances between 1–3 m is used [61–66]. Figure 8a shows 2D data obtained with GISAXS measurements. The vertical cut (Fig. 8b) provides information about the vertical film composition relative to the sample plane, for example, the correlated roughness [79, 81]. The horizontal cut (Fig. 8c) on the 2D data, on the other hand, provides information on the lateral structure of the film and gives information on the relevant length scales for OPVs [64, 65, 82]. Together with the vertical information obtained via XRR measurements, the morphology of, say, a bulk heterojunction (BHJ) system can be reconstructed as shown for the model system P3HT:PCBM [76]. Degradation of OPV devices, for instance, is morphology driven as shown by in operando GISAXS [64] and IV characterization [83]. Wang et al. [84] further proved the dependence on device geometry.

After exciton separation, the generated charge carriers need to be transported to the device's electrodes. The charge transport and conductivity in small-molecule, polymeric, and hybrid systems are strongly correlated to the crystallization of the materials [85–87]. In addition, the conductivity in polymers is anisotropic. In active layers and selective interface layers of OPV devices, the vertical conductivity

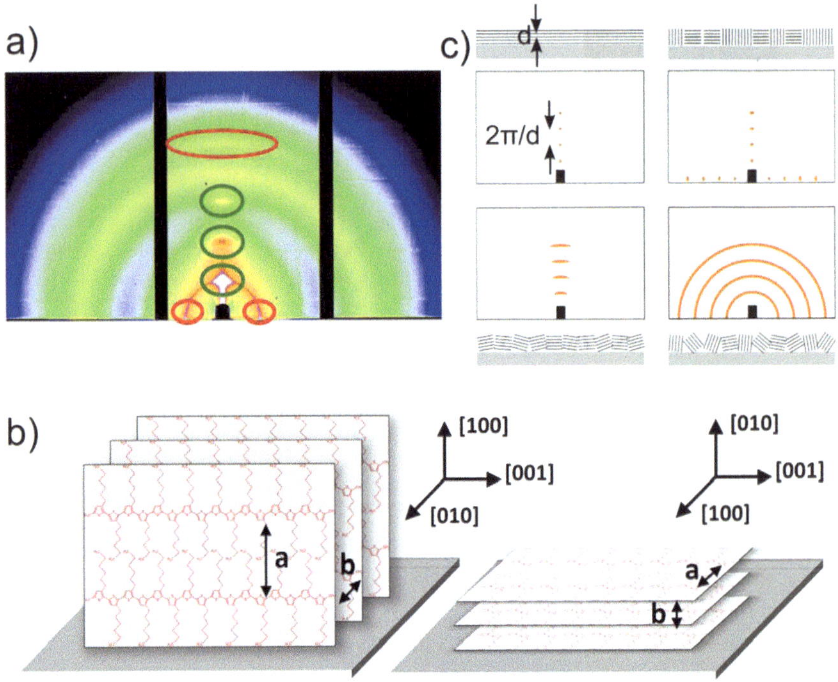

Fig. 9 (**a**) Sample two-dimensional grazing-incidence wide-angle X-ray scattering (2D GIWAXS) data of P3HT:PCBM microstructured bilayers with the scattering from repeating lattice vectors of the film crystals highlighted. (**b**) Schematic representation of P3HT crystals in an edge-on (*left*) and a face-on (*right*) orientation with respect to the substrate. The lattice constants a and b as well as the crystal directions [100], [010], and [001] are depicted. (**c**) Sketch of film crystallinity and corresponding 2D GIWAXS data in case of vertical lamellar stacking (*upper left*), crystallites with vertical and horizontal orientation (*upper right*), oriented domains (*lower left*), and full rotational disorder of crystallites (*lower right*). (Parts of images (**b**) and (**c**) are reproduced with permission from [62]. Copyright 2015, Wiley-VCH Verlag GmbH & Co. KGaA)

is important, whereas, for example, polymeric electrodes require a good lateral conductivity. In organic thin films, the intramolecular conductivity as well as the intermolecular conductivity within crystal grains are important [87]. Therefore, the molecular orientation of crystals needs to be taken into account. GIWAXS provides information about molecular stacking distances [64, 76, 87–89], crystal orientations [82, 86, 90], and even the lower limit of crystal sizes [76, 87, 89, 91–94]. Figure 9a shows the 2D data obtained with GIWAXS measurements with the crystal lattice stacking highlighted. In Fig. 9b two polymer crystal orientations are shown. The stacking distances marked *a* and *b* are repeated and hence result in diffraction peaks. Different orientations of the crystals appear at different positions on the diffraction semicircle as shown in Fig. 9c. Further, the imperfect orientation of crystals results in widening of the diffraction peak along the semicircle, as shown in Fig. 9c.

The importance of the molecular orientations, stacking distances, and evolution of the latter on the conductivity of films was shown in detail for PEDOT:PSS polymeric electrodes for use in indium tin oxide (ITO)-free OPVs [86, 87]. With the model system P3HT:PCBM [76] as an example, solvents have been shown to significantly impact crystal sizes and molecular stacking distances. In addition, it has been shown that with nano-imprinting and the resulting confinement of the polymers, the molecular orientation of P3HT can be controlled [95, 96].

Despite all the advantages of XRR, GISAXS, and GIWAXS, especially for all-polymeric systems, the contrast between materials can be insufficient. In such cases the contrast can be enhanced by selective deuteration for neutron experiments. Hard X-rays are scattered from the electron shell of atoms and the contrast between materials is determined by their electron densities. Neutrons are scattered from the core of atoms, which leads to a fundamentally different scattering contrast between the materials [71]. As an alternative to neutron scattering, polarized resonant soft X-ray scattering (R-SoXS) was developed as a strong tool for tuning the contrast between materials via the chemical sensitivity of this technique [73, 74, 97, 98]. Soft X-ray scattering uses an energy range typically below 2 keV. These energies are in the range where electronic transitions in lighter atoms, such as carbon, take place [74]. With this, the scattering can be tuned to specific bond types providing chemical sensitivity. Figure 10 shows a fundamental carbon transition, the C1s-π^* in aromatic carbon rings. In case the energy is resonant with this transition, aromatic molecules collectively act as dipole antennas and the scattering is dependent on the relative alignment of the electric field [74]. Thus, the scattering signal can correspond to scales where the molecular orientation is preserved and then can expand over several crystal grains and even to noncrystalline parts [74]. As Pfadler et al. [99] showed, the contrast in the all-polymeric system P3HT:poly([N,N′-bis(2-octyldodecyl)-naphthalene-1,4,5,8-bis(dicarboximide)-2,6-

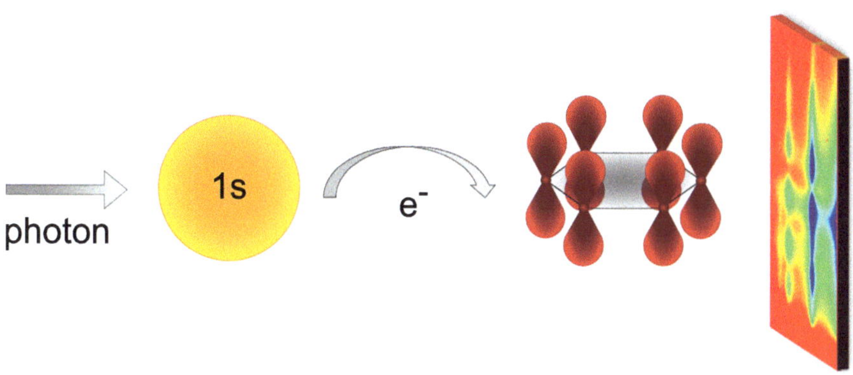

Fig. 10 Basic principle of chemical-sensitive polarized resonant soft X-ray scattering, where, for instance, the photon energy excites an electron from the 1s orbital to an unoccupied molecular orbital

diyl]-*alt*-5,5′-(2,2′-bithiophene)) [P(NDI2OD-T2)] is too low for hard X-ray investigation. The interface between the two materials in this work was artificially nanostructured and therefore is well defined (see Fig. 17). With grazing-incidence R-SoXS, the reason for the absence of scattering was proven to be the lack of material contrast at the interface between both materials [99].

Ultimately, the variety of grazing-incidence scattering techniques provides room for structural investigations ranging from molecular size (e.g., molecular stacking distances) to mesoscopic size in the nanometer range (e.g., active layer domain sizes and artificial structures). With the help of the methods discussed here, one can investigate the vertical and lateral structures that are relevant for the exciton separation and charge-transport mechanisms in organic thin films. Further, additional parameters that influence the conductivity in such films, such as molecular orientation, crystallite sizes, and stacking distances, can be investigated in wide-angle scattering. Finally, with the possibility of applying the principles of XRR and GISAXS with hard X-rays to neutron or soft X-ray scattering, the contrast between materials can be tuned such that even materials with similar electron densities can be distinguished. In the following section the influence of film structures and charge carrier dynamics on the device performance of OPVs is discussed in detail.

4 Device Characterization

In the last section we described the active layer morphology characterization with grazing-incidence scattering techniques. Previously, the self-organization of discotic LCs was discussed in detail. We reviewed how to obtain a well-defined supramolecular assembly from small-molecule HBCs as a result of self-organization, and we also explained the importance of molecular building block alignment for oriented charge carrier mobility. However, efficient charge carrier separation upon photoexcitation in excitonic solar cells can be exclusively realized if a second organic semiconductor is added to energetically form a type II heterointerface [100]. In general, the external quantum efficiency (EQE) of an organic (i.e., excitonic) solar cell can be expressed as a product of the efficiencies (η) of single processes involved from light absorption to current extraction [101, 102]:

$$\begin{aligned}\eta(\text{EQE}) &= \eta(\text{absorption}) \times \eta(\text{IQE}) = \eta(\text{absorption}) \times \eta(\text{exciton diffusion}) \\ &\times \eta(\text{exciton dissociation}) \times \eta(\text{charge collection}),\end{aligned} \tag{1}$$

where η(IQE) describes the internal quantum efficiency, which is determined by processes governed by the electronic properties of the photoactive film. An ideal D–A couple for solar light harvesting shows complementary absorption and suitable energy levels. Furthermore, a bicontinuous D–A morphology providing direct percolation pathways to the external electrodes is crucial for efficient charge collection. Here, in a first step, phase-separated BHJ solar cells were fabricated from solution-processable derivatives of the nano-graphene HBC donors blended with small-molecule perylene-diimide (PDI) acceptors. High-orientation–dependent bulk charge carrier mobilities up to 0.2 cm^2/Vs (PDI) and 1 cm^2/Vs (HBC) have been shown for both materials and for several derivatives [10, 103, 104]. In 2001, Schmidt-Mende et al. [105] successfully implemented the small-molecule D–A couple in an organic solar cell and demonstrated high EQEs for solution-processed BHJ devices based on HBC-Ph-C_{12} and PDI. Subsequent studies on this system and the pristine materials investigated the critical dependency of the supramolecular assembly on processing conditions and how slight changes in molecular design strongly influence the film morphology [27, 106–108]. We investigated the influence of altered HBC donor molecules on fundamental processes that affect the overall photovoltaic device performance and correlated the results to morphological properties that we obtained in a separate X-ray analysis study on thin blend films [109, 110]. The chemical structures of the solar cell materials we used are shown in Fig. 11, and the thermotropic properties of the chemical compounds along with the characteristic stacking parameters are summarized in Table 1.

Alkyl chains with different lengths (6, 8, 12, and 16 carbon atoms), a triple-bond linker between the HBC core and the residual phenyl group, and a swallow-tailed dialkylphenyl chain are attached to the HBC core.

Altered molecules were synthesized following Fechtenkötter et al.'s technique [112]; the modified procedure is described in detail in the supporting information of Hesse et al. [109]. In short, the high sterical demand of the hexadecyl alkyl chain and its limited reactivity make a more complex synthetic route necessary to obtain the HBC-Ph-C_{16}. Furthermore, because of the introduction of an additional triple bond between the HBC core and the substituents at the periphery (for the HBC≡Ph-C_{12} and the HBC≡Ph-$C_{12,12}$), we used another synthetic route, which was based on a sixfold Sonogashira–Hagihara coupling reaction in the last synthetic step [109]. We have shown photovoltaic devices assembled from various phenyl-substituted HBCs blended with PDI using an inverted device structure. Long-term air-stable devices were assembled with an EQE of up to 27% (inset of Fig. 12).

It was concluded that short alkylic side chains attached *to* the discotic core of the HBC molecules allow more efficient photocurrent generation as a result of an improved exciton separation yield. However, low fill factors, which can possibly be explained by bimolecular recombination and space charge formation, limited the overall power conversion efficiency (PCE) of these devices [109]. In order to elucidate the supramolecular assembly and the stacking of the discotic molecules used in the active layer of the photovoltaic devices, we measured the X-ray diffraction. The studies revealed the morphology of these mixed D–A blend films and supported evidence for the formation of D–A complexes in the blend films. For blend mixtures of PDI and HBC molecules with short alkylic side chains, these complexes are packed into a layered structure lying edge-on (parallel) to the substrate (as sketched in Fig. 13).

The layered structure was found to be disrupted when increasing the side chain length of the HBC molecule, and eventually a disordered structure is formed for long side chains ($n > 12$). We attributed this behavior to the size difference between

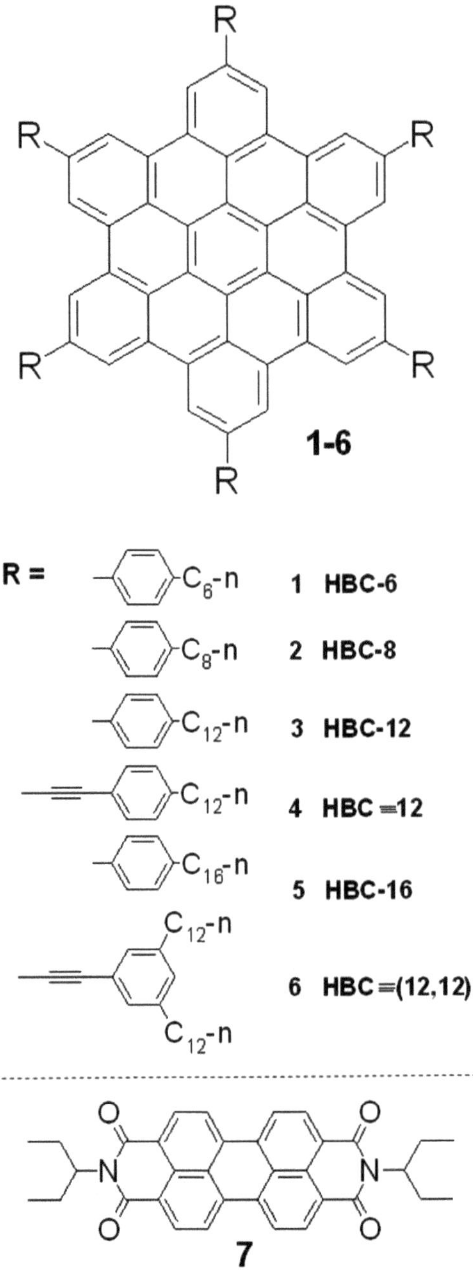

Fig. 11 Chemical structure of organic materials used in solar cells. 1–6: electron acceptor hexa-*peri*-hexabenzocoronene with different residues at the periphery. 7: electron donor perylene diimide. (Reprinted from Hesse et al. [109] with permission from Elsevier)

the aromatic parts of the HBC and PDI molecules. Based on these results, the OPV device performance using these blend mixtures can be understood in more detail. Intimate contact is provided for HBC–PDI blends when short side chains are used,

Table 1 Characteristic stacking parameters of extruded films. Characteristic distances and phase transitions have been derived by two-dimensional powder X-ray diffraction measurements and differential scanning calorimetry: *Cr* crystalline, *LC* liquid crystalline, *I* isotropic. Packing structure: (1) = hexagonal, (2) = cubic, (3) = tilted herringbone. All HBC compounds show a liquid-crystalline phase transition with at least one accessible phase transition

Material	Core–core (nm)	d (π–π) (nm)	Phase transitions
PDI	3.39 [111]	0.35 [105]	68 °C, Cr–I [107]
HBC-6	2.76	0.35	161 °C, LC–LC
HBC-8	2.97	0.35	153 °C, LC–LC
HBC-12	3.42	0.35	−22 °C, 62 °C, LC–LC
HBC-16	3.70	0.35	9 °C, LC–LC
HBC≡Ph-C_{12}	$3.37^{(1)}$–$3.18^{(2)}$	0.35	11 °C, 72 °C, LC–LC
HBC≡Ph-$C_{12,12}$	3.63	$0.44^{(3)}$	59 °C, LC–LC

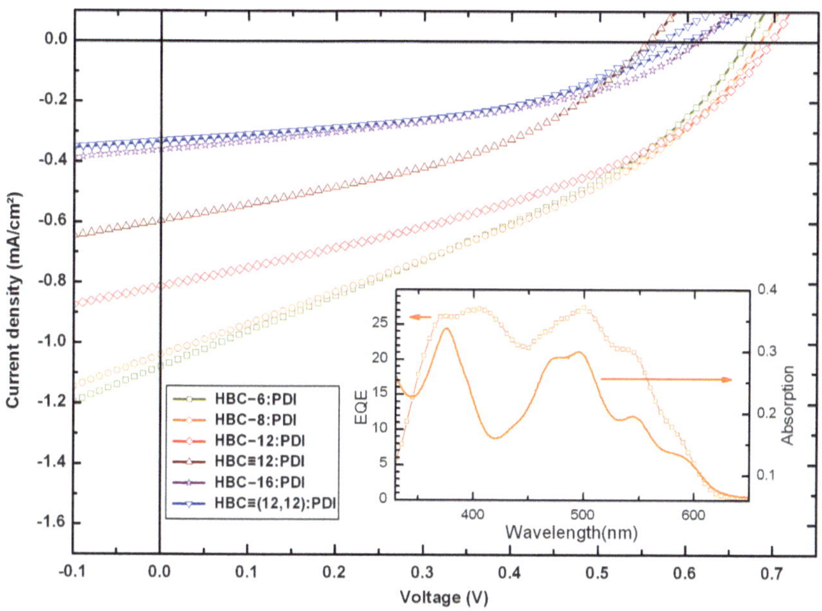

Fig. 12 Hexa-*peri*-hexabenzocoronene:perylene diimide (HBC:PDI) blend *J–V* characteristics. The best photovoltaic devices are obtained for blends containing short side chain HBCs. The low fill factor of the devices can be explained by strong bimolecular recombination. (Reprinted from Hesse et al. [109] with permission from Elsevier)

resulting in higher current generation. On the other hand, the two molecules with a very similar and planar core shape tend to intermix on a molecular scale; that is, unfavorable morphologies for charge extraction are formed. The results are low fill factors as well as device performance that strongly depends on the illumination intensity. Such a strongly suppressed charge extraction for HBC–PDI blends

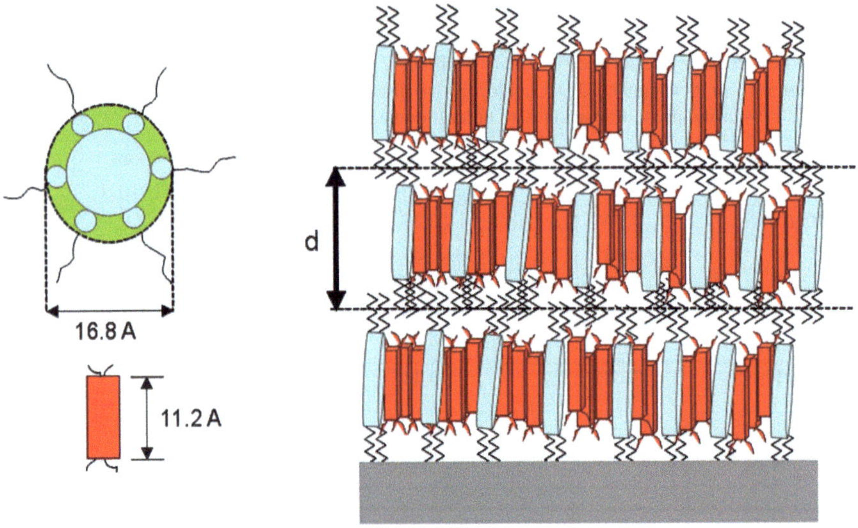

Fig. 13 Schematic representation of the layered structure formed upon blending solution-processable hexa-*peri*-hexabenzocoronene (*blue*) and perylene diimide (*red*) molecules as supported by X-ray diffraction measurements. The layered structure parallel to the substrate plane leads to strong bimolecular recombination and limits the incorporation of these molecules into OPV devices. (Reprinted from Al-Hussein et al. [110] with permission from Elsevier)

featuring a layered structure parallel to the substrate (Fig. 13) is fully consistent with the hole current transients of pristine HBCs presented in Fig. 7. There, anisotropic hole mobilities are determined: Going from a homeotropic organization of pristine HBC to an unorganized orientation and then to an edge-on orientation, the respective mobilities are decreased from 10^{-3} cm^2/Vs to 10^{-5} cm^2/Vs to ultimately being completely suppressed for an edge-on organization if alkyl side chains hinder the charge transport between the layers. In evaporated HBC films featuring no side chains, more suitable supramolecular assembly and crystalline packing of HBC donor molecules were evidenced using X-ray diffraction measurements. For these films our studies revealed a long exciton diffusion length of $L_D \sim 25$ nm for layers of the pristine unsubstituted discotic material (Fig. 14) [113]. As such, when aligned and assembled appropriately, discotic small molecules can very well compete with the highly conjugated polymeric compounds commonly used in OPV devices.

Advanced photon-harvesting concepts suitable for decoupling exciton harvesting and charge transport are of fundamental interest and might potentially allow us to further increase the efficiency of OPV devices. We added an energy relay dye to fullerenes; it resulted in increased light harvesting and a significantly improved PCE for OPV devices. This new method was analyzed in detail and showed that the sensitization allows decoupling of light absorption and charge transport in the thin-film OPV devices [114]. Although fullerenes exhibit excellent properties as electron acceptors, their visible light absorption is limited. Thus,

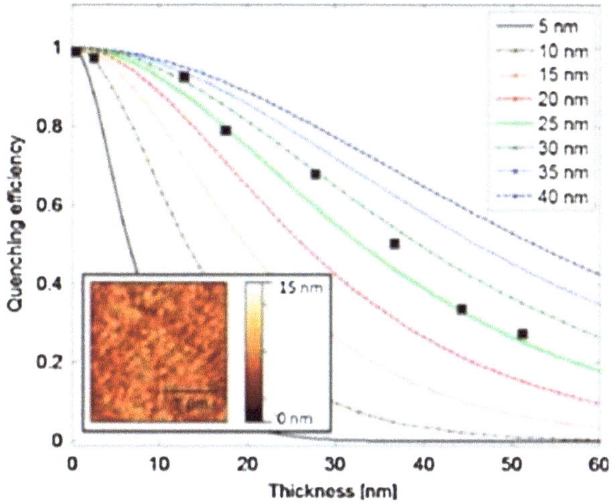

Fig. 14 Photoluminescence studies reveal high exciton diffusion length ($L_D \sim 25$ nm) for flat layers of unsubstituted hexa-*peri*-hexabenzocoronene (HBC) layers grown from the vapor phase. The *solid curves* describe the simulated quenching efficiency expected for different exciton diffusion lengths in HBC. The *squares* describe the measured values. These indicate an exciton diffusion length of ~25 nm. Crystalline packing and a suitable supramolecular assembly are attributed to allow for this unperturbed exciton migration in the pristine layers of the disc-shaped molecule. (Reprinted from Hesse et al. [113]. Copyright 2012, Wiley-VCH Verlag GmbH & Co. KGaA, Weinheim)

strongly light-absorbing donor materials are needed for fullerene acceptors to efficiently harvest light in the thin active layer of OPV devices. The photocurrent generation and thus PCE of this type of solar cell are confined by the overlap of the relatively narrow absorption band of commonly used donor molecules with the solar spectrum. Here we present the concept of fullerene dye sensitization, which allows increased light harvesting on the electron acceptor side of the heterojunction. The concept was shown for the UV-absorbing small-molecule HBC (Fig. 15) and a near-infrared absorbing polymer, poly[2,1,3-benzothiadiazole-4,7-diyl [4,4-bis(2-ethylhexyl)-4H-cyclopenta [2,1-b:3,4-b′] dithiophene-2,6-diyl]], respectively.

In both systems, remarkably higher PCE values were achieved via perylene sensitization of the fullerene acceptor. Steady-state photoluminescence, transient absorption, and transient photocurrent decay studies revealed pathways of the additionally generated excited states at the sensitizer molecule. The findings suggest fluorescence resonance energy transfer from the photo-excited dye to the fullerene, allowing enhanced photon harvesting of the thin-film devices in the spectral region of pronounced dye absorption.

As discussed above, a suitable supramolecular assembly of pristine HBC material can be obtained and exhibits attractive properties for OPV devices. Instead, for intermixed D–A blends of discotic molecules we found an unsuitable stacking suffering from pronounced recombination (i.e., low fill factors). The use of nanostructured interfaces at the D–A heterojunction potentially allows us to combine

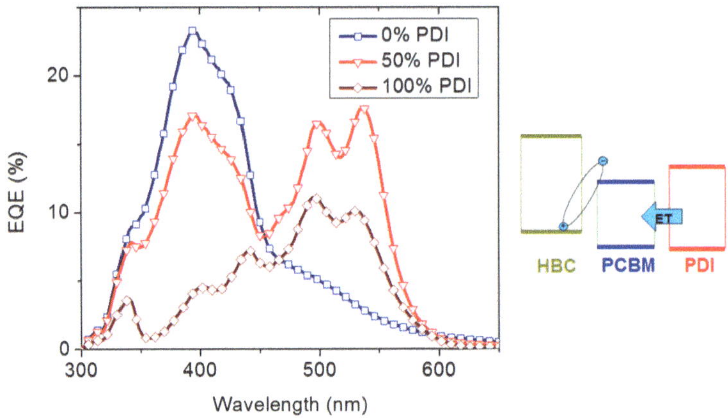

Fig. 15 Dye sensitization of fullerenes allows decoupling of photoabsorption and charge separation/transport in organic photovoltaic devices. An energy-transfer mechanism is evidenced from the photoexcited dye molecule toward the fullerene derivative followed by charge separation at the heterojunction with the donor molecule, causing enhanced photon harvesting in the spectral region of pronounced dye absorption. (Depicted from Hesse et al. [114])

the benefits of bilayered and intermixed device geometries: An increased interfacial area and consequently enhanced exciton separation yield can be combined with pristine material properties in the respective bicontinuous material phases. Using nano-imprint lithography (NIL), we fabricated interdigitated geometries on a nanometer-length scale for organic materials [115, 116]. Hexagonally ordered anodized aluminum oxide (AAO) templates were used as stamps, and an HBC derivative with acrylate moieties was synthesized that can be polymerized in situ in the confinement of the AAO template [117]. An exact pattern transfer of the AAO structures into the organic material was shown, and film fabrication occurred for a variety of nanowire dimensions on square-centimeter areas. The fabrication directly on conductive glass, support, and control over the formation of a dense barrier layer render this approach appealing for the fabrication of fully organic nanostructured photovoltaic devices [116] (Fig. 16).

However, the HBC derivative used in these studies did not seem ideal for photovoltaic applications. In fact, the cross linking leads to charge-trapping effects that were expected to occur at the acrylate moieties and that were attributed to strongly suppressed charge transport in the organic layer.

Recently, Png and coworkers [118] demonstrated a simple and versatile photo-cross-linking methodology based on sterically hindered bis(fluorophenyl azide)s. The photo-cross-linking efficiency is shown to be high and to be dominated by alkyl side chain insertion reactions, which do not degrade the semiconducting properties of polymers in the interconnected, that is, undissolvable, network. Highly efficient interpenetrating heterostructure devices can be obtained if a second organic semiconductor is back-infiltrated into the lightly cross-linked polymer network. Liu et al. [119] demonstrated highly reproducible and state-of-the-art

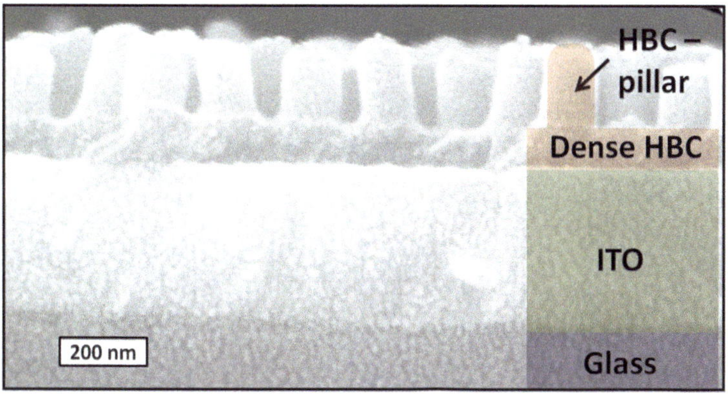

Fig. 16 A cross-sectional scanning electron microscopy graph shows an imprinted organic film with precisely controlled nanostructure dimensions. Highly ordered pillars bound to a dense barrier layer consisting of cross-linked hexa-*peri*-hexabenzocoronene material are fabricated on indium tin oxide–coated class support. (Depicted from Hesse et al. [116])

efficiencies (above 4%) for the regioregular P3HT:PCBM D–A model system with IQEs up to 90% [Eq. (1)]. The controlled back-infiltrated heterostructure is prepared, for instance, by infiltrating an acceptor material into a photo-cross-linked polymer donor matrix. Islands of trapped donor as well as acceptor phases (that can occur in BHJ devices) can be excluded since this approach intrinsically features a controlled built-in continuity of donor and acceptor phases, respectively. Going one step further, a bilayer heterojunction device featuring a controlled nanostructured interface is considered the perfect D–A morphology [120–123]. He et al. [124] demonstrated a solvent-assisted double nano-imprint approach that allowed the formation of nanostructured poly((9,9-dioctylfluorene)-2,7-diyl-*alt*-[4,7-bis(3-hexylthien-5-yl)-2,1,3-benzothiadiazole]-2′,2″-diyl) (F8TBT)/P3HT all-polymer bilayer devices with a PCE of 1.9%. The pre-patterned P3HT layer itself was used as an imprinting mold to structure the F8TBT layer in a dichloromethane (orthogonal solvent for P3HT) vapor-saturated atmosphere. We developed a new approach to fabricate all solution-processed, nano-embossed D–A heterostructures for organic solar cells [97]. In our corresponding study, we investigated the obtained nanostructured bilayer system consisting of the *p*-type polymer P3HT and the *n*-type polymer P(NDI2OD-T2) [125–129]. Most importantly, the complementary absorption of P(NDI2OD-T2) and P3HT facilitated the elucidation of the donor and acceptor contributions in spectrally resolved measurements. A distinct D–A interface in this bottom-up solution process was realized via photo-cross linking of the P(NDI2OD-T2) network and the subsequent deposition of P3HT from solution. AAO membranes, featuring self-organized hexagonal nanopore arrays of different sizes, were used in a NIL process to tailor the topography of the P(NDI2OD-T2) layer [115, 117] (Fig. 17). In addition to common local real-space characterization techniques, such as scanning electron microscopy and atomic force microscopy, a

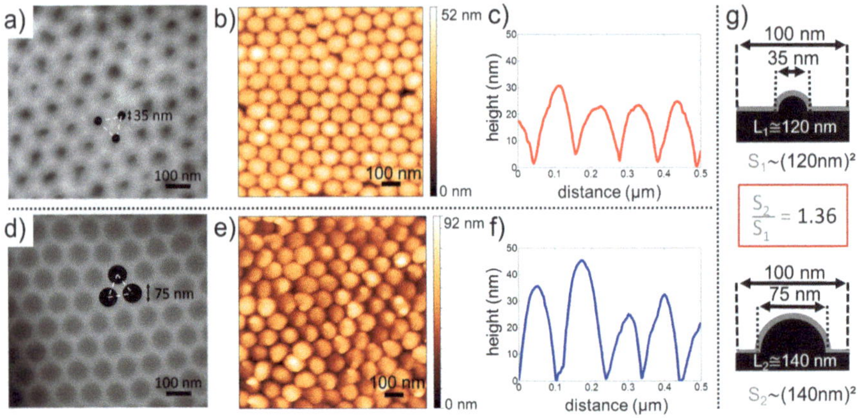

Fig. 17 Comparison of extracted topographies. The 35-nm structures are shown in the first row, and the 75-nm structures are shown in the second row. Each column offers a direct comparison of the respective nanostructures. (**a**, **d**) Scanning electron microscopy pictures of the master mold structures used in the nano-imprint lithography process. Both anodized aluminium oxide membranes feature a honeycomb lattice with a periodicity of 100 nm and only differ from each other in terms of the specific pore diameter. (**b**, **e**) Atomic force microscopy (AFM) topographic height profiles of imprinted P(NDI2OD-T2) films. (**c**, **f**) Corresponding AFM line sections are shown. (**g**) Simplified model to give a quantitative starting point for the difference in surface area of the two employed Master mold structures. The 75-nm structure is estimated to feature 36% surface enhancement over the 35-nm structure. (Reprinted with permission from Pfadler et al. [97]. Copyright 2014, American Chemical Society)

statistically averaged reciprocal space analysis method, namely grazing-incidence resonant soft X-ray scattering (GI-RSoXS), was applied to probe the quality of the nano-imprinting process and the final interdigitating structure.

For our investigation, nanostructured bilayer devices were found to show a direct correlation between an enhanced D–A interfacial area (Figs. 17 and 18) and an enhanced EQE. As found from EQE spectra, both excitons generated in the donor as well as in the acceptor polymer equally contribute to the relative improvement in EQE.

The enhancement in EQE is subsequently visible in the corresponding *J*–*V* measurements with an increased J_{SC}. More importantly, we focused on the influence of the systematically enhanced D–A interfacial area on polaron recombination processes, which we elucidated for different solar cell working regimes. Therefore, our fundamental study included complementary measurement techniques, such as diode dark current characteristics, transient photovoltage, and impedance spectroscopy. Our investigation showed that the controlled enhancement of the D–A interfacial area on a nanometer scale has been beneficial for the harvesting of photoexcitons, while simultaneously enhancing the undesired polaron recombination losses in excitonic solar cells. Based on the investigation of our model system for an excitonic solar cell, and according to findings concerning the importance of interfacial multiphase structures in BHJ solar cells, we draw an updated hypothetic picture

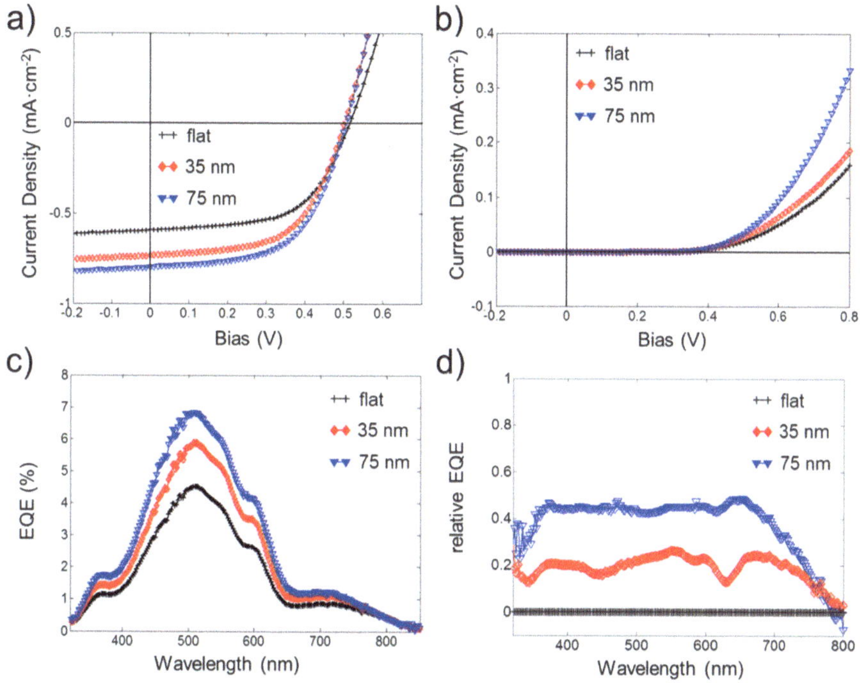

Fig. 18 Representative *J–V* and external quantum efficiency (EQE) characteristics of comb-like bilayer devices featuring an enhancement in the interfacial area. (**a**) *J–V* characteristics recorded under simulated AM 1.5G solar illumination. (**b**) Diode dark current characteristics emphasizing polaron recombination in the forward direction. (**c**) Corresponding EQE of nanostructured devices. (**d**) Relative EQE enhancement: [EQE(structured) − EQE(flat)]/EQE(flat)]. (Reprinted with permission from Pfadler et al. [97]. Copyright 2014, American Chemical Society)

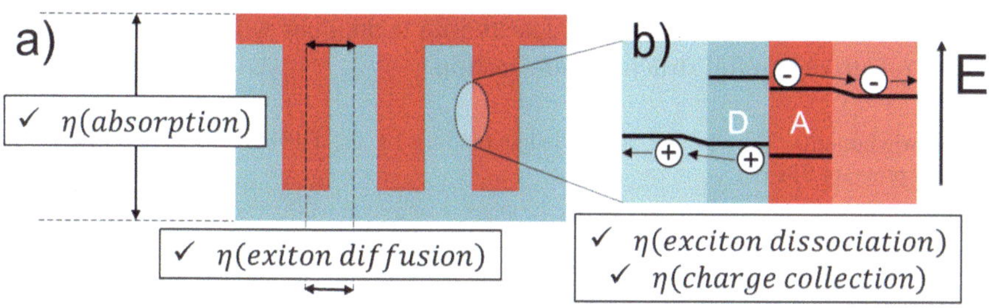

Fig. 19 Illustration of the ideal comb-like morphology for (**a**) a high-efficiency excitonic solar cell [Eq. (1)] with (**b**) energetically optimized interface for efficient charge separation. (Reprinted with permission from Pfadler et al. [97]. Copyright 2014, American Chemical Society)

for the perfect comb-like morphology for an excitonic solar cell by additionally considering the energy landscape with respect to the interface to suppress the measured enhanced recombination [130, 131]. The obtained structure, which is discussed with respect to Eq. (1), is shown in Fig. 19. A vertically optimized spacing guarantees optimized η(absorption), whereas a horizontal spacing of the D–A interfaces with respect to the exciton diffusion length guarantees optimized η(exciton diffusion). In addition to the bicontinuous spatial optimization (Fig. 19a), the perfect morphology needs to be energetically optimized (Fig. 19b). In the ideal energy landscape, both holes and electrons are funneled away from the interface immediately after exciton splitting through slight changes in the levels of the highest occupied molecular orbital and the lowest unoccupied molecular orbital of the donor and acceptor, respectively. For instance, this can be realized via changes in the local crystallinity or different phases as discussed by Jamieson and Burke and coworkers [130, 131] for polymer–fullerene BHJs.

5 Conclusion

In recent years there has been tremendous progress in analyzing and understanding the role of structural alignment and crystallization in BHJ solar cells. Nevertheless, controlling the alignment down to a molecular level in large-area devices remains a huge challenge. While molecular alignment and assembly of discotic materials is possible, it is very challenging to realize ideal structures in D–A blends. Therefore, one possible solution for ideal alignment on a molecular level and high interfacial area might be realized by nanoimprint. The key issue for high-performance solar cells is not only a large interfacial area, but also design of the interface so that it promotes exciton separation and suppresses recombination. Here the differences in energy levels of a crystalline versus amorphous phase can already play a decisive role affecting the final device performance. Therefore, future challenges involve combined synthetic and device engineering efforts to allow full control over the alignment and crystallization in mixtures of two donor and acceptor materials at the right places in the device. Advanced scattering techniques will continue to play a major role to obtain full structural information.

Acknowledgments The authors acknowledge funding by the DFG in the framework of the priority program SPP1355.

References

1. Hirschberg JHKK Brunsveld L, Ramzi A, Vekemans JAJM, Sijbesma RP, Meijer EW (2000) Nature 407:167–170
2. Davis JT, Spada GP (2007) Chem Soc Rev 36:296–313
3. Antonietti M, Förster S (2003) Adv Mater 15:1323–1333
4. Goodby JW, Saez IM, Cowling SJ, Görtz V, Draper M, Hall AW, Sia S, Cosquer G, Lee S-E, Raynes EP (2008) Angew Chem Int Ed 47:2754–2787
5. Pisula W, Zorn M, Chang JY, Müllen K, Zentel R (2009) Macromol Rapid Commun 30:1179–1202

6. O'Neill M, Kelly SM (2003) Adv Mater 15:1135–1146
7. Demus JGD, Gray GW, Spiess H-W, Vill V (1998) Handbook of liquid crystals. Wiley VCH, Weinheim
8. Boden N, Bushby RJ, Clements J, Movaghar B, Donovan KJ, Kreouzis T (1995) Phys Rev B 52:13274–13280
9. Laschat S, Baro A, Steinke N, Giesselmann F, Hägele C, Scalia G, Judele R, Kapatsina E, Sauer S, Schreivogel A, Tosoni M (2007) Angew Chem Int Ed 46:4832–4887
10. Sergeyev S, Pisula W, Geerts YH (2007) Chem Soc Rev 36:1902–1929
11. Bushby RJ, Lozman OR (2002) Curr Opin Colloid Interface Sci 7:343–354
12. Warman JM, Piris J, Pisula W, Kastler M, Wasserfallen D, Müllen K (2005) J Am Chem Soc 127:14257–14262
13. Schouten PG, Warman JM, Gelinck GH, Copyn MJ (1995) J Phys Chem 99:11780–11793
14. Warman JM, Van De Craats AM (2003) Mol Cryst Liq Cryst 396:41–72
15. Kastler M, Pisula W, Wasserfallen D, Pakula T, Müllen K (2005) J Am Chem Soc 127:4286–4296
16. Bengs H, Karthaus O, Ringsdorf H, Baehr C, Ebert M, Wendorff JH (1991) Liq Cryst 10:161–168
17. Brown SP, Schnell I, Brand JD, Müllen K, Spiess HW (2000) J Mol Struct 521:179–195
18. Crispin X, Cornil J, Friedlein R, Okudaira KK, Lemaur V, Crispin A, Kestemont G, Lehmann M, Fahlman M, Lazzaroni R, Geerts Y, Wendin G, Ueno N, Brédas J-L, Salaneck WR (2004) J Am Chem Soc 126:11889–11899
19. Fischbach I, Pakula T, Minkin P, Fechtenkötter A, Müllen K, Spiess HW, Saalwächter K (2002) J Phys Chem B 106:6408–6418
20. Hunter CA, Sanders JKM (1990) J Am Chem Soc 112:5525–5534
21. Fischbach I, Ebert F, Spiess HW, Schnell I (2004) ChemPhysChem 5:895–908
22. Arikainen EO, Boden N, Bushby RJ, Clements J, Movaghar B, Wood A (1995) J Mater Chem 5:2161–2165
23. Motoyanagi J, Yamamoto Y, Saeki A, Alam MA, Kimoto A, Kosaka A, Fukushima T, Seki S, Tagawa S, Aida T (2009) Chem Asian J 4:876–880
24. McKeown NB, Helliwell M, Hassan BM, Hayhurst D, Li H, Thompson N, Teat SJ (2007) Chem Eur J 13:228–234
25. Shearman GC, Yahioglu G, Kirstein J, Milgrom LR, Seddon JM (2009) J Mater Chem 19:598–604
26. Bengs H, Closs F, Frey T, Funhoff D, Ringsdorf H, Siemensmeyer K (1993) Liq Cryst 15:565–574
27. Pisula W, Tomović Ž, Simpson C, Kastler M, Pakula T, Müllen K (2005) Chem Mater 17:4296–4303
28. Bisoyi HK, Kumar S (2008) J Phys Org Chem 21:47–52
29. Gearba RI, Bondar AI, Goderis B, Bras W, Ivanov DA (2005) Chem Mater 17:2825–2832
30. Wasserfallen D, Kastler M, Pisula W, Hofer WA, Fogel Y, Wang Z, Müllen K (2006) J Am Chem Soc 128:1334–1339
31. Pisula W, Kastler M, Wasserfallen D, Pakula T, Müllen K (2004) J Am Chem Soc 126:8074–8075
32. Pisula W, Tomović Ž, El Hamaoui B, Watson MD, Pakula T, Müllen K (2005) Adv Funct Mater 15:893–904
33. Pisula W, Kastler M, Wasserfallen D, Mondeshki M, Piris J, Schnell I, Müllen K (2006) Chem Mater 18:3634–3640
34. Brunsveld L, Folmer BJB, Meijer EW, Sijbesma RP (2001) Chem Rev 101:4071–4098
35. De Greef TFA, Smulders MMJ, Wolffs M, Schenning APHJ, Sijbesma RP, Meijer EW (2009) Chem Rev 109:5687–5754
36. Ghosh S, Li X-Q, Stepanenko V, Würthner F (2008) Chem Eur J 14:11343–11357
37. Keizer HM, Sijbesma RP (2005) Chem Soc Rev 34:226–234
38. Dou X, Pisula W, Wu J, Bodwell GJ, Müllen K (2008) Chem Eur J 14:240–249

39. Wasserfallen D, Fischbach I, Chebotareva N, Kastler M, Pisula W, Jäckel F, Watson MD, Schnell I, Rabe JP, Spiess HW, Müllen K (2005) Adv Funct Mater 15:1585–1594
40. Levitsky IA, Kishikawa K, Eichhorn SH, Swager TM (2000) J Am Chem Soc 122:2474–2479
41. Foster EJ, Jones RB, Lavigueur C, Williams VE (2006) J Am Chem Soc 128:8569–8574
42. Feng X, Pisula W, Takase M, Dou X, Enkelmann V, Wagner M, Ding N, Müllen K (2008) Chem Mater 20:2872–2874
43. Feng X, Marcon V, Pisula W, Hansen MR, Kirkpatrick J, Grozema F, Andrienko D, Kremer K, Mullen K (2009) Nat Mater 8:421–426
44. Wu J, Fechtenkötter A, Gauss J, Watson MD, Kastler M, Fechtenkötter C, Wagner M, Müllen K (2004) J Am Chem Soc 126:11311–11321
45. Pisula W, Tomović Ž, Watson MD, Müllen K, Kussmann J, Ochsenfeld C, Metzroth T, Gauss J (2007) J Phys Chem B 111:7481–7487
46. Hill JP, Jin W, Kosaka A, Fukushima T, Ichihara H, Shimomura T, Ito K, Hashizume T, Ishii N, Aida T (2004) Science 304:1481–1483
47. Jin W, Yamamoto Y, Fukushima T, Ishii N, Kim J, Kato K, Takata M, Aida T (2008) J Am Chem Soc 130:9434–9440
48. Hansen MR, Schnitzler T, Pisula W, Graf R, Müllen K, Spiess HW (2009) Angew Chem Int Ed 48:4621–4624
49. Bjornholm T, Hassenkam T, Reitzel N (1999) J Mater Chem 9:1975–1990
50. Becerril HA, Roberts ME, Liu Z, Locklin J, Bao Z (2008) Adv Mater 20:2588–2594
51. Wang S, Kivala M, Lieberwirth I, Kirchhoff K, Feng X, Pisula W, Müllen K (2011) ChemPhysChem 12:1648–1651
52. Pisula W, Kastler M, El Hamaoui B, García-Gutiérrez M-C, Davies RJ, Riekel C, Müllen K (2007) ChemPhysChem 8:1025–1028
53. De Cupere V, Tant J, Viville P, Lazzaroni R, Osikowicz W, Salaneck WR, Geerts YH (2006) Langmuir 22:7798–7806
54. Grelet E, Bock H (2006) EPL (Europhysics Letters) 73:712
55. Thiebaut O, Bock H, Grelet E (2010) J Am Chem Soc 132:6886–6887
56. Pouzet E, Cupere VD, Heintz C, Andreasen JW, Breiby DW, Nielsen MM, Viville P, Lazzaroni R, Gbabode G, Geerts YH (2009) J Phys Chem C 113:14398–14406
57. Liu C-Y, Fechtenkötter A, Watson MD, Müllen K, Bard AJ (2003) Chem Mater 15:124–130
58. Schweicher G, Gbabode G, Quist F, Debever O, Dumont N, Sergeyev S, Geerts YH (2009) Chem Mater 21:5867–5874
59. Cisse L, Destruel P, Archambeau S, Seguy I, Jolinat P, Bock H, Grelet E (2009) Chem Phys Lett 476:89–91
60. Kastler M, Pisula W, Laquai F, Kumar A, Davies RJ, Baluschev S, Garcia–Gutiérrez MC, Wasserfallen D, Butt HJ, Riekel C, Wegner G, Müllen K (2006) Adv Mater 18:2255–2259
61. Hexemer A, Müller-Buschbaum P (2015) IUCrJ 2:106–125
62. Müller-Buschbaum P (2014) Adv Mater 26:7692–7709
63. Ruderer MA, Müller-Buschbaum P (2011) Soft Matter 7:5482
64. Chou KW, Yan B, Li R, Li EQ, Zhao K, Anjum DH, Alvarez S, Gassaway R, Biocca A, Thoroddsen ST, Hexemer A, Amassian A (2013) Adv Mater 25:1923–1929
65. Kozub DR, Vakhshouri K, Orme LM, Wang C, Hexemer A, Gomez ED (2011) Macromolecules 44:5722–5726
66. Renaud G, Lazzari R, Leroy F (2009) Surf Sci Rep 64:255–380
67. Smilgies DM, Busch P, Papadakis CM, Posselt D (2002) Synchrotron Radiat News 15:35–42
68. Babonneau D (2010) J Appl Crystallogr 43:929–936
69. Chourou ST, Sarje A, Li XS, Chan ER, Hexemer A (2013) J Appl Crystallogr 46:1781–1795
70. Lazzari R (2002) J Appl Crystallogr 35:406–421
71. Guo S, Cao B, Wang W, Moulin J-F, Müller-Buschbaum P (2015) ACS Appl Mater Interfaces 7:4641–4649
72. Rawolle M, Sarkar K, Niedermeier MA, Schindler M, Lellig P, Gutmann JS, Moulin J-F, Haese-Seiller M, Wochnik AS, Scheu C, Müller-Buschbaum P (2013) ACS Appl Mater Interfaces 5:719–729

73. Collins BA, Cochran JE, Yan H, Gann E, Hub C, Fink R, Wang C, Schuettfort T, McNeill CR, Chabinyc ML, Ade H (2012) Nat Mater 11:536–543
74. Mannsfeld SCB (2012) Nat Mater 11:489–490
75. Gruber M, Rawolle M, Wagner J, Magerl D, Hörmann U, Perlich J, Roth SV, Opitz A, Schreiber F, Müller-Buschbaum P, Brütting W (2013) Adv Energy Mater 3:1075–1083
76. Ruderer MA, Guo S, Meier R, Chiang H-Y, Körstgens V, Wiedersich J, Perlich J, Roth SV, Müller-Buschbaum P (2011) Adv Funct Mater 21:3382–3391
77. Rawolle M, Braden EV, Niedermeier MA, Magerl D, Sarkar K, Fröschl T, Hüsing N, Perlich J, Müller-Buschbaum P (2012) ChemPhysChem 13:2412–2417
78. Schmidt-Hansberg B, Sanyal M, Klein MFG, Pfaff M, Schnabel N, Jaiser S, Vorobiev A, Müller E, Colsmann A, Scharfer P, Gerthsen D, Lemmer U, Barrena E, Schabel W (2011) ACS Nano 5:8579–8590
79. Rawolle M, Ruderer MA, Prams SM, Zhong Q, Magerl D, Perlich J, Roth SV, Lellig P, Gutmann JS, Müller-Buschbaum P (2011) Small 7:884–891
80. Rawolle M, Niedermeier MA, Kaune G, Perlich J, Lellig P, Memesa M, Cheng Y-J, Gutmann JS, Müller-Buschbaum P (2012) Chem Soc Rev 41:5131
81. Guo S, Ruderer MA, Rawolle M, Körstgens V, Birkenstock C, Perlich J, Müller-Buschbaum P (2013) ACS Appl Mater Interfaces 5:8581–8590
82. Wu W-R, Jeng U-S, Su C-J, Wei K-H, Su M-S, Chiu M-Y, Chen C-Y, Su W-B, Su C-H, Su A-C (2011) ACS Nano 5:6233–6243
83. Schaffer CJ, Palumbiny CM, Niedermeier MA, Jendrzejewski C, Santoro G, Roth SV, Müller-Buschbaum P (2013) Adv Mater 25:6760–6764
84. Wang W, Pröller S, Niedermeier MA, Körstgens V, Philipp M, Su B, Moseguí González D, Yu S, Roth SV, Müller-Buschbaum P (2015) ACS Appl Mater Interfaces 7:602–610
85. Guldin S, Hüttner S, Tiwana P, Orilall MC, Ülgüt B, Stefik M, Docampo P, Kolle M, Divitini G, Ducati C, Redfern SAT, Snaith HJ, Wiesner U, Eder D, Steiner U (2011) Energy Environ Sci 4:225–233
86. Palumbiny CM, Heller C, Schaffer CJ, Körstgens V, Santoro G, Roth SV, Müller-Buschbaum P (2014) J Phys Chem C 118:13598–13606
87. Palumbiny CM, Liu F, Russell TP, Hexemer A, Wang C, Müller-Buschbaum P (2015) Adv Mater 27:3391–3397
88. Huang Y-C, Tsao C-S, Chuang C-M, Lee C-H, Hsu F-H, Cha H-C, Chen C-Y, Lin T-H, Su C-J, Jeng U-S, Su W-F (2012) J Phys Chem C 116:10238–10244
89. Verploegen E, Miller CE, Schmidt K, Bao Z, Toney MF (2012) Chem Mater 24:3923–3931
90. Treat ND, Shuttle CG, Toney MF, Hawker CJ, Chabinyc ML (2011) J Mater Chem 21:15224
91. Treat ND, Brady MA, Smith G, Toney MF, Kramer EJ, Hawker CJ, Chabinyc ML (2011) Adv Energy Mater 1:82–89
92. Kohn P, Rong Z, Scherer KH, Sepe A, Sommer M, Müller-Buschbaum P, Friend RH, Steiner U, Hüttner S (2013) Macromolecules 46:4002–4013
93. Niedermeier MA, Rawolle M, Lellig P, Körstgens V, Herzig EM, Buffet A, Roth SV, Gutmann JS, Fröschl T, Hüsing N, Müller-Buschbaum P (2013) ChemPhysChem 14:597–602
94. Rogers JT, Schmidt K, Toney MF, Kramer EJ, Bazan GC (2011) Adv Mater 23:2284–2288
95. Aryal M, Trivedi K, Hu W (2009) ACS Nano 3:3085–3090
96. Hlaing H, Lu X, Hofmann T, Yager KG, Black CT, Ocko BM (2011) ACS Nano 5:7532–7538
97. Pfadler T, Coric M, Palumbiny CM, Jakowetz AC, Strunk KP, Dorman JA, Ehrenreich P, Wang C, Hexemer A, Png RQ, Ho PK, Muller-Buschbaum P, Weickert J, Schmidt-Mende L (2014) ACS Nano 8:12397–12409
98. Ruderer MA, Wang C, Schaible E, Hexemer A, Xu T, Müller-Buschbaum P (2013) Macromolecules 46:4491–4501
99. T Pfadler, M Coric, CM Palumbiny, AC Jakowetz, K-P Strunk, JA Dorman, P Ehrenreich, C Wang, A Hexemer, R-Q Png, PKH Ho, P Müller-Buschbaum, J Weickert, L Schmidt-Mende (2014) ACS Nano 8(12):12397–12409. doi:10.1021/nn5064166
100. Gregg BA (2003) J Phys Chem B 107:4688–4698
101. Deibel C, Dyakonov V (2010) Rep Prog Phys 73:096401

102. Park SH, Roy A, Beaupre S, Cho S, Coates N, Moon JS, Moses D, Leclerc M, Lee K, Heeger AJ (2009) Nat Photonics 3:297–302
103. van de Craats AM, Warman J, Fechtenkötter A, Brand JD, Harbison M, Müllen K (1999) Adv Mater 11:1469–1472
104. Struijk CW, Sieval AB, Dakhorst JE, van Dijk M, Kimkes P, Koehorst RB, Donker H, Schaafsma TJ, Picken SJ, van de Craats AM (2000) J Am Chem Soc 122:11057–11066
105. Schmidt-Mende L, Fechtenkötter A, Müllen K, Moons E, Friend RH, MacKenzie J (2001) Science 293:1119–1122
106. Schmidtke JP, Friend RH, Kastler M, Müllen K (2006) J Chem Phys 124:174704
107. Li J, Kastler M, Pisula W, Robertson JWF, Wasserfallen D, Grimsdale AC, Wu J, Müllen K (2007) Adv Funct Mater 17:2528–2533
108. May F, Marcon V, Hansen MR, Grozema F, Andrienko D (2011) J Mater Chem 21:9538–9545
109. Hesse HC, Weickert J, Al-Hussein M, Dössel L, Feng X, Müllen K, Schmidt-Mende L (2010) Sol Energy Mater Sol Cells 94:560–567
110. Al-Hussein M, Hesse HC, Weickert J, Dössel L, Feng X, Müllen K, Schmidt-Mende L (2011) Thin Solid Films 520:307–313
111. Provencher F, Laprade JF, Côté M, Silva C (2009) Phys Status Solidi (c) 6:93–96
112. Fechtenkötter A, Saalwächter K, Harbison MA, Müllen K, Spiess HW (1999) Angew Chem Int Ed 38:3039–3042
113. Hesse HC, Schaffer C, Hundschell C, Narita A, Feng X, Müllen K, Nickel B, Schmidt-Mende L (2012) Phys Status Solidi (a) 209:785–789
114. Hesse HC, Weickert J, Hundschell C, Feng X, Müllen K, Nickel B, Mozer AJ, Schmidt-Mende L (2011) Adv Energy Mater 1:861–869
115. Guo LJ (2007) Adv Mater 19:495–513
116. Hesse HC, Lembke D, Dössel L, Feng X, Müllen K, Schmidt-Mende L (2011) Nanotechnology 22:055303
117. Lee W, Ji R, Gosele U, Nielsch K (2006) Nat Mater 5:741–747
118. Png RQ, Chia PJ, Tang JC, Liu B, Sivaramakrishnan S, Zhou M, Khong SH, Chan HS, Burroughes JH, Chua LL, Friend RH, Ho PK (2010) Nat Mater 9:152–158
119. Liu B, Png R-Q, Zhao L-H, Chua L-L, Friend RH, Ho PKH (2012) Nat Commun 3:1321
120. Watkins PK, Walker AB, Verschoor GLB (2005) Nano Lett 5:1814–1818
121. Weickert J, Dunbar RB, Hesse HC, Wiedemann W, Schmidt-Mende L (2011) Adv Mater 23:1810–1828
122. Coakley KM, McGehee MD (2004) Chem Mater 16:4533–4542
123. Venkataraman D, Yurt S, Venkatraman BH, Gavvalapalli N (2010) J Phys Chem Lett 1:947–958
124. He X, Gao F, Tu G, Hasko D, Hüttner S, Steiner U, Greenham NC, Friend RH, Huck WTS (2010) Nano Lett 10:1302–1307
125. Yan H, Chen Z, Zheng Y, Newman C, Quinn JR, Dotz F, Kastler M, Facchetti A (2009) Nature 457:679–686
126. Rivnay J, Toney MF, Zheng Y, Kauvar IV, Chen Z, Wagner V, Facchetti A, Salleo A (2010) Adv Mater 22:4359–4363
127. Rivnay J, Steyrleuthner R, Jimison LH, Casadei A, Chen Z, Toney MF, Facchetti A, Neher D, Salleo A (2011) Macromolecules 44:5246–5255
128. Dang MT, Hirsch L, Wantz G (2011) Adv Mater 23:3597–3602
129. Mori D, Benten H, Okada I, Ohkita H, Ito S (2014) Energy Environ Sci 7:2939–2943
130. Burke TM, McGehee MD (2014) Adv Mater 26:1923–1928
131. Jamieson FC, Domingo EB, McCarthy-Ward T, Heeney M, Stingelin N, Durrant JR (2012) Chem Sci 3:485–492

Tuning Side Chain and Main Chain Order in a Prototypical Donor–Acceptor Copolymer: Implications for Optical, Electronic, and Photovoltaic Characteristics

Marcel Schubert, Johannes Frisch, Sybille Allard, Eduard Preis, Ullrich Scherf, Norbert Koch, and Dieter Neher

Contents

M. Schubert (✉)
Soft Matter Photonics, School of Physics and Astronomy, University of St Andrews, KY16 9SS St Andrews, UK
e-mail: ms293@st-andrews.ac.uk

J. Frisch
Institut für Physik and IRIS Adlershof, Humboldt-Universität zu Berlin, Brook-Taylor-Straße 6, 12489 Berlin, Germany
e-mail: johannes.frisch@helmholtz-berlin.de

S. Allard • E. Preis • U. Scherf
Macromolecular Chemistry and Institute for Polymer Technology, Bergische Universität Wuppertal, Gaußstraße 20, 42119 Wuppertal, Germany
e-mail: sallard@uni-wuppertal.de; preis@uni-wuppertal.de; scherf@uni-wuppertal.de

N. Koch
Institut für Physik and IRIS Adlershof, Humboldt-Universität zu Berlin, Brook-Taylor-Straße 6, 12489 Berlin, Germany

Helmholtz Zentrum Berlin für Materialien und Energie GmbH, Albert-Einstein-Straße 15, 12489 Berlin, Germany
e-mail: norbert.koch@physik.hu-berlin.de

D. Neher
Physics of Soft Matter, School of Physics and Astronomy, University of Potsdam, 14476 Potsdam, Germany
e-mail: neher@uni-potsdam.de

K. Leo (ed.), *Elementary Processes in Organic Photovoltaics*, Advances in Polymer Science 272, DOI 10.1007/978-3-319-28338-8_10

Abstract The recent development of donor–acceptor copolymers has led to an enormous improvement in the performance of organic solar cells and organic field-effect transistors. Here we describe the synthesis, detailed characterisation, and application of a series of structurally modified copolymers to investigate fundamental structure–property relationships in this class of conjugated polymers. The interplay between chemical structure and optoelectronic properties is investigated. These are further correlated to the charge transport and solar cell performance, which allows us to link their chemical structure to the observed physical properties.

Keywords Aggregate states • All-polymer heterojunctions • Alternating copolymers • Ambipolar charge transport • Ambipolar materials • Backbone modifications • Bilayer solar cells • Charge separation • Conformational disorder • Crystalline phases • Donor–acceptor copolymers • Electron traps • Energetic disorder • Energy-level alignment • Fermi-level alignment • Fermi-level pinning • Interface dipole • Interlayer • Intrachain order • Intragap states • Microscopic morphology • Mobility imbalance • Mobility relaxation • Monte Carlo simulation • Multiple trapping model • Nonradiative recombination • OFET • Open-circuit voltage • Optoelectronic properties • Partially alternating copolymers • Photo-CELIV • Photocurrent • Photovoltaic gap • Polymer intermixing • Recombination losses • Spectral diffusion • Statistical copolymers • Stille-type cross-coupling • Structure–property relationships • Time-dependent mobility • Time-of-flight (TOF) • Transient photocurrent • Ultraviolet photoelectron spectroscopy • Vacuum-level alignment • X-ray photoelectron spectroscopy

1 Introduction to Donor–Acceptor Copolymers

The development and optimization of donor–acceptor (D–A) copolymers has greatly improved the performance of conjugated polymers over the past decade. Comprising an electron-rich (donor) and an electron-deficient (acceptor) molecular building block, these copolymers allow the systematic and easy modification of the material's optoelectronic properties [1] (Fig. 1). They can be widely tuned with respect to their optical bandgap and molecular frontier orbitals. Most impressively, the charge carrier mobility of the latest generation of D–A copolymers exceeds values of 10 cm^2/Vs [2, 3]. Being on a par with the performance of amorphous silicon and other inorganic semiconductors renders them the ideal materials for applications such as organic thin-film transistors, inverters, and circuitry. Furthermore, they dis-

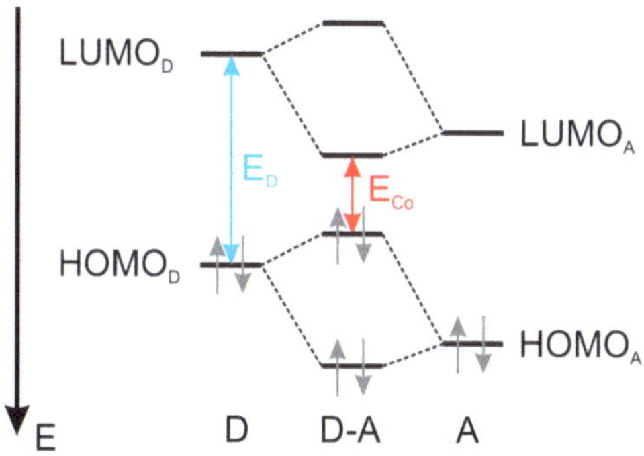

Fig. 1 Energy scheme illustrating the working principle of copolymers built up by the covalent binding of an electron-deficient acceptor (A) unit with an electron-rich donor (D) fragment. Due to the hybridization of the frontier molecular orbitals, the bandgap of the copolymer (E_{Co}) is strongly reduced compared to those of the donor (E_D) or acceptor block. For a rough estimate, E_{Co} is given by the energy difference between the highest occupied molecular orbital of the donor and the lowest unoccupied molecular orbital of the acceptor, while there are more chemical and physical parameters that can affect the bandgap [1]

play a class of organic semiconductors that comprises a large number of ambipolar materials, which possess comparable mobilities for electrons and holes [4].

The concept of combining chemically and structurally different monomer units into one backbone promoted the development of an enormous number of copolymer structures with an equal variety of physical and chemical properties. Consequently, D–A copolymers are now the first choice for the donor component in organic solar cells, where being able to precisely adjust the energetic and optical properties is of fundamental importance. On the other hand, they can also be incorporated as electron acceptors because of their high electron mobilities [5–8]. So-called all-polymer solar cells, which combine both donor and acceptor polymers, have recently demonstrated efficiencies above 5 % [9–11]. Thus, electron-transporting D–A copolymers have the potential to rival the dominating role of fullerene-based acceptors, which have been the predominant acceptors in organic solar cells for more than 20 years. Even copolymers with very similar chemical structures can now be combined as donor and acceptor within the same active layer [12], potentially avoiding difficulties arising from the different mechanical properties of polymers and small molecules.

In addition to tuning the electronic properties, D–A copolymers offer a wide variability in the nature and sequence of the side chains, thereby controlling packing and interchain interaction and opening exciting opportunities for the control of the blend morphology [13].

Another approach that is not yet in widespread use is breaking the strict alternation of the donor and acceptor units along the copolymer backbone, thereby altering the intrachain order. It is expected that the transition from strictly alternating to partially alternating or even statistical copolymer will result in subtle changes to the optical, electronic, and charge transport properties, with important consequences for their application in photovoltaic devices. Theoretical work and numerical simulations have indeed proposed that imbalanced electron and hole mobilities as well as an increased degree of energetic disorder of the donor and acceptor material may assist the split of electron–hole pairs at D–A heterojunctions [14–16]. Manipulating the energetic disorder by changing the intrachain order is an attractive approach to experimentally investigate this correlation.

Here we summarize the synthesis and investigation of a series of D–A copolymers with 9,9-dialkylfluorene (F) and 4,7-di(thiophene-2-yl)-2,1,3-benzothiadiazole (TBTT) building blocks in different arrangements. Poly(FTBTT) (PFTBTT) copolymers are prototypical, given that they were among the first D–A-type copolymers designed especially for use in organic solar cells [17]. Since then various studies have investigated the charge generation and charge recombination processes, ultrafast photophysics, and morphology of PFTBTT-based solar cells [18–26].

2 Synthesis of Alternating and Partially Alternating PFTBTT Copolymers

The aim of the synthetic part was generating a series of structurally modified D–A copolymers. Additional goals were having a sufficiently high electron mobility to apply them as the acceptor component in all-polymer solar cells and investigating the influence of the degree of intrachain order on the charge carrier mobility and photovoltaic performance.

First, two alternating copolymers were investigated, alt-PF8TBTT with linear octyl and alt-PF8/12TBTT with branched octyldodecyl substituents attached to the fluorene unit. The change from linear to branched side chains was expected to alter the interchain order in the solid state, thus also affecting the charge carrier mobility and device performance.

The copolymers were synthesized from 2,7-dibromo-9,9-dialkyl fluorene [27] and distannylated TBTT in a Stille-type cross-coupling reaction following procedures described in the literature [28, 29] (Fig. 2). Copolymer alt-PF8TBTT was isolated with a molecular weight M_n of 5000 g/mol and a polydispersity index (PDI) of 2.0–2.5. Copolymer alt-PF8/12TBTT was generated with a molecular weight M_n of 8000 g/mol and a PDI of 2.1.

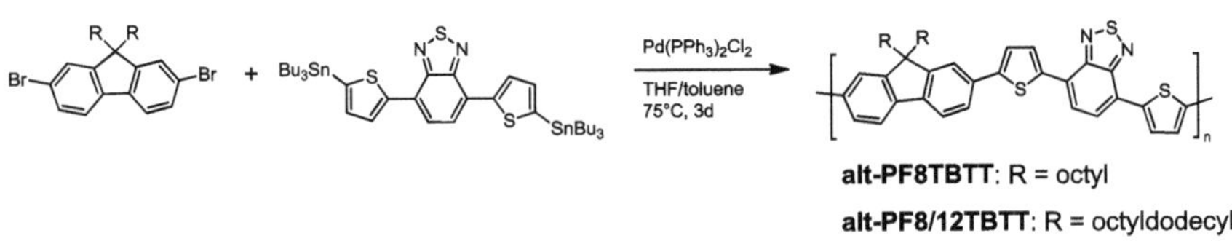

Fig. 2 Synthesis of alternating PFTBTT copolymers via Stille-type coupling reactions

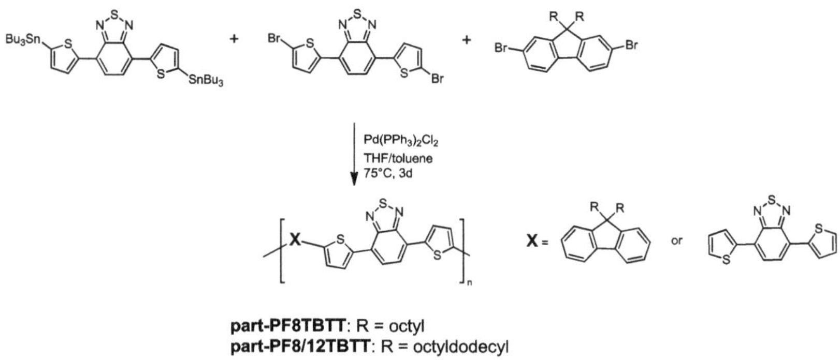

Fig. 3 Synthesis of partially alternating copolymers via a Stille-type coupling reaction that introduces a higher TBTT content into the backbone. The enrichment is estimated to be about 5 % compared to the alternating copolymer, yielding blocks of several directly connected TBTT units

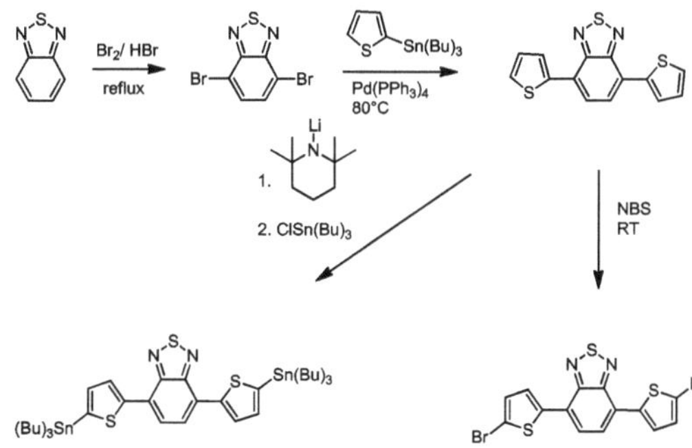

Fig. 4 Monomer synthesis of the distannylated TBTT (*left*) and the dibrominated TBTT (*right*)

To increase the intrachain disorder, we also synthesized two partially alternating copolymers, part-PF8TBTT and part-PF8/12TBTT, in a Stille-type cross-coupling reaction starting with 2,7-dibromo-9,9-dialkyl fluorene, distannylated TBTT, and dibromo TBTT, in which the TBTT monomer unit was randomly connected to TBTT or fluorene units (Fig. 3).

The monomer syntheses followed literature procedures, while the concept of partially alternating copolymers displayed a novel attempt to gradually degrade the otherwise perfect alternation of D–A copolymers [28–33]. The monomer syntheses of dibromo or distannylated TBTT starting from benzothiadiazole are illustrated in Fig. 4. The stoichiometry was adjusted to equimolar amounts of bromo and stannyl functions. Further analysis revealed that these partially alternating copolymers contain a TBTT content of about 50 %. Thus, the overall composition is comparable to those of the alternating copolymers. However, according to Fig. 3, the partially alternating copolymers contain blocks of several directly connected TBTT units although the probability for the formation of these blocks is rather small. Table 1 summarizes the molecular weights and optical properties of all copolymers.

Table 1 Number- (M_n) and weight-averaged (M_w) molecular masses, polydispersity (PDI), and maximum absorbance (Abs_{max}) in chloroform of all copolymers used in this study. Copolymer structures: *alt,* alternating; *part,* partially alternating; *stat,* statistical. Side chains: 8 = octyl; 8/12 = octyldodecyl; 2/6 = ethylhexyl; 3/12 = farnesyl

	M_n (g/mol)	M_w (g/mol)	PDI (M_w/M_n)	Abs_{max} ($CHCl_3$) (nm)
alt-PF8TBTT	5000	10,000	2.00	384, 536
alt-PF8/12TBTT	8000	17,000	2.12	384, 536
alt-PF3/12TBTT	4900	13,300	2.71	357, 530
part-PF8TBTT	3000	5000	1.66	378, 536
part-PF8/12TBTT	8200	13,000	1.58	378, 542
stat-PF2/6TBTT (30 % TBTT)	10,000	20,000	2	355, 522
stat-PF3/12TBTT (2 % TBTT)	10,000	15,000	1.5	–

To alter the copolymer structure even stronger, we synthesized statistical PFTBTTs according to Yamamoto-type homo-coupling of the two dibrominated monomers [33, 34]. The resulting copolymers showed a higher molecular weight but only a maximum TBTT content of about 30 % by using a monomer molar ratio of 60:40 (TBTT:F) (see Table 1). A series of Yamamoto-type reactions showed that the incorporation ratio of TBTT depends less on the monomer ratio used, revealing a lower reactivity of the electron-poor building block into the polymer chains. Because of their gradually increasing TBTT content (from about 2–30 %), statistical copolymers were used to investigate the effect of the polymer backbone composition on the copolymers' optoelectronic properties.

3 Energetic and Optical Properties of Pristine Copolymers

In order to characterize the alteration of the physical properties introduced by the chemical modifications, we first present a thorough investigation of the energetic structure and optical properties of a series of different PFTBTT copolymers. Structurally different backbone modifications are introduced with so-called partially alternating copolymers. Finally, we compare alternating and partially alternating copolymers with different side chains.

3.1 Influence of Acceptor Concentration in D–A Copolymers on the Energy-Level Alignment

To determine the effects of different acceptor concentrations on the optical and electronic properties, we investigated spin-coated films of different PFTBTT copolymers on poly(3,4-ethylene dioxy thiophene):poly(styrene sulfonic acid) (PEDT:PSS) substrates. The PEDT:PSS was used to ensure the same polymer

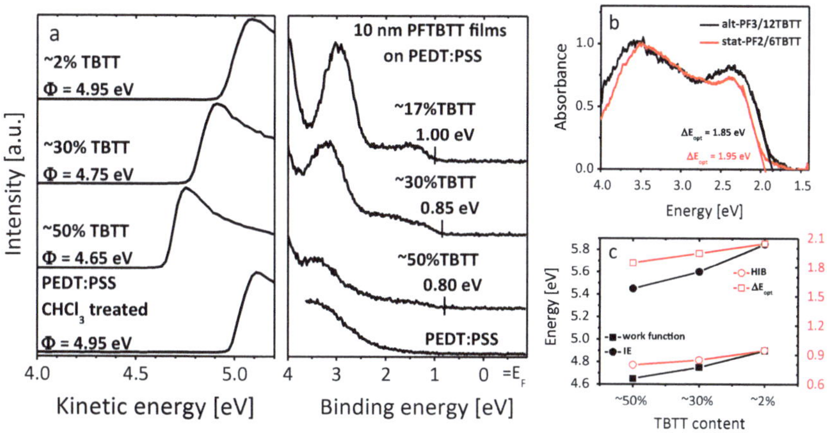

Fig. 5 (**a**) Secondary electron cutoff (*left*) and valence band spectra (*right*) of spin-coated 10-nm PFTBTT films on PEDT:PSS substrates. Polymer films differ in the acceptor (TBTT) content value, from ~50 % for alternating copolymers to ~30 and 2 % for statistical copolymers. (**b**) Ultraviolet–visible spectroscopy (UV-Vis) absorption spectra of spin-coated 10-nm alt-PF3/12TBTT (*black*) and stat-PF2/6TBTT (*red*) films on PEDT:PSS substrates. The absorption spectra of PEDT:PSS were subtracted. (**c**) Work function, ionization energy, hole injection barrier, and ΔE_{opt} (optical gap) of PFTBTT films with varying TBTT content on top of PEDT:PSS substrates determined by ultraviolet photoelectron spectroscopy and UV-Vis measurements

layer morphology as in the electrical and optoelectronic studies. The acceptor concentration was varied among ~50 % (alt-PF3/12TBTT), ~30 % (stat-PF2/6TBTT), and ~2 % (stat-PF3/12TBTT), and the energy-level alignment was probed with ultraviolet photoelectron spectroscopy (UPS). From these measurements, we determined the secondary electron cutoff (SECO) and valence band (VB) spectra, which reveal the work function (Φ) and valence band onset of the copolymers on top of PDOT:PSS, respectively.

SECO and VB spectra of the different copolymers and of the PEDT:PSS electrode are shown in Fig. 5a. As the acceptor (TBTT) content increases, the emission feature at 2.7 eV shifts to higher binding energy and broadens, whereas the VB onset elongates toward the Fermi energy (E_F). However, Φ and the hole injection barrier (HIB) decrease with increasing acceptor content. Consequently, the ionization energy (IE) changes from 5.45 eV (~50 % TBTT) to 5.60 eV (~30 % TBTT) and to 5.85 eV (~2 % TBTT). The value for low TBTT content is close to that found in pure polyfluorene, where the IE was determined to be 5.75 eV [35]. Note that the substrate work function did not change after deposition of a PFTBTT film with ~2 % acceptor content, suggesting an established vacuum-level alignment. This situation changes for higher acceptor concentrations and, thus, smaller IEs. The substrate Φ decreases almost linearly with increasing acceptor content to 4.75 eV (~30 % TBTT) and 4.65 eV (~50 % TBTT). Accordingly, the Φ value where the transition between vacuum-level alignment and Fermi-level pinning occurs is far below 4.95 eV for higher acceptor content in PFTBTT copolymers. Consequently, pinning of the Fermi level occurs at intragap states between the VB maximum and E_F, whose density is expected to be well below the sensitivity of the present UPS experimental setup [36–38]. Such high densities of gap states could potentially hinder the charge transport within the PFTBTT phase and could also negatively affect the photovoltaic performance.

Similar results can be found for the optical gap (ΔE_{opt}) determined by the onset of the optical absorption spectra (see Fig. 5b). The onset is shifted by 0.1 eV to higher energies upon decreasing the acceptor content from ~50 to ~30 %. The optical gap is further increased to 2.05 eV for lower acceptor concentrations (see Fig. 5c). By subtracting ΔE_{opt} from the IE, we can estimate the electron affinity (EA) of the polymers. Accordingly, the acceptor concentration changes both the IE and the EA. The characteristic values (Φ, HIB, ΔE_{opt} and IE) are summarized in Fig. 5c as a function of the acceptor concentration. Note that the HIB does not change in parallel to the IE. Therefore, calculating the VB onset position with respect to the Fermi level, assuming vacuum-level alignment at the polymer/PEDT:PSS interface, fails for PFTBTT films with ~30 and ~50 % acceptor concentrations because of the pinning of the Fermi level at intragap states.

3.2 Aggregation Effects in Partially Alternating Copolymers

Partially alternating copolymers were synthesized in order to have a lower degree of structural disorder compared to the statistical copolymers discussed above. Although the chemical modifications we introduced are only minor, the direct comparison between partially alternating copolymers and their alternating counterparts reveals distinct changes to the optical properties. Absorption spectra for copolymers with two different side chain substitutions in solution and in thin spin-coated films are displayed in Fig. 6. In solution, partially alternating copolymers show broadened absorption spectra, again indicating an increase in the energetic disorder. In addition,

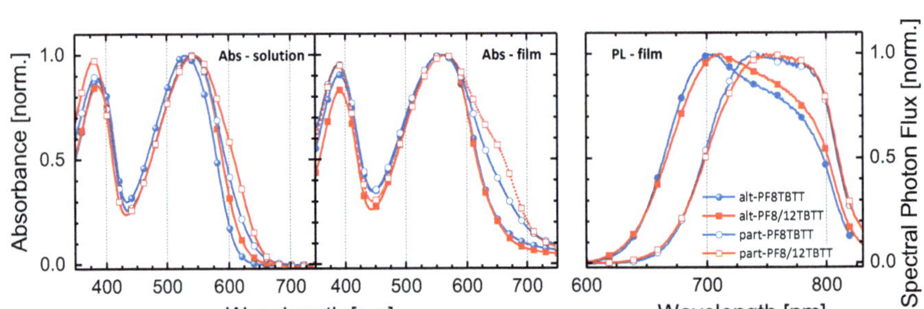

Fig. 6 Normalized absorbance (Abs) of alternating (*closed symbols*) and partially alternating (*open symbols*) PFTBTT copolymers measured in solution (*left*) and in thin film (*middle*). Also shown are normalized photoluminescence film spectra (*right*)

the thin-film absorption of the partially alternating copolymers is redshifted by about 0.15–0.2 eV. This shift is caused by the appearance of an additional absorption shoulder centered at about 650 nm, while the two main absorption bands around 375 and 550 nm remain clearly visible in all four derivatives. A redshift is also observed in the photoluminescence spectra, which proves that the partially alternating copolymers comprise a reduced optical bandgap. The appearance of such spectral features has been assigned to aggregate states that form upon the aggregation or crystallization of conjugated polymers; they are redshifted because of an improved planarization of the backbone [39, 40]. Furthermore, these states dominate the emission characteristics because of their fast spectral diffusion into these lower-energy states. The formation of aggregate states is further supported by the absorption spectra recorded in a good solvent (chloroform) and at low concentration (0.1 g/L), where the formation of aggregates is suppressed [20, 40, 41]. Most likely, aggregation of the partially alternating moieties is induced by multiple connected TBTT segments that induce stacking of neighbored chains. On the other hand, we attribute the absence of the red-edge absorbance in alternating and statistical copolymers to a rather disordered and amorphous film morphology, which is consistent with the reported absence of X-ray diffraction signals in alt-PF8TBTT [20].

4 Charge Transport Properties of Alternating, Partially Alternating, and Statistical Copolymers

Ambipolar charge transport for the PFTBTT-type copolymer F8TBT was first demonstrated in a light-emitting organic field-effect transistor (OFET) structure [21]. Figure 7 displays OFET characteristics measured in a bottom gate–top electrode geometry with a silanized silicon dioxide gate insulator. The latter was introduced to prevent electron trapping at the interface between the gate dielectric and the conjugated polymer [42]. Ambipolar charge transport in alt-PF8TBTT is clearly visible with mobilities for holes and electrons of $\mu_h = 2 \times 10^{-3}$ cm^2/Vs and $\mu_e = 4 \times 10^{-4}$ cm^2/Vs, respectively. Thus, alt-PF8TBTT and related copolymers can be regarded as ambipolar materials with an almost balanced electron and hole transport. This is consistent with the excellent acceptor properties of alt-PF8/12TBTT in organic solar cells [25].

Transient photocurrent measurements were performed to address the bulk electron transport in the three different copolymer structures. The technique of photo-generated charge extraction by linearly increasing voltage (photo-CELIV) is applied because it allows us to probe time-dependent charge transport phenomena in organic layers with active layer thicknesses of about 100 nm [43]. A charge-generation layer was introduced in the devices to ensure that the transport of electrons is measured [44]. Extracted electron mobilities are displayed in Fig. 8. The mobility differs by almost two orders of magnitude and is clearly related to the degree of structural

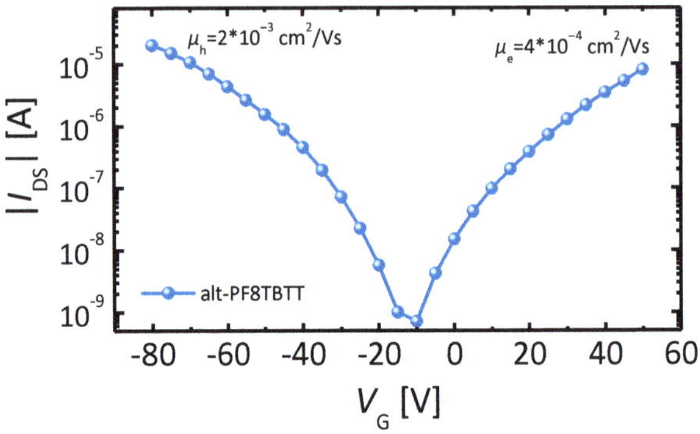

Fig. 7 Transfer characteristic of alt-PF8TBTT in bottom gate field-effect transistor at a source-drain voltage of −80 V

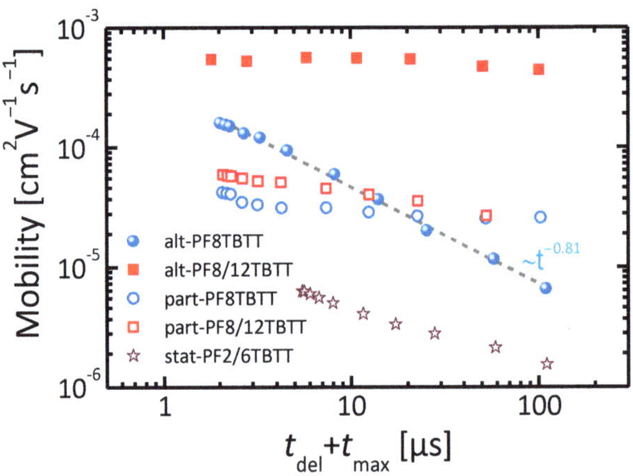

Fig. 8 Photogenerated charge extraction by linearly increasing voltage (photo-CELIV) results for the electron mobility of alternating (*filled squares* and *spheres*), partially alternating (*open squares* and *circles*), and statistical (*open stars*, 32 % TBTT) PFTBTT copolymers. To improve charge generation, all devices featured a 3-nm-thin P3HT charge-generation layer. Complete device structure: ITO/PEDT:PSS/P3HT-IL/PFTBTT/Sm/Al; device area: 1 mm^2. Layer thickness: P3HT-IL = 3 nm, PFTBTT = 50–200 nm. Details about the device preparation and analysis, including calculation of the mobility, were published earlier [44]

disorder in the copolymer backbone. Upon a short delay, the highest mobility is measured for the two alternating copolymers, followed by the partially alternating copolymers. The statistical copolymer has by far the lowest mobility. This could be a result of the lower TBTT content, which effectively reduces the number of low-energy transport sites. In addition, the high structural disorder is likely to introduce an increased energetic disorder and could result in the formation of electron traps, as described ahead.

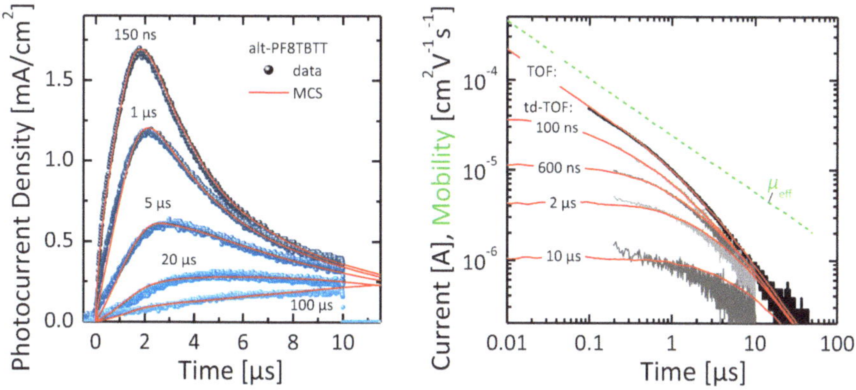

Fig. 9 (*left*) Photogenerated charge extraction by linearly increasing voltage (photo-CELIV) transients for alt-PF8TBTT measured in a charge-generation layer device. All transients are shifted along the time axis to set the beginning of the extraction pulse to $t = 0$. The delay between photoexcitation and the beginning of the linearly increasing extraction voltage is indicated on each transient. (*right*) Time-of-flight (TOF) and time-delayed-TOF transients, where the delay time is indicated for the latter. In both graphs, *red lines* display results from Monte Carlo simulations assuming a trapping and release-–dominated electron transport. The effective mobility (μ_{eff}) in such a system is time dependent and undergoes a power law decay as it is displayed (*green dashed line*)

In addition to the large differences in the magnitude of the mobilities, we observe that the charge carrier mobility for several copolymers is not constant in time but instead displays a severe relaxation. Although described before, time-dependent mobilities are still a poorly understood phenomenon and were thus further investigated. The fastest mobility decay is observed for alt-PF8TBTT, where it follows a power law decay. This is accompanied by a change in the shape of the photo-CELIV current transients (Fig. 9), which shows a decreasing and shifting photocurrent signal that lacks a clear maximum after prolonged delay. The charge-generation-layer technique also allowed us to perform time-of-flight (TOF) and a newly developed technique called time-delayed TOF (td-TOF) measurements on the same devices. Introducing a delayed extraction renders the characteristic kink in the TOF transient to disappear.

The anomalous shape of the photocurrent transients and how it is related to the mobility relaxation was further analyzed with Monte Carlo–type computer simulations. A time-dependent mobility has been observed in systems where the charge transport is affected by an exponential density of trap states [45]. Within this so-called multiple trapping model, trapped charges must be thermally activated to a manifold of transport states. To account for the high electrical fields that are applied at the end of a photo-CELIV pulse, an additional field-dependent detrapping mechanism has been introduced [44]. With this model, all transient photocurrent experiments conducted on alt-PF8TBTT could be simulated with a single parameter set as can be seen in Fig. 9. Thus, we may conclude that the observed changes in the characteristics of the transients are related to the continuous relaxation of the electron mobility caused by the presence of an exponential distribution of electron traps.

Trap-dominated electron currents are a common phenomenon in steady-state current voltage measurements of conjugated polymers. It has been assumed previously that this might be the result of chemical defects that are unintentionally introduced during the synthesis and processing of organic semiconductors [46]. In contrast to this theory, we observe strong differences in the transport characteristics of alt-PF8TBTT and PF8/12TBTT, in a way that one is trap-controlled while the other is trap-free. However, both have been synthesized by following the same procedures and both copolymers are different only in the length of the solubilizing side chains. Also, the device preparation and characterization were carried out under the same conditions. We therefore propose that trapping in alt-PF8TBTT is caused by effects related to the molecular structure or microscopic morphology rather than originating from chemical defects [44]. Conformational disorder in amorphous poly(fluorene) aggregates has been found to increase upon octyl-substitution by combined molecular dynamics and density functional theory–based simulations [47]. This increased energetic disorder might also be the origin of the mobility relaxation of the statistical copolymer stat-PF2/6TBTT. This raises the question of the missing trapping effects in partially alternating copolymers that were shown before to comprise a significantly broadened energy landscape. However, transport in these copolymers might be dominated by the additionally formed aggregate states (Fig. 6), which are lower in energy [48]. The important role of aggregate states for electron transport was recently described for a high-mobility D–A copolymer [13].

The various effects observed in electron transport characteristics underline the importance of the backbone regularity of D–A -copolymers. Even small imperfections in the alternation of D–A units will have negative effects on the charge transport because of a significant increase in energetic disorder. However, the formation of aggregate or crystalline phases might compensate for the increased disorder, which increases the complexity of the interplay between energetic and transport properties. Moreover, side chains provide an important parameter for optimizing the copolymers.

5 Application in Bilayer Solar Cells

To investigate the correlation between the electronic parameters of the newly synthesized acceptor polymers and the final photovoltaic properties, we prepared bilayer solar cells using the interlayer approach. We chose this rather simple device structure to avoid the complexity that goes along with the intermixed morphology in bulk heterojunction polymer blends. We formed bilayer all-polymer solar cells by first preparing a thin insoluble donor layer consisting of poly(3-hexylthiophene) on which a thicker layer of the PFTBTT copolymers is spin coated [44]. The investigation starts again with a detailed analysis of the interlayer, followed by the description of the energy-level alignment at the P3HT/PFTBTT heterojunction. Finally, we studied the effect of depositing a metal top contact to determine the energetic structure of the complete device. This section presents the photovoltaic properties and interprets them in view of the measured optical, energetic and charge transport properties.

5.1 PFTBTT Copolymers as Acceptors Combined with P3HT

5.1.1 Energy-Level Alignment at the P3HT/PFTBTT Interface

The P3HT-interlayer (P3HT-IL) is an insolubilized P3HT film with a layer thickness of 3–6 nm [49]. As can be seen in Fig. 10, Φ and the position of the VB onset of the P3HT-IL are identical compared to a spin-coated P3HT film with a thickness of 10 nm. Unfortunately, this only holds for the same material and solvent. Different energy-level positions are observed for P3HT with different molecular weights, regioregularities, solvents, or preparation conditions or that come from different providers [50, 51]. All these parameters influence the order of the individual

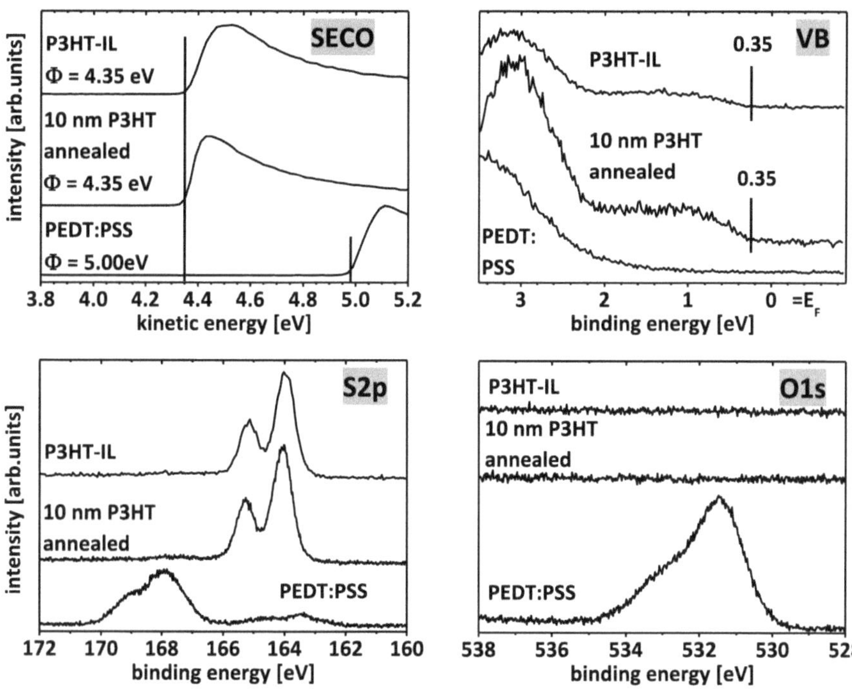

Fig. 10 Secondary electron cutoff, valence band, S2p, and O1s core-level spectra of the pristine PEDT:PSS substrate, a 10-nm P3HT film annealed in ultrahigh vacuum, and those of the P3HT-IL

polymer chains within the film and the micrometer morphology of the polymer layer. In particular, the tilt angle of the polymer backbone plane with respect to the substrate influences the IE of the films [52, 53]. Here, as shown in Fig. 10, Φ of the pristine PEDT:PSS layer below was decreased by 0.65 eV because of the formation of the P3HT-IL. The formed interface dipole at the anode–donor interface defines the effective Φ of this material combination [54]. It is thus important when correlating material parameters and device performance to note that the effective Φ is the relevant value for the anode side in the device in contrast to Φ of the bare PEDT:PSS anode. Additionally, no holes or pinholes can be observed in the P3HT-IL. This is indicated by the complete attenuation of all substrate core-level features due to the interlayer formation (see core-level region spectra in Fig. 10). Therefore, the P3HT-IL can be used to study the energy-level alignment at the intimate P3HT-IL/PF8TBTT and P3HT-IL/PF8/12TBTT interface, preventing direct contact between the acceptor polymer and the anode material as well as intermixing of the two polymers.

The VB spectra of the alternating and partially alternating PF8TBTT and PF8/12TBTT spin coated on the P3HT-IL (Fig. 11) show no more intensity close to E_F originating from the underlying P3HT-IL. The absence of P3HT features is evidence for complete coverage of the P3HT-IL by the PFTBTT copolymers and the absence of pronounced polymer intermixing. The film thickness of all polymers was ~20 nm. The sample Φ (see SECO in Fig. 11) increased after film formation for all copolymers up to 4.80 eV. Adding the measured Φ values to the determined VB onset, we found that the resulting IEs range from 5.40 and 5.45 eV for part- and alt-PF8TBTT to 5.50 eV and 5.60 eV for part- and alt-PF8/12TBTT, respectively. It is important to note that all calculated IEs are much higher compared to the effective Φ of the PEDT:PSS/P3HT-IL substrate. Therefore, the Φ increase caused by the deposition of copolymers was unexpected. In contrast to the measured Φ shifts at the P3HT-IL/copolymer interface, vacuum-level alignment was predicted for this material combination [55, 56]. However, as noted before (see Sect. 3.1),

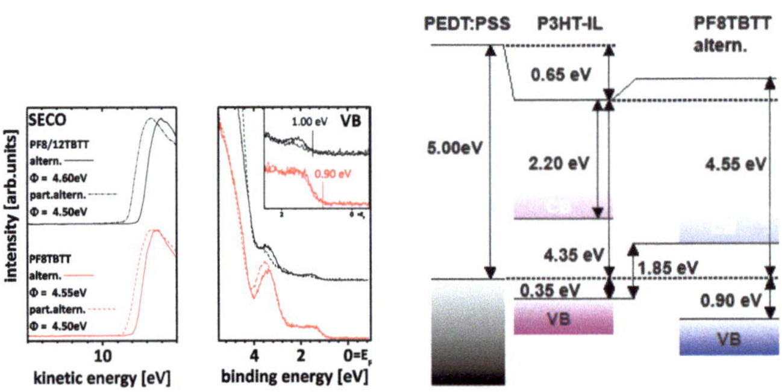

Fig. 11 (*left*) Secondary electron cutoff and valence band (VB) of the alternating and partially alternating PF8TBTT (*red*) and PF8/12TBTT (*black*) spin coated from chloroform solution on P3HT-IL. (*right*) Energy-level scheme at the P3HT-IL/alt-PF8TBTT interface. Work function (Φ), VB onset, and photovoltaic gap (Δ_{PVG}) for all copolymers investigated are summarized in Table 2

the low density of gap states in PF8TBTT and PF8/12TBTT does not show up in photoemission (because of their low density) but can cause pinning of E_F far in the gap [37].

UPS results for the PEDT:PSS/P3HT-IL/alt-PF8TBTT interfaces are summarized in the energy-level diagram in Fig. 11. The position of the P3HT conduction band (CB) onset was calculated by subtracting the transport gap of 2.5 eV from the VB onset position [57]. The transport gap of alt-PF8TBTT was estimated to be 2.4 eV (optical gap of 1.9 eV plus exciton binding energy of 0.5 eV). It is well accepted in the literature that the upper limit for the V_{OC} is given by the photovoltaic gap (Δ_{PVG}; the difference between the VB onset of the donor and the CB onset of the acceptor) in organic heterojunction solar cells [58, 59]. In the present case of P3HT and alt-PF8TBTT, a photovoltaic gap of 1.85 eV was determined from the energy-level diagram in Fig. 11. This value is, however, significantly larger than the energy difference between the ionization energy of P3HT (4.70 eV) and the electron affinity of the alt-PF8TBTT (3.05 eV) assuming vacuum-level alignment, which would yield 1.65 eV. This indicates the importance of directly determining the interfacial D–A energy-level alignment. However, compared to the measured V_{OC} of 1.21 eV in the device (see Table 3), the UPS-derived value of the photovoltaic gap was 0.64 eV larger. Losses on the order of 0.5 eV and above, as deduced directly from the UPS experiments, are in agreement with previously determined energetic losses in bulk heterojunction solar cells of alt-PF8TBTT and PCBM and equal those found in some of the best photovoltaic D–A combinations, which show a typical loss of 0.5–0.7 eV measured from the energy of the charge-transfer state [60, 61]. The energy difference between V_{OC} and the photovoltaic gap is evidence for nonradiative recombination processes and energetic disorder.

5.1.2 Full Electronic Structure

In order to acquire the energetic structure of a complete solar cell device, we used UPS to analyze the deposition of the samarium (Sm) top electrode. Figure 12 shows SECO and VB spectra for different Sm thicknesses. The sample Φ drastically decreased to 2.3 eV because of the deposition of 10 Å Sm and stabilized at 2.60 eV for a higher Sm coverage. The intensity of all copolymer-related VB features decreased from the early stage of interface formation. For increasing Sm coverage, the alt-PF8TBTT features shifted to a higher binding energy (BE) (Fig. 12) and the emission intensity from Sm close to the Fermi level increased. The PFTBTT features are not broadened but are clearly visible up to a Sm coverage of 2 Å. The energy-level shift of the alt-PF8TBTT features to a higher BE represents the magnitude of the built-in field because of Fermi-level alignment between the two electrodes in the device. As Fig. 12 shows, the maximum of the localized peak (at ~3.5 eV BE) was clearly shifted to a higher BE by ~0.85 eV. Unfortunately, the final position of the VB onset cannot be determined because of the rapid attenuation of the features above 2 Å Sm coverage. Because the EA of the alt-PF8TBTT of 3.35 eV (see previous section) is much larger than the Sm Φ, we expect an interfacial dipole

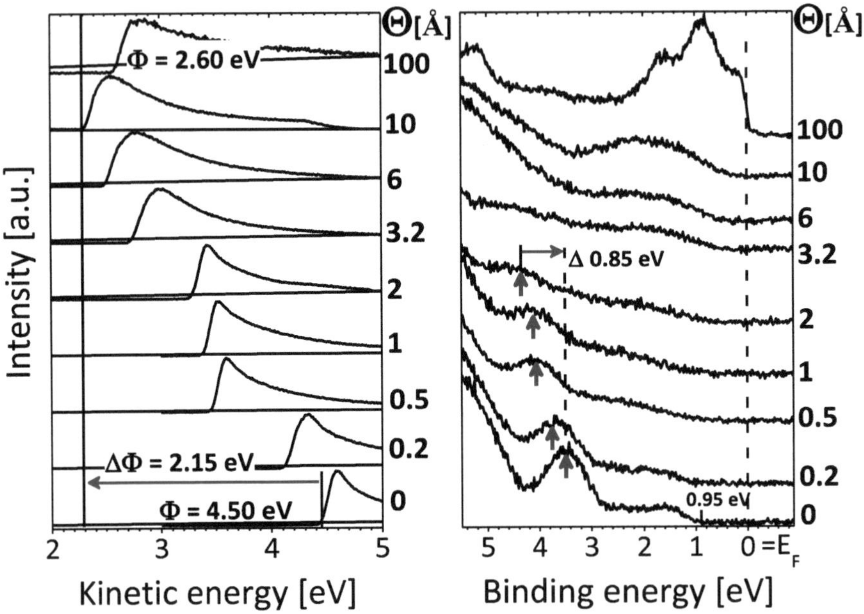

Fig. 12 Secondary electron cutoff (*left*) and photoemission spectra of the valence region (*right*) and for increasing Sm coverage (Φ) on a P3HT-IL/alt-PF8TBTT heterojunction. *Black arrows* indicate the position of the lowest localized π band of the alt-PF8TBTT

(ID) to be formed at the polymer/Sm interface as a result of unoccupied states of the polymer being pinned at E_F of the Sm electrode.

To determine the strength of the ID at the alt-PF8TBTT/Sm interface, we performed the reverse experiment—spin coating an alt-PF8TBTT film onto a pristine Sm layer (see Fig. 13). We first investigated the influence of the deposition method on the pristine Sm film by spin coating the pure solvent on a freshly evaporated Sm film. After the spin coating, the C and O concentrations on the Sm surface drastically increased [see X-ray photoelectron spectroscopy (XPS) survey spectra in Fig. 13]. In parallel, Φ of the Sm surface decreased by 0.2 eV and the VB region spectrum (see Fig. 13) became similar to that of SmO_X [62], as clearly indicated by the vanished photoelectron intensity at E_F. Both UPS and XPS clearly showed a complete oxidation of the Sm surface and the copolymer was deposited on a thin SmO_X film rather than on pristine Sm. Nevertheless, investigations of the energy-level alignment at the SmO_X/alt-PF8TBTT interface include important information about the pinning position of the Sm Fermi level with respect to the occupied states of the alt-PF8TBTT, where the E_F pinning was again induced by unoccupied states of the copolymer. These become occupied by electrons close to the interface and cause the formation of an interface dipole and consequently the observed Φ increase to 3.40 eV. The levels to which electrons are transferred could be either the CB minimum or states below if tail states in the gap close to

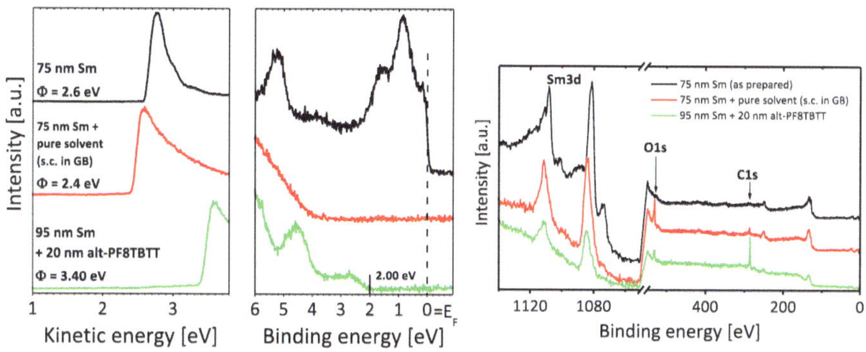

Fig. 13 Ultraviolet photoelectron spectroscopy spectra of the secondary electron cutoff (*left*) and valence region (*center*) as well as X-ray photoelectron spectroscopy (XPS) survey spectra (*right*) of a 75-nm Sm film vapor deposited in ultrahigh vacuum (UHV) before (*black*) and after direct transfer to an N_2-filled glove box (GB) in low-vacuum conditions (10^{-2} mbar) (*red line*). The *green* spectrum represents the XPS survey spectrum of a 20-nm alt-PF8TBTT film spin coated in an N_2-filled GB on a pristine 95-nm Sm film vapor deposited in UHV

the CB exist. The energy value at which the CB minimum becomes pinned above E_F depends on the effective density of states, the distribution of tail states, and the number of charges that needed to be transferred to reach an electronic equilibrium. However, the resulting interface dipole at the SmO_X/alt-PF8TBTT interface was determined to 1.00 eV. As shown in Fig. 13, the VB onset of the copolymer is located 2.00 eV below E_F. This is 0.25 eV higher than the last measurable position of the alt-PF8TBTT VB onset (at 1.75 eV BE) in the reverse deposition sequence described earlier.

In conclusion, Fermi-level alignment between the ITO/PEDT:PSS substrate and the Sm film is established independently of the polymer heterojunction in between. This is clearly indicated by the position of the Sm E_F at 0 eV BE. Note that Fermi-level alignment across the heterojunction is comparable to short-circuit conditions in a device. By combining the results of the photoemission experiments, one can derive the full electronic structure across all interfaces in the P3HT-IL/alt-PF8TBTT solar cell, as shown in Fig. 14 [63]. The established built-in field across the heterojunction at closed-circuit conditions can be determined to 1.10 eV. As can be seen in the energy-level diagram, because of the built-in field, the energy levels of the alt-PF8TBTT are tilted and Fermi-level alignment is established across the layers.

5.1.3 Bilayer Solar Cells with PFTBTT Copolymers as Acceptors

Investigating the effects of altered energetic disorder and charge transport properties on the photovoltaic performance of the different PFTBTT copolymers was a central aim of this study.

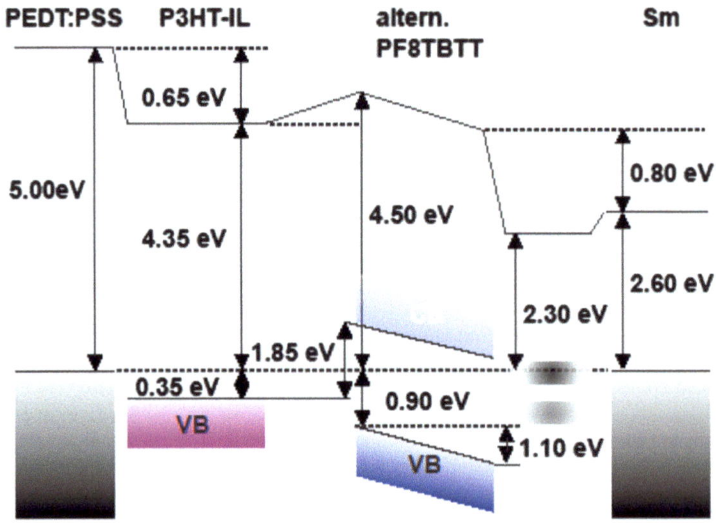

Fig. 14 Schematic energy-level diagram of a P3HT-IL/alt-PF8TBTT bilayer solar cell

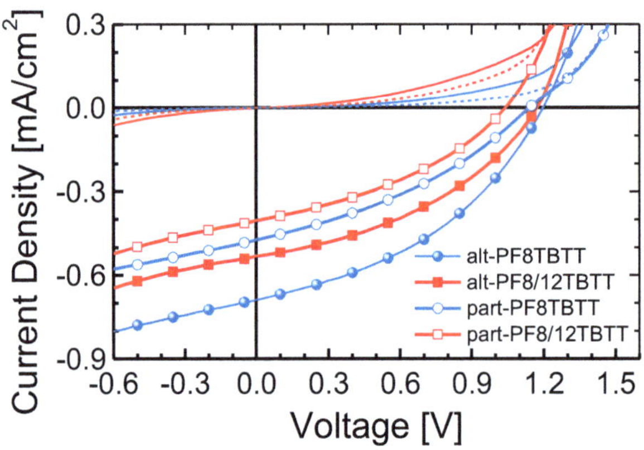

Fig. 15 Current–voltage characteristics of bilayer solar cells comprising PFTBTT copolymers as acceptor on top of a P3HT-IL measured at 100 mA/cm^2 (*symbols*) and in the dark (*lines*, *solid* = alt, *dashed* = part). Devices were annealed for 10 min at 140 °C. Device structure: ITO/PEDT:PSS/P3HT-interlayer(∼3 nm)/PFTBTT(40 nm)/Ca(20 nm)/Al(100 nm)

Typical current–voltage characteristics and solar cell parameters of P3HT-IL–based bilayer solar cells comprising different PFTBTT copolymers as acceptor are displayed in Fig. 15 and Table 3, respectively. From all copolymers, solar cells with statistical PFTBTT copolymer exhibit very low short-circuit currents, about an order of magnitude smaller than the current measured for alternating copolymers. Additionally, the solar cells suffer from very low fill factors of about 20 % and small open-circuit voltages. Hereby, we found that neither the structure of the side chains or that of the backbone (via varying TBTT fractions) considerably affects the solar cell performance of statistical copolymers. Since the statistical copolymer had the lowest electron mobility measured, the mobility imbalance between donor (P3HT) and acceptor (PFTBTTs) polymer was the largest [14]. However, the low photocurrent, fill factor, and open-circuit voltage—together all a clear sign for recombination-limited solar cells—clearly show that an increased mobility imbalance does not improve the photovoltaic performance. In fact, the severe recombination losses of free charge carriers are likely to be introduced by the accumulation of charges in the device, a direct consequence of the low electron mobility determined for stat-PF2/6TBTT (Fig. 8) [64]. Thus, we conclude that the potentially positive effect of a mobility imbalance is removed by an increased recombination. This concept might indeed only work if a high-mobility donor and acceptor component are combined to ensure an efficient separation of the electron and hole densities inside the active layer.

Compared to the statistical copolymers, partially alternating PFTBTTs show a much better photovoltaic response, but they are clearly not as good as the alternating derivatives. This is true for both side chain variants. Overall, the photocurrent follows almost the same trend as the electron mobility shown in Fig. 8. Only alt-PF8TBTT seems to fall away from this trend; however, as a result of the mobility relaxation, alt-PF8TBTT can be expected to have the highest mobility on timescales below a picosecond, which are more relevant for charge carrier generation [23, 24, 65]. Consequently, our results give some evidence that the charge separation efficiency benefits from a high electron mobility. On the other hand, our results show that the larger degree of disorder observed for the two partially alternating copolymers (Fig. 6) does not improve the photovoltaic performance as predicted by theory for ideal bilayer structures.

In addition to the impact of the mobility and disorder, we analyzed the interplay between the energetic structure and the photovoltaic parameters. We expected the open-circuit voltage to have a direct correlation to the experimentally determined energetic structure. In particular, we were investigating if the reduced optical bandgap of the partially alternating copolymers and the experimentally determined photovoltaic gap (Table 2) could be directly correlated to the open-circuit voltage. Indeed, the open-circuit voltage and the photovoltaic gap decreased in parallel when we compared alternating with partially alternating polymers, as can be seen in Table 3, although the correlation of the open-circuit voltage as a function of

Table 2 Work function F, valance band (VB) onset, and photovoltaic gap (Δ_{PVG}) for the different copolymer films deposited on P3HT-IL

Copolymer	Φ (eV)	VB onset (eV)	Δ_{PVG} (eV)
alt-PF8TBTT	4.55	0.90	1.85
part-PF8TBTT	4.50	0.90	1.65
alt-PF8/12TBTT	4.60	1.00	1.75
part-PF8/12TBTT	4.50	1.00	1.55

Table 3 Solar cell performance of alternating (alt), partially alternating (part), and statistical (stat) PFTBTT copolymers. All devices were annealed at 140 °C for 10 min before evaporation of the top contact. The structure of the solar cells was ITO/PEDT:PSS/P3HT-interlayer(∼3 nm)/PFTBTT(40 nm)/Ca(20 nm)/Al(100 nm)

PFTBTT	V_{OC} (V)	J_{SC} (mA/cm^2)	FF (%)	η_{PCE} (%)	$\Delta_{PVG}-V_{OC}$ (eV)
alt-PF8TBTT	1.21	0.690	40	0.33	0.64
alt-PF8/12TBTT	1.17	0.530	40	0.24	0.58
part-PF8TBTT	1.15	0.470	35	0.19	0.50
part-PF8/12TBTT	1.04	0.400	30	0.15	0.51
stat-PF2/6TBTT (30 % TBTT)	0.89	0.056	20	0.01	–

FF fill factor, *PCE* power conversion efficiency
Values for Δ_{PVG} were taken from Table 2

the photovoltaic gap was not linear. Because of different recombination kinetics at the different polymer heterojunctions and the relatively high dark currents observed in the interlayer devices, this nonlinear correlation was unexpected. A direct comparison of Δ_{PVG} and V_{OC} for alt-PF8TBTT and part-PF8TBTT revealed that the loss term ($\Delta_{PVG} - V_{OC}$) was 0.14 eV larger for alt-PF8TBTT than for part-PF8TBTT. Interestingly, this difference corresponds well with the reduction in the optical gap for the partially alternating copolymer. Consequently, the loss mechanism in the device seemed to be similar for alt-PF8TBTT and part-PF8TBTT, and the V_{OC} was directly limited by the energy-level alignment at the heterojunction, as we determined with UPS. We observed the same trend for the two PF8/12TBTT copolymers although the comparison was obscured by the higher dark currents, which resulted in a smaller difference of only 0.07 eV between the loss terms. Thus, we were able to link the experimentally determined energetic structure of several all-polymer heterojunctions to their photovoltaic performance, providing a new and more precise way to analyze the interplay between energy-level alignment and solar cell performance.

6 Summary

Our results show that a larger degree of disorder and mobility imbalance—as we introduced to the photovoltaic systems through chemical modifications of the acceptor polymers—does not improve the photovoltaic performance. The main reason for this lack of improvement is a decrease in the charge carrier mobility for polymers that do not comprise a strictly alternating copolymer structure. Ultimately, this inefficient electron transport increases the bimolecular recombination of free charges and drastically reduces device performance. This finding is in accordance with recent computer simulations and analytical results that predict the charge carrier mobilities of organic solar cells will have a tremendous influence on the *J–V* characteristics [14, 64]. Moreover, we observed that mainly the photocurrent benefits from a high electron mobility, which could be related to improved charge carrier dissociation at the heterojunction. Finally, we were able to correlate the energetic structure measured with UPS to the open-circuit voltage of the complete solar cell devices. Overall, our results will help guide researchers toward making further improvements to this successful class of synthetic semiconducting copolymers.

Acknowledgment The authors thank the DFG for funding within the DFG Priority Program 1355 "Elementary Processes of Organic Photovoltaics." This report is based on results obtained by the collaboration of project no. 15 ("Tuning the Optical and Charge-Transporting Properties of the Electron-Accepting Phase in Polymer Solar Cells") and project number 11 ("Electronic Properties of Interfaces with Conjugated Polymers and Polyelectrolytes").

References

1. Zhang Z-G, Wang J (2012) J Mater Chem 22:4178
2. Kang I, Yun H-J, Chung DS, Kwon S-K, Kim Y-H (2013) J Am Chem Soc 135:14896
3. Tseng H-R, Phan H, Luo C, Wang M, Perez LA, Patel SN, Ying L, Kramer EJ, Nguyen T-Q, Bazan GC, Heeger AJ (2014) Adv Mater 26:2993
4. Yi Z, Wang S, Liu Y (2015) Adv Mater 27:3589
5. Guo X, Facchetti A, Marks TJ (2014) Chem Rev 114:8943
6. Lin Y, Zhan X (2014) Mater Horiz 1:470
7. Steyrleuthner R, Schubert M, Jaiser F, Blakesley JC, Chen Z, Facchetti A, Neher D (2010) Adv Mater 22:2799
8. Yan H, Chen Z, Zheng Y, Newman C, Quinn JR, Dotz F, Kastler M, Facchetti A (2009) Nature 457:679
9. Kang H, Uddin MA, Lee C, Kim K-H, Nguyen TL, Lee W, Li Y, Wang C, Woo HY, Kim BJ (2015) J Am Chem Soc 137:2359
10. Mori D, Benten H, Okada I, Ohkita H, Ito S (2014) Energy Environ Sci 7:2939
11. Ye L, Jiao X, Zhou M, Zhang S, Yao H, Zhao W, Xia A, Ade H, Hou J (2015) Adv Mater. doi:10.1002/adma.201503218
12. Li W, Roelofs WSC, Turbiez M, Wienk MM, Janssen RAJ (2014) Adv Mater 26:3304
13. Steyrleuthner R, Di Pietro R, Collins BA, Polzer F, Himmelberger S, Schubert M, Chen Z, Zhang S, Salleo A, Ade H, Facchetti A, Neher D (2014) J Am Chem Soc 136:4245
14. Marsh RA, Groves C, Greenham NC (2007) J Appl Phys 101:083509
15. Offermans T, Meskers SCJ, Janssen RAJ (2005) Chem Phys 308:125
16. Yang F, Forrest SR (2008) ACS Nano 2:1022
17. Svensson M, Zhang F, Veenstra SC, Verhees WJH, Hummelen JC, Kroon JM, Inganäs O, Andersson MR (2003) Adv Mater 15:988
18. Groves C, Marsh RA, Greenham NC (2008) J Chem Phys 129:114903
19. Huang Y, Westenhoff S, Avilov I, Sreearunothai P, Hodgkiss JM, Deleener C, Friend RH, Beljonne D (2008) Nat Mater 7:483
20. Liu J, Choi H, Kim JY, Bailey C, Durstock M, Dai L (2012) Adv Mater 24:538
21. McNeill CR, Abrusci A, Zaumseil J, Wilson R, McKiernan MJ, Burroughes JH, Halls JJM, Greenham NC, Friend RH (2007) Appl Phys Lett 90:193506
22. McNeill CR, Halls JJM, Wilson R, Whiting GL, Berkebile S, Ramsey MG, Friend RH, Greenham NC (2008) Adv Funct Mater 18:2309
23. McNeill CR, Westenhoff S, Groves C, Friend RH, Greenham NC (2007) J Phys Chem C 111:19153
24. Mori D, Benten H, Ohkita H, Ito S (2015) Adv Energy Mater 5:1500304

25. Mori D, Benten H, Ohkita H, Ito S, Miyake K (2012) ACS Appl Mater Interfaces 4:3325
26. Yan H, Collins BA, Gann E, Wang C, Ade H, McNeill CR (2012) ACS Nano 6:677
27. Kodomari M, Satoh H, Yoshitomi S (1988) J Org Chem 53:2093
28. Baillargeon VP, Stille JK (1986) J Am Chem Soc 108:452
29. Li J-H, Liang Y, Wang D-P, Liu W-J, Xie Y-X, Yin D-L (2005) J Org Chem 70:2832
30. Ellinger S, Ziener U, Thewalt U, Landfester K, Möller M (2007) Chem Mater 19:1070
31. Letizia JA, Salata MR, Tribout CM, Facchetti A, Ratner MA, Marks TJ (2008) J Am Chem Soc 130:9679
32. Mulherin RC, Jung S, Huettner S, Johnson K, Kohn P, Sommer M, Allard S, Scherf U, Greenham NC (2011) Nano Lett 11:4846
33. Nothofer H-G (2005) Dissertation, Universität Potsdam
34. Yamamoto T (1992) Prog Polym Sci 17:1153
35. Hwang J, Kim E-G, Liu J, Brédas J-L, Duggal A, Kahn A (2007) J Phys Chem C 111:1378
36. Koch N, Elschner A, Rabe JP, Johnson RL (2005) Adv Mater 17:330
37. Lange I, Blakesley JC, Frisch J, Vollmer A, Koch N, Neher D (2011) Phys Rev Lett 106:216402
38. Sueyoshi T, Fukagawa H, Ono M, Kera S, Ueno N (2009) Appl Phys Lett 95:183303
39. Garreau S, Leclerc M, Errien N, Louarn G (2003) Macromolecules 36:692
40. Steyrleuthner R, Schubert M, Howard I, Klaumünzer B, Schilling K, Chen Z, Saalfrank P, Laquai F, Facchetti A, Neher D (2012) J Am Chem Soc 134:18303
41. Schubert M, Dolfen D, Frisch J, Roland S, Steyrleuthner R, Stiller B, Chen Z, Scherf U, Koch N, Facchetti A, Neher D (2012) Adv Energy Mater 2:369
42. Chua L-L, Zaumseil J, Chang J-F, Ou EC-W, Ho PK-H, Sirringhaus H, Friend RH (2005) Nature 434:194
43. Bange S, Schubert M, Neher D (2010) Phys Rev B 81:035209
44. Schubert M, Preis E, Blakesley JC, Pingel P, Scherf U, Neher D (2013) Phys Rev B 87:024203
45. Orenstein J, Kastner M (1981) Phys Rev Lett 46:1421
46. Nicolai HT, Kuik M, Wetzelaer GAH, de Boer B, Campbell C, Risko C, Brédas JL, Blom PWM (2012) Nat Mater 11:882
47. Kilina S, Dandu N, Batista ER, Saxena A, Martin RL, Smith DL, Tretiak S (2013) J Phys Chem Lett 4:1453
48. Noriega R, Rivnay J, Vandewal K, Koch FPV, Stingelin N, Smith P, Toney MF, Salleo A (2013) Nat Mater 12:1038
49. Huang DM, Mauger SA, Friedrich S, George SJ, Dumitriu-LaGrange D, Yoon S, Moulé AJ (2011) Adv Funct Mater 21:1657
50. Chang J-F, Sun B, Breiby DW, Nielsen MM, Sölling TI, Giles M, McCulloch I, Sirringhaus H (2004) Chem Mater 16:4772
51. Zen A, Pflaum J, Hirschmann S, Zhuang W, Jaiser F, Asawapirom U, Rabe JP, Scherf U, Neher D (2004) Adv Funct Mater 14:757
52. Hao XT, Hosokai T, Mitsuo N, Kera S, Okudaira KK, Mase K, Ueno N (2007) J Phys Chem B 111:10365
53. Heimel G, Salzmann I, Duhm S, Rabe JP, Koch N (2009) Adv Funct Mater 19:3874
54. Frisch J, Vollmer A, Rabe JP, Koch N (2011) Org Electron 12:916
55. Braun S, Salaneck WR, Fahlman M (2009) Adv Mater 21:1450
56. Hwang J, Wan A, Kahn A (2009) Mater Sci Eng R Rep 64:1
57. Kanai K, Miyazaki T, Suzuki H, Inaba M, Ouchi Y, Seki K (2010) Phys Chem Chem Phys 12:273
58. Dennler G, Scharber MC, Brabec CJ (2009) Adv Mater 21:1323
59. Rand BP, Burk DP, Forrest SR (2007) Phys Rev B 75:115327
60. Burke TM, Sweetnam S, Vandewal K, McGehee MD (2015) Adv Energy Mater 5:1500123
61. Vandewal K, Tvingstedt K, Gadisa A, Inganäs O, Manca JV (2010) Phys Rev B 81:125204
62. Kuriyama T, Kunimori K, Kuriyama T, Nozoye H (1998) Chem Commun 501. doi:10.1039/A707932J

63. Frisch J, Schubert M, Preis E, Rabe JP, Neher D, Scherf U, Koch N (2012) J Mater Chem 22:4418
64. Würfel U, Neher D, Spies A, Albrecht S (2015) Nat Commun 6:6951
65. Schubert M, Collins BA, Mangold H, Howard IA, Schindler W, Vandewal K, Roland S, Behrends J, Kraffert F, Steyrleuthner R, Chen Z, Fostiropoulos K, Bittl R, Salleo A, Facchetti A, Laquai F, Ade HW, Neher D (2014) Adv Funct Mater 24:4068

Charge Carrier Generation, Recombination, and Extraction in Polymer–Fullerene Bulk Heterojunction Organic Solar Cells

Frédéric Laquai, Denis Andrienko, Carsten Deibel, and Dieter Neher

Contents

Abstract In this chapter we review the basic principles of photocurrent generation in bulk heterojunction organic solar cells, discuss the loss channels limiting their efficiency, and present case studies of several polymer–fullerene blends. Using

F. Laquai
King Abdullah University of Science and Technology (KAUST), Thuwal, Saudi Arabia
e-mail: frederic.laquai@kaust.edu.sa

D. Andrienko
Max Planck Institute for Polymer Research, Ackermannweg 10, 55128 Mainz, Germany
e-mail: denis.andrienko@mpip-mainz.mpg.de

C. Deibel (✉)
Institut für Physik, Technische Universität Chemnitz, 09126 Chemnitz, Germany
e-mail: deibel@physik.tu-chemnitz.de

D. Neher
Institute of Physics and Astronomy, University of Potsdam, Karl-Liebknecht-Strasse 24-25, 14476 Potsdam, Germany
e-mail: neher@uni-potsdam.de

K. Leo (ed.), *Elementary Processes in Organic Photovoltaics*,
Advances in Polymer Science 272, DOI 10.1007/978-3-319-28338-8_11

steady-state and transient, optical, and electrooptical techniques, we create a precise picture of the fundamental processes that ultimately govern solar cell efficiency.

Keywords Charge extraction • Charge generation • Charge recombination • Organic solar cells • PBT7 • PBTTT • PCPDTBT

1 Photophysics of an Organic Solar Cell

Organic photovoltaic (OPV) devices have recently exceeded power conversion efficiencies (PCEs) of 10 % in single-junction cells [1] and 13.2 % in a multijunction device using small molecules [2]. The tenfold efficiency increase has been achieved in less than a decade by fine-tuning material processing and bandgap engineering. To further improve photovoltaic performance and to move forward to commercial applications, we are now urged to critically assess and revise the current understanding of photophysical processes occurring in OPV devices, in particular those that limit efficiency. For instance, the mechanism of charge carrier generation via dissociation of photogenerated excitons is still debated: In the past it was often assumed that *bound* charge-transfer (CT) states are formed at the donor–acceptor (D–A) interfaces and dissociate into free-charge carriers in a field-*dependent* separation process. Recent studies, however, indicate that several systems exhibit a field-*independent* charge generation with an internal quantum efficiency close to unity. A multitude of explanations for the efficient charge separation has been proposed, such as hot CT-state dissociation [3, 4], delocalization of charges during the ultrafast CT process [5], high initial values of charge mobilities and the presence of energy cascades [6], an interplay of entropic and enthalpic contributions [7], and long-range electrostatic forces at intermixed D–A interfaces [8, 9].

The controversial interpretation of experimental results hints at an approaching paradigm shift. New models that try to account for coherent CT, heterogeneity of blends, and nonequilibrium processes urgently require further development and validation. Once established, they will help us to understand the interplay among the chemical composition, bulk and interfacial morphologies, energetic landscapes for excitons, as well as CT and charge-separated states. This understanding is the key for the successful design of new materials.

The purpose of this review is threefold: first, to summarize the current understanding of processes that take place during photocurrent generation in an organic solar cell; second, to link them to the morphology and chemical composition for several polymer–fullerene blends; and finally, to generalize the results to other bulk heterojunction solar cells.

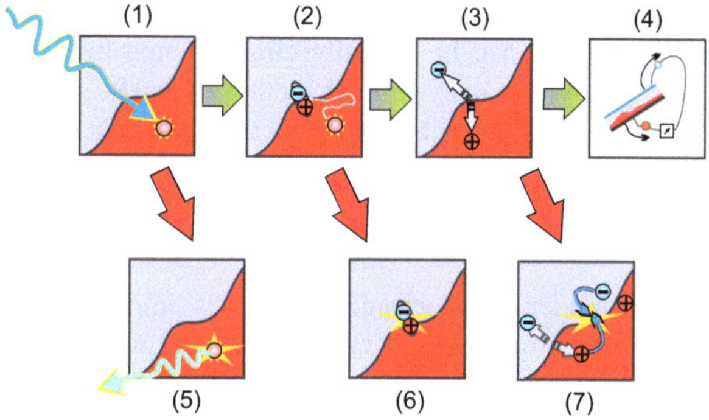

Fig. 1 Photophysical processes in organic solar cells leading to photocurrent generation (*green arrows*) and photocurrent loss (*red arrows*): (1) exciton generation by photon absorption in the donor and/or acceptor material; (2) exciton diffusion-limited and ultrafast charge transfer (CT) at the interface, potentially creating an intermediate CT state at the interface; (3) separation into free-charge carriers, followed by charge transport through the bulk materials; (4) charge carrier extraction at the electrodes; (5) exciton recombination; (6) geminate recombination of CT states; and (7) nongeminate recombination of free charges

1.1 Charge Carrier Photogeneration

We start by listing the steps leading to photocurrent generation. After photon absorption in the donor/acceptor materials and consequent exciton generation, the exciton diffuses toward the D–A interface, where an intermediate CT state is formed after an ultrafast CT reaction. The CT state then splits into free charges, which drift-diffuse in the bulk materials and are eventually extracted at the electrodes. These steps are depicted in Fig. 1. Each step is accompanied by loss processes, which are also shown in Fig. 1. The different loss channels include the following:

1. Incomplete photon absorption in the photoactive layer, mainly for three reasons:
   a. The optical gap of organic materials is typically larger than 1.4 eV; hence, photons with lower energies are not absorbed by the photoactive layer. This issue can be addressed in part by using low-bandgap materials with an absorption in the near-infrared (NIR) region. Such materials, however, often exhibit lower quantum efficiencies [10, 11].
   b. The photoactive layer is rather thin (100 nm) and thus captures only a fraction of the above-bandgap photons because of its finite optical density. To enhance absorption, the thickness of the photoactive layer can be increased. This, however, often leads to a lower fill factor and to a reduction in the PCE, mainly as a result of the low charge carrier mobility of organic materials [12].
   c. The fullerene derivatives, frequently used as acceptors, contribute only little to the total absorbance of the photoactive layer, as their absorption cross sections are small in the visible spectrum. Moreover, low-bandgap polymers often require an excess of the fullerene component for optimum

device performance, which further dilutes the absorbance of the photoactive layer. This issue can be partially circumvented by using (much more costly) [6,6]-phenyl C_{71}-butyric acid methyl ester ($PC_{70}BM$), which has a stronger absorption in the visible spectrum than [6,6]-phenyl-C_{61}-butyric acid methyl ester ($PC_{60}BM$). Another alternative is to use non-fullerene acceptors. Here, efficiencies approaching those obtained with fullerene acceptors were achieved only recently [13].

2. The exciton diffusion length in organic materials is limited to about 10 nm [14]. This is a consequence of the short exciton lifetime, which is typically in the range of several hundred picoseconds, and the incoherent exciton motion process, which is best described by the hopping of the exciton between chromophores with different energy values. Hence, exciton harvesting is most efficient if the exciton has to diffuse only a short distance before encountering the interface, which requires a *nanoscale* phase separation in bulk heterojunction solar cells. The intrinsically short exciton diffusion length also limits the performance of *bilayer* solar cells. Here, the organic materials with an enhanced exciton diffusion length could help to reduce the recombination of free charges compared with bulk heterojunction devices [15] since the bilayer device has a much smaller area of D–A interfaces.
3. Interfacial CT will not necessarily yield free charges but may lead to the formation of bound electron–hole pairs, in which the two oppositely charged particles have not entirely escaped their mutual Coulomb attraction. Such pairs are often called *geminate pairs* (GPs), where "geminate" reflects the fact that the electron and the hole are the product of the same primary photoexcitation. As charges are further separated from each other across the heterojunction, the binding energy of the GP reduces.
4. The mechanism of charge photogeneration is still debated. Either it is directly by excitons in the neat materials in process (3), or it is via the GP dissociation. An important question is whether "hot charge transfer states" dominate the charge photogeneration or thermally relaxed GPs. Also, tightly bound CT states may represent an ultimate loss channel because of geminate recombination, which is still discussed controversially.
5. Charge carrier recombination during the transport to the electrodes is often the main loss mechanism for state-of-the-art solar cells [15, 16]. In disordered organic semiconductor thin films, charge transport occurs via hopping of charge carriers between transport sites within the density of states [17]. In a device it is often described by drift-diffusion models using the classical semiconductor band theory as a simplified starting point. The work function difference of the device electrodes creates an internal (built-in) electric field in the organic layer, which causes the charge carriers to drift to their respective electrodes. However, charge carriers may get trapped either in tail states of the density of states or in deeper-lying trap states. Trapping leads to both slower charge transport and trap-assisted recombination equivalent to Shockley–Read–Hall recombination in inorganic semiconductors. Furthermore, free-charge carriers can encounter

each other again at the D–A interface and undergo nongeminate recombination during the transport to the electrodes if not extracted prior to their encounter and recombination. Therefore, the competition between recombination and extraction rates needs to be tuned in favor of the latter. Recently, this competition was analyzed by the groups of Koster, Würfel, and Neher [18, 19]. It was shown that nongeminate recombination affects the device efficiency even for reasonably high mobilities of 10^{-3} cm^2/Vs and that electron and hole mobilities in excess of 10^{-2} cm^2/Vs are needed when aiming at an efficient extraction of charges out of thick active layers.

6. Finally, at the electrodes the charge carriers are extracted from the photoactive layer. Extraction of the wrong carrier type (which is equivalent to recombination [20]) can be countered by adding hole- or electron-blocking layers. Note that energy barriers at the metal–organic interface (or materials of low conductivity for the majority carrier type) can hinder charge extraction, leading to charge accumulation and to so-called S-shaped current–voltage characteristics [21, 22].

The aforementioned loss channels reduce the photocurrent of the solar cell, the fill factor of the device (at least if charge carriers are involved), and the open-circuit voltage, V_{OC}. In fact, the open-circuit voltage of an organic solar cell is typically much lower than the energy of the absorbed photons. Several loss mechanisms contribute to the V_{OC} loss: 1. Absorption of above-bandgap photons creates losses because of the fast internal conversion of the exciton and the dissipation of the excess energy into heat. 2. Charge transfer at the interface adds a loss because of the energy-level offset between donor and acceptor. 3. The nongeminate recombination of charges [23, 24] determines the charge density and with that the Fermi-level splitting under open-circuit conditions. Together with the broadening of the density of states, this recombination results in a V_{OC} loss of at least 0.5 V [8].

The low open-circuit voltage is a significant drawback of organic solar cells; improving V_{OC} would definitely boost the commercialization of organic solar cells [25].

1.2 Charge-Transfer States and Geminate Charge Recombination

In this section we discuss in more detail the role of CT states. At the D–A interface, only a fraction of electrons and holes escape their Coulomb attraction and split into free charges. Electron–hole pairs that have not managed to fully separate will remain bound and form interfacial GPs. These electron–hole pairs might either reside directly at the interface, forming CT states [26], or be at a larger distance with negligible wave function overlap. In the latter case, terms such as “bound” and “loosely bound polaron pairs,” or “dark states,” are often used [27, 28]. Interfacial electron–hole pairs that have relaxed into CT states can undergo geminate recombination to the ground state. Geminate recombination of CT states has been experimentally observed by time-resolved photoluminescence

(PL) spectroscopy and exhibits characteristic PL at an energy corresponding roughly to the energy gap between the ionization potential of the donor and the electron affinity of the acceptor, both in a solid state. The lifetime of the emission has been determined to be on the order of several hundred picoseconds up to several nanoseconds depending on the investigated material system [29]. Another way to excite CT states is by driving the solar cell as a light-emitting diode [29]. In this case, the radiative recombination of CT states leads to a low-energy emission, with a significant redshift of the emission peak with regard to the emission from the pure donor and acceptor components. Interestingly, charge-transfer photoluminescence (CT-PL) and electroluminescence (CT-EL) spectra do not fully overlap, with the EL peaking at a lower energy. It has therefore been proposed that the PL is from CT states on D–A pairs within mixed domains (molecular heterojunctions) while CT-EL is caused by the recombination of CT states formed at the interface between donor and acceptor domains (domain heterojunction) [30]. Due to a detailed balance, free-charge generation upon illumination, which is the inverse of the EL process, must involve the same CT states that are formed in EL [24]. Indeed, it has been demonstrated that quenching of the CT-state PL by an electric field requires fields that are substantially larger than those usually present in operating photovoltaic devices [30, 31], which questions whether the CT states leading to PL are precursors to free charges. The upper limit of the binding energy of the CT state in poly[2-methoxy-5-(3′,7′-dimethyloctyloxy)-1,4-phenylenevinylene]:1-(3-methoxycarbonyl)propyl-1-phenyl[6,6]C_{61} (MDMO-PPV:PCBM) has been experimentally determined to be on the order of 200 meV [32], which corresponds to several times $k_B T$ at room temperature and thus makes thermally assisted splitting unlikely. Therefore, the interpretation of CT-PL data with regard to the efficiency of free-charge generation should be considered with care. On the other hand, Vandewal and coworkers [33] were able to demonstrate efficient free-charge generation even when exciting D–A blends at the energy maximum of CT-EL, suggesting a thermally relaxed interfacial CT state as the precursor to free charges.

Further evidence of geminate recombination has been obtained using transient absorption (TA) spectroscopy (TAS) [34, 35]. In TA pump-probe experiments, geminate recombination can be identified as a pump-intensity–independent signal decay of both the ground-state bleach and charge-induced absorption that can be fitted by a single exponential with inverse rate constants on the order of picoseconds to nanoseconds [36]. The intensity-independent character of the recombination process is a consequence of the localization of the CT states at the interface since they are much less mobile than free charges. Thus, interactions between CT states or between CT states and other excited species occur only at very high excitation densities. The quasi single-exponential recombination dynamics of CT states, often observed in TA experiments, is, however, not straightforward to interpret. In principle, the disordered nature of bulk heterojunction D–A blends should cause a distribution of electron–hole pair distances across the interface, in turn causing a distribution of CT-state lifetimes, resulting in a stretched exponential rather than a single exponential decay. However, one may argue that the spectroscopically observed CT states are relaxed and similar in nature, resulting in a narrow distribution of lifetimes, which can be approximated by a single exponential.

1.3 Free Charges and Nongeminate Recombination

Electron–hole pairs, which escaped their mutual Coulomb attraction, form charge-separated states. Charge-separated states contribute to the device's photocurrent if their spatially separated carriers are extracted at the electrodes. During charge transport to the electrodes, however, free electrons and holes can encounter and recombine. Since the recombining charge carriers originate from different photoexcitations, this mechanism is termed nongeminate recombination. The rate of nongeminate recombination, R, depends on both the carrier concentration n and the effective lifetime $\tau(n)$. Usually, the higher the charge carrier concentration is, the faster the recombination, $R \sim n/\tau(n)$, will be. For most state-of-the-art organic solar cells, the nongeminate recombination of already separated electrons and holes is the dominant loss mechanism.

The simplest description of charge carrier recombination is Langevin's model. In this model, the Coulomb attraction between electrons and holes results in a recombination rate of $R = e/\epsilon\epsilon_0 (\mu_e + \mu_h) np$ [37]. Here e is the elementary charge, $\epsilon\epsilon_0$ is the material dielectric constant, and n/p and $\mu_{\mathrm{e/h}}$ are the electron and hole densities and mobilities, respectively. If the electron and hole concentrations under illumination are similar (for instance, in systems without doping), then $n \approx p$, the recombination process is of the second order, and the carrier concentration decay is inversely proportional to the time, $n \sim t^{-1}$.

In polymer–fullerene systems, deviations from Langevin's recombination have been observed. For instance, Pivrikas et al. [38] noted a reduction in the recombination rate by orders of magnitude for poly(3-hexylthiophene-2,5-diyl): 1-(3-methoxycarbonyl)propyl-1-phenyl[6,6]C_{61} (P3HT:PCBM). This reduction is often taken into account by multiplying Langevin's recombination rate by a prefactor varying within the range from 10^{-3} to 10^{-1}, depending on the material system, and is termed a reduced Langevin recombination. For P3HT:PCBM the reported values are of the order of 0.02 or less, whereas almost no reduction is measured in MDMO-PPV:PCBM [39]. Reduced recombination can be explained by the redissociation of bound pairs [40], the presence of the charge carrier concentration gradients [41], and spatial localization of recombination zones to D–A interfaces [42, 43].

Apart from the recombination rate reduction, time-dependent experiments, for example, transient absorption, demonstrated that the decay of the carrier concentration can be different from the t^{-1} behavior predicted by Langevin's model [44], leading to a power law decay rate, $R \sim n^{\lambda+1}$. It was found that the exponent λ increases with lower temperatures [45, 46], indicating a relationship to thermal activation. Inspired by transient absorption experiments on pristine polymers [47] and using random walk modeling [48], the power law decay has been related to the trapping of charge carriers in the tail states of the density of states [17]. This effect can be accounted for by assuming the carrier concentration-dependent mobility, $R \sim \mu_{\mathrm{eff}}(n)n^2$. Other effects influencing the recombination exponent include the injection of majority carriers by electrodes with small injection barriers [49] and doping [50].

To conclude this section, we emphasize that fast charge carrier extraction from the photoactive layer and the suppression of nongeminate recombination are prerequisite in order to obtain a high performance from photovoltaic devices. In a working solar cell, the nongeminate recombination of spatially-separated charge carriers always competes with charge carrier extraction during the drift–diffusion of charges toward the electrodes [51]. If charge generation is efficient but charge extraction is slower than nongeminate recombination, the buildup of high charge carrier concentrations can lead to the device having a reduced fill factor and lower PCE [18].

2 Case Studies

In what follows, we present a detailed photophysical characterization of four polymer blends, the chemical structures of which are shown in Fig. 2.

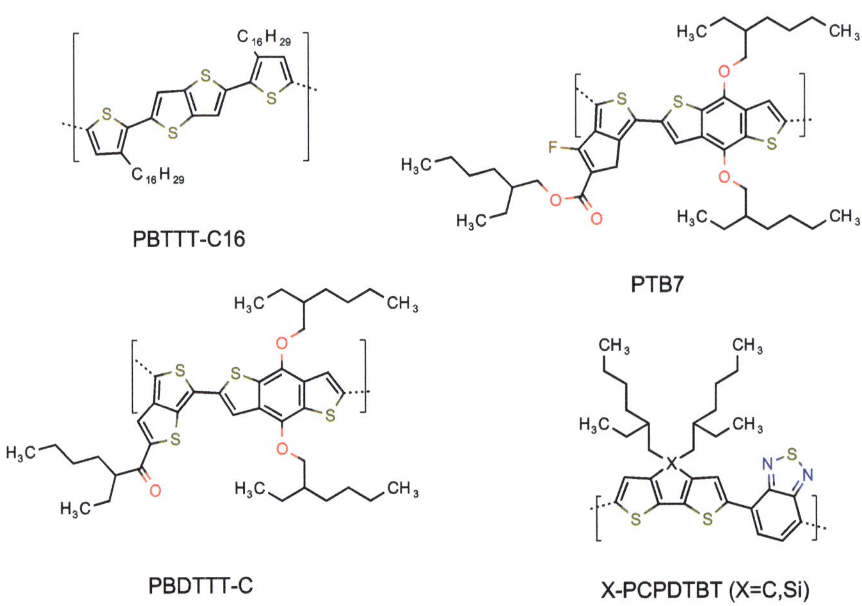

Fig. 2 Structures of polymers discussed in this chapter: poly(2,5-bis(3-hexadecylthiophen-2-yl)thieno[3,2-*b*] thiophene (PBTTT); poly((4,8-bis((2-ethylhexyl)oxy)benzo (1,2-*b*:4,5-*b*')dithiophene-2,6-diyl) (3-fluoro-2-((2-ethylhexyl)carbonyl) thieno(3,4-b)thiophenediyl)) (PTB7); poly[(4,8-bis-(2-ethylhexyloxy)-benzo(1,2-*b*:4,5-*b*')dithiophene)-2,6-diyl-*alt*-(4-(2-ethylhexanoyl)-thieno[3,4-*b*]thiophene)-2-6-diyl)] (PBDTTT-C); poly[2,6-(4,4-bis-(2-ethylhexyl)-4H-cyclopenta[2,1-*b*:3,4-*b*']dithiophene)-*alt*-4,7-(2,1,3-benzothiadiazole)] (C-PCPDTBT), and poly[2,6-(4,4-bis-(2-ethylhexyl)dithieno[3,2-*b*:2',3'-d]silole)-*alt*-4,7-(2,1,3-benzothiadiazole)] (Si-PCPDTBT). Blends were made using either [6,6]-phenyl-C_{61}-butyric acid methyl ester ($PC_{60}BM$) or [6,6]-phenyl-C_{71}-butyric acid methyl ester ($PC_{70}BM$)

2.1 *Effect of Solvent Additives on Charge Carrier Recombination in PTB7:PCBM*

Blends of the donor polymer thieno[3,4-*b*]thiophene-*alt*-benzodithiophene (PTB7) and the fullerene acceptor $PC_{70}BM$ are among the most efficient OPV materials, with a reported record PCE of 9.2 % [52]. The photovoltaic performance of these blends improves dramatically upon adding 1,8-diiodooctane (DIO) to the solution. DIO has a strong impact on the active layer morphology [53–56]: Pristine blends comprise large (100-nm) clusters consisting almost entirely of fullerene embedded in mixed polymer-rich domains. DIO suppresses the formation of large fullerene clusters without affecting the overall crystallinity or composition of domains. As a consequence, the short-circuit current and the fill factor of the photovoltaic cell increase [52] because of the more efficient generation of free charges or/and reduced nongeminate recombination.

To disentangle these two contributions, the effect of DIO on the generation and recombination of charges was investigated by a combination of transient photovoltage, charge extraction, and time-delayed collection field (TDCF) experiments [15]. Two sets of PTB7:$PC_{70}BM$ 1:1.5 blend solar cells were compared: one prepared from pure chlorobenzene (CB) solution and another from CB with a small amount of DIO as co-solvent. The current–voltage characteristics for both sets are shown in Fig. 3. As expected, the corresponding PCE increases from 3.6 to 7.0 % when DIO is added.

Pre-bias-dependent TDCF measurements [57, 58] were performed to investigate the field dependence of the photogeneration yield, that is, the efficiency of the CT-state dissociation. Devices processed without DIO show a pronounced field

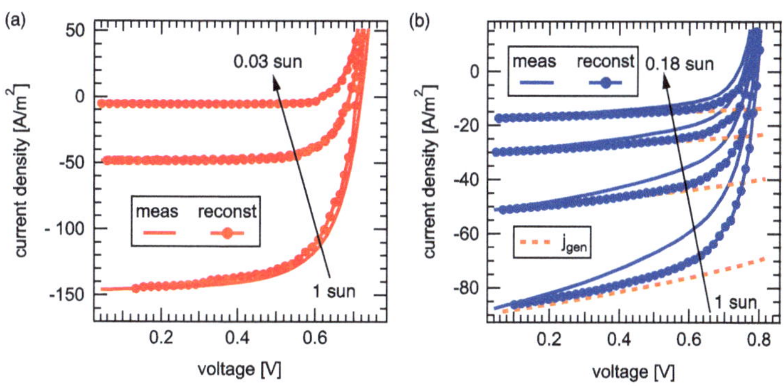

Fig. 3 Measured and reconstructed current–voltage response of PTB7:$PC_{70}BM$ 1:1.5 solar cells processed with (**a**) and without (**b**) 1,8-diiodooctane (DIO) as cosolvent for different illumination intensities at 300 K. The solar cell with DIO (**a**) is only limited by nongeminate recombination, which is responsible for the entire shape of the current–voltage characteristics. The device without DIO (**b**) is additionally limited by field-dependent photogeneration, yielding the generation current j_{gen} as determined from time-delayed collection field measurements. [Adapted from Adv. Funct. Mater. 24, 1306 (2014), reprinted by permission of John Wiley & Sons, Inc.]

dependence in conjunction with a lower carrier yield, whereas devices prepared with DIO exhibit a higher carrier yield in line with a weak field dependence. It appears that the well-balanced presence of mixed domains and neat material domains in the blend prepared with DIO assists the charge photogeneration by promoting a field-independent CT separation. This observation is also in agreement with theoretical calculations predicting that well-ordered domains of the donor and acceptor phases with intermixed interfaces are required in order to split CT states [9].

In contrast, the nongeminate recombination in both sets of devices (see Fig. 4) shows comparable charge carrier lifetimes at low charge carrier densities, resulting in a recombination order of about 3.5 at room temperature. If the density of tail state distribution were exponential, this would correspond to a characteristic energy of the tail states of about 50 meV. For carrier concentrations generated at 1 sun illumination, however, only the blend with DIO shows a recombination order of 2, that means a second-order recombination mechanism.

The relative contributions of geminate and nongeminate recombinations are shown in Fig. 5, in the reconstructed current–voltage characteristics. The device prepared with DIO is limited only by nongeminate recombination, while the device prepared without DIO also exhibits a field-dependent photogeneration as described

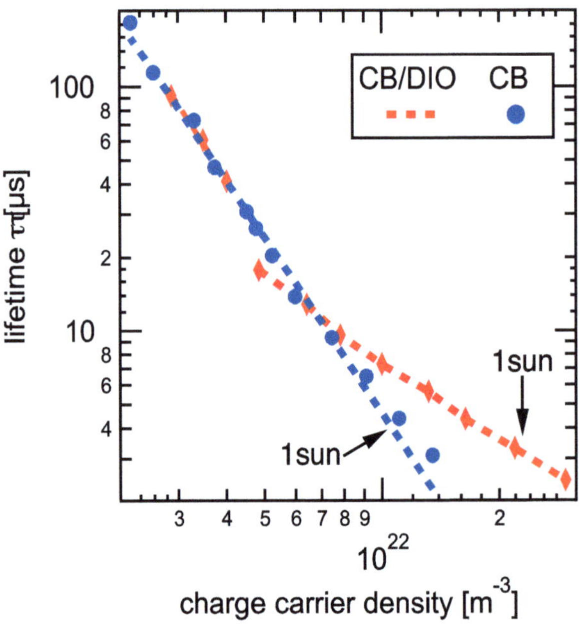

Fig. 4 Effective lifetime in dependence of charge carrier concentration of PTB7:$PC_{70}BM$ 1:1.5 solar cells processed with and without 1,8-diiodooctane (DIO). Both devices show trap-limited losses with a recombination order of about 3.5 at low carrier concentrations. Around one sun illumination, only the device with DIO shows second-order recombination, as described in the text. [Adapted from Adv. Funct. Mater. 24, 1306 (2014), reprinted by permission of John Wiley & Sons, Inc.]

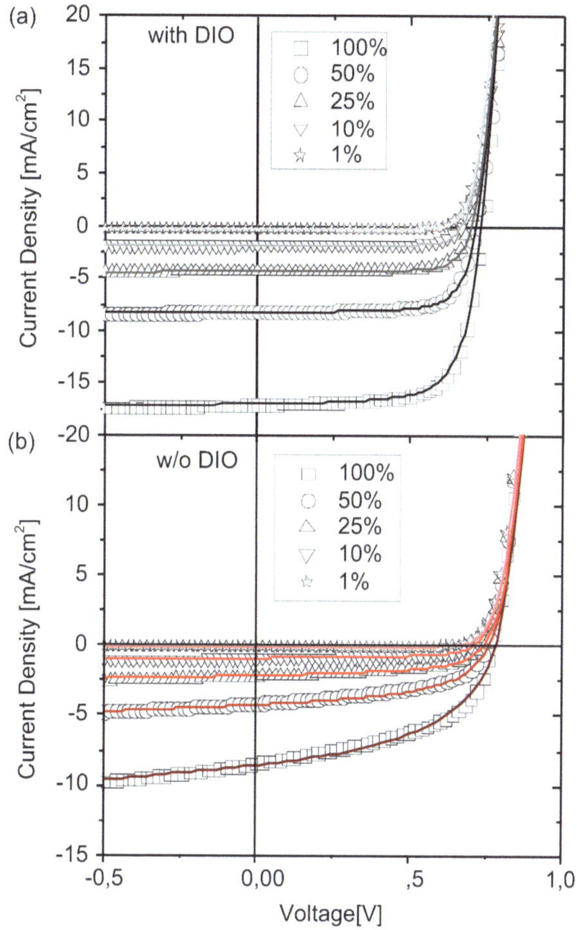

Fig. 5 *J–V* characteristics at different light intensities relative to one sun for the device prepared with (**a**) and without (**b**) 1,8-diiodooctane. *Symbols* are the experimental data and *solid lines* are drift-diffusion simulations based on experimentally determined parameters. [Reprinted with permission from Phys. Chem. C 119, 8310 (2015). Copyright American Chemical Society]

above. We propose that the discrepancy between the measured and reconstructed curves in the device prepared without DIO is caused by the trapping of electrons in isolated fullerene clusters from which they cannot escape but from which they can still contribute to recombination. This scenario is consistent with the lower electron mobility reported for devices prepared without DIO [59].

To address this point in more detail, we combined TDCF with bias-assisted charge extraction (BACE) [60], space charge limited current measurements, and numerical device simulations [61]. In the BACE technique, charges are extracted at a reverse bias, thus reducing the amount of charge being lost by nongeminate recombination or trapping.

It turned out that the nongeminate recombination coefficient of the pristine blend, $\gamma = 4 \times 10^{-17}$ m^3/s, is larger than in the DIO-processed blend, $\gamma_{DIO} = 1.5 \times 10^{-17}$ m^3/s. This can be attributed to the presence of large intermixed regions in the blend prepared without the additive. Interestingly, the electron mobility is an order of magnitude lower in the pristine blend, which seems to be counterintuitive to the fact that this blend comprises large fullerene domains, with diameters comparable to the active layer thickness. However, a recent study by Hedley et al. [62] suggested that these domains consist of smaller fullerene spheres, of the order of 20–60 nm. While the electron mobility within these spheres may be quite high, the macroscopic transport across the entire fullerene domain can be limited by transitions between small spheres, explaining the low mobilities and charge trapping.

Given the fact that the use of the additive affects all relevant parameters (mobilities, the nongeminate recombination coefficient, the field dependence on photogeneration), drift-diffusion simulations of the *J–V* characteristics were performed to identify the main reason for the large difference in performance. The simulation results are compared to the experimental *J–V* curves in Fig. 5. The detailed analysis of these simulations revealed that the low performance of the devices prepared without DIO cannot be explained solely on the basis of a field-dependent generation and a higher coefficient for nongeminate recombination. Instead, the effect of the additive can only be quantified when the reduced electron mobility in the device processed without DIO is taken into account, which causes inefficient electron extraction and a poor fill factor (see Fig. 5b). This result emphasizes the importance of efficient charge extraction for reaching high currents and fill factors.

In fact, a similar situation is encountered in as-prepared P3HT:PCBM blends. These devices suffer from poor performance, mainly because of a low fill factor. The efficiency can be largely improved by thermal annealing or solvent treatment. Our recent experimental and simulation work revealed that annealing affects all relevant parameters, but it is mainly the low mobility of holes in the as-prepared blend that causes its poor performance [63].

Bartesaghi et al. [18] recently showed that the fill factor of organic bulk heterojunction devices can be related to a dimensionless quantity, θ, which is proportional to the total loss current caused by nongeminate recombination divided by the extraction current at short-circuit conditions:

$$\theta = \frac{\gamma G d^4}{\mu_\mathrm{h}\mu_\mathrm{e} V_\mathrm{int}^2} \sim \frac{J_\mathrm{rec}}{J_\mathrm{extr}}$$

where θ itself is a function of the nongeminate recombination coefficient γ, the generation rate G, the active layer thickness d, the mobilities of electrons and holes, μ_e and μ_h, and the built-in bias V_int. We find that $\theta = 0.16$ for the pristine PTB7:PCBM blend and $\theta_\mathrm{DIO} = 0.018$ for the device processed with DIO. An order-of-magnitude reduction upon the addition of DIO is in agreement with the observed increase in fill factor from 49.7 to 71.5 % and is the combined effect of a reduced nongeminate recombination coefficient and a higher electron mobility.

2.2 Photophysics of C- and Si-PCPDTBT Blends

Poly[2,6-(4,4-*bis*-(2-ethylhexyl)-4*H*-cyclopenta[2,1-*b*:3,4-*b'*] dithiophene)-*alt*-4,7-(2,1,3-benzothiadiazole)] (C-PCPDTBT) and poly[2,6-(4,4-bis(2-ethylhexyl) dithieno[3,2-b:2,3-d]silole)-*alt*-4,7-(2,1,3 benzothiadiazole)] (Si-PCPDTBT or PSBTBT) are D–A low-bandgap polymers that have demonstrated photovoltaic efficiencies exceeding 5 %. Peet et al. [64] showed that the efficiency of the carbon-bridged polymer C-PCPDTBT blended with $PC_{70}BM$ increases drastically when a small amount of an alkanedithiole such as 1,8-octanedithiol (ODT) is added to the solution prior to spin coating. It was proposed that ODT selectively dissolves fullerenes and promotes the aggregation of polymer chains into more ordered supramolecular structures prior to complete drying of the photoactive layer [65]. Substitution of the bridging carbon by silicon increased the interchain order of Si-PCPDTBT. As a result, Si-PCPDTBT/$PC_{70}BM$ solar cells could reach conversion efficiencies over 5 % without the use of any additives.

The performance differences caused by different film preparation conditions and the substitution of the carbon by silicon atoms have been addressed by several research groups. For instance, Durrant and coworkers [66] used TAS to determine the amplitude of the charge-induced absorption signal and thereby the free-charge carrier yield in the absence and presence of an electric field. They concluded that charge generation in both C- and Si-PCBTBT with PCBM is field independent at room temperature. Furthermore, based on the observation of a lower charge carrier yield in TAS experiments, the authors assigned the poorer performance of the C-PCPDTBT:PCBM blend processed without ODT to a lower efficiency for photogeneration, caused by geminate recombination of interfacial CT states in highly intermixed blends. The polaron dynamics at longer timescales was explained by the nongeminate recombination of free carriers competing with charge extraction. Interestingly, the polaron dynamics did not show any appreciable dependence on bias even at longer timescales, where the carrier dynamics is mostly determined by the competition between recombination and extraction. As some of these data have been recorded for a very high excitation density of 85 $\mu J/cm^2$, screening of the external electric field by space charge effects likely occurred. In fact, in a later study Neher's group [67, 68] showed that in C-PCPDTBT:PCBM and Si-PCBTBT:PCBM charge generation is field dependent and the magnitude of the field dependence decreases in the order C-PCPDTBT:PCBM (without additive) > C-PCPDTBT:PCBM (with additive) > Si-PCBTBT:PCBM, explaining the pronounced differences in fill factor observed for these systems. The field dependence on generation was shown to scale inversely with the polymer domain size, highlighting the need for sufficiently large and preferably pure domains for efficient free-charge generation [69, 70].

Blom and coworkers [71] also studied the effect of solvent additives on the charge carrier yield in C-PCPDTBT:PCBM photovoltaic devices. They succeeded to model the entire device's *J*–*V* curves using Braun–Onsager's model, including a field-assisted separation of rather long-lived CT states. They proposed that adding

ODT reduces the rate of geminate recombination of the CT state from $1.7 \cdot 10^7$ s^{-1} to $3 \cdot 10^6$ s^{-1}. However, the rate constants are very sensitive to the underlying model and thus may lag behind physical interpretation. In fact, a recent paper by Jarzab et al. [72] indicates that the lifetime of photogenerated CT states in C-PCPDTBT:PCBM is only about 500 ps as measured by time-resolved fluorescence spectroscopy.

The early time charge-generation processes in C-PCPDTBT:PCBM have also been studied by Heeger et al. [73] using ultrafast pump-probe spectroscopy. Their results indicate ultrafast exciton dissociation and charge generation from CT states on a timescale of several picoseconds. Rao et al. and Chow et al. [74, 75] recently presented ns-μs TA studies on C-PCPDTBT:PCBM blends and demonstrated that the polymer's triplet state is populated by the nongeminate recombination of free charges created by exciton dissociation after photoexcitation. Interestingly, the blend processed with a solvent additive showed more triplet excitons, supporting the assignment of triplet formation to a nongeminate recombination mechanism; in other words, the more free charges that were created, the higher was the yield of triplets. In fact, for optimized blends an almost quantitative recombination of charges into the triplet state was observed under pulsed laser excitation as typically used in TA experiments [75].

In our own studies on C-PCPDTBT:PCBM blends processed with and without solvent additives, we confirmed that processing with solvent additives leads to enhanced demixing of the blend's components [76]. Whereas in the blend processed without additive the CT upon excitation of the polymer was exclusively ultrafast, the blend processed with an additive showed ultrafast charge separation as well as a diffusion-limited component extending to several picoseconds. Interestingly, a diffusion-limited CT component becomes visible in the blend processed without DIO when exciting the fullerene acceptor, and it is even more pronounced in the blend processed with an additive, indicating enhanced demixing. Additionally, processing with an additive was found to increase the yield of free charges by reducing the geminate recombination of interfacial CT states. As mentioned above, parallel work by the Cambridge group demonstrated that both blends create triplet states upon the nongeminate recombination of free charges, with the extent to which triplets were created depending on the microstructure of the blend controlled by the processing conditions. In our own work we compared the creation of triplets in C/Si-PCPDTBT:PCBM blends by a combination of TAS and multivariate curve resolution alternating least-squares (MCR-ALS) data analysis [77]. Typical ps-ns NIR TA data of a Si-PCPDTBT:PCBM blend along with the component-associated spectra and dynamics of singlet excitons, charge carriers, and triplet states (obtained by MCR-ALS) are shown in Fig. 6. We have also looked into the triplet-state formation in C-PCPDTBT:PCBM blends processed with and without adding DIO. The triplet yield increases upon processing the C-PCPDTBT:PCBM with cosolvents because of increased free-charge generation, in line with the results reported by the Cambridge group. However, triplet generation was found to be reduced in Si-PCBTBT:PCBM blends despite a further increased yield of free charges compared

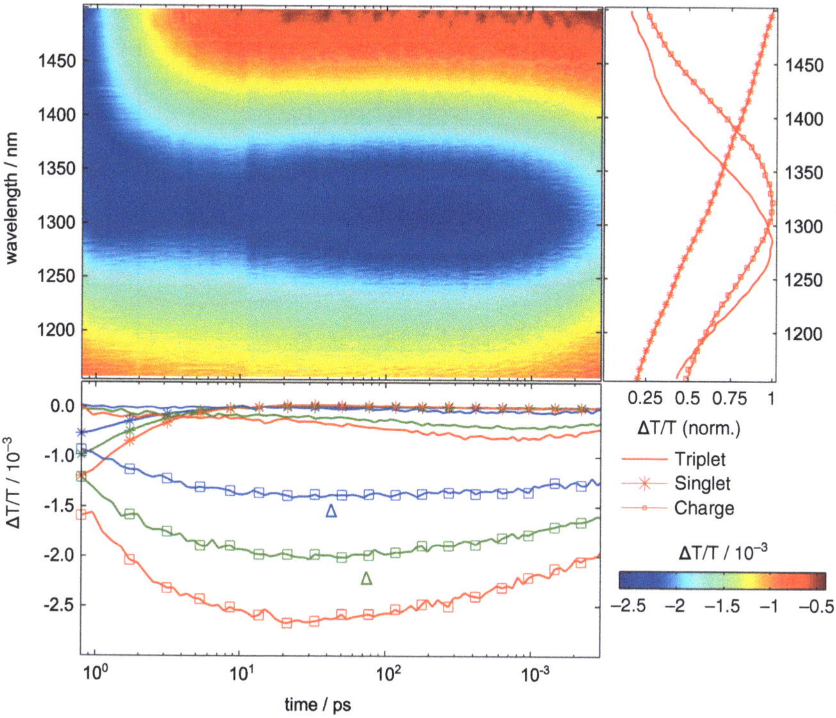

Fig. 6 Contour plot of ps-ns near-infrared transient absorption data of a Si-PCPDTBT:PCBM blend, component spectra (*right panel*), and component dynamics (*bottom panel*) obtained by multivariate curve resolution analysis. [Reproduced from Energy Environ. Sci. 8, 1511 (2015) with permission from the Royal Society of Chemistry]

to C-PCPDTBT:PCBM (see Fig. 7). The reason for the decreased triplet formation is not yet fully understood; it could be related to the blend's microstructure and interfacial morphology as well as the slightly different triplet energy levels of both material systems. Further experiments on related systems are ongoing to elucidate the interplay among energetics, morphology, and triplet-state formation.

2.3 *Effect of Morphology on Charge Generation in PBTTT:PCBM*

The conjugated polymer poly(2,5-bis(3-hexadecylthiophen-2-yl)thieno[3,2-b] thiophene (PBTTT-C16) is a versatile model system for OPVs, as it allows a systematic variation of the blend morphology when the type and fraction of the fullerene acceptor are changed. Depending on the size of the fullerene, either the acceptor molecules can intercalate between the polymer side chains, leading to an intimate mixing of donor and acceptor on a local scale, or they are too large to intercalate,

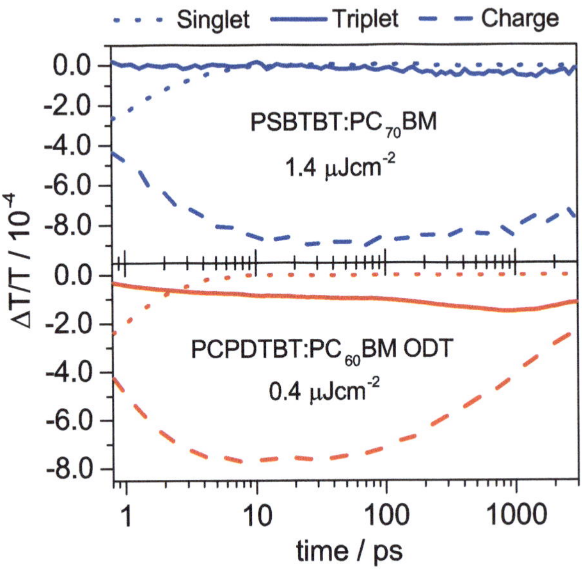

Fig. 7 Singlet-, charge carrier, and triplet-state dynamics in Si-PCPDTBT:PCBM (*upper panel*) and optimized C-PCPDTBT:PCBM blends (*lower panel*) obtained by multivariate curve resolution analysis of ps-ns near-infrared transient absorption data. Note that under similar excitation conditions charge carrier recombination (*dashed lines*) is slower and fewer triplet states (*solid lines*) are created in the Si-PCPDTBT:PCBM system compared to PCPDTBT:PCBM. [Reproduced from Energy Environ. Sci. 8, 1511 (2015) with permission from the Royal Society of Chemistry]

so that separate material domains are formed. These properties explain why this material system is interesting for studying the relationship between morphology and the fundamental processes in organic bulk heterojunction solar cells. The intercalation of fullerene molecules between polymer side chains and the formation of a bimolecular crystal (cocrystal) were previously investigated on PBTTT:fullerene blends by McGehee's group [78, 79]. PBTTT polymer chain spacing increases when blended with $PC_{60}BM$ relative to the neat polymer, indicating the formation of a closely intermixed cocrystal in which fullerene molecules intercalate polymer side chains. For excess concentrations of fullerene molecules beyond a 1:1 ratio, pure fullerene phases are formed in addition to the intercalated/cocrystal phase. In contrast, bis-$PC_{60}BM$–based blends do not exhibit increased chain spacing, as the fullerene molecules bearing two side chains cannot intercalate for steric reasons. We recently confirmed these results using photothermal deflection spectroscopy [80]: Here a clear sub-bandgap absorption below 1.65 eV is an exclusive feature of the polymer–fullerene blend—that is, it is not seen in either of the neat materials—and corresponds to the CT-state absorption, which is created by the close interaction between the donor and acceptor molecules. In comparison, the CT-state absorption of PBTTT:bis-PCBM is reduced by at least one order of magnitude compared to PBTTT:PCBM. This confirms the reduced interaction of the donor and acceptor molecules at the interface because of the lack of intercalation in the bis-PCBM–based blend.

TDCF measurements were performed on these model systems at very low laser fluence to exclusively study the competition between free-charge generation and geminate recombination while avoiding any unwanted carrier losses from nongeminate recombination [80]. Between open-circuit conditions and −4V bias photogenerated charges are extracted, where their field dependence is a measure of the competition between field-dependent charge generation and its counterprocess, namely, geminate recombination. For 1:0.7 PBTTT:$PC_{60}BM$, the field dependence changes by a factor of 10, indicating that the separation yield at low internal voltages is very weak and cannot compete with geminate recombination. For the fully intercalated 1:1 blend, the photogeneration yield still changes by a factor of 3. In contrast, the 1:4 blend shows a comparably weak field dependence by a factor of about 1.3. Only in the latter case (namely, the intercalated blend with excess fullerene molecules that create a separate pure phase) is a reasonable photovoltaic performance achieved. Hence, for an effective dissociation of bound charge carrier pairs in an intercalated system, connected to a high photogeneration yield and weak field dependence, an excess fullerene phase is necessary to ensure charge generation. This allows the spatial separation of the GP, which is directly related to delocalization [81, 82].

Concerning the already-mentioned debate on the role of hot CT complexes on the photogeneration yield, we compared spectrally resolved quasi–steady-state photocurrent spectroscopy to pulsed ns-laser excitation by a TDCF technique on PBTTT:$PC_{60}BM$ blends. In the time-resolved experiment, two different photon energies were used, leading to excitation of singlet excitons in the donor and acceptor at 2.34 eV as well as direct subgap excitation at 1.17 eV, which is exclusively absorbed by the CT complexes [80]. With both experimental techniques we found that the influence of the excitation energy (above- or subgap) on the overall field dependence for photogeneration is very weak and cannot be traced to an effect of hot CT complexes.

Another aspect related to morphology becomes important. Comparing the best of the above intercalated systems, the 1:4 PBTTT:$PC_{60}BM$ blend to a nonintercalated 1:1 bis-PCBM blend, we found that the overall amount of photogeneration is more favorable in the former, whereas the latter has a more efficient extraction of photogenerated charge carriers [80]. Between short- and open-circuit conditions, the normalized photogenerated charge of PBTTT:bis-PCBM exceeds PBTTT:PCBM, resulting in an overall field dependence of about 15 % for the bis-PCBM-based solar cell. We expect that the lower extraction yield in the PBTTT:$PC_{60}BM$ device is caused by the extraction of electrons from the intercalated phase into the pure fullerene phase. This extra step is not required in the bis-PCBM–based blend, where the bound charge carriers, which can only be separated at the D–A interface, can more easily delocalize into the respective nonintercalated material phases. Laquai's group [83] recently performed ps-μs TA experiments on PBTTT:$PC_{60}BM$ blends using a blending ratio of 1:1, corresponding to the cocrystal, and a 1:4 ratio as optimized for device performance. The TA experiments confirmed substantial fast

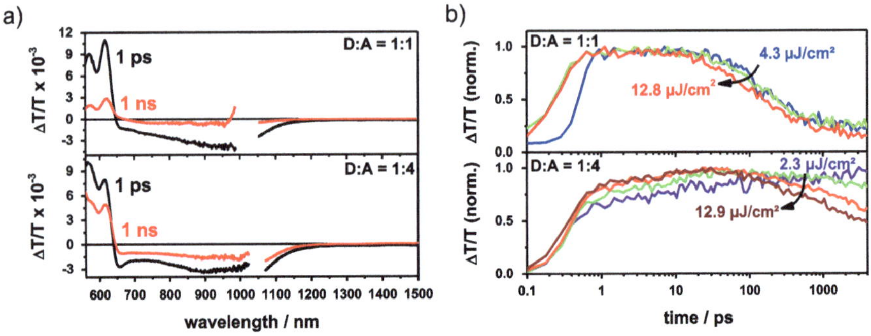

Fig. 8 (**a**) Transient absorption spectra at 1 ps and 1 ns after photoexcitation of a PBTTT:PCBM blend with a 1:1 (*upper panel*) and 1:4 (*lower panel*) blending ratio and (**b**) corresponding kinetics of the ground-state bleach for a 1:1 (*upper panel*) and 1:4 (*lower panel*) blending ratio. Note the fast and intensity-independent signal decay in the 1:1 blend indicating geminate recombination of charge-transfer states. [From Macromol. Rapid Commun. 36, 1054 (2015), reprinted by permission of John Wiley & Sons, Inc.]

sub-ns geminate recombination in the 1:1 blend, in line with the low photocurrent observed in devices (Fig. 8). Furthermore, geminate recombination was found to be substantially reduced in 1:4 blends, in line with higher photocurrents. These experiments complement the TDCF experiments presented by Zusan et al. [85] and support their findings regarding the importance of extended fullerene-rich regions for efficient charge carrier separation.

2.4 *Charge Carrier Photogeneration and Triplet Exciton Formation in PBDTTT-C:PCBM*

PBDTTT-C is another example from the family of benzodithiophene-based low-bandgap polymers, to which PTB7 also belongs. Laquai's group [84] recently reported fs-μs TA experiments on PBDTTT-C:PC_{60}BM blends. Ultrafast CT, following photoexcitation of the polymer, was observed. The assignment of the ps TA spectrum to charges was confirmed by comparing the TA signal to the absorption spectrum of a chemically oxidized polymer film. Interestingly, the NIR photoinduced absorption spectrum showed a pronounced redshift on the ps-ns timescale, indicating the formation of a second component. By MCR-ALS analysis, as outlined above for the PCPDTBT:PC_{60}BM material system, two components were identified that constituted the experimentally measured TA data matrix (see Fig. 9). Here, the second, spectrally redshifted, component was identified as the polymer's triplet state by comparison of the component spectrum to the triplet absorption obtained independently on a metal-porphyrin–sensitized polymer. Furthermore, MCR analysis showed that triplets were the dominant excited state at 1 ns after photoexcitation and the analysis revealed a strong fluence dependence of

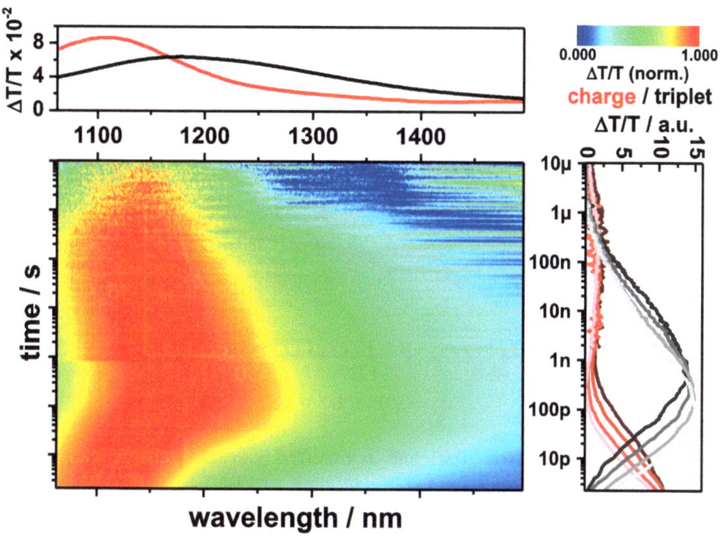

Fig. 9 Normalized contour plot of the ps-μs near-infrared transient absorption data obtained on a PBDTTT-C:PCBM blend at an excitation density of 11.5 μJ/cm^2. The panel on *top* shows the component spectra and the panel on the *right side* the ps-μs component dynamics obtained by multivariate curve resolution analysis of the experimental data. Note the fast and intensity-dependent recombination of charge carriers and the concomitant population of the triplet state on the sub-ns timescale. On the ns-μs timescale, triplet states undergo annihilation and repopulate the pool of charge carriers. [Reprinted with permission from J. Phys. Chem. C 119, 13509 (2015). Copyright American Chemical Society]

the charge carrier decay dynamics and triplet-state formation. This clearly points to a nongeminate recombination of free charges as the origin of triplet formation in this system, which is in line with results obtained on other low-bandgap polymer–fullerene blends [74, 75, 77]. However, on the ns-μs timescale, the MCR analysis showed a regeneration of the charge carrier population. We attributed this to the annihilation of triplet states, which leads to the generation of higher-excited singlet and triplet excitons that now have sufficient energy to undergo charge separation at the polymer–fullerene interface and thereby recreate a small pool of charges. This study demonstrates that triplet-state formation is not exclusive to the aforementioned low-bandgap systems but may be a more general phenomenon in polymer–fullerene blends in which the polymer's triplet level is sufficiently low to be populated from interfacial triplet CT states that are created during nongeminate charge recombination. In fact, as there may be an equilibrium between free charges and CT states in working devices as previously discussed by Burke et al. [25], any additional channel such as triplet energy transfer from the CT state to the polymer's triplet state could have an impact on device performance even under solar illumination conditions, where the carrier concentration is lower than under pulsed laser excitation as typically used in TA experiments.

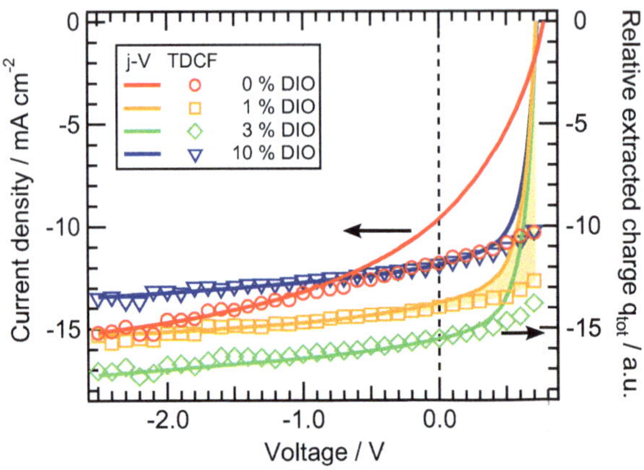

Fig. 10 *J–V* characteristics (*left axis*) and normalized extracted charge carrier density q_{tot} obtained using time-delayed collection field (*right axis*) of PBDTTT-C:$PC_{71}BM$ solar cells with varying DIO content. The illumination intensity was 1 sun. The difference between *J–V* and q_{tot} corresponds to nongeminate recombination losses as indicated by the *shaded area* using the example of 1 % DIO. [Figure reprinted by permission from Macmillan Publishers Ltd., Sci. Rep. 5, 8286 (2015)]

In a related investigation, the fraction of the cosolvent DIO in the preparation of PBDTTT-C:$PC_{71}BM$ solar cells was systematically increased. Zusan et al. [85] found that DIO has effects on the blend microstructure which influence charge photogeneration and recombination. The impact of the cosolvent on the current–voltage characteristics, including the contributions of geminate and nongeminate recombination, is shown in Fig. 10. Without the cosolvent, the blend morphology constitutes large $PC_{71}BM$ agglomerates in a PBDTTT-C–rich matrix, in accordance with reports by Collins et al. [86]. The photogeneration was poor because of the accordingly limited interfacial area, also leading to singlet exciton losses within the large fullerene domains. Already small amounts of DIO (0.6 %) led a reduction in the $PC_{71}BM$ domain size and a higher interfacial area between the donor and acceptor. The PCE is approximately doubled from 2.6 to 5.3 % because of the more efficient exciton dissociation; as a result of a strongly enhanced D–A interface, the PCE of the 0.6 % DIO device reaches a moderate level of only 5.3 %. However, the field-dependent photogeneration of the CT complexes and the inefficient collection of free-charge carriers in the too finely intermixed D–A matrix are problematic. Only an increase in the DIO content up to 3 % allows the charge photogeneration and collection to be improved by further decreasing the $PC_{71}BM$ domains and growing pure PBDTTT-C regions. With similar light harvesting, the resulting maximum PCE is 6.9 %. It is limited by a dominant nongeminate recombination and by substantially reduced but still present geminate losses. This study demonstrates that the photovoltaic performance is controlled by a crucial balance between local photogeneration and charge carrier transport due to the blend morphology.

3 Summary and Discussion

The study presented of four polymer blends supports the accepted viewpoint that the morphology of the blend, which is determined by the (post)processing conditions, is crucial to its photophysical properties. This is consistent with the view that low-energy (thermalized) rather than hot CT states are the main precursors to free charges [33]. As a consequence, the efficiency for free-charge photogeneration is inherently connected to the energy of the relaxed CT state and that of the charge-separated state, with these energies being influenced by the molecular packing and orientation at the D–A heterojunction and within the charge-transporting domains. Given the complexity of the systems, it is also clear that the adequate understanding of photophysics can only be achieved by combining several experimental techniques and theoretical modeling. It is, however, significantly more difficult to provide a clear link between morphology changes and photophysics at an atomic level, which is needed to design new materials. Here we identify several such structure–morphology–property relationships.

First, we shall argue that the role of the additive (or a good solvent) is to provide better local packing (π stacking) of polymer chains as well as to achieve the required sizes of polymer/fullerene domains. This conclusion has been supported by other studies, which also point out that most conjugated polymers are polymorphs [87–90], and additives can help to select the required polymorph. Better local ordering naturally improves charge transport because of reduced paracrystallinity and stronger electronic coupling elements [90–93]. A less trivial observation is that long-range ordering also influences the relative energy-level alignment as well as can help to reduce geminate and nongeminate recombination by contributing to the electrostatic forces acting on CT and charge-separated states at the interface [8, 9]. Detailed analysis of the long-range electrostatic effects teaches us that this comes at a price: While helping to split CT states and reduce nongeminate recombination, electrostatic effects also reduce the open-circuit voltage [9]. Since electrostatic forces, to a large extent, are determined by the quadrupole moment of a polymer repeat unit and its orientation with respect to the interface, one should look for a suitable compromise when tuning the chemistry of the polymer backbone.

Second, a far less trivial observation is that one should still allow for a certain degree of intermixing at the D–A interface. The argument here is that the localized electron and hole at an ideally flat interface are always bound electrostatically, with a binding energy on the order of 0.3 eV. If, however, a donor molecule protrudes into the acceptor phase (or vice versa), its gas-phase ionization potential and electron affinity become “dressed” by the electrostatic fields typical for the acceptor side of the interface. This reduces the binding energy of the CT state and makes it possible to split this state thermally [9].

With these design rules in mind, we can outline the future targets for OPV research. It has already been mentioned that the two main obstacles that prevent OPV commercialization are (1) limited by a thin photoactive layer with low external quantum efficiency and (2) low open-circuit voltages, in spite of large photovoltaic gaps. In both cases the potential strategies can be readily identified: To improve the external quantum efficiency we need materials with higher charge carrier mobilities, which facilitate extraction prior to recombination and thus allow for thicker photoactive layers. Furthermore, acceptors with an absorption stronger than that of fullerenes should be actively investigated.

The problem of the low open-circuit voltage is more intricate since arguments that account only for the relative energy-level alignment are insufficient to formulate appropriate design rules. It seems that we need to gain a better control of interfacial ordering (including intermixing) in order to balance the binding energy of CT states and losses in the open-circuit voltage. In addition, we need to look for materials with a small energetic disorder since the significant part of reducing the photovoltaic gap to the open-circuit voltage is caused by pinning the chemical potentials of holes and electrons in (rather broad) densities of states of the donor and the acceptor.

Overall, we now better understand the subtle interplay among molecular architecture, morphology, and electrostatic effects that jointly produces a high-performing solar cell, and we are ready to move forward assisted by the design rules we formulated.

Acknowledgments This work was partially supported by Deutsche Forschungsgemeinschaft (DFG) under the Priority Program "Elementary Processes of Organic Photovoltaics" (SPP 1355), BMBF grant MESOMERIE (FKZ 13N10723) and MEDOS (FKZ 03EK3503B), and DFG program IRTG 1404. The project has received funding from the NMP-20-2014—"Widening Materials Models" program under grant agreement number 646259 (MOSTOPHOS). F. Laquai thanks the Max Planck Society for funding the Max Planck Research Group.

References

1. Liu Y, Zhao J, Li Z, Mu C, Ma W, Hu H, Jiang K, Lin H, Ade H, Yan H (2014) Nat Commun 5:5293
2. Heliatek (2013) http://www.heliatek.com/de/presse/pressemitteilungen/details/heliatek-erreicht-neuen-organischen-photovoltaik-weltrekord-mit-einer-effizienz-von-132
3. Bakulin A, Rao A, Pavelyev V, van Loosdrecht P, Pshenichnikov M, Niedzialek D, Cornil J, Beljonne D, Friend RH (2012) Science 335:1340
4. Grancini G, Maiuri M, Fazzi D, Petrozza A, Egelhaaf H-J, Brida D, Cerullo G, Lanzani G (2013) Nat Mater 12:29
5. Gagorik AG, Mohin JW, Kowalewski T, Hutchison GR (2015) Adv Funct Mater 25:1996
6. Burke TM, McGehee MD (2014) Adv Mater 26:1923
7. Monahan NR, Williams KW, Kumar B, Nuckolls C, Zhu X-Y (2015) Phys Rev Lett 114:247003
8. Poelking C, Tietze M, Elschner C, Olthof S, Hertel D, Baumeier B, Würthner F, Meerholz K, Leo K, Andrienko D (2015) Nat Mater 14:434
9. Poelking C, Andrienko D (2015) J Am Chem Soc 137:6320
10. Scharber MC, Koppe M, Gao J, Cordella F, Loi MA, Denk P, Morana M, Egelhaaf H-J, Forberich K, Dennler G, Gaudiana R, Waller D, Zhu Z, Shi X, Brabec CJ (2010) Adv Mater 22:367
11. Vandewal K, Ma Z, Bergqvist J, Tang Z, Wang E, Henriksson P, Tvingstedt K, Andersson MR, Zhang F, Inganäs O (2012) Adv Funct Mater 22:3480

12. Bartelt JA, Lam D, Burke TM, Sweetnam SM, McGehee MD (2015) Adv Energy Mater 5:1500577
13. Zhao J, Li Y, Lin H, Liu Y, Jiang K, Mu C, Ma T, Lin Lai JY, Hu H, Yu D, Yan H (2015) Energy Environ Sci 8:520
14. Mikhnenko OV, Azimi H, Scharber M, Morana M, Blom PWM, Loi MA (2012) Energy Environ Sci 5:6960
15. Foertig A, Kniepert J, Gluecker M, Brenner T, Dyakonov V, Neher D, Deibel C (2014) Adv Funct Mater 24:1306
16. Kirchartz T, Pieters BE, Kirkpatrick J, Rau U, Nelson J (2011) Phys Rev B 83:115209
17. Baranovskii SD (2014) Phys Status Solidi 251:487
18. Bartesaghi D, Pérez IDC, Kniepert J, Roland S, Turbiez M, Neher D, Koster LJA (2015) Nat Commun 6:7083
19. Würfel U, Neher D, Spies A, Albrecht S (2015) Nat Commun 6:6951
20. Strobel T, Deibel C, Dyakonov V (2010) Phys Rev Lett 105:266602
21. Wagenpfahl A, Rauh D, Binder M, Deibel C, Dyakonov V (2010) Phys Rev B 82:115306
22. Tress W, Leo K, Riede M (2011) Adv Funct Mater 21:2140
23. Rau U (2007) Phys Rev B 76:085303
24. Vandewal K, Tvingstedt K, Gadisa A, Inganäs O, Manca JV (2009) Nat Mater 8:904
25. Burke TM, Sweetnam S, Vandewal K, McGehee MD (2015) Adv Energy Mater 5
26. Deibel C, Strobel T, Dyakonov V (2010) Adv Mater 22:4097
27. Shoaee S, Subramaniyan S, Xin H, Keiderling C, Tuladhar PS, Jamieson F, Jenekhe SA, Durrant JR (2013) Adv Funct Mater 23:3286
28. Dimitrov SD, Durrant JR (2014) Chem Mater 26:616
29. Scharber MC, Lungenschmied C, Egelhaaf H-J, Matt G, Bednorz M, Fromherz T, Gao J, Jarzab D, Loi MA (2011) Energy Environ Sci 4:5077
30. Tvingstedt K, Vandewal K, Zhang F, Inganäs O (2010) J Phys Chem C 114:21824
31. Inal S, Schubert M, Sellinger A, Neher D (2010) J Phys Chem Lett 1:982
32. Kern J, Schwab S, Deibel C, Dyakonov V (2011) Phys Status Solidi—Rapid Res Lett 5:364
33. Vandewal K, Albrecht S, Hoke ET, Graham KR, Widmer J, Douglas JD, Schubert M, Mateker WR, Bloking JT, Burkhard GF, Sellinger A, Fréchet JMJ, Amassian A, Riede MK, McGehee MD, Neher D, Salleo A (2014) Nat Mater 13:63
34. Hodgkiss JM, Campbell AR, Marsh RA, Rao A, Albert-Seifried S, Friend RH (2010) Phys Rev Lett 104:177701
35. Howard IA, Mauer R, Meister M, Laquai F (2010) J Am Chem Soc 132:14866
36. Howard IA, Laquai F (2010) Macromol Chem Phys 211:2063
37. Langevin PMP (1903) Ann Chim Phys 28:433
38. Pivrikas A, Sariciftci NS, Juška G, Österbacka R (2007) Prog Photovolt Res Appl 15:677
39. Mingebach M, Walter S, Dyakonov V, Deibel C (2012) Appl Phys Lett 100:193302
40. Koster LJA, Mihailetchi VD, Ramaker R, Blom PWM (2005) Appl Phys Lett 86:123509
41. Deibel C, Wagenpfahl A, Dyakonov V (2009) Phys Rev B 80:075203
42. Koster LJA, Mihailetchi VD, Blom PWM (2006) Appl Phys Lett 88:052104
43. Heiber MC, Baumbach C, Dyakonov V, Deibel C (2015) Phys Rev Lett 114:136602
44. Shuttle CG, O'Regan B, Ballantyne AM, Nelson J, Bradley DDC, de Mello J, Durrant JR (2008) Appl Phys Lett 92:093311
45. Foertig A, Baumann A, Rauh D, Dyakonov V, Deibel C (2009) Appl Phys Lett 95:052104
46. Gorenflot J, Heiber MC, Baumann A, Lorrmann J, Gunz M, Kämpgen A, Dyakonov V, Deibel C (2014) J Appl Phys 115:144502
47. Montanari I, Nogueira AF, Nelson J, Durrant JR, Winder C, Loi MA, Sariciftci NS, Brabec C (2002) Appl Phys Lett 81:3001
48. Nelson J (2003) Phys Rev B 67:155209
49. Deibel C, Rauh D, Foertig A (2013) Appl Phys Lett 103:043307
50. Deledalle F, Kirchartz T, Vezie MS, Campoy-Quiles M, Shakya Tuladhar P, Nelson J, Durrant JR (2015) Phys Rev X 5:011032

51. Baranovski S (2006) Charge transport in disordered solids with applications in electronics, 1 edn. Wiley, 498 p. ISBN: 978-0-470-09504-1
52. He Z, Zhong C, Su S, Xu M, Wu H, Cao Y (2012) Nat Photonics 6:593
53. Liang Y, Xu Z, Xia J, Tsai S-T, Wu Y, Li G, Ray C, Yu L (2010) Adv Mater 22, E135
54. Chen W, Xu T, He F, Wang W, Wang C, Strzalka J, Liu Y, Wen J, Miller DJ, Chen J, Hong K, Yu L, Darling SB (2011) Nano Lett 11:3707
55. Yan H, Collins BA, Gann E, Wang C, Ade H, McNeill CR (2012) ACS Nano 6:677
56. Lou SJ, Szarko JM, Xu T, Yu L, Marks TJ, Chen LX (2011) J Am Chem Soc 133:20661
57. Mort J, Chen I, Troup A, Morgan M, Knights J, Lujan R (1980) Phys Rev Lett 45:1348
58. Kniepert J, Schubert M, Blakesley JC, Neher D (2011) J Phys Chem Lett 2:700
59. Foster S, Deledalle F, Mitani A, Kimura T, Kim K-B, Okachi T, Kirchartz T, Oguma J, Miyake K, Durrant JR, Doi S, Nelson J (2014) Adv Energy Mater 4:1400311
60. Lange I, Kniepert J, Pingel P, Dumsch I, Allard S, Janietz S, Scherf U, Neher D (2013) J Phys Chem Lett 4:3865
61. Kniepert J, Lange I, Heidbrink J, Kurpiers J, Brenner TJK, Koster LJA, Neher D (2015) J Phys Chem C 119:8310
62. Hedley GJ, Ward AJ, Alekseev A, Howells CT, Martins ER, Serrano LA, Cooke G, Ruseckas A, Samuel IDW (2013) Nat Commun 4:2867
63. Kniepert J, Lange I, van der Kaap NJ, Koster LJA, Neher D (2014) Adv Energy Mater 4:1301401
64. Peet J, Kim JY, Coates NE, Ma WL, Moses D, Heeger AJ, Bazan GC (2007) Nat Mater 6:497
65. Peet J, Cho NS, Lee SK, Bazan GC (2008) Macromolecules 41:8655
66. Jamieson FC, Agostinelli T, Azimi H, Nelson J, Durrant JR (2010) J Phys Chem Lett 1:3306
67. Albrecht S, Schindler W, Kurpiers J, Kniepert J, Blakesley JC, Dumsch I, Allard S, Fostiropoulos K, Scherf U, Neher D (2012) J Phys Chem Lett 3:640
68. Albrecht S, Vandewal K, Tumbleston JR, Fischer FSU, Douglas JD, Fréchet JMJ, Ludwigs S, Ade H, Salleo A, Neher D (2014) Adv Mater 26:2533
69. Albrecht S, Janietz S, Schindler W, Frisch J, Kurpiers J, Kniepert J, Inal S, Pingel P, Fostiropoulos K, Koch N, Neher D (2012) J Am Chem Soc 134:14932
70. Albrecht S, Tumbleston JR, Janietz S, Dumsch I, Allard S, Scherf U, Ade H, Neher D (2014) J Phys Chem Lett 5:1131
71. Moet DJD, Lenes M, Morana M, Azimi H, Brabec CJ, Blom PWM (2010) Appl Phys Lett 96:213506
72. Jarzab D, Cordella F, Gao J, Scharber M, Egelhaaf H-J, Loi MA (2011) Adv Energy Mater 1:604
73. Hwang I-W, Cho S, Kim JY, Lee K, Coates NE, Moses D, Heeger AJ (2008) J Appl Phys 104:033706
74. Rao A, Chow PCY, Gélinas S, Schlenker CW, Li C-Z, Yip H-L, Jen AK-Y, Ginger DS, Friend RH (2013) Nature 500:435
75. Chow PCY, Gélinas S, Rao A, Friend RH (2014) J Am Chem Soc 136:3424
76. Etzold F, Howard IA, Forler N, Cho DM, Meister M, Mangold H, Shu J, Hansen MR, Müllen K, Laquai F (2012) J Am Chem Soc 134:10569
77. Etzold F, Howard IA, Forler N, Melnyk A, Andrienko D, Hansen MR, Laquai F (2015) Energy Environ Sci 8:1511
78. Miller NC, Gysel R, Miller CE, Verploegen E, Beiley Z, Heeney M, McCulloch I, Bao Z, Toney MF, McGehee MD (2011) J Polym Sci Part B Polym Phys 49:499
79. Miller NC, Cho E, Junk MJN, Gysel R, Risko C, Kim D, Sweetnam S, Miller CE, Richter LJ, Kline RJ, Heeney M, McCulloch I, Amassian A, Acevedo-Feliz D, Knox C, Hansen MR, Dudenko D, Chmelka BF, Toney MF, Brédas J-L, McGehee MD (2012) Adv Mater 24:6071
80. Zusan A, Vandewal K, Allendorf B, Hansen NH, Pflaum J, Salleo A, Dyakonov V, Deibel C (2014) Adv Energy Mater 4:1400922
81. Veldman D, Ipek O, Meskers SCJ, Sweelssen J, Koetse MM, Veenstra SC, Kroon JM, van Bavel SS, Loos J, Janssen RAJ (2008) J Am Chem Soc 130:7721
82. Deibel C, Strobel T, Dyakonov V (2009) Phys Rev Lett 103:036402

83. Gehrig DW, Howard IA, Sweetnam S, Burke TM, McGehee MD, Laquai F (2015) Macromol Rapid Commun 36:1054
84. Gehrig DW, Howard IA, Laquai F (2015) J Phys Chem C 119:13509
85. Zusan A, Gieseking B, Zerson M, Dyakonov V, Magerle R, Deibel C (2015) Sci Rep 5:8286
86. Collins BA, Li Z, Tumbleston JR, Gann E, McNeill CR, Ade H (2013) Adv Energy Mater 3:65
87. Fischer FSU, Trefz D, Back J, Kayunkid N, Tornow B, Albrecht S, Yager KG, Singh G, Karim A, Neher D, Brinkmann M, Ludwigs S (2015) Adv Mater 27:1223
88. Fischer FSU, Kayunkid N, Trefz D, Ludwigs S, Brinkmann M (2015) Macromolecules 48:3974
89. Scharsich C, Fischer FSU, Wilma K, Hildner R, Ludwigs S, Köhler A (2015) J Polym Sci Part B Polym Phys 53:1416
90. Poelking C, Daoulas K, Troisi A, Andrienko D (2014) P3HT revisited – from molecular scale to solar cell devices Adv Polym Sci 265:139–180
91. Poelking C, Andrienko D (2013) Macromolecules 46:8941
92. Gemünden P, Poelking C, Kremer K, Daoulas K, Andrienko D (2015) Macromol Rapid Commun 36:1047
93. Andrienko D (2014) Supramolecular materials for opto-electronics. Royal Society of Chemistry, Cambridge

Controlling the Electronic Interface Properties in Polymer–Fullerene Bulk Heterojunction Solar Cells

T. Stubhan, N. Wolf, J. Manara, V. Dyakonov, and C.J. Brabec

Contents

Abstract This work covers the use of solution-processed metal oxides as interface layers for organic solar cells. To study the interface properties, intrinsic and Al-doped ZnO_x were chosen as reference systems. From the class of n-type metal oxides, ZnO_x was chosen because it can be doped when it is solution processed.

T. Stubhan (✉)
Materials for Electronics and Energy Technology (i-MEET), Friedrich-Alexander-Universität Erlangen-Nürnberg, Martensstrasse 7, 91058 Erlangen, Germany
e-mail: tobias.stubhan@fau.de

N. Wolf • J. Manara • V. Dyakonov
Bavarian Center for Applied Energy Research (ZAE Bayern), Am Galgenberg 87, 97074 Würzburg, Germany
e-mail: jochen.manara@zae-bayern.de

C.J Brabec
Materials for Electronics and Energy Technology (i-MEET), Friedrich-Alexander-Universität Erlangen-Nürnberg, Martensstrasse 7, 91058 Erlangen, Germany

Bavarian Center for Applied Energy Research (ZAE Bayern), Haberstr. 2a, 91058 Erlangen, Germany

K. Leo (ed.), *Elementary Processes in Organic Photovoltaics*,
Advances in Polymer Science 272, DOI 10.1007/978-3-319-28338-8_12

Furthermore, the influence of thin modification layers applied on top of the metal oxides is investigated.

Keywords Aluminum-doped zinc oxide • Barium hydroxide • Charge transport in nanoparticular films • Conjugated polyelectrolyte • Device structure • Diketopyrrolopyrrole-quinquethiophene • Doped zinc oxide • Electronic surface properties • Energy diagram • Energy-level diagrams • EQE spectra • Inferface layers for organic solar cells • Interface layers • Interface modification layers • Inverted organic solar cells • *J-V* characteristics • Low temperature • Molybdenum oxide • Mott–Schottky capacitance • Oxygen • P3HT • PCDTBT • Phosphonic acid anchored fullerene SAM • Polyvinylpyrrolidone • Poole–Frenkel effect • Reference solar cells • SAM modification • Stability of interface materials • Stabilization of AZO nanoparticles • Summary • Thickness dependence • Trioxadecanoic acid • Water • Zinc oxide

1 Introduction

Organic solar cells have been processed in two different geometries: (1) in the inverted geometry, where the ZnO layer is directly processed on the transparent conducting oxide electrode as well as (2) in the normal cell structure, where the ZnO is deposited on top of the semiconductor and beneath the top cathode. This was possible because in addition to the classical sol-gel-processed ZnO_x layers we were able to process intrinsic and Al-doped ZnO_x layers via a nanoparticle suspension without high-temperature annealing processes that harm the polymeric active layer. The main objectives were to systematically vary the doping level and to build a correlation between the device functionality and the work function of the interface layer: its conductivity and carrier concentration, respectively.

Finally, these investigations provided valuable insight into the origin of the second diode and surface recombination in organic solar cells. Second diodes are internal diodes in solar cells that are connected in series to the photovoltaic junction and cause a kink in the *J–V* curve in the fourth quadrant and a delayed opening of the diode in the first quadrant [1].

The main aspect of this work was the development, advancement, design, characterization, and understanding of interface layers for a solution-processed polymer–fullerene solar cell. Although the development of the full solar cell (including the semiconductor, top electrode, and packaging) was not the focus of this work, this research delivered further important input to the architecture and design criteria for the rest of the solar cell's stack.

The first part of this work focused on the development and characterization of solution-processed ZnO at various bulk doping levels. The second part focused on modifying the electronic surface properties of doped and intrinsic ZnO. A description of the work's progress and the results obtained are summarized in the following sections.

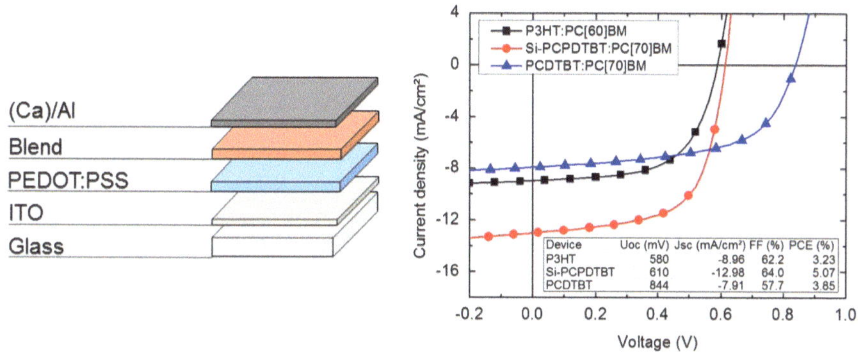

Fig. 1 *Left:* layer stack of the normal architecture solar cells. *Right:* *J–V* characteristics of the representative polymer solar cells with different donor polymers

1.1 *Reference Solar Cells*

Reference solar cells were built in the normal architecture. Figure 1 displays representative *J–V* curves of solar cells and the standard architecture used with different donor polymers in the active layer.

1.2 *Fundamentals and Mechanisms*

1.2.1 Charge Transport in Nanoparticle Thin Films of ZiZnO and Aluminum-Doped ZnO

According to the first part of the work package, the synthesis, characterization, and testing of reference materials, namely ZnO, TiOx, and a first aluminum-doped ZnO (AZO) precursor according to known routes taken from the literature [2–4], started. The lattice and crystallite structure of the materials was investigated through X-ray diffraction measurements. We obtained anatase TiOx with a crystallite size of 12.5 nm and wurtzite ZnO and AZO with crystallite sizes of 4.2 and 4.0 nm, respectively. A slight shift in the peak of the 002 plane of the AZO in comparison to the ZnO is a sign for interstitial doping. With optical measurements, the transmission and bandgap of the materials were investigated and found to be in a suitable range for a window layer in solar cell applications. The transmission of all materials is over 90 % in the visible part of the spectrum. This is a result of the wide optical bandgaps: 3.85 eV for TiOx, 3.44 eV for ZnO, and 3.34 eV for AZO. Electrical characterization showed conductivities on the order of 10^{-4} S/cm for TiOx and 10^{-6} S/cm for ZnO, whereas AZO had the highest conductivity, on the order of 10^{-3} S/cm.

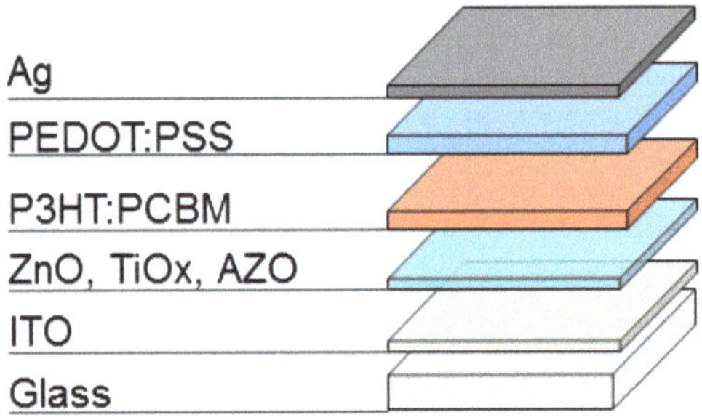

Fig. 2 Layer stack of the investigated inverted polymer solar cells

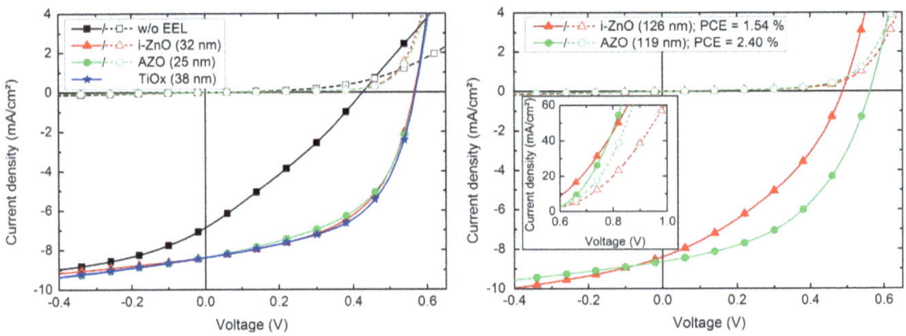

Fig. 3 *Left: J–V* characteristics of the inverted polymer solar cells with thin electron extraction layers. *Right*: *J–V* characteristics of the inverted polymer solar cells with thick electron extraction layers

Next, these materials were incorporated in inverted polymer solar cells with a usual thickness of a few tens of nanometers (20–40 nm). It was found that all three materials worked comparably and as well as electron extraction layers (EELs) in polymer solar cells (Figs. 2 and 3, right). Interestingly, doped ZnO turned out to be superior to intrinsic ZnO in the thick-film limit, that is, for extraction layers with a thickness of 100 nm and above (Fig. 3, right). The solar cells with thick AZO EELs maintained their performance compared with the thin EELs, but the efficiency of the devices with thick intrinsic ZnO EELs deteriorated (Table 1). Thick and robust films are a prerequisite to reliable large-scale EEL printing.

In subsequent work, the thickness of the interface layer was increased to even 600 nm without limiting the device performance.

As one of our goals was to build more environmentally stable devices, it was necessary to replace the commonly used acidic and hygroscopic PEDOT:PSS as the p-type interlayer, which is known to decrease device lifetimes. Therefore, we developed a solution process to deposit the more stable MoO_3 as the anode interlayer, which yields solar cells with a performance comparable to that of PEDOT:PSS.

Table 1 Key parameters of the inverted solar cells with different materials and thicknesses of the electron extraction layer

	V_{OC} (mV)	J_{SC} (mA/cm^2)	PCE (%)	FF (%)	R_S (Ω/cm^2)	R_{Sh} (kΩ/cm^2)	R (kΩ/cm^2)
w/o EEL	426	−6.86	0.85	29.2	23.6	3.0	0.29
TiOx	571	−8.38	2.59	54.1	–	–	–
i-ZnO (32 nm)	565	−8.41	2.56	53.9	1.1	6.2	0.59
i-ZnO (126 nm)	486	−8.40	1.54	37.7	3.2	3.1	0.33
AZO (25 nm)	569	−8.36	2.42	50.8	1.1	4.9	0.44
AZO (119 nm)	558	−8.69	2.40	49.4	1.2	3.9	0.50

V_{OC} open-circuit voltage, J_{SC} short-circuit current density, *PCE* power conversion efficiency, *FF* fill factor, R_S series resistance, R_{Sh} parallel resistance, *R* resistance

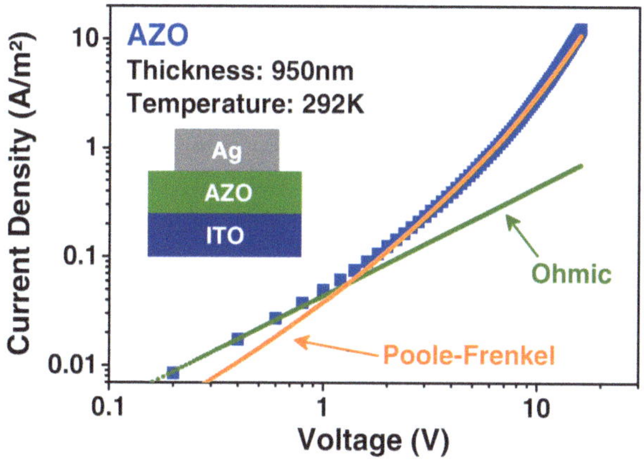

Fig. 4 *J–V* characteristics of aluminum-doped ZnO in a double-logarithmic plot. Two fit functions demonstrate that the low-voltage data are nicely described by Ohmic behavior, while the high-voltage data follow the Poole–Frenkel effect

1.2.2 Charge Transport in Nanoparticle Thin Films of ZnO and Aluminum-Doped ZnO

To improve our understanding of the above results, we carried out a charge-transport study on intrinsic and doped zinc oxide with commercially available ZnO/AZO dispersions from Nanograde Ltd. While the *J–V* characteristics at low voltage obey Ohm's law, transport in the high-voltage data regime is controlled by the Poole–Frenkel effect. The occurrence of the Poole–Frenkel effect is related to coulombically bound electrons, which have to overcome a field-dependent barrier to the next nanoparticle. Also, the conduction mechanism was demonstrated to be equivalent for the higher conductive Al-doped ZnO particles, where Al acts as a shallow donor. To the best of our knowledge, this is the first time that the Poole–Frenkel effect was unambiguously demonstrated as the dominant transport mechanism in a nanoparticle thin film (Fig. 4).

1.3 Processing of Interface Materials

Following our work program, the formulation design of doped metal oxide inks was started. As the primary prototype, the metal oxide ZnO was chosen since it has been shown that ZnO [4–8]

- can be processed from solution
- can be doped (e.g., with In, Ag, Ga, Mg, but especially Al)
- has been demonstrated as a functional layer in organic photovoltaic (OPV) cells when it was processed as an intrinsic semiconductor from solution
- has been demonstrated as a functional electrode for organic light-emitting diodes when it was processed from vacuum (atomic layer deposition, sputtering, etc.)

A variety of coating procedures have been well established for many different kinds of applications. In this work, we used sol-gel processes [9] because of their various advantages. Sol-gel precursors are easy to handle in atmospheric conditions, a defined band-doping of the processed materials is possible [10], they are cost-efficient, and they can be deposited by various techniques, including dip coating, spin coating, or several printing and coating processes [11]. There are two different sol-gel processes: the so-called high-temperature and low-temperature paths [12].

In the high-temperature method, the ZnO_x sol is prepared using zinc-acetate-2-hydrate, diethanolamine, and isopropyl alcohol (ISP). Aluminum-nitrate-9-hydrate is added as dopant [13]. The functional layer is formed by annealing the thin-film layer at temperatures over 250 °C [14]. This classic sol-gel process resulted in AZO functional layers with a charge carrier density of 4.2×10^{19} cm^{-3}, a specific resistivity of 3.1×10^{-3} Ω/cm, and a transmittance of up to 0.88 at a layer thickness of 1250 nm.

The low-temperature path is based on nanoparticle suspensions [15]. These suspensions consist of crystalline ZnO or AZO nanoparticles with a primary particle size between 15–40 nm. The nanoparticles are redispersed in solvents such as ethanol, diethylene glycol, or ISP and stabilized with surface-modifying agents [e.g., carbon acids such as pentanoic or trioxadecanoic acid (TODS)] [14]. The surface-modifying agents are added to the suspension to prevent aggregation and agglomeration of the nanoparticles. Additionally, an ultrasonic dissolver and stirrer are used for deagglomeration. In addition to the surface-modifying agents, a binder phase is necessary to provide a sufficient mechanical stability of the coating by interconnecting the particles on the one hand and to adhere the coating to the substrate on the other hand [16].The binder phase may also include functional groups that enable a curing of the coating at temperatures below 130 °C by using, for instance, ultraviolet radiation and photostarter. The state-of-the-art are polymeric binder phases based on glycidoxypropyltrimethoxysilane or methacryloxypropyltrimethoxysilane [17].

The nanoparticles themselves can, for example, be prepared by a surfactant-assisted sol-gel method using zinc nitrate and citric acid as starting materials [18]. A polyesterification reaction due to the Pechini process can also be used for the preparation of ZnO and AZO powders. This synthesis process is based on the chelation of metal ions by an α-hydroxyl acid such as citric acid in order to form stable homogeneous solutions. The powder can be obtained via calcination [19]. The various processes described were completed by a high-temperature calcination (300 °C) and a subsequent particle surface modification to result in stable suspensions with a particle size below 100 nm. These systems can be processed at low temperatures and are compatible with temperature-sensitive substrates such as polymers.

To date there has been only one paper describing the stabilization of AZO nanoparticles in ethylene glycol [20]. As ethylene glycol affects the performance of the solar cells negatively, it could not be used in our work. We have managed to stabilize AZO nanoparticles with at least four different surface-modifying agents in at least three solvents, respectively, solvent mixtures. Therefore, a series of ZnO solutions has been prepared and was successfully integrated in polymer solar cells.

A solution containing AZO particles, TODS surfactant, and ISP was identified as one candidate for yielding suitable films for solar cell integration. Only 140 °C is necessary to anneal these films, so they can be incorporated onto the temperature-sensitive organic active layer. The electrical functionality of the obtained AZO film was confirmed in standard solar cells with the layer stack displayed in Fig. 5 (left). The *J–V* characteristics in Fig. 5 (right) show that the AZO forms the electron-selective contact with the proper work function to achieve the expected high open-circuit voltage, V_{OC}. The high fill factor (FF) shows the good quality of the AZO EEL.

Further investigations have been performed on redispersed AZO nanoparticle solutions with different stabilizers, and we researched mixtures of stabilizers to find a solution with long-term stability and improved properties within organic solar cells.

Stable suspensions in IPS could be prepared with the surfactants TODS and polyvinylpyrrolidone (PVP) as well as with a mixture of TODS and PVP and with acetylacetone (AcAc), 1,4-butanediol, hexamethylenetetramine, and ε-

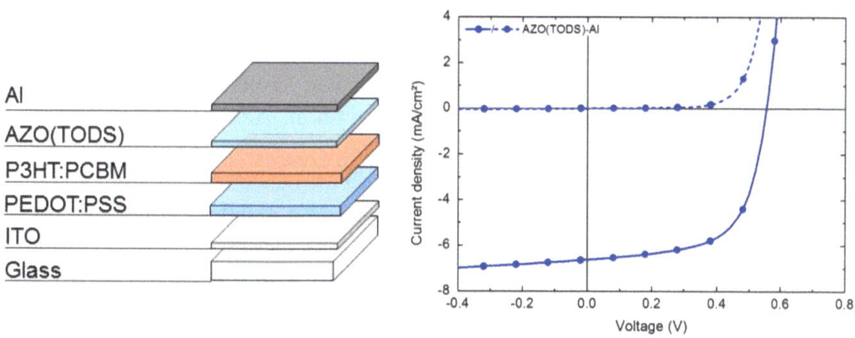

Fig. 5 *Left*: layer stack of the investigated solar cells. *Right*: *J–V* characteristics of standard solar cells with aluminum-doped ZnO (trioxadecanoic acid) electron extraction layer

Table 2 Hydrodynamic particle sizes of the prepared redispersed AZO nanoparticle solutions using trioxadecanoic acid (TODS), TODS and polyvinylpyrrolidone (PVP), and acetylacetone (AcAC) measured by dynamic light scattering directly after and 3 months after the preparation

Surfactant	TODS (nm)	TODS + PVP (nm)	AcAc (nm)
Particle sizes directly after dispersing	155	145	142
Particle sizes 3 months after dispersing	161	146	155

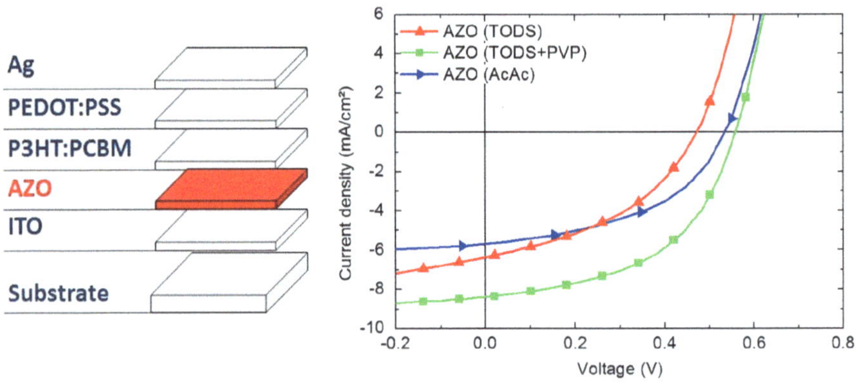

Fig. 6 *Left:* layer stack of inverted polymer–fullerene solar cell with aluminum-doped ZnO (AZO) interface layer. *Right*: *J–V* characteristics of devices containing AZO interface layers built with trioxadecanoic acid (TODS), TODS and polyvinylpyrrolidone, and acetylacetone nanoparticle suspensions

caprolactone. The smallest particle sizes could be reached by using the surfactants TODS, TODS + PVP, and AcAc, which are shown in Table 2. Usually, the hydrodynamic particle sizes are higher than the primary particle sizes due to agglomeration and aggregation of the primary particles. The particle sizes are similar, which confirms the long-term stability of the prepared AZO nanoparticle solutions.

The surface roughness of the thin films prepared at low temperatures reached values down to 20 nm with an electrical conductivity of up to 10^{-2} S/cm for layers prepared directly after dispersing the AZO nanoparticles as well as for layers prepared with a three-month-old AZO nanoparticle suspension. Finally, inverted polymer–fullerene solar cells have been built with an AZO interface layer using these AZO nanoparticle solutions (Fig. 6, left). The performance of the devices built was found to depend on the surfactant used, with the best obtained with TODS and PVP (Fig. 6, right). The effect can be attributed to a smaller particle size in the dispersion and better layer morphology.

Finally, redispersed AZO nanoparticle solutions with long-term stability have been developed and can successfully be applied to organic solar cells to investigate their properties and improve the performance.

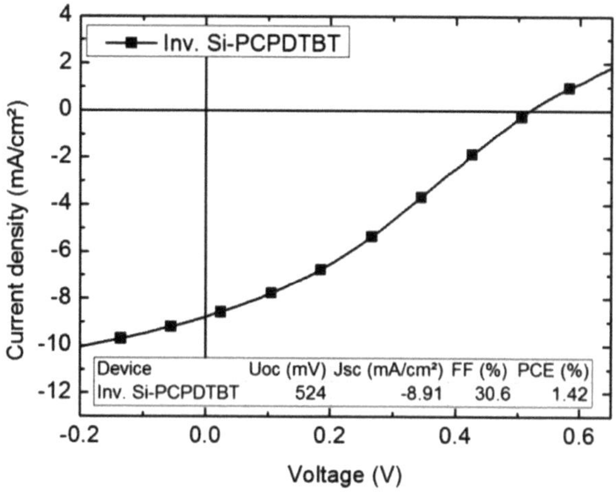

Fig. 7 Inverted solar cell with an aluminum-doped ZnO electron extraction layer and an Si-PCPDTBT:PC_{70}BM active layer (structure similar to Fig. 4)

1.4 Interface Modification Layers

One of the weak points of metal oxide–based EELs frequently discussed is their reliability and reproducibility when processed at low temperature (i.e., at 140 °C). Low processing temperatures may yield incomplete ligand removal and thus result in bad control of the electronic surface properties. For our investigated EEL materials, the performance gap between standard and inverted architecture is small. However, for active layers containing other high-performance donor polymers (e.g., poly[2,6-(4,4-bis-(2-ethylhexyl)dithieno[3,2-*b*:2′,3′-d]silole)-*alt*-4,7-(2,1,3-benzothiadiazole)] (Si-PCPDTBT); Fig. 7) or small molecules, the gap can be large.

1.4.1 Self-assembled Monolayer Modification

Decorating metal oxides with self-assembled monolayers (SAMs) is one way to achieve predefined surface properties. The experiment depicted in Fig. 8 with the key parameters listed in Table 3 demonstrates how varying the interface properties influences the device performance of inverted solar cells based on poly(3-hexylthiophene-2,5-diyl) (P3HT). The phosphonic acid-anchored aliphatic SAM (SAM-C6) forms a very unfortunate surface on the AZO, which decreases the device performance significantly and inflicts a second diode behavior in the *J–V* curve. In contrast, the fullerene-functionalized SAMs [SAM(C6-C60) and SAM(C18-C60)] as termination groups for the surface of the doped ZnO films improve the device efficiency [21].

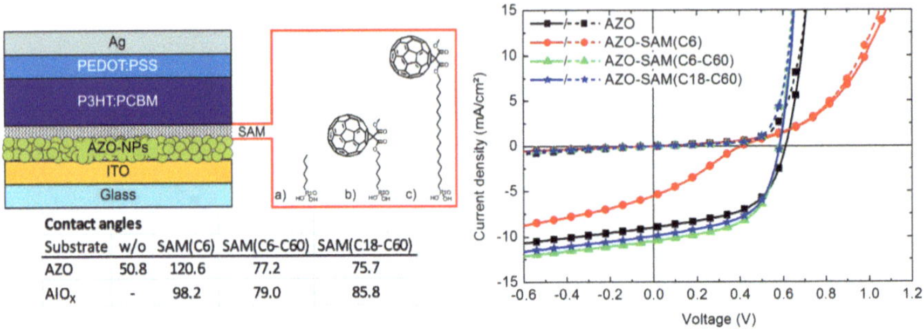

Fig. 8 *Left:* layer stack of the investigated solar cells including the chemical structure of the self-assembled monolayers (SAMs) and the results of the contact angle measurements to confirm the proper SAM deposition on aluminum-doped ZnO (AZO) and reference AlO_x layers. *Right:* *J–V* characteristics of solar cells with SAM-decorated AZO electron extraction layers

Table 3 Key parameters of solar cells with aluminum-doped ZnO electron extraction layers decorated with self-assembled monolayers

	V_{OC} (mV)	J_{SC} (mA/cm^2)	FF (%)	PCE (%)	R_S (Ω/cm^2)	R_P (Ω/cm^2)
AZO	602	−9.23	51.9	2.88	1.2	655
AZO-SAM(C6)	354	−3.94	25.3	0.38	9.5	1011
AZO-SAM(C6-C60)	583	−9.92	57.4	3.32	0.6	1473
AZO-SAM(C18-C60)	584	−9.24	53.5	2.88	0.9	1353

V_{OC} open-circuit voltage, J_{SC} short-circuit current density, *PCE* power conversion efficiency, *FF* fill factor, R_S series resistance, R_P polarization resistance

1.4.2 Barium Hydroxide Modification

Inverted organic solar cells with air-stable interface materials and top electrodes and an efficiency of 6.01 % are achieved by inserting a barium hydroxide [$Ba(OH)_2$] layer between the AZO EEL and the active layer. A low-bandgap diketopyrrolopyrrole-quinquethiophene alternating copolymer (pDPP5T-2) and phenyl-C61-butyric acid methyl ester ($PC_{60}BM$) were chosen as the active layer compounds (Fig. 9). Compared to the control device without $Ba(OH)_2$, inserting a 10-nm-thick $Ba(OH)_2$ layer results in an enhanced V_{OC} of 10 %, a short-circuit current density, J_{SC}, of 28 %, an FF of 28 %, and a power conversion efficiency (PCE) of 80 % (Fig. 10 and Table 4). Modifying AZO with a solution-processed low-cost $Ba(OH)_2$ layer increased the efficiency of the inverted device by dominantly reducing the energy barrier for electron extraction from PCBM (see schematic in Fig. 11); consequently, a higher built-in voltage and increased surface charge density are observed (see Fig. 12). The effects of a $Ba(OH)_2$ layer on the improvement in the device performance stem from the potential surface doping by diffusion of Ba^{2+} ions through the surface of the active layer, the reduced charge recombination, the decreased energy barrier for electron extraction and transport

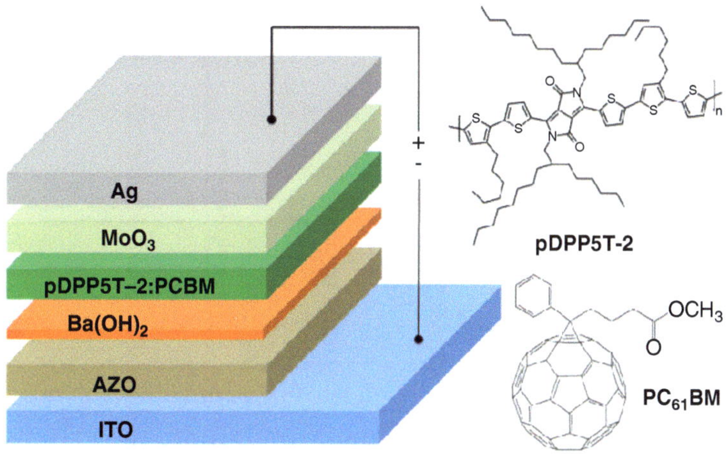

Fig. 9 Schematic of the inverted organic solar cells' structure and chemical structures of the pDPP5T-2 and $PC_{60}BM$

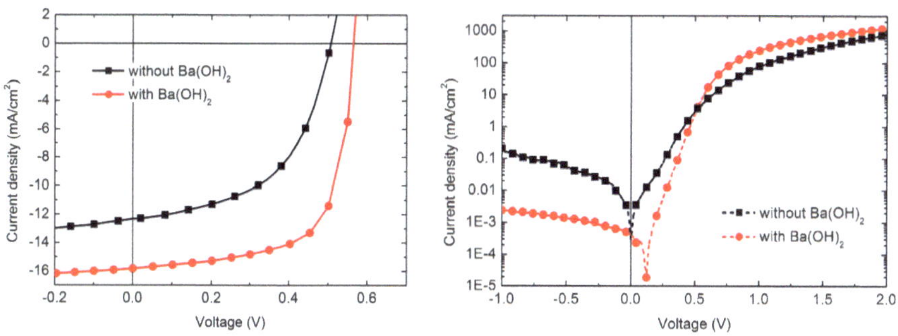

Fig. 10 *Left*: *J–V* characteristics of inverted organic solar cells without and with the barium hydroxide layer. *Right:* corresponding logarithmic plot of dark *J–V* characteristics

Table 4 Key values of the *J–V* characteristics of inverted organic solar cells without and with the $Ba(OH)_2$ layer

	V_{OC} (V)	J_{SC} (mA/cm^2)	FF (%)	PCE (%)	R_S (Ω/cm^2)	R_{Sh} (kΩ/cm^2)	n	J_0 (mA/cm^2)
Without $Ba(OH)_2$	0.50	−12.31	53.90	3.33	1.05	11.58	1.87	1.97×10^{-5}
		−11.78[a]						
With $Ba(OH)_2$	0.55	−15.79	69.18	6.01	0.88	376.41	1.65	4.70×10^{-7}
		−15.19[a]						

V_{OC} open-circuit voltage, J_{SC} short-circuit current density, *FF* fill factor, *PCE* power conversion efficiency, R_S series resistance, R_{Sh} parallel resistance, n, J_0
[a]Current density values derived from the external quantum efficiency measurement

via inserting an interface dipole layer, and the increased built-in potential from a corresponding increase in the surface charge density [22].

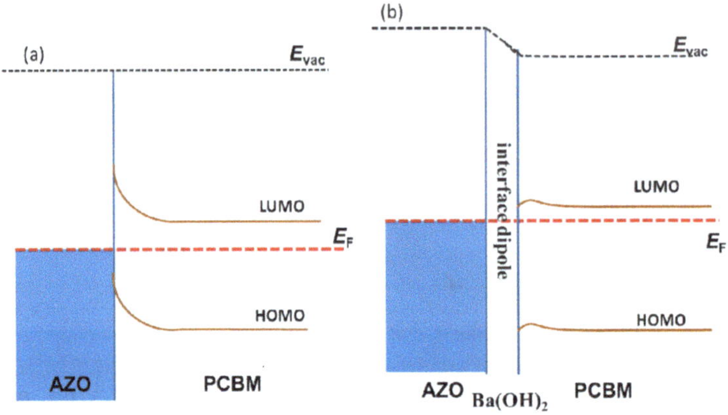

Fig. 11 Schematic energy diagrams for band conditions at the AZO/PCBM junctions (**a**) without and (**b**) with barium hydroxide on the AZO layer

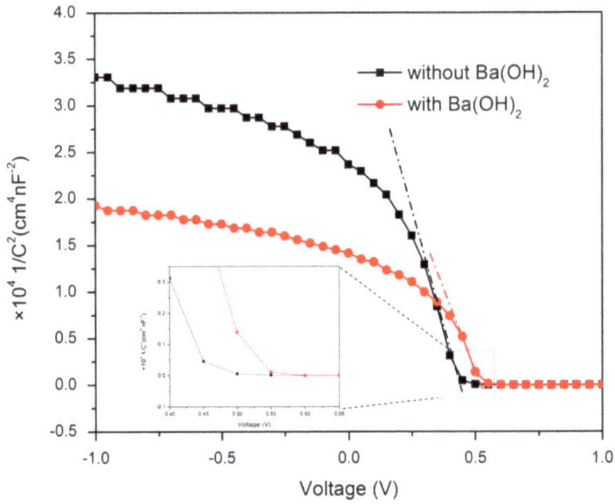

Fig. 12 Mott–Schottky capacitance plots of the voltage data of the device without and with the barium hydroxide layer. Inset: The intercept with voltage axis yields the built-in voltage

1.4.3 Conjugated Polyelectrolyte Modification

The device structure, the molecular structure of poly[3-(6-trimethylammonium-hexyl)thiophene] (P3TMAHT), and the energy levels of the components used in the devices are shown in Fig. 13. Detailed methods for the syntheses of N(Ph-2T-DCN-Et)$_3$ and P3TMAHT are described in the literature [23, 24]. Through application of the P3TMAHT layer on top of the AZO, the work function of AZO is reduced to from 4.2 to 3.8 eV. Indeed, the increased polarity due to the accumulation of ions

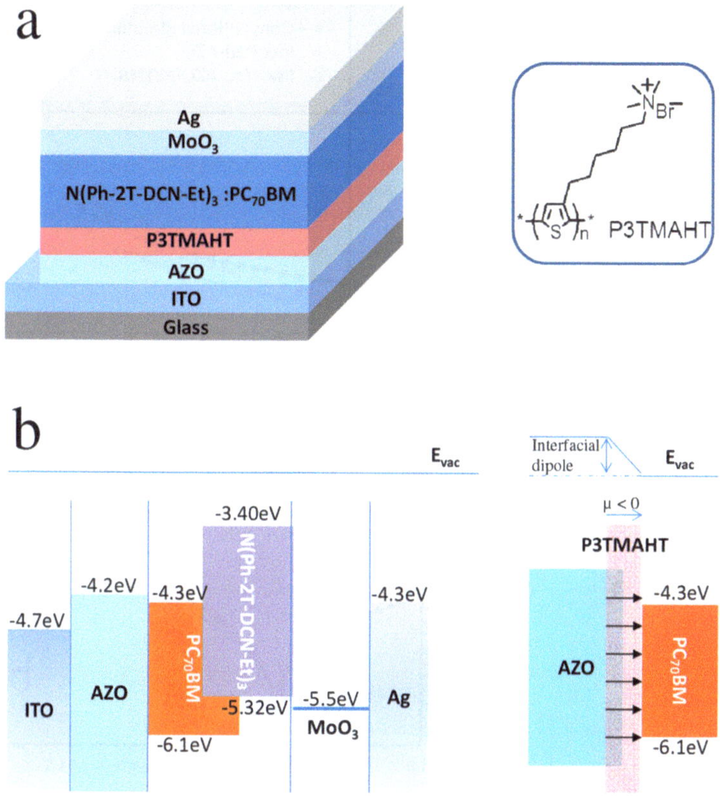

Fig. 13 *Top*: the device structure of the inverted N(Ph-2T-DCN-Et)$_3$:PC$_{70}$BM solar cells. Inset shows the molecular structures of P3TMAHT. *Bottom:* schematic energy-level diagrams for flat band conditions at the AZO/(P3TMAHT)/active layer junction with and without the P3TMAHT layer

and the formation of a dipole layer at the interface of AZO is observed as described in the references [25, 26].

The current density vs. voltage (*J–V*) characteristics of N(Ph-2T-DCN-Et)3:PC70BM–based conventional [ITO/PEDOT:PSS/N(Ph-2T-DCN-Et)3:PC70 BM/Ca/Ag] and inverted [ITO/AZO/(P3TMAHT)/N(Ph-2T-DCN-Et)3:PC70BM/ MoO3/Ag] devices under simulated AM 1.5 and in the dark are shown in Fig. 14 (top and middle), respectively. Device parameters are deduced from the *J–V* characteristics in the light and dark (summarized in Table 5). The reference conventional cells based on N(Ph-2T-DCN-Et)3:PC[70]BM as active layer have a V_{OC} of 0.939 V, a J_{SC} of 7.79 mA/cm^2, an FF of 45.9 %, and a PCE of 3.36 %, while the series resistance (R_S) is 1.29 Ω/cm^2 and the parallel resistance (R_{Sh}) is 1.84 kΩ/cm^2. In contrast, the inverted cells based on AZO as the electron-transporting layer show a PCE of 3.17 % with a V_{OC} of 0.840 V, a J_{SC} of 8.47 mA/cm^2, and an FF of 44.5 %. The surface recombination, work function

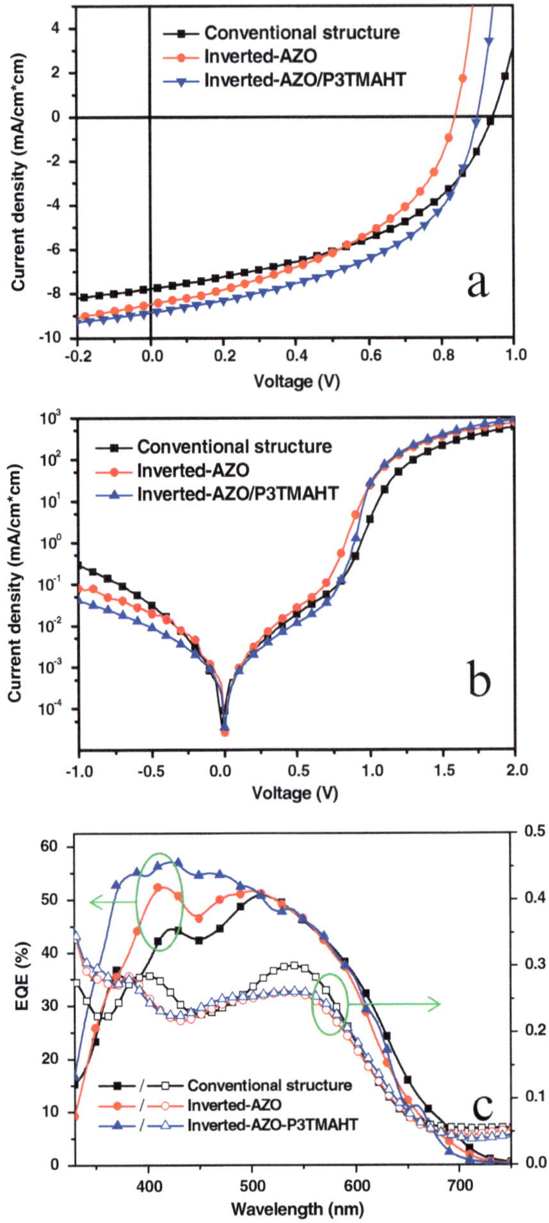

Fig. 14 *Top: J–V* characteristics of the conventional and inverted devices based on various interface layers under the illumination of AM 1.5G, 100 mW/cm^2. *Middle: J–V* characteristics of the devices in *dark*. *Bottom*: External quantum efficiency spectra of the conventional device and the inverted devices based on various interface layers, and absorption spectra of these blend films of PEDOT/active layer, AZO/active layer, and AZO/P3TMAHT/active layer on the ITO substrates

mismatch, or interface barrier might result in a reduced V_{OC} from 0.939 to 0.840 V, thus resulting in a slightly lower PCE of 3.17 %. Consequently, we introduced P3TMAHT as the interface layer on the AZO layer, which facilitates electron injection and transport by reducing the energy barrier between the transparent cathode and active layer. This directly resulted in an increased V_{OC} (from 0.840

Table 5 Performance parameters of N(Ph-2T-DCN-Et)$_3$:PC$_{70}$BM solar cells with conventional and inverted structures based on various interface layers

	V_{OC} (V)	J_{SC} (mA/cm^2)	FF (%)	PCE (%)	R_S (Ω/cm^2)	R_P (Ω/cm^2)
Conventional	0.939	7.79	45.9	3.36	1.29	1.84
Inverted-AZO	0.840	8.47	44.5	3.17	1.21	0.85
Inverted-AZO/P3TMAHT	0.898	8.85	49.2	3.91	0.96	2.34

V_{OC} open-circuit voltage, J_{SC} short-circuit current density, *FF* fill factor, *PCE* power conversion efficiency, R_S series resistance, R_P polarization resistance

to 0.898 V). In addition, compared with the AZO-based inverted devices, the combination of AZO and the P3TMAHT layer slightly enhanced the J_{SC} (from 8.47 to 8.85 mA/cm^2) and FF (from 44.5 to 49.2 %) and thus resulted in a higher PCE of 3.91 %.

Considering the improvement in the J_{SC}, Heeger et al. [27] speculated that the increased built-in potential and internal electric field might help the charge carriers to escape shallow traps and reduce trap-assisted recombination. In addition, Kim et al. [28] thought that the smoother and more hydrophobic surface of the buffer layer (AZO/P3TMAHT layer, as shown in Fig. 14, bottom) improved the compatibility between the organic active layer and the AZO layer and that the better contact induced the enhanced J_{sc}. The external quantum efficiency (EQE) spectra of various devices are shown in Fig. 14 (bottom). The EQE results agree with the J_{SC} of various devices mentioned above.

The improved PCE of the device with the AZO/P3TMAHT layer are also consistent with the higher EQE value (>55 %) exhibited in the PC$_{70}$BM absorption region between 350–500 nm. Although the absorption spectra of blends on the AZO and AZO/P3TMAHT layer are similar, as shown in Fig. 14 (bottom), the higher EQE values of the AZO/P3TMAHT layer-based device in the PC$_{70}$BM absorption region prove that AZO/P3TMAHT is more efficiently extracting electrons from PC$_{70}$BM and thus successfully reduces the recombination.

1.5 Stability of Interface Materials

We synthesized and prepared two different types of Al^{3+}-doped ZnO thin transparent films by using a low-cost sol-gel doctor-blade technique. The effect of temperature, adsorption, and diffusion of moisture or moisture vapor on the electrical and optical properties of the films was investigated and correlated to device degradation. The films were placed in a climatic chamber (type C-40/200) for different deposition times. The storage of the samples took place under the following conditions: 40 °C, 90 % relative humidity, 0–140 h. Our results show that the optical properties such as transmittance and bandgap are not influenced by damp-heat testing. Scanning electron microscopy and X-ray diffraction measurements clearly revealed the more amorphous structure of low-temperature AZO thin films

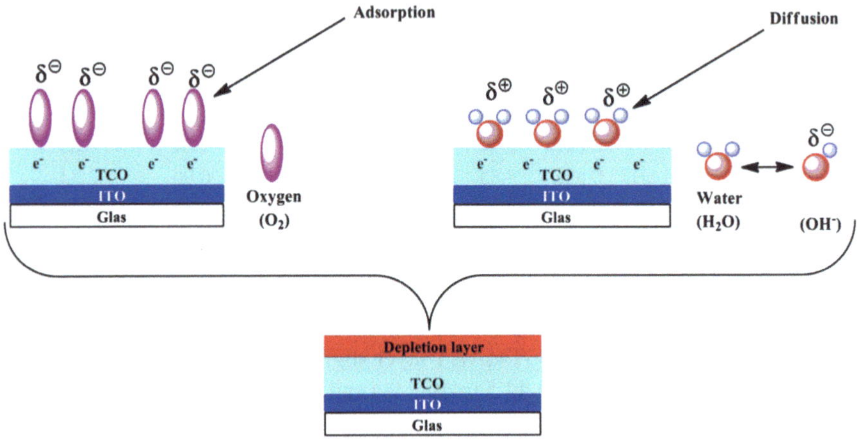

Fig. 15 Schematic view of induced adsorption of oxygen, and diffusion of water or vapor from the ambient atmosphere in the climatic chamber. (Image modified and adapted from [29])

with a nanoporous surface morphology. The analyzed Fourier transform infrared spectra confirmed the impact of both adsorbed oxygen and diffused water on AZO layers, promoting the degradation process (see Fig. 15). The electrical properties of both high- and low-temperature AZO films, that is, work function and conductivity, showed evident degradation upon damp-heat exposure, indicating the strong interaction between the metal surface and the ambient conditions that leads to the formation of a depletion layer. Accelerated device degradation under damp-heat testing underlines the formation of a barrier or depletion layer at the interface, reducing V_{OC}, J_{SC}, and FF. The intrinsic stability of metal oxide–based interlayers as well as that of their nanostructure is an essential criterion for the stability of organic electronic devices. Investigations of different metal oxide surfaces are required to better understand how to further improve the electrical integrity of the surface under damp-heat conditions [30].

1.6 Summary

- Representative solar cells with P3HT, Si-PCPDTBT, and poly[2,6-(4,4-bis(2-ethylhexyl)-4H-cyclopenta[2,1-*b*;3,4-*b′*]dithiophene)-*alt*-4,7(2,1,3-benzothiadiazole)] with power efficiencies between 3–5 % were processed.
- The synthesis, characterization, and benchmarking of the reference materials, namely TiOx, ZnO, and AZO for EELs in polymer solar cells, were completed. The relationship between material properties and solar cell performance could be established. Applied in thin films, all three materials performed comparably in the solar cells. AZO performed superior to ZnO in the thick film limit. It is

now possible to compare the materials developed in this work to known interface layers.

- A series of AZO formulations with different compositions was developed and the prepared thin films were characterized. A formulation that yields films with suitable properties for the solar cell application could be identified, and the first solar cells that contain these films were built, with promising results.
- Furthermore, the experiments focused on the influence of modification layers on device performance. Three different approaches for surface modifications were applied and investigated:
  - SAMs: PA-anchored aliphatic and fullerene-functionalized SAMs were used as termination groups for doped metal oxide films. The surface properties can be modified over a large interval with a huge impact on device performance.
  - $BaOH_2$: A thin modification layer of $BaOH_2$ was applied to AZO. This led to ...., which improved the device performance of the respective devices by a large margin.
  - CPE: The conductive polyelectrolyte P3MAHT was deposited on top of AZO to improve the performance of inverted small-molecule–based organic solar cells. The P3MAHT reduced the work function, thus enhancing the contact formation to PCBM.
- Additionally, stable redispersed AZO nanoparticle suspensions for a low-temperature coating process that provided long-term stability together with customized electrical properties (such as the electrical conductivity), a small surface roughness, and a high transmittance were developed.
- Conclusions
  - Doping of ZnO allowed us to increase the layer thickness of these interface layers to more than 500 nm without deteriorating device performance.
  - Surface modification has a huge impact on device performance. Many different active layer systems can yield improved performance with surface modifications; however, there is no universal solution. A toolbox based on a suitable doped metal oxide with a variety of modifications proved to be the best approach to adjust the interface properties to the active layers' needs.

Acknowledgment This work is supported by the German Science Foundation (DFG), grant numbers BR 4031/1-1, DY 18/7-1, BR 4031/1-2, and DY 18/7-2.

References

1. Wagenpfahl A, Rauh D, Binder M, Deibel C (2010) Phys Rev B 82:115306
2. Wang J, Polleux J, Lim J, Dunn B (2007) J Phys Chem C 111:14925
3. Alam MJ, Cameron DC (2001) J Vac Sci Technol A 19:1642
4. Park Y-S, Park HK, Cho SW, Jeong JA, Choi KH, Kim HK, Lee JY, Cho WJ (2008) Electrochem Solid-State Lett 11:J85

5. Huang H-H, Chu SY, Kao PC, Chen YC (2008) Thin Solid Films 516(16):5664
6. Yoshida Y, Tanaka S, Hiromttsu I, Fujita Y, Yoshino K (2008) Jpn J Appl Phys 47:867
7. Park YR, Nam E, Kim YS (2008) Jpn J Appl Phys 47:468
8. Tsai C-L, Lin YJ, Wu PH, Chen SY, Liu DS, Hong HJH, Liu CJ, Chang HC (2007) J Appl Phys 101(11):113713
9. Al-Dahoudi N, Aegerter MA (2006) Thin Solid Films 502:193–197
10. Dislich H (1986) J Non-Cryst Solids 80:115–121
11. Rydzek M, Reidinger M, Arduini-Schuster M, Manara J (2011) Prog Org Coat 70:369–375
12. Reidinger M, Rydzek M, Scherdel C, Arduini-Schuster M, Manara J (2009) Thin Solid Films 517:3096–3099
13. Rydzek M, Reidinger M, Arduini-Schuster M, Manara J (2012) Thin Solid Films 520:4114–4118
14. Rydzek M, Reidinger M, Scherdel C, Arduini-Schuster M, Manara J (2009) High Temp–High Pressures 38:277–293
15. Puetz J, Al-Dahoudi N, Aegerter MA (2004) Adv Eng Mater 6:733–737
16. Al-Dahoudi N (2003) Dissertation, Universität des Saarlandes
17. Al-Dahoudi N, Aegerter MA (2006) Thin Solid Films 520:193
18. Zhang YL, Yang Y, Zhao JH, Tan RQ, Cui P, Song WJ (2009) J Sol-Gel Sci Technol 51:198
19. Serier H, Gaudon M, Ménétrier M (2009) Solid State Sci 11:1192
20. Zhao J, Tan R, Zhang Y, Yang Y, Guo Y, Zhang X, Wang W, Song W (2011) J Am Ceram Soc 94:725–728
21. Stubhan T, Salinas M, Ebel A, Krebs FC, Hirsch A, Halik M, Brabec CJ (2012) Adv Energy Mater 2:532
22. Zhang H, Stubhan T, Li N, Turbiez M, Matt GJ, Ameri T, Brabec CJ (2014) J Mater Chem A 2:18917
23. Min J, Luponosov YN, Ameri T, Elschner A, Peregudova SM, Baran D, Heumüller T, Li N, Machui F, Ponomarenko S, Brabec CJ (2013) Org Electron 14:219–229
24. Scherf U, Gutacker A, Koenen N (2008) Acc Chem Res 41:1086–1097
25. Seo JH, Gutacker A, Sun Y, Wu H, Huang F, Cao Y, Scherf U, Heeger AJ, Bazan GC (2011) J Am Chem Soc 133:8416
26. He ZC, Zhong CM, Su SJ, Xu M, Wu HB, Cao Y (2012) Nat Photonics 6:591–595
27. Kyaw AKK, Wang DH, Gupta V, Zhang J, Chand S, Bazan GC, Heeger AJ (2013) Adv Mater 25:2397–2402
28. Choi H, Park JS, Jeong E, Kim G-H, Lee BR, Kim SO, Song MH, Woo HY, Kim JY (2011) Adv Mater 23:2759–2763
29. Jeong JK, Won Yang H, Jeong JH, Mo Y-G, Kim HD (2008) Appl Phys Lett 93:123508
30. Litzov I, Azimi H, Matt G, Kubis P, Stubhan T, Popov G, Brabec CJ Org Electron 15(2):569–576
31. Oh H, Krantz J, Stubhan T, Litzov I, Brabec CJ (2011) Sol Energy Mater Sol Cells 95:2194
32. Stubhan T, Oh H, Pinna L, Krantz J, Litzov I, Brabec CJ (2011) Org Electron 12:1539
33. Stubhan T, Ameri T, Salinas M, Krantz J, Machui F, Halik M, Brabec CJ (2011) Appl Phys Lett 98:253308
34. Wolf N, Rydzek M, Gerstenlauer D, Arduini-Schuster M, Manara J (2013) Thin Solid Films 532:60–65
35. Wolf N, Stubhan T, Manara J, Dyakonov V, Brabec CJ (2014) Thin Solid Films 564:213
36. Sista S, Park M-H, Hong Z, Wu Y, Hou J, Kwan WL, Li G, Yang Y (2010) Adv Mater 22:380

Near-Infrared Sensitization of Polymer/Fullerene Solar Cells: Controlling the Morphology and Transport in Ternary Blends

Tayebeh Ameri, Michael Forster, Ullrich Scherf, and Christoph J. Brabec

Contents

T. Ameri (✉)
Department of Materials Science and Engineering, Materials for Electronics and Energy Technology (i-MEET), Friedrich-Alexander University Erlangen-Nuremberg, Martensstrasse 7, 91058 Erlangen, Germany
e-mail: tayebeh.ameri@fau.de

M. Forster • U. Scherf
Mathematik and Naturwissenschaften Fachgebiet Makromolekulare Chemie and Institut für Polymertechnologie, Bergische Universität Wuppertal, Gaußstraße 20, 42097 Wuppertal, Germany

C.J. Brabec
Department of Materials Science and Engineering, Materials for Electronics and Energy Technology (i-MEET), Friedrich-Alexander University Erlangen-Nuremberg, Martensstrasse 7, 91058 Erlangen, Germany

Bavarian Center for Applied Energy Research (ZAE Bayern), Haberstr. 2a, 91058 Erlangen, Germany

K. Leo (ed.), *Elementary Processes in Organic Photovoltaics*,
Advances in Polymer Science 272, DOI 10.1007/978-3-319-28338-8_13

Abstract The concept of near-infrared (NIR) sensitization can be used as an alternative strategy to extend the spectral sensitivity of wide-bandgap polymers in polymer/fullerene solar cells. In ternary systems consisting of a conjugated polymer donor, a fullerene acceptor, and a sensitizer, the fullerene needs to act as an electron acceptor as well as an electron-transport matrix, the polymeric donor should provide a sufficiently high hole mobility, and the sensitizers should sensitize the bulk heterojunction solar cell in the red/NIR region. So far we have used various optoelectrical and structural techniques to investigate the possible mechanisms of the charge transfer and charge transport among the three components and microstructure of the ternary blends. In this review-like chapter, we present our recent achievements on developing the concept of NIR sensitization for polymer/fullerene solar cells by mainly addressing the important aspect of the relationship between morphology and transport.

Keywords Cascade alignment • Differential scanning calorimetry • Fullerene multiadducts • GIWAXS • Hansen solubility parameters • Morphology agent • Near-IR sensitization • OPV • Phase diagram • SCLC • Surface energy • Ternary solar cell • Time-resolved pump-probe spectroscopy • Transport

1 Introduction

The growing world energy demand has urged people to seek new alternatives to conventional fossil resources. Photovoltaics offers the ability to generate electricity in a clean and renewable way. Among existing solar cells, organic photovoltaics (OPVs) is one of the most promising technologies for low-cost energy production, with the advantages of semitransparency, flexibility, and solution processing. Because of these beneficial and attractive features, a great deal of attention has been given to research and development for higher OPV efficiency over the past few years. A significant breakthrough of power conversion efficiencies (PCEs) over 10 % was recently reported on both polymer- and small-molecule–based OPVs [1–5] (http://www.heliatek.com/en/press/press-releases/details/heliatek-sets-new-organic-photovoltaic-world-record-efficiency-of-13-2). Though impressive, further performance development is required to ensure a bright industrial future for OPVs.

Two of the main attractive features of conjugated polymers are their strong absorption coefficient together with their ability to tune the bandgap. As such, this material class is ideal for photovoltaics. However, since conjugated polymers are predominantly molecular excitonic absorbers, their absorption is characterized by a spectrally narrow absorption of a few 100 nm instead of a flat absorption plateau seen with inorganic semiconductors. This is especially unsatisfying for solar cell operations, where one wants to absorb the whole spectrum resonant to the bandgap with equal oscillator strength. As one of the most well-known strategies to address this limitation, polymers with different bandgaps and absorptions can be coupled into tandem cells. Tandem cells allow photons resonant to the bandgap of both polymers to be collected with minimal thermalization losses. However, they are based on a complicated multilayer stack with serious technical challenges. Among those challenges are processing a robust intermediated layer, coupling appropriate absorbers, as well as optimizing the active layers' thicknesses. In this chapter we present an elegant alternative strategy to extend the spectral sensitivity of wide-bandgap polymers into the near-infrared (NIR) region, the so-called organic ternary solar cell. In a ternary organic solar cell, an infrared sensitizer is simply added to the host system consisting of a wide-bandgap polymer blended with a fullerene derivative. The ternary approach improves the photon harvesting in thickness-limited photoactive layers, resulting in higher short-circuit current densities (J_{sc}) and therefore higher PCEs in a simple single junction. As such, the ternary solar cell concept avoids the demanding challenges of multijunction solar cell processing for spectrally broad light harvesting. Nevertheless, charge transfer and charge transport among the three components, as well as the complex morphology of the ternary blends, are considered to be the main challenges of ternary solar cells [6]. The concentration of the sensitizer decisively affects the microstructure formation, while the respective energy-level alignment is also decisive for the sensitization mechanism. Cascade charge-transfer, parallel-like charge-transfer, and energy-transfer reactions are among the most relevant mechanisms for successful sensitization. The first demonstration of a ternary solar cell including a low-bandgap polymer sensitizer was presented by Koppe et al. [7], who sensitized poly(3-hexylthiophene):[6,6]-phenyl-C_{61}-butyric acid methyl ester (P3HT:PCBM) with 20 wt% of the NIR absorber poly[2,6-(4,4-bis(2-ethylhexyl)-4H-cyclopenta[2,1-*b*;3,4-*b′*]dithiophene)-*alt*-4,7(2,1,3-benzothiadiazole)] (PCPDTBT). Optimized ternary devices obtained PCEs of 2.8 %, which is 10 % higher than the performance of the relevant binary reference. To date, the most efficient ternary organic solar cells exhibit PCEs of over 8 %, showing a wide absorption spectrum [8–11].

So far, we have carried out a detailed study on the ink formulation and phase diagram of the ternary blend. In 2011, when just a few groups were working on the ternary solar cells, we focused on the development of ternary solar cells and proving the potential of the NIR sensitization concept, for the first time achieving an improvement of over 30 % compared to the reference binary device. Afterward, charge-transfer/transport and morphology investigations were our main focus, so that an initial design rule for the organic ternary solar cells was established. Our work included polymeric as well as small-molecule sensitizers. A description of the progress of our work and the results obtained are summarized in the following sections.

2 Results and Discussion

2.1 Solubility and Ink Formulation

Two different conjugated polymers, the semicrystalline P3HT and dominantly amorphous C-PCPDTBT, and the fullerene derivative of PCBM were selected as

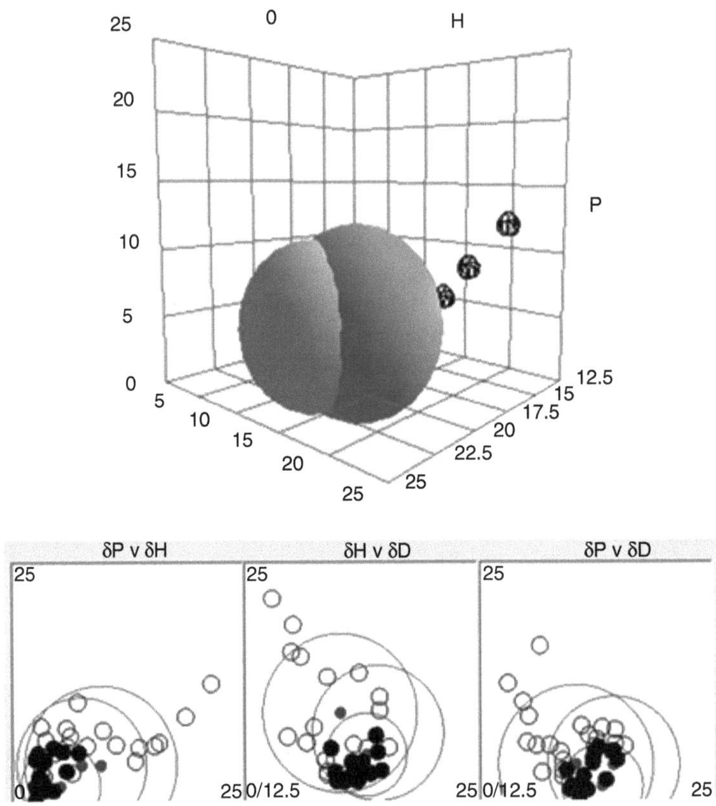

Fig. 1 Hansen solubility parameter diagram for solutes at 60 °C with four solvents, 2.5 g/L, for P3HT, PCPDTBT, and PCBM

prototype candidates to evaluate the applicability of the Hansen solubility parameter (HSP) [12–14] concept for organic semiconductors and ternary systems.

P3HT was certainly the most difficult component to dissolve. The significantly better solubility of PCPDTBT and PCBM is immediately recognized by the bigger radius of their spheres. The lower and upper limits of the solubility regime are the diameter of the solubility sphere. All solvents within this regime are expected to dissolve all three semiconductors; this is visualized in Fig. 1. Finally, we analyzed the ternary system of P3HT:PCPDTBT:PCBM with the following solubility range: $\delta_D = 14.4–23.0\ MPa^{1/2}$, $\delta_P = 0.0–5.7\ MPa^{1/2}$, and $\delta_H = 0.5–8.8\ MPa^{1/2}$ at 60 °C.

Overall, we could demonstrate that HSP is a representative method to describe and predict the solubility of dominantly amorphous organic semiconductors. Crystalline or dominantly aggregating semiconductors require further in-depth investigations. This opens the opportunity to use HSP for the design of organic semiconductor inks. More details of this work can be found in [15].

2.2 Phase Diagram Investigation

We used differential scanning calorimetry (DSC) to visualize the mixing behavior of multicomponent systems and extract phase diagrams for binary and ternary organic semiconductor blends. The P3HT:PCBM host system showed only one eutectic point, at around 60 wt% P3HT (40 wt% PCBM) at an eutectic temperature of about 220 °C. Figure 2a summarizes the phase behavior of the multicomponent blends from the first DSC heating curves. We can divide the mixtures into two types: semicrystalline (melting peaks observed during heating) and amorphous (no melting peaks observed during heating). The addition of small amounts of C-PCPDTBT (<20 wt%) did not significantly change the eutectic temperature of the ternary blends at a polymer:fullerene ratio of 60:40 wt%. This was demonstrated by the existence of a eutectic channel (dark) along the composition of 40 wt% PCBM in Fig. 2b. A dramatic reduction in the PCBM crystallinity was observed by adding more than 20 wt% PCPDTBT.

The phase diagram showed excellent agreement with the solar cell performance. On the basis of the hypereutectic composites (40–60 wt% total polymer) with a low concentration of the sensitizer (<20 wt% C-PCPDTBT), the most efficient devices were achieved. Further increasing the C-PCPDTBT content in the ternary blend reduced the PCBM semicrystallinity and therefore the device performance as well. Far from the eutectic region, where we have a high concentration of PCBM and over 20 wt% sensitizer (amorphous trapezoid area), no functional devices were achieved.

In summary, we established the phase diagram on the ternary blends for the first time and found a good correlation between the thermal behavior of the semiconductor blends and their photovoltaic properties. The details of this section are presented in [16].

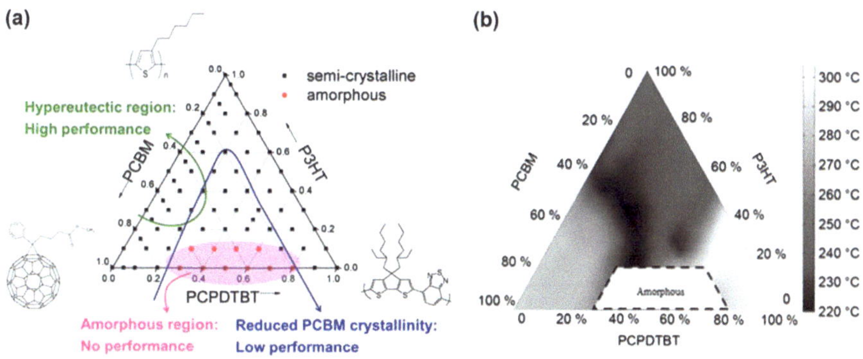

Fig. 2 (**a**) Overview of ternary blend mixtures from differential scanning calorimetry (DSC) first heating curves. *Full squares* stand for the crystalline mixtures, with melting peaks in DSC heating curves. *Full circles* represent amorphous mixtures. (**b**) Two-dimensional ternary phase diagram with the end melting temperature of the DSC first heating curves as *liquidus lines*

2.3 *Proving the Potential of the NIR Sensitization Concept*

To prove the potential of the NIR sensitization concept, we gradually introduced the small-bandgap polymer of poly[(4,40-bis(2-ethylhexyl)dithieno[3,2-*b*:20,30-d]silole)-2,6-diyl-*alt*-(4,7-bis(2-thienyl)-2,1,3-benzothiadiazole)-5,50-diyl] (Si-PCPDTBT) [17, 18] into the binary blend of P3HT:PCBM as an NIR sensitizer. All devices were fabricated on the basis of the structure presented in Fig. 3a. Adding from 10 up to 40 wt% of Si-PCPDTBT resulted in a continuous increase in the external quantum efficiency (EQE) in the NIR region (Fig. 3b). A significant performance improvement of over 30 % was achieved compared to the P3HT:PCM

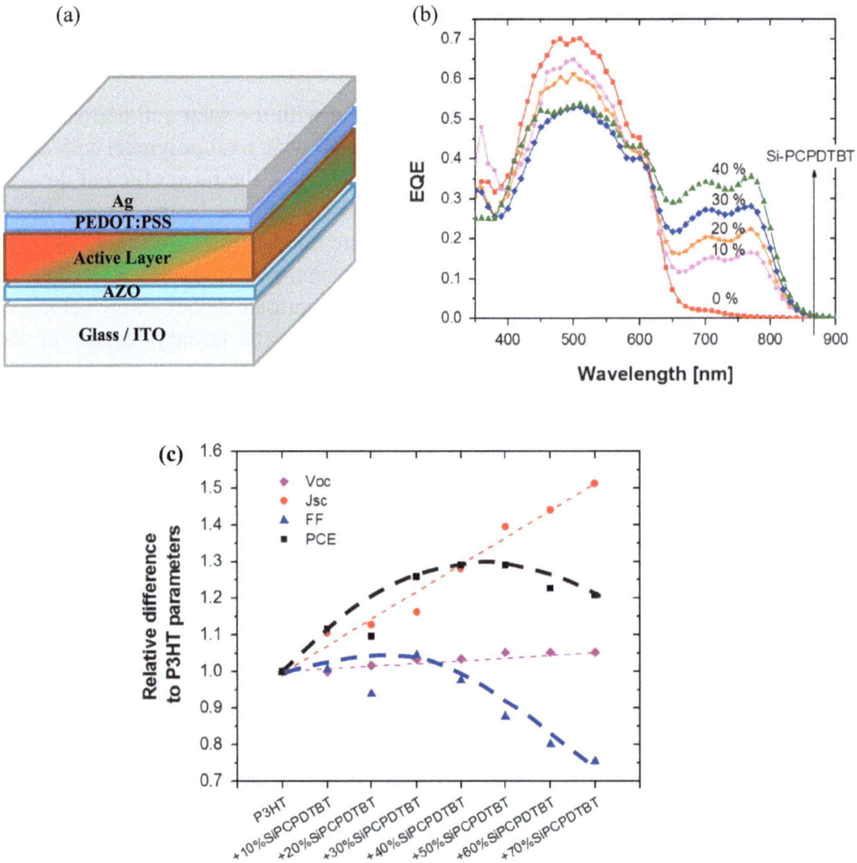

Fig. 3 (**a**) Schematic illustration of the inverted architecture of the binary/ternary solar cells. (**b**) External quantum efficiency spectra of P3HT:Si-PCPDTBT:PCBM ternary devices with different sensitizer concentrations. (**c**) Relative changes of the device parameters with increasing Si-PCPDTBT content. The performance of binary P3HT:PCBM is chosen as reference and set to 100 %. All other device data are normalized to that reference

binary device. Interestingly, even up to 40 wt% Si-PCPDTBT content, the average fill factor (FF) stayed around 60 % and the J_{sc} still increased (Fig. 3c). For further details, see [19].

2.4 *Investigation of Charge-Transfer Mechanisms*

Interestingly, two similar low-bandgap polymers of C-PCPDTBT and Si-PCPDTBT revealed a completely different functionality in the ternary devices [7, 19, 20]. The performance of ternary solar cells based on P3HT:PCBM was slightly improved in the presence of just 20 wt% C-PCPDTBT, mainly showing an enhanced J_{sc}. The addition of larger amounts of the sensitizer led, however, to a significant decrease in the FF relative to reference binary devices. In contrast, the presence of Si-PCPDTBT even up to 40 wt% enhanced the PCE of the device without compromising the FF. These two analogous copolymers possess very similar chemical structures with comparable highest occupied molecular orbitals (HOMO) and lowest unoccupied molecular orbitals (LUMO), optical bandgaps, and absorption spectra. However, we carried out a detailed exciton/charge-transfer study by means of an ultrafast time-resolved pump probe to elucidate the origin of the observed difference in these two ternary systems.

Differential absorption spectra revealed that in ternary P3HT:C-PCPDTBT: PCBM 0.9:0.1:1 films, upon photoexciting C-PCPDTBT charges are created on P3HT. Importantly, the charges are the result of a hole transfer from C-PCPDTBT to P3HT (Fig. 4a). On the one hand, hole transfer starts at times faster than 1 ps, which is consistent with the mechanism of direct hole transfer, shown in Fig. 4b. Hole transfer extended, nevertheless, throughout the investigated timescale of 7500 ps, containing a dominant component of 140 ps. This observation supports, on the other hand, the mechanism of hole transfer upon diffusion, with a calculated diffusion length of around 1.4 nm (Fig. 4c).

Very similar results were achieved for the Si-PCPDTBT–based ternary systems. Picosecond time-resolved pump-probe spectroscopy revealed that the sensitization of P3HT:PCBM with Si-PCPDTBT involves the transfer of photogenerated positive polarons from the low-bandgap polymer to P3HT within sub-picoseconds up to a few hundreds of picoseconds. Furthermore, intensity-dependent experiments in combination with global fitting showed that the charge transfer from Si-PCPDTBT to P3HT competes with the nongeminate charge carrier recombination of the holes in the Si-PCPDTBT phase with electrons in the PCBM phase, both diffusive processes. At excitation densities corresponding to steady-state conditions under one sun, modeling predicted hole-transfer efficiencies exceeding 90 %, in accordance with internal quantum efficiency measurements. At higher pump intensities, bimolecular recombination suppressed the hole-transfer process effectively Details are presented in [21, 22].

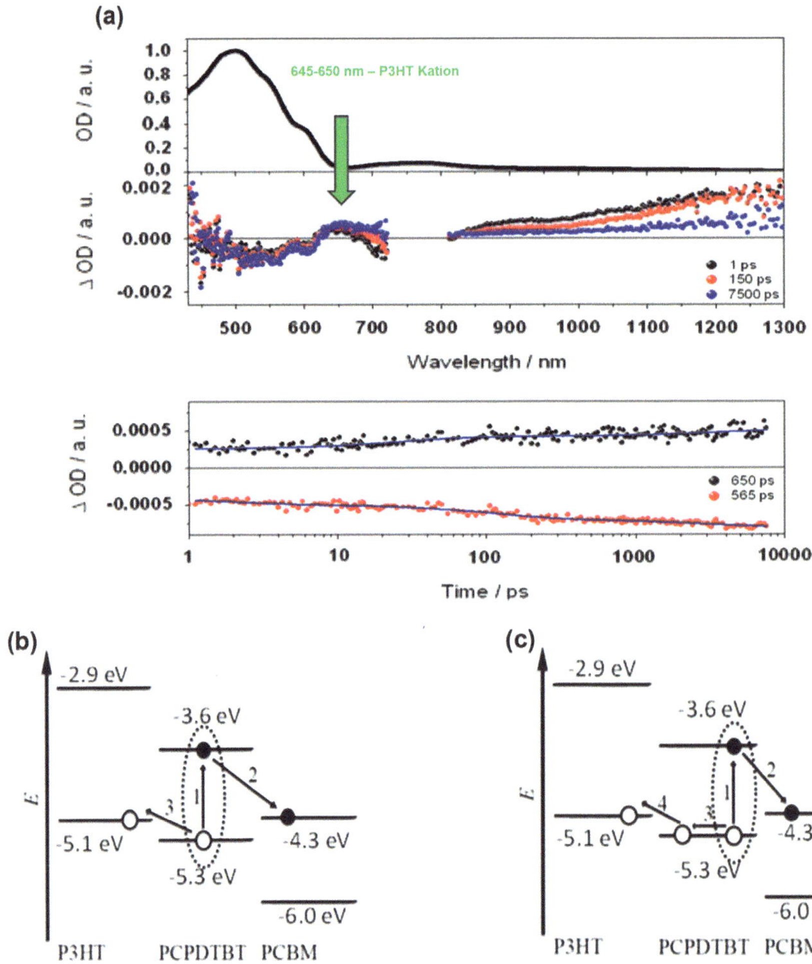

Fig. 4 (**a**) Absorption spectrum (*solid line*, *upper part*), differential absorption spectra upon excitation at 775 nm (100 nJ) recorded with different time delays, and time absorption profiles at 565 nm (*red*, *lower part*) and 650 nm (*black*, *lower part*) of a P3HT:PCPDTBT:PCBM 0.9:0.1:1 film. (**b**) Proposed mechanism of charge separation in ternary films without diffusion in C-PCPDTBT domains—including, firstly, generation of an exciton in C-PCPDTBT, secondly, electron transfer from PCPDTBT to PCBM, and, thirdly, hole transfer from PCPDTBT-positive polaron to P3HT. (**c**) Proposed mechanism of charge separation in ternary films with diffusion in C-PCPDTBT domains—including, firstly, generation of an exciton in C-PCPDTBT, secondly, electron transfer from PCPDTBT to PCBM, thirdly, hole diffusion to an interface between C-PCPDTBT and P3HT, and, fourthly, hole transfer from C-PCPDTBT-positive polaron to P3HT

2.5 Morphology Investigation

Since our detailed exciton/charge-transfer study revealed similar transfer mechanisms in both ternary systems, we hypothesized that the performance difference is likely to originate from the fact that the addition of C-PCPDTBT and Si-PCPDTBT sensitizer imposes different types of impact on the morphology of the host P3HT:PCBM blend. We performed a full-fledged characterization by ultraviolet–visible NIR spectroscopy, atomic force microscopy, surface energy measurements, DSC, grazing-incidence wide-angle X-ray scattering (GIWAXS), photoluminescence, and space charge limited current (SCLC) to get insights into the correlation between device performance and morphology. The most important characterizations are presented here; further details can be seen in [20, 23].

Surface energy has been shown to play a decisive role in the segregation of components. The surface energy of C-PCPDTBT ($\gamma_{\text{C-PCPDTBT}} = 40.5 \pm 1.5$ mN/m) is much higher than that of Si-PCPDTBT ($\gamma_{\text{Si-PCPDTBT}} = 26.4 \pm 2.1$ mN/m). On the basis of the interfacial surface energies between the pristine materials and the calculated wetting coefficient, the location of the analogous sensitizers in the host matrix of P3HT:PCBM was predicted. The wetting coefficient of C-PCPDTBT in P3HT:PCBM (-0.4 ± 0.2) suggested that C-PCPDTBT is most likely located at the P3HT:PCBM interfaces, with the tendency to interact with the semicrystalline (or aggregated) domains of fullerene. In contrast, the calculated wetting coefficient of 1.1 ± 0.2 for Si-PCPDTBT in the P3HT:PCBM blends suggests that Si-PCPDTBT is located in the amorphous P3HT domains at the P3HT:PCBM interfaces, with a high tendency to interact with the crystalline domains of P3HT (presented in the schematic of Fig. 5). To prove this prediction, we carried out further studies.

According to the DSC measurements, the normalized enthalpy of the binary P3HT:C-PCPDTBT system showed a linear relationship, with a slope of nearly 1, with increasing C-PCPDTBT content (Fig. 6a). This shows that C-PCPDTBT does not influence the P3HT crystallinity in the binary system and could be a sign that in the ternary system C-PCPDTBT affects mainly the PCBM crystalline domains rather than P3HT ones. C-PCPDTBT obviously does not intercalate into the crystalline P3HT domains but rather mixes with the amorphous phase. In contrast, a strong, superlinear decrease in the PCBM enthalpy was already observed at a 5 wt% C-PCPDTBT content for binary blends of PCBM:C-PCPDTBT, and there were no PCBM melting peaks for blends with >30 wt% C-PCPDTBT content. This decreasing PCBM crystallinity indicates that PCBM does indeed intercalate in C-PCPDTBT. These results were inconsistent with the GIWAXS results, where adding more than 20 wt% C-PCPDTBT disturbed the PCBM crystallinity peak much more strongly than the P3HT crystallinity peak in the ternary blends (Fig. 6b). On the other hand, by increasing the Si-PCPDTBT fraction beyond 20 wt%, we found that the relative reduction in the PCBM peak intensity is notably lower compared to that observed for the polymer interlayer peak (compare Fig. 6c-c and c-d). This suggests that the intensity reduction stems not only from PCBM

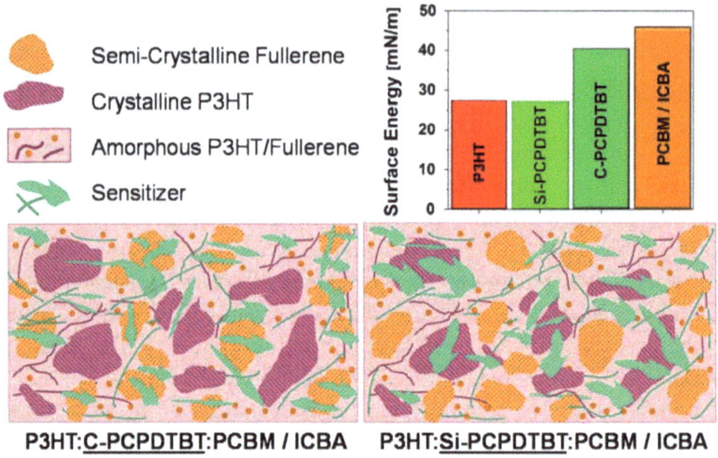

Fig. 5 Illustration of the ternary morphology based on the calculated wetting coefficient, which predicted that the C-PCPDTBT is most likely located at the P3HT:PCBM interfaces with the tendency to interact with the semicrystalline (or aggregated) domains of fullerene. In contrast, the Si-PCPDTBT is located in the amorphous P3HT domains at the P3HT:PCBM interfaces with a high tendency to interact with the crystalline domains of P3HT

incorporation but also from the incorporation of Si-PCPDTBT domains into the P3HT phase.

In agreement with the surface tension prediction and GIWAXS results, the SCLC results (Fig. 6d) confirmed that Si-PCPDTBT exhibits a tighter interaction with the amorphous P3HT:PCBM interfaces and semicrystalline domains of P3HT than the semicrystalline, aggregated domains of PCBM. In contrast, the C-PCPDTBT sensitizer has a detrimental effect on the semicrystalline domains of the fullerene derivatives, impairing the device performance drastically, even at concentrations over 20 wt%.

This study shows how slight structural modifications can significantly impact the behavior of a complex, multiphase system. As an initial set of design rules, we showed that the surface energy of the sensitizer is a driving force for its location in the host matrix and is an important parameter to control the morphology of ternary blends. By designing sensitizers bearing fine-tuned surface energy, using appropriate functional groups, it is possible to incorporate these materials into the donor or acceptor phase. For instance, in systems with low electron mobility [such as P3HT:indene-C60 bisadduct (ICBA)], it is beneficial to use a sensitizer that prefers to mix with the polymer, as this will have fewer negative effects on the electron mobility. For these systems the disruption of the fullerene semicrystalline domains is more detrimental to the device performance than is the disruption of the polymer domains.

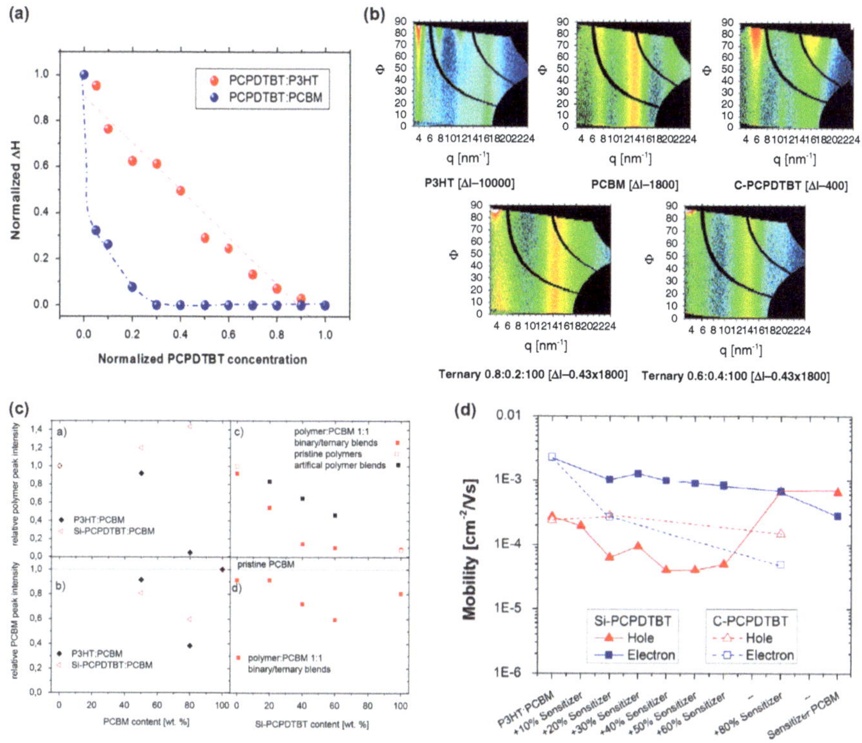

Fig. 6 (**a**) Normalized enthalpy change determined by integral P3HT and PCBM of melting peak area in binary blends of C-PCPDTBT:P3HT and C-PCPDTBT:PCBM. (**b**) Two-dimensional grazing-incidence wide-angle X-ray scattering (GIWAXS) detector of P3HT (*left*, *upper part*), PCPDTBT (*middle*, *upper part*), and PCBM (*right*, *upper part*) as well as P3HT:PCPDTBT:PCBM 0.8:0.2:1 (*left*, *lower part*) and 0.6:0.4:1 (*right*, *lower part*). (**c**) GIWAXS intensity of the main polymer interlayer peak (c-a and c-c) and main PCBM peak (c-b and c-d) normalized to the peak intensity of the corresponding pristine polymer and PCBM film, respectively, as a function of the Si-PCPDTBT (c-c and c-d) and PCBM contents (c-a and c-b). (**d**) Space charge limited current electron and hole mobility vs. sensitizer (Si-PCPDTBT and C-PCPDTBT) content added to the P3HT:PCBM host system

2.6 *Near-infrared Sensitization of Fullerene Multiadducts*

Our morphology findings, presented in the previous section, encouraged us to use the ternary components approach for the NIR sensitization of fullerene multiadducts by adding the low-bandgap copolymer Si-PCPDTBT into the P3HT:ICBA binary blend. We successfully overcame the limitations of ICBA:Si-PCPDTBT–based organic solar cells and demonstrated efficient organic solar cells (OSCs) with a high FF for the ternary composite P3HT:Si-PCPDTBT:ICBA. A significant efficiency improvement of over 30 % was achieved by adding 20 wt% of Si-PCPDTBT to a P3HT/fullerene multiadduct blend (Fig. 7a). We have successfully demonstrated

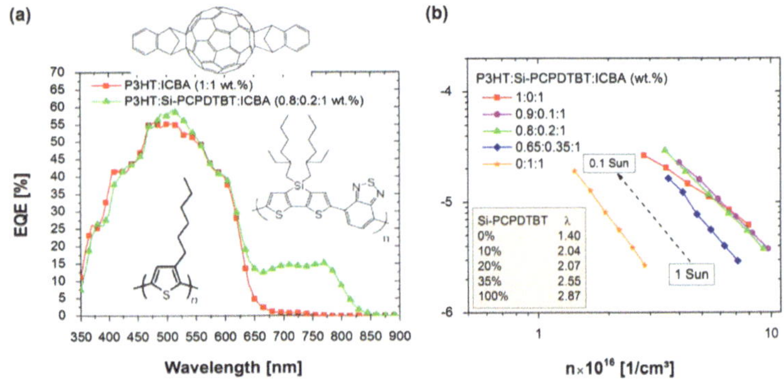

Fig. 7 (**a**) External quantum efficiency spectra of P3HT:ICBA and 0.8:0.2:1 wt% P3HT:Si-PCPDTBT:ICBA. (**b**) Charge carrier's lifetime vs. the charge carrier density of binary P3HT:ICBA and Si-PCPDTBT:ICBA and ternary systems with different content of Si-PCPDTBT content. The inset table presents the recombination exponent of every system

that the concept of ternary blend–based solar cells can be used to tackle many problems, such as narrow absorption window of the polymers and low open-circuit voltage, V_{OC}. Difficulties in blending nonpolar ICBA systems into low-bandgap polymers can be overcome. Moreover, the amount of laboriously synthesized and expensive polymer components such as Si-PCPDTBT is reduced when applying ternary blends that still rely on the favorable recombination dynamics and transport properties of the P3HT matrix (Fig. 7b).

2.7 *Small-Molecule–Based Sensitizers*

2.7.1 Benzannulated Aza-BODIPY Dyes

Ternary composite inverted OSCs were fabricated based on P3HT:PCBM blended with two different NIR-absorbing benzannulated aza-BODIPY dyes, difluoro-bora-bis-(1-phenyl-indoyl)-azamethine (10c) or difluoro-bora-bis-(1-(5-methylthiophen)-indoyl)-azamethine (10f). Although the addition of both aza-BODIPY dyes enhanced the absorption of the blends, only the addition of 10c resulted in enhanced NIR spectral photosensitivity (Fig. 8a).

According to photoinduced absorption results presented in Fig. 8c, d, the photoinduced absorption between 600–700 nm ascribed to the P3HT delocalized polarons was discernible in the transient absorption spectra of 10c-based ternary film as a result of hole transfer from 10c to P3HT. Additionally, a broad photoinduced absorption ranging from 780–1200 nm was recognizable. In contrast, differential absorption spectra of P3HT:10f:PCBM failed to reveal any relevant signals upon NIR sensitization. As a matter of fact, the low HOMO level of 10f is unsuitable to power a cascade hole transfer to P3HT (Fig. 8b). As a consequence, the

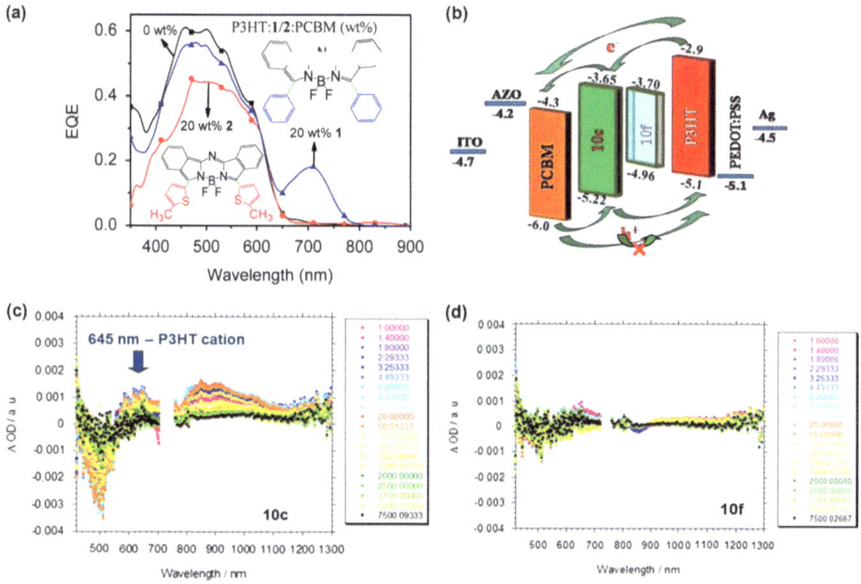

Fig. 8 (**a**) External quantum efficiency spectra of host P3HT:PCBM, P3HT:10c:PCBM, and P3HT:10f:PCBM with 20 wt% sensitizer. (**b**) Schematic energy band diagram of P3HT:10c/10f:PCBM–based solar cells. Transient absorption spectra of (**c**) P3HT:10c:PCBM 0.9:0.1:1 wt% film and (**d**) P3HT:10f:PCBM 0.9:0.1:1 wt% film upon laser excitation at 720 nm at room temperature, recorded with different time delays

10f radical cation, generated by charge transfer to PCBM, is more likely to undergo geminate or nongeminate recombination since hole transfer to P3HT is energetically unfavorable. Therefore, if the sensitizer concentration is not high enough to form a parallel channel for the charge carrier transport, in order to avoid such undesired charge trapping, it is important to have a cascade alignment between the energy levels of the sensitizer and host donor and acceptor compounds. For further details, see [24].

2.7.2 Dithienylthienothiadiazole- and Dithienylthienoselenadiazole-Based Small Molecule

Some novel small molecules based on dithienylthienothiadiazole- and Dithienylthienoselenadiazole units were synthesized following well-known literature procedures (see Supplementary Information). The chemical structures of these compounds are presented in Fig. 9.

The ternary solar cells based on the P3HT:PCBM host system and aforementioned small molecules as sensitizer were fabricated in an inverted structure. Table S1 (in Supplementary Information) presents the performance of the devices, incorporating 10, 20, and 30 wt% sensitizers. When 10 wt% AE906 and AE910

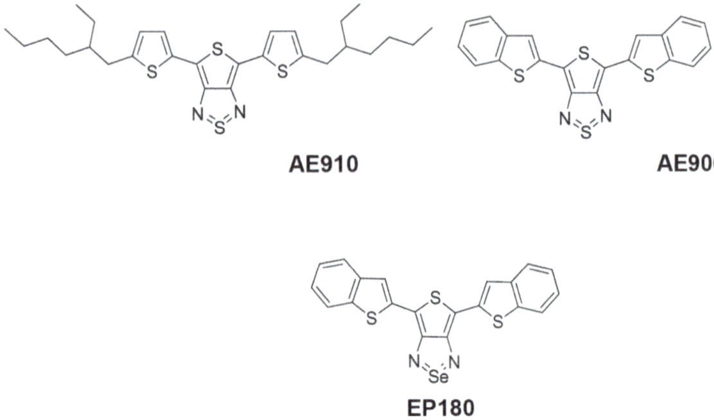

Fig. 9 Chemical structures of the synthesized dithienylthienothiadiazole- and Dithienylthienoselenadiazole–based small molecules AE906/AE910/EP180

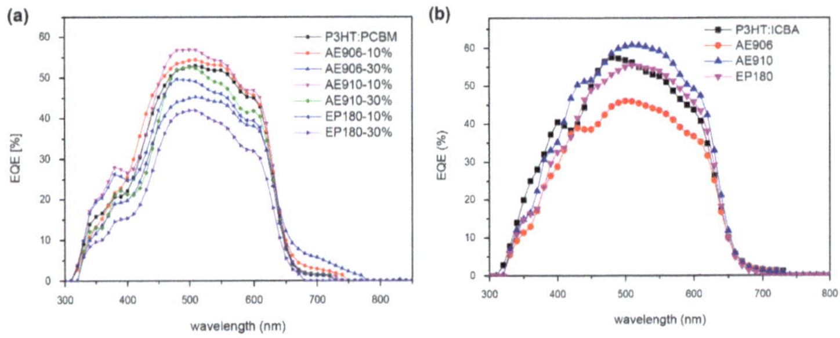

Fig. 10 External quantum efficiency spectra of (**a**) P3HT:PCBM reference and ternary systems with 10 and 30 wt% dithienylthienothiadiazole- and Dithienylthienoselenadiazole–based sensitizers. (**b**) P3HT:ICBA reference and ternary systems with 20 wt% dithienylthienothiadiazole- and Dithienylthienoselenadiazole–based sensitizers

were added, the J_{sc} was slightly improved compared to the host P3HT:PCBM. As Fig. 10a shows, this improvement is mainly based on the improved photocurrent collection over the P3HT absorption region, 400–650 nm, rather than the NIR region. On the other hand, the V_{oc} and FF were improved up to 5 % and 15 %, respectively, for different sensitizer contents.

Very similar results were achieved for the P3HT:ICBA system, where AE910 was used as the sensitizer. In this case, the EQE again showed an increase all over the spectrum, from 400 up to 650 nm (Fig. 10b) and 15 % FF and 5 % V_{oc} improvement were achieved compared to the reference P3HT:ICBA.

These results suggest that the ternary concepts can also have other functionalities than infrared sensitization and even act as a morphology agent. Further morphological and transport studies are required to shed light on the influence of these ternary compounds on the host matrix.

3 Conclusion

In summary, the potential of NIR sensitization is proven to enhance the EQE by over 30 %. Furthermore, ternary sensitization enables full performance of bis-fullerenes (with P3HT as matrix). Based on our initial design rules for the ternary systems, the surface energy of the NIR sensitizer is an important driver for the sensitizer location in the host matrix. Disruption of the fullerene semicrystalline domains is more detrimental than disruption of the polymer domains. This implies our need for novel fullerenes with improved properties. According to the results achieved from ultrafast pump-probe spectroscopy, for the ternary systems with a cascade energy-level alignment charge transfer happens in two stages of ps and ns, consistent with the mechanism of direct hole transfer and hole transfer upon diffusion.

However, high-efficiency host references are required to boost the performance of ternary solar cells beyond 10 %, and the morphology and charge transfer/transport still remain the big challenges for ternary systems, requiring further systematic studies to shed light on them.

Acknowledgments This work has been funded by the DFG project (SPP1355), grant numbers. BR 4031/2 − 1 and BR 4031/2-2.

T.A and C.J.B. gratefully acknowledge the support of the Cluster of Excellence "Engineering of Advanced Materials" at the University of Erlangen–Nuremberg, which is funded by the German Research Foundation (DFG) within the framework of its "Excellence Initiative" and the project "Solar Technologies Go Hybrid" (SolTech). U.S. and M.F. gratefully acknowledge the support of the DFG project, grant number SCHE 410/23-2.

References

1. Green MA, Emery K, Hishikawa Y, Warta W, Dunlop ED (2012) Prog Photovolt Res Appl 20:606
2. He Z, Xiao B, Liu F, Wu H, Yang Y, Xiao S, Wang C, Russell TP, Cao Y (2015) Nat Photonics 9:174–179
3. Liu Y, Zhao J, Li Z, Mu C, Ma W, Hu H, Jiang K, Lin H, Ade H, Yan H (2014) Nat Commun 5:5293
4. Kan B, Li MM, Zhang Q, Liu F, Wan XJ, Wang YC, Ni W, Long GK, Yang X, Feng HR, Zuo Y, Zhang MT, Huang F, Cao Y, Russell TP, Chen YS (2015) J Am Chem Soc. doi:10.1021/jacs.5b00305
5. Mohd Yusoff ARB, Kim D, Kim HP, Shneider FK, da Silva WJ, Jang J (2015) Energy Environ Sci 8:303
6. Ameri T, Khoram P, Min J, Brabec CJ (2013) Adv Mater 25:4245
7. Koppe M, Egelhaaf HJ, Dennler G, Scharber MC, Brabec CJ, Schilinsky P, Hoth CN (2010) Adv Funct Mater 20:338
8. Lu L, Xu T, Chen W, Landry ES, Yu L (2014) Nat Photonics 8:716
9. Cheng P, Li Y, Zhan X (2014) Energy Environ Sci 7:2005
10. Yang YM, Chen W, Dou L, Chang W, Duan H, Bob B, Li G, Yang Y (2015) Nat Photonics 9:190
11. Liu S, You P, Li J, Li J, Lee C-S, Ong BS, Suryae C, Yan F (2015) Energy Environ Sci 8:1463

12. Hansen CM (2007) Hansen solubility parameters—a user's handbook, 2nd edn. CRC Press, Boca Raton, FL, Chapter 1
13. Hansen CM, Smith AL (2004) Carbon 42:1591
14. Walker B, Tamayo A, Duong DT, Dang X-D, Kim C, Granstrom J, Nguyen T-Q (2011) Adv Energy Mater 1(2):221
15. Machui F, Abbott S, Waller D, Koppe M, Brabec CJ (2011) Macromol Chem Phys 212:2159
16. Li N, Machui F, Waller D, Koppe M, Brabec CJ (2011) Sol Energy Mater Sol Cells 95:3465
17. Scharber MC, Koppe M, Gao J, Cordella F, Loi MA, Denk P, Morana M, Egelhaaf HJ, Forberich K, Dennler G, Gaudiana R, Waller D, Zhu Z, Shi X, Brabec CJ (2010) Adv Mater 22:367
18. Morana M, Azimi H, Dennler G, Egelhaaf HJ, Scharber M, Forberich K, Hauch J, Gaudiana R, Waller D, Zhu Z, Hingerl K, van Bavel SS, Loos J, Brabec CJ (2010) Adv Funct Mater 20:1180
19. Ameri T, Min J, Li N, Machui F, Baran D, Forster M, Schottler KJ, Dolfen D, Scherf U, Brabec CJ (2012) Adv Energy Mater 2:1198
20. Machui F, Rathgeber S, Li N, Ameri T, Brabec CJ (2012) J Mater Chem 22:15570
21. Löslein H, Ameri T, Matt GJ, Koppe M, Egelhaaf HJ, Troeger A, Sgobba V, Guldi DM, Brabec CJ (2013) Macromol Rapid Commun 34:1090
22. Koppe M, Egelhaaf HJ, Clodic E, Morana M, Lüer L, Troeger A, Sgobba V, Guldi DM, Ameri T, Brabec CJ (2013) Adv Energy Mater 3:949
23. Ameri T, Khoram P, Heumüller T, Baran D, Machui F, Troeger A, Sgobba V, Guldi DM, Halik M, Rathgeber S, Scherf U, Brabec CJ (2014) J Mater Chem A 2:19461
24. Min J, Ameri T, Gresser R, Lorenz-Rothe M, Baran D, Troeger A, Sgobba V, Leo K, Riede M, Guldi DM, Brabec CJ (2013) ACS Appl Mater Interfaces 5:5609

Critical Dimensions in Small-Molecule Plasmonic Particle Solar Cells

Till Jägeler-Hoheisel, Johannes Benduhn, Christian Körner, and Karl Leo

Contents

Abstract In this review, we summarize design principles of organic solar cells with plasmonic nanostructures. A process for scalable vacuum processing of silver nanoparticles is presented. Approximations for losses inside plasmonic structures are derived with respect to particle size and absorber material. We evaluate the characteristic length scales of plasmonic near-field enhancement and backscattering. The thickness of the absorber layers can be significantly reduced in plasmonic devices showing increased power conversion efficiency. The strongest plasmonic effects are observed in coupling particle structures.

Keywords Absorber layer thickness • Absorption effciency • Backscattering • Charge extraction • Coupling plasmonic particles • Coupling of plasmonic particles to absorbing media • Device stack • Dipolar plasmonic resonance in dispersive media • Dipolar resonance • Extinction efficiency • Fill factor • For-

---

T. Jägeler-Hoheisel (✉) • J. Benduhn • C. Körner • K. Leo
Institut für Angewandte Photophysik, Technische Universität Dresden, George-Bähr-Straße 1, 01069 Dresden, Germany
e-mail: Till.Hoheisel@iapp.de; leo@iapp.de

K. Leo (ed.), *Elementary Processes in Organic Photovoltaics*,
Advances in Polymer Science 272, DOI 10.1007/978-3-319-28338-8_14

ward scattering • Infra-red • Inter-particle spacing • Modified long-wavelength approach • Morphology in plasmonic organic solar cells • Nanotechnology • Near-field effect • Near-field enhancement • Particle aspect ratio • Particle density • Particle growth • Particle growth after seeding • Particle position • Particle size • Passivation of plasmonic particles • Phase of plasmonic resonance • Photocurrent • Plasmonic device concepts • Plasmonic solar cell • Polarizability • Reflection • Reflection at plasmonic particles • Scattering efficiency • Semiconductor bandgap • Silver particle deposition • Spectral overlap • Splitting of plasmonic resonance • Vollmer-weber growth

1 Motivation

Organic solar cells (OSCs) have made tremendous progress in recent years, achieving more than 10 % power conversion efficiency (PCE) for various device concepts and also making huge progress in upscaling, including roll-to-roll processing and commercial applications on plastic foils [1].

In OSCs the absorption of photons does not directly result in charge carrier generation. Actual state-of-the-art OSCs use mixed layers of donor and acceptor materials to allow for short exciton diffusion paths and efficient creation of charge carriers. Although the exact morphology of such bulk heterojunction (BHJ) cells depends on materials involved and also on the preparation conditions, the intermixing of the donor and acceptor materials in almost all cases leads to increased recombination between opposite charge carriers. This process typically increases with increasing BHJ thickness. This behavior results in a trade-off in the thickness of the absorber layer. Thicker layers allow one to harvest a larger fraction of the incident photons and increase the short-circuit photocurrent J_{sc}. On the other hand, reduced fill factors (FFs) in thick BHJ layers prescribe that only a fraction of the photocurrent can be extracted at a maximum power point.

The absorber layer thicknesses of efficient organic devices are in the range of 120 nm in typical solution-processed OSCs [2, 3], while in vacuum-processed OSCs the absorber layer thickness is even less, with typical values of 40 nm [4, 5]. Such thin devices can significantly profit from engineering thin-film optics because the device thickness is lower than the coherence length of the solar irradiation, which is in the order (about) of roughly 600 nm [6]. OSCs can benefit from utilizing the interference pattern of incident light and light reflected at the back contact or even multiple passes in weak cavities. Doped organic transport layers or other transparent conductive materials are not only a basis for optimizing the electrical properties of such devices, but also serve to adjust the thin-film optics to achieve maximum photocurrent from the active layers.

The absorber layer trade-off also motivates the search for additional subwavelength light manipulation techniques to enhance absorption inside the intrinsic absorber layers. Subwavelength-sized plasmonic nanostructures can confine light on a subwavelength scale in near fields around their surface. Plasmonic structures can show scattering cross sections multiple times larger than their geometrical areas. Both properties, combined with their small size, make them promising for integration in thin-film devices that have dimensions of typically only a few wavelengths. Thin-film OSCs can also be viewed as detectors that allow us to coherently probe the local field within the solar cell stack and reveal optical phenomena connected to plasmonic particles.

Significant interest in the subject will also arise from technological steps in upscaling of organic photovoltaic technologies. From the viewpoint of device manufacturing, metal structures such as metal nanowire meshes [7, 8] or thin metal contacts [9–11] are very likely to be used as contacts for flexible thin-film devices, allowing roll-to-roll processes. Thus, a detailed understanding of plasmonic effects in those devices is needed both to use the full potential of nanostructured electrodes such as metal grids and nanowire meshes and to avoid loss channels such as unwanted reflection, fluorescence quenching to plasmonic contacts, or parasitic absorption. Furthermore, plasmonic absorption enhancement has the potential to significantly reduce the amount of absorber material needed [12].

2 Plasmonic Device Design and Nanotechnology

Plasmonic solar cells can be categorized according to their working principle [13] or to their device structure [14].

Plasmonic particles or gratings could be used to scatter light into waveguide modes within the device. This technique has been extensively studied in high-refractive-index inorganic solar cells [12]. Backscattering, which depends on the phase of the induced field within the particles, can also serve to engineer the local light intensity for different wavelengths independently, as is discussed in Sect. 6. Another strategy involves confining light on a subwavelength scale within the near field of the particle; the light can then be used to enhance absorption when the active layer is placed within the extension of this near field. This concept is discussed in detail in Sect. 5. A third plasmonic approach is to couple light into propagating modes confined to a metal surface parallel to the absorber layer. This approach requires complicated coupling schemes such as grating [15] or prism coupling [16, 17] for momentum matching, which in principle can be achieved with scattering particles—however, at the price of combining two lossy structures.

Another classification often used is the position of plasmonic particles within the device stack, which is also more related to the manufacturing of the devices. Although absorption enhancement can be observed in various structures and is accessible in optical simulations, a main technological task is the implementation of suitable plasmonic nanostructures in OSC devices without reducing their electrical performance. Assuming a transparent substrate, particles deposited on the substrate before depositing the absorber layer are in a particle *in front of* or *below* the absorber layer configuration. Accordingly, particles *inside* the absorber layer and particles *at the back* of the absorber layer are distinguished [14].

When the particles or plasmonic structures are predeposited before the sensitive organic layers, there is a free choice of processing conditions. This may include very harsh conditions such as high temperatures, enabling Ostwald ripening of metal films [13, 18], patterning of the substrate via sputtering [19], the use of capillary forces in drying solvent films [20–22], laser interference patterning [23], among others. A drawback of particles in front of the cell is that the stack subsequently deposited onto the particles may suffer from rough morphologies or even shunts that can deteriorate device performance or limit the feasible particle size [24–26]. The thin layers of the OSC stack are prone to roughness of the substrate when deposited because devices require the formation of closed functional layers that can have thickness down to only a few nanometers. This might lead to shunts or require smoothing [8, 27] of the surface, which typically leads to a large distance between the plasmonic structures and the absorber layers, impeding near-field enhancement. Still the problems following from the roughness of the structures may limit the maximum particle size or more fundamentally lead to losses because light can be scattered away from the solar cell or absorbed in the particles before passing the active layers [24]. In classical high-index inorganic solar cell materials, the application of particles in front of the solar cell has been demonstrated because the strong contrast of the refractive index at the air–semiconductor interface leads to mainly forward scattering of plasmonic particles [12, 13]. Forward scattering has been demonstrated in OSCs using the refractive index contrast between PEDOT:PSS and organic absorbers [26, 28] or conformal coating of plasmonic structures with MoO_3 below solution-processed absorber layers [29].

A tempting approach is to mix particles with the absorber layers in solution-processed OSCs. However, charge carrier recombination at the metal particles within the intrinsic semiconductor can reduce the device performance [30], and special care has to be taken with respect to the surface passivation [31] of the particles. This approach is often used within the absorber layers of solution-processed OSCs, which allow one to spin-coat particles together with functional layers of the solar cell device. While the shape of the particles can be chosen to have even further increased field-enhancement [32], the size of particles within the absorber layer is limited [25, 26] and reflection and absorption losses may occur. Furthermore, a comparison of organic BHJ layers spin coated with and without particles is not straightforward, even when the layer thickness or the amount of absorber is kept constant. There are some reports to the contrary for low concentrations of particles in a special material combination [33], but generally the addition of particles and surface-modifying chemicals may lead to unexpected effects in the BHJ morphology [34], which is reflected, for example, in reports of improved electrical characteristics [14, 31]. However, this approach seems to be less suitable in our vacuum-processed OSCs, and we would like to refer to [14], who also highlight the conceptual difficulties in the interpretation of these experiments.

The approach of placing particles or plasmonic structures [25, 35] behind but close to the organic absorber layer is the most challenging with respect to suitable processing techniques that should not affect the organic stack below. However, this can avoid roughness problems in the device stack because most of the layers are produced on a flat substrate. Light reaching the particles already had a certain probability of getting absorbed while passing the absorber layer, and light reflected at the plasmonic structures at least has one more pass through the absorber layer before leaving the device. In our interpretation of the experiments, we may assume that the morphology of the absorber layer is not affected by the plasmonic structure and that the layer thickness below the particles is well defined. We focused on this approach because we believe it is most compatible with small-molecule OSCs in the following sections, starting with techniques for deposition of the particles.

3 Particle Deposition in Small-Molecule OSCs

Small-molecule OSCs are very sensitive to air and moisture. Our devices are typically processed under ultrahigh vacuum (UHV) at pressures below 10^{-6} mbar from temperature gradient vacuum sublimation–purified materials and encapsulated with glass lids in a nitrogen atmosphere. These requirements of high purity and inert conditions make the solution processing of particles or nanowires onto existing organic stacks a complicated task [7].

We thus focused on techniques for processing silver nanoparticles in small-molecule cells in UHV. Thermal evaporation in UHV is a widely used technique in the research and production of organic devices, and the evaporation of metallic back contacts is used even in most solution-processed devices. The typical parameters that can be controlled in vacuum deposition are the rate at which materials are evaporated , the temperature of the substrate , and the material of the substrate itself. These parameters influence the probability of the formation of seed clusters [36]. Using an analogy with water droplets, we may speak of *wetting* and *nonwetting* substrates to describe the interfacial energy and contact angles involved. A reduced deposition rate and elevated substrate temperatures should lead to an overall reduction in the density of seed clusters on a nonwetting substrate. A solar cell stack underneath, however, limits the temperature of the substrate during silver deposition. Figure 1a shows that silver deposition on the typical OSC material C_{60} even at very low rates of 0.003 nm s^{-1} and elevated substrate temperatures of 80 °C lead to the formation of rather small silver particles that have insufficient sizes for efficient light manipulation, as discussed in Sect. 4.

To improve particle formation, we investigated the wetting properties of various transparent conducting oxides [10]. Thin interlayers of inorganic transparent conductive oxides were then evaporated onto the predeposited OSC stack to engineer the wetting properties [37]. Particles obtained from 3 nm of silver deposited with the same rate and substrate temperature (0.003 nm/s, 80 °C) are shown in Fig. 1b, c for dewetting layers formed from 1.5 nm of WO_3 and 2 nm of V_2O_5, respectively.

Once a suitable dewetting layer is found, which in an ideal case should align to the transport level of the adjacent layers, the particle density n, that is, particles per area, can be controlled by adjusting the substrate temperature T. Figure 2a presents the particle densities found in devices comprising a dewetting layer of V_2O_5 for the

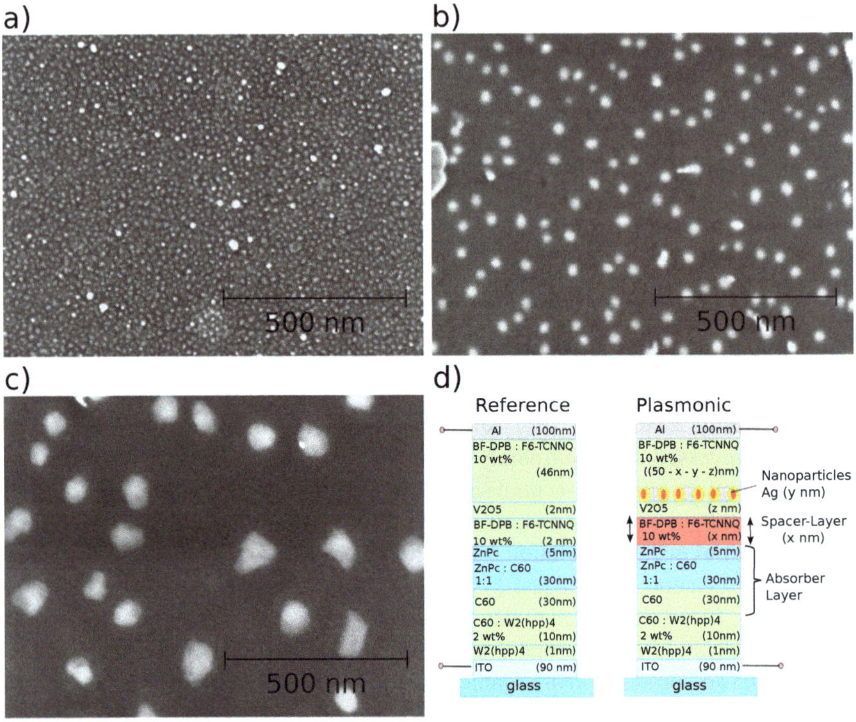

Fig. 1 Scanning electron microscopy scans of silver growth on different substrates at a fixed deposition rate (0.003 nm/s) and substrate temperature (80 °C) and sketched device stack. (**a**) 4 nm of silver on C_{60}. (**b**) 3 nm of silver on 1.5 nm of WO_3. (**c**) 3 nm of silver on 2 nm of V_2O_5. (**d**) Typical device stack including silver nanoparticles in the hole-transport layer at the back of the absorber layer

fixed deposition rate (0.003 nm/s) plotted against the inverse absolute temperature. The fit (solid curve) can be described by $n = 10^{-7.5} \times 10^{(\frac{3090\,\text{K}}{T})}$. Within the moderate temperature range of 30°–95 °C, the surface density varies within two orders of magnitude from up to 650 μm^{-2} to less than 10 μm^{-2}.

Once a sufficient number of seed clusters at a given temperature and deposition rate have formed on the surface, new silver atoms are efficiently captured by the existing particles during diffusion [36], reducing the probability of the formation of new seed clusters. As diffusion is temperature activated, higher temperatures increase the diffusion length of silver atoms and reduce the number of clusters needed to efficiently capture impinging atoms, thus reducing the particle density. Interparticle spacing of more than 200 nm reveals that silver atoms can diffuse large distances on the surface of V_2O_5.

When atom trapping at seed clusters is the dominating effect, the additional deposition of silver only results in growth of the existing clusters. Figure 2b shows particle in-plane diameters for various mass thicknesses of silver obtained at

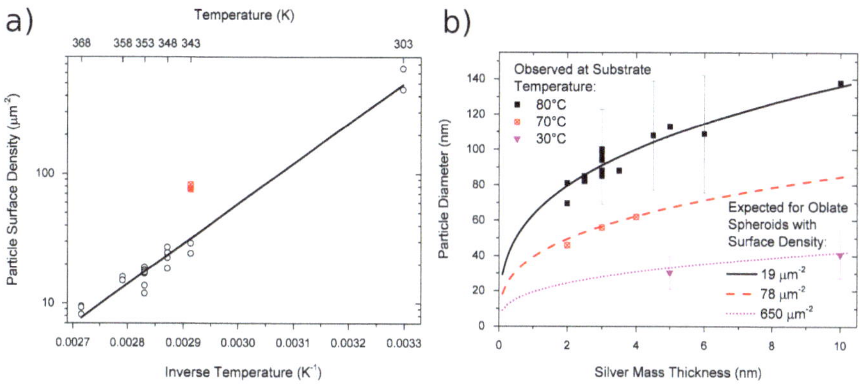

Fig. 2 Effect of substrate temperature on particle surface density and the relationship between the amount of silver deposited and the particle long-axis diameter. The deposition rate was fixed (0.003 nm/s) and a thin (2- and 4- nm) V_2O_5 interlayer was used on the organic thin-film stack to promote particle growth. (**a**) Particle surface density plotted on a logarithmic scale against inverse absolute temperature. (**b**) Particle diameters obtained from different amounts of silver at a given substrate temperature. The lines show the geometrically expected long-axis diameter for prolate spheroids with a short-to-long-axis ratio of 0.4 at surface densities determined experimentally

different substrate temperatures. Scanning electron microscopy (SEM) and atomic force microscopy (AFM) investigations of the shape of particles produced on V_2O_5 reveal that the shape of the particles in terms of the ratio of particle height to diameter is around 0.4 independent of the particle diameter and seems to be a property of the material system used and the contact angle involved. The lines in the same graph show the long-axis diameters geometrically expected for prolate ellipsoids with the same aspect ratio and the volume of each particle determined by the total amount of silver and the given surface densities. However, for larger particles (see Fig. 7c,d) the relative contribution of the surface energy is reduced, allowing for more irregular shapes, and even thin metal sheets extending from the particles suggest growth along terraces of crystallites similar to observations made by Metois et al. [38] in lead particles.

The whole process allows particles to be placed behind the absorber layer and also introduces means of adjusting the particle position inside the stack by depositing spacer layers between the absorber layer and the particle dewetting layer, which provides insight into plasmonic processes by determining specific length scales. A typical vacuum-deposited OSC stack is sketched in Fig. 1d. We use a glass substrate covered with transparent indium tin oxide (ITO) contacts. The overlap between the structured ITO electrode and a perpendicular aluminum (100- nm) back contact defines the device area of 6.44 mm^2. In this study we use fullerene (C_{60}) n-doped with 10 wt % of tetrakis-[1,3,4,6,7,8-hexahydro-2H-pyrimido[1,2-a]pyrimidinato] ditungsten(II) ($W_2(hpp)_4$) as electron-transport layers at the ITO electrode side and an intrinsic C_{60} hole-blocking layer, which also contributes to the photocurrent. BHJ absorber layers were produced by coevaporation of zinc phthalocyanine (ZnPc) and fullerene (C_{60}) in a one-to-one ratio. As electron-blocking layer, intrinsic ZnPc is used. At the metal back contact hole-transport layers (HTLs) made from N,N-((diphenyl-N,N-bis)9,9,-dimethyl-fluoren-2-yl)-benzidine) (BF-DPB) p-doped with 2,2'-(perfluoronaphthalen-2,6-diylidene)dimalononitrile (F6-TCNNQ) (10 wt %) are used.

Devices presented in Sects. 4–6 were fabricated with the following stack: ITO (90 nm), $W_2(hpp)_4$ (1 nm), n-C_{60} (10 nm), C_{60} (30 nm), ZnPc:C_{60} (30 nm), ZnPc (5 nm), p-BF-DPB (x nm), V_2O_5 (z nm), Ag (y nm), p-BF-DPB ($50 - x - y - z$ nm), F6-TCNNQ (1 nm), Al (100 nm). The reference sample contains no silver, and the best reference device was found for $x = 2$ nm of the spacer layer [39]. During particle deposition the substrate was heated to 80 °C. A silver thickness of $y = 3$ nm resulted in particle diameters and densities of 87 nm and 19 μm^{-2} on $z = 2$ nm of a V_2O_5 dewetting layer. A dewetting layer of $z = 4$ nm and $y = 3$, 5, and 10 nm of Ag resulted in diameters as given in Fig. 4b.

Devices varying the thickness h of the BHJ absorber layer in Sect. 7 were fabricated as follows: ITO (90 nm), $W_2(hpp)_4$ (1 nm), C_{60} (5 nm), ZnPc:C_{60} (h nm), ZnPc (3 nm), p-BF-DPB (2 nm), V_2O_5 (2 nm), Ag (2.5 nm), p-BF-DPB (55 nm), F6-TCNNQ (1 nm), Al (100 nm), resulting in particle diameters of (83 ± 13) nm and densities of 18 μm^{-2}.

Devices for the investigation of different particle densities in Sect. 8 were fabricated using the following stack sequence: ITO (90 nm), $W_2(hpp)_4$ (1 nm), C_{60} (10 nm), ZnPc:C_{60} (30 nm), ZnPc (3 nm), p-BF-DPB (2 nm), V_2O_5 (2 nm), Ag (j nm), p-BF-DPB ($48-j$ nm), F6-TCNNQ (1 nm), Al (100 nm). High-particle-density samples were produced at 75 °C using $j = 3$ nm, 6 nm, and 8 nm of Ag for particle formation, which resulted in the diameters and densities given in Fig. 7. For the low-density sample described in Fig. 7, 95 °C and 1.5 nm Ag were used.

Absorptance spectra were recorded with a fiber-coupled reflection setup and a 100-nm Al layer at the back of the glass substrate as reference. External quantum efficiency (EQE) spectra were recorded using chopped illumination and a lock-in amplifier.

For further details of device characterization, refer to [10, 39, 40].

4 Particle Size as a Critical Parameter

The goal of producing large, well-separated particles is driven by the attempt to reduce losses within the particles. The effect of the particle diameter on losses and tuning of the plasmonic resonance with respect to the absorption coefficient of the active absorber materials is discussed in this section, while the interparticle distance is revisited in Sect. 8.

Plasmonic particle resonances or localized surface plasmons are often described by the induced dipole of the lowest-order dipolar mode thought of as being located in the center of the particle. The complex amplitude of this induced dipole is the product of the exciting electrical field multiplied by the dielectric constant of vacuum and the medium and the polarizability α of the particle. A practical approximation [41] of this polarizability, including the effect of particle size as the particle radius a and size parameter $x = ka$, is recalled in Eq. (1). This equation describes spherical particles, but similar modified long-wavelength approaches have been successfully evaluated numerically and experimentally with spheroidal particles [41–43]. These approaches could be used to even further increase the agreement with the experimental data obtained with prolate ellipsoids. However, this would introduce additional parameters and would not change the general trends.

$$\alpha = a^3 \frac{\epsilon' - 1}{\epsilon' + 2 - (6\epsilon' - 12)\frac{x^2}{10} - i\frac{2x^3}{3}(\epsilon' - 1)}. \tag{1}$$

The complex polarizability includes the wavelength-dependent dielectric constants of both the particle ϵ and the environment ϵ_m, which enter in the form of a reduced dielectric constant $\epsilon' = \frac{\epsilon}{\epsilon_m}$. For several metals, the real part of the dielectric constant is negative for frequencies lower than the plasma frequency, and so the real part of the denominator in α can vanish, producing strong induced dipoles with their magnitude only limited by the dissipative imaginary component of the dielectric functions.

The application of this analytical approach in the context of absorbing media, which is the relevant case for most solar cell applications, is, however, new. We evaluate the absorption efficiency within the particle as the difference $Q_{\text{abs}} = Q_{\text{ext}} - Q_{\text{sca}}$ of extinction and scattering efficiencies. The efficiency factors are the respective cross sections normalized to the geometrical area of the particles according to Eqs. (2) and (3), which have suitable powers in the wavenumber $k = \frac{2\pi\sqrt{\epsilon_m}}{\lambda_0}$ to obtain real-valued numbers by taking the real part and the absolute value, respectively:

$$Q_{\text{ext}} = -\frac{4}{a^2}\Re\{\mathrm{i}\, k\, \alpha\}, \tag{2}$$

$$Q_{\text{sca}} = \frac{8}{3a^2}|k^2\alpha|^2. \tag{3}$$

Unlike numerical approaches, which often focus on special geometries, we chose to use analytical approximations of the energy flow across the surface of the particle in absorbing media without concerning the actual boundaries of the medium. However, light scattered by the particle can be absorbed within the medium, as scattering only describes the energy flow calculated at the surface of the particle. Although very simplistic, this approach is fast and can be used to quantify the effects of particle size and dispersion of the medium as well as the relative phase of the particle polarizability with respect to the incident field, which can be very important for the device design and is usually not directly accessible in numerical approaches.

We use the optical constants of silver from [44] and the optical constants of a one-to-one mixture of ZnPc and C_{60} similar to in-plane data in [45] to plot a typical

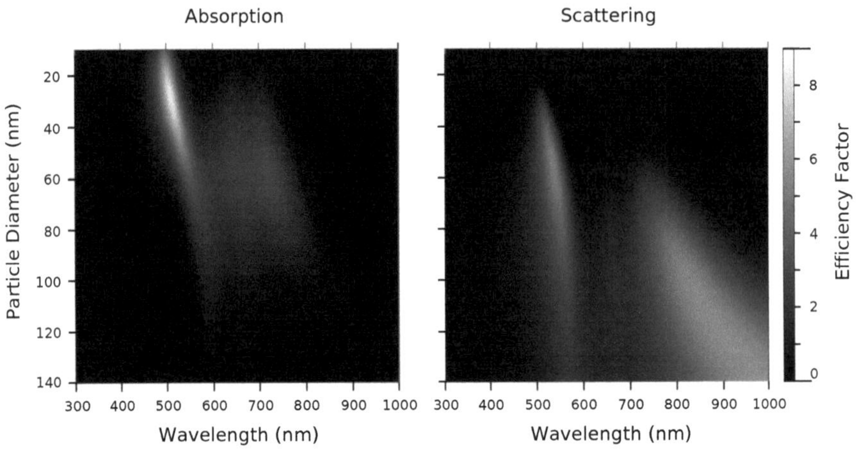

Fig. 3 Absorption and scattering efficiency factors (*color coded*) of the dipolar mode of plasmonic silver particles embedded in an absorbing medium corresponding to a 1:1 mixture of ZnPc:C_{60}. The efficiency factors are absorption and scattering cross sections normalized to the geometrical cross section of the particles. Wavelength is varied along the x-axis and particle diameter is varied along the y-axis

example of absorption and scattering efficiency for different particle diameters in Fig. 3. Small particles show extinction at short wavelengths, and a huge fraction of the light is dissipated as expressed by the large absorption coefficient of the particles, which reaches values of more than eight times the projected geometrical area of the particle. Similar to nondispersive media, the proportion of light manipulated but not absorbed by particles expressed in the scattering efficiency is higher in larger particles. For large particle diameters, the absorptive component is negligible and extinction of the particles is mainly because of scattering.

In contrast to a nonabsorbing medium, where the plasmonic resonance would continuously redshift with increasing particle size [42, 43], the coupling of the dispersive medium and the plasmonic resonance in regions of strong medium absorption result in a splitting of the plasmonic resonance into two branches around each flank of the ZnPc absorption band. The relative weight of the two subpeaks, however, continuously shifts to the red with increasing particle size. The splitting of the resonance peak into two subpeaks is an interesting effect [46], which persists to very low absorber concentrations when the interface of ZnPc:C_{60} and the wide-bandgap BF-DPB where the particles are located in our devices is treated in an effective medium approach, which explains why the spectra of additional absorptance and EQE in our devices reproduce this behavior. However, in the context of absorption enhancement in OSCs, we are most interested in photocurrent generation. Concerning plasmonic photocurrent enhancement, this implies that plasmonic resonances coupling to the absorber layer lead to a broadening of the region of photocurrent production.

For efficient light manipulations, large metal particles with long wavelength resonances and low losses are desirable. However, limitations of the particle size are implied by technical requirements for device preparation but more fundamentally also by the bandgap and thus the absorption coefficient of the semiconductors, which have to significantly overlap the particle resonance. For example, numerical investigations of typical semiconductors and particle diameter [47] reveal that within amorphous silicon solar cells, no effective near-field enhancement can be obtained because the resonance of low loss particles is well beyond the bandgap of the high-refractive-index semiconductor. Accordingly, in several materials, plasmonic enhancement is only possible with scattering particles are embedded in a low-refractive-index environment and far away from the absorber.

Figure 4a shows absorptance (1 minus direct reflectance) of the reference devices and OSCs including nanoparticles of different diameters placed as close as possible to the absorber layer. With respect to the reference device, new plasmonic peaks arise at the long-wavelength side of the ZnPc main absorption band. These peaks shift to the red and broaden with increasing particle diameter. In Fig. 4b spectra of additional absorption and EQEs as derived by subtracting the reference including V_2O_5 from the plasmonic devices are compared with the EQE response of the reference. Additional photocurrent is produced in spectral regions where the additional absorptance related to the particle resonance and the EQE of the device overlap. This also emphasizes that plasmonic resonances can be used to broaden the absorption spectrum of organic absorbers. This indicates a trade-off between scattering efficiency and overlap with the semiconductor absorption, which in this case favors smaller particles. However, the EQE enhancement in the short-wavelength region (below 550 nm is not related to a corresponding absorptance

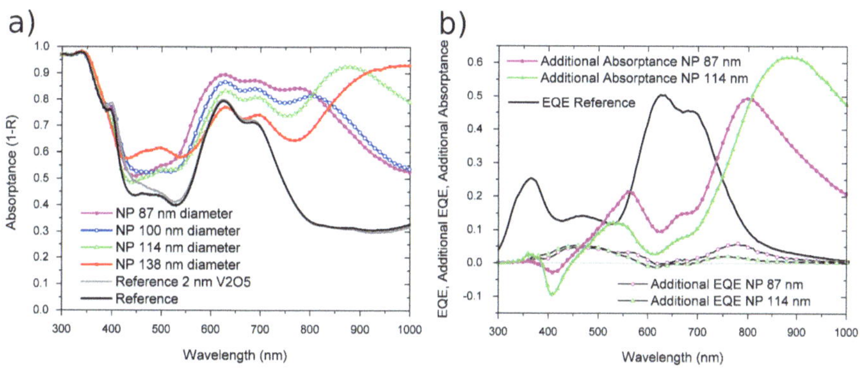

Fig. 4 Effect of particle size on absorptance and photocurrent enhancement. (**a**) Absorptance of the reference devices and devices including nanoparticles of different diameters formed from different amounts of silver. (**b**) Comparison of the external quantum efficiencies (EQEs) of the reference device, additional absorptance, and additional EQE calculated by subtracting the spectrum of the V_2O_5 reference from spectra observed in the plasmonic devices, including nanoparticles of different size

enhancement. This short-wavelength enhancement will be explained in Sect. 6 by considering the thin-film optics of the device and the fact that the induced particle field is increasingly out of phase from the driving field for wavelengths shorter than those of the resonance peaks.

5 Length Scales for Near-Field Enhancement

Depositing particles onto predeposited organic thin-film device stacks allows the systematic study of how the position of the particles with respect to the absorber layer influences the EQE of the devices. By varying the position of the particles, we can identify different mechanisms leading to enhanced photocurrents relevant for the device design. For example, we use spacer layers of doped HTL to adjust the distance between plasmonic particles and the intrinsic layers. Because the dewetting layer already has a thickness of 2 nm, this minimum particle-to-absorber distance corresponds to direct contact of the intrinsic layer and the dewetting layer. EQE enhancement spectra are calculated by normalizing the EQE of plasmonic devices by the EQE of the reference devices. EQE enhancement spectra for plasmonic devices comprising spacer layers of different thicknesses are plotted in Fig. 5a. A short- and a long-wavelength contribution can be distinguished according to their dependence on the particle to absorber layer distance.

EQE enhancement spectra reveal that the photocurrent in the long-wavelength region for wavelengths larger than 550 nm varies strongly with the particle-to-absorber distance. This part of the spectra can be directly related to the plasmonic resonances, which are in phase with the incident field. However, this near-field

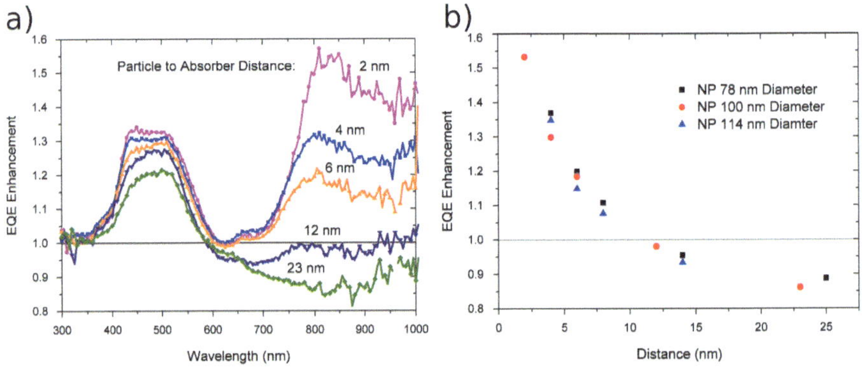

Fig. 5 Distance dependence of external quantum efficiency (EQE) enhancement relative to the reference device. (**a**) EQE enhancement spectra of plasmonic solar cells with different distances between particles and absorber layer. (**b**) Mean enhancement factors between 800–850 nm plotted vs. the particle-to-absorber distance

enhancement strongly decreases with increasing distance of the particles from the absorber. Mean enhancement factors in the wavelength range of 800–850 nm are plotted in Fig. 5b for several sets of ZnPc:C_{60} samples with comparable particle size and varying the particle-to-absorber distance. Although the magnitude of the maximum enhancement depends on the specific stack design, the effects of near-field enhancement and supposed parasitic absorption consistently balance at distances around 8 nm. A description of the peak enhancement assuming exponential decay and a base level of 0.86 accounting for some absorption of the particles visualizes that the near-field enhancement falls to half its magnitude every 4 nm [39]. This is in agreement with distance dependencies found in the absorption enhancement by Rand et al. [48] and Wang et al. [24] and values obtained in the photocurrent of dye-sensitized solar cells when particles are passivated with TiO_2 [49]. The dielectric constant of the coating material and the semiconducting absorbers decrease the spread of the near field with respect to air and water [50] but allow for much larger working distances than in inorganic semiconductors.

This confinement of the near field close to the particles is strongly connected to the question of suitable passivation layers [15, 29] or coating [49] of the particles. It is understandable that such low distances pose a significant issue in finding suitable combinations of buffer and dewetting layers between the particles and the absorber layer that maintain good electrical characteristics of the devices. We reproduced the near-field photocurrent enhancement with various absorber materials; the effect was most pronounced in long-wavelength-absorbing semiconductors such as aza-BODIPY [51]. However, to maintain reasonable electrical characteristics with more efficient absorber materials, combinations of buffer layers thicker than 7 nm were required.

The range at which the near field of the particles penetrates the absorber layer also plays an important role for device optimization. The relative strength of the near-field EQE enhancement (see Fig. 6f) is higher in thinner absorber layers, in agreement with the plasmonic near field penetrating a larger fraction of the absorber layer.

6 Reflection of Particles

The EQE enhancement (Fig. 5a) at a wavelength shorter than 550 nm shows little resemblance to the absorption enhancement (Fig. 4) and the mere scattering coefficient (Fig. 3). The explanation of this effect, which is not directly connected to an absorptance enhancement of the whole device, requires considering the thin-film optics of the whole device as well as the phase of the particle resonance with respect to the incident field.

For such short wavelengths, the particle resonance is nearly completely out of phase. Although the polarizability at short wavelengths is much smaller than that at the resonance position, the induced field reduces the electrical field at the particle position. The apparent node in the total field strength leads to a constructive

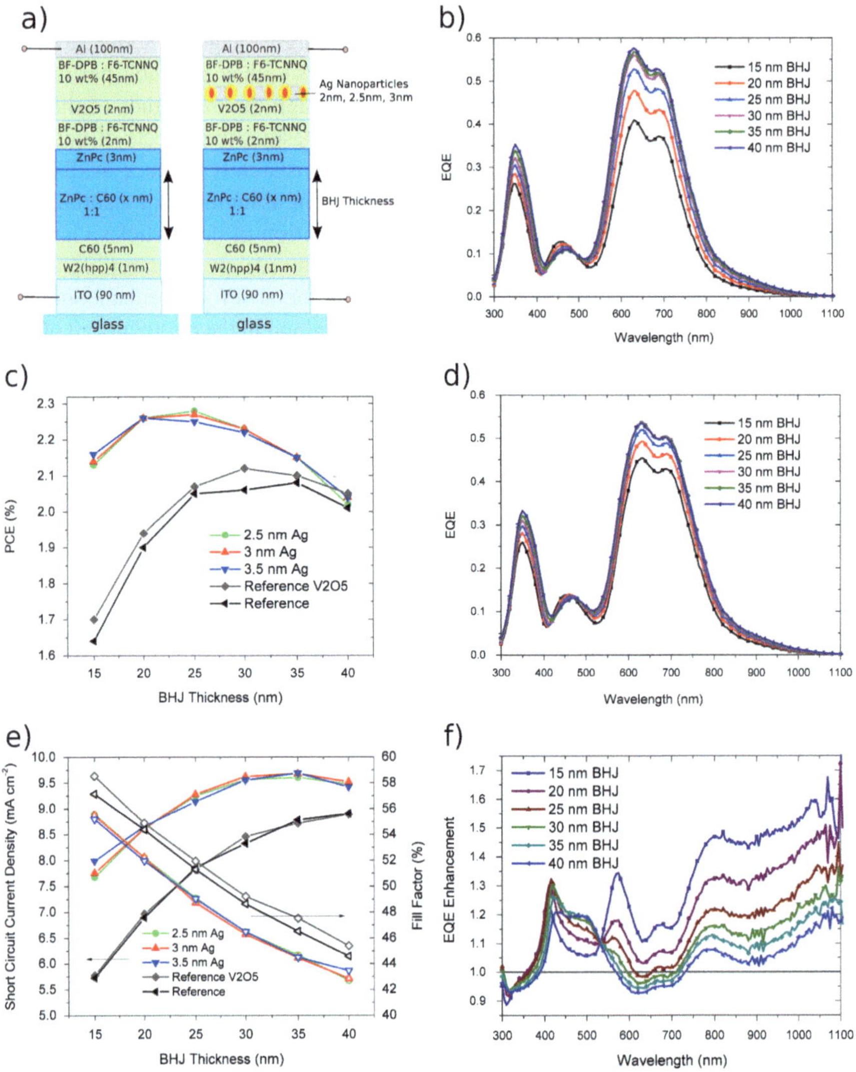

Fig. 6 Bulk heterojunction (BHJ) absorber layer thickness variation in plasmonic and reference ZnPc and C_{60} OSCs. (**a**) Organic thin-film stack of reference and plasmonic devices used to investigate different BHJ thicknesses. (**b**) External quantum efficiency (EQE) spectra of nonplasmonic reference devices for different thicknesses of the BHJ absorber layer. (**c**) Power conversion efficiency vs. BHJ thickness for reference and plasmonic devices. (**d**) EQE spectra of plasmonic devices for different thicknesses of the BHJ absorber layer, including particles with (83 ± 13) nm diameter and a surface density of $18\,\mu m^{-2}$ formed from 2.5 nm of silver. (**e**) Short-circuit current density (*filled symbols*) and fill factor (*open symbols*) vs. BHJ thickness for reference and plasmonic devices. (**f**) EQE enhancement spectra calculated by normalizing the EQE of the plasmonic device to the reference device of corresponding BHJ thickness

interference of short-wavelength light at a position located a quarter of a wavelength away from the particles. The particles work like a selective mirror, enhancing photocurrent when the absorber layer is sufficiently thick to include a significant part of the interference maximum. The enhancement in the short-wavelength region for samples with increasing absorber layer thickness in Fig. 6f at the left of the isosbestic point thus broadens and shifts to longer wavelength. Absorption within the particles is assumed to be very low in this region (see Fig. 3). The effect shows a weak distance dependence (see Fig. 5a), which is connected to the quarter-wavelength criterion for constructive interference, and its relative contribution is also more independent of the absorber thickness (Fig. 6f). Light reflected at the particle layer seems to be related to the total amount of silver deposited [39] (see Fig. 7a, b). A significant fraction of the overall photocurrent increase can be attributed to this scattering effect, which is less dependent on the position of the particles with respect to the absorber layer, making device preparation easier. The effect observed here using the particles behind the absorber as reflectors located at a quarter of a wavelength behind the absorber and detuned to longer wavelength also points to the equivalent of using short-wavelength plasmonic particles in front of the absorber as directors. Forward scattering has been proven to be useful in plasmonic devices especially when the local dielectric environment favors forward scattering and minimizes reflection losses [28, 29].

7 Optimized Absorber Thickness

With plasmonic absorption enhancement found as a versatile tool to engineer the optics within subwavelength thin-film layer stacks, the absorption-transport trade-off is revisited. The stack design used to investigate a variation of the absorber layer thickness is depicted in Fig. 6a.

To begin with the reference devices, the typical absorption–transport trade-off in the PCE [40] of amorphous thin-film solar cells is found. With respect to the BHJ thickness, PCE (Fig. 6c) has a maximum for a BHJ thickness of around 30 nm. With an open-circuit voltage of around 0.51 V, which is independent of the BHJ thickness, this can be understood as the maximum of the product of the short-circuit current, which is increasing, and the FF, which drops with increasing BHJ thickness (Fig. 6e). Photocurrent production does not linearly scale with absorber layer thickness. This effect is visualized in the peak EQE values, which seem to approach an upper limit in the reference devices for increasing BHJ thickness (Fig. 6b). Absorption in thick layers is almost perfect in spectral regions of strong absorption, and improvements by thicker layers can only be realized by broadening the absorption bands. With increasing BHJ thickness, the extraction of charge carriers shows a stronger dependence on the applied field, and the recombination of charge carrier pairs and increased series resistance within the BHJ deteriorate the device performance at the maximum power point, which is expressed by the lowered FF for higher BHJ thickness as plotted in Fig. 6e.

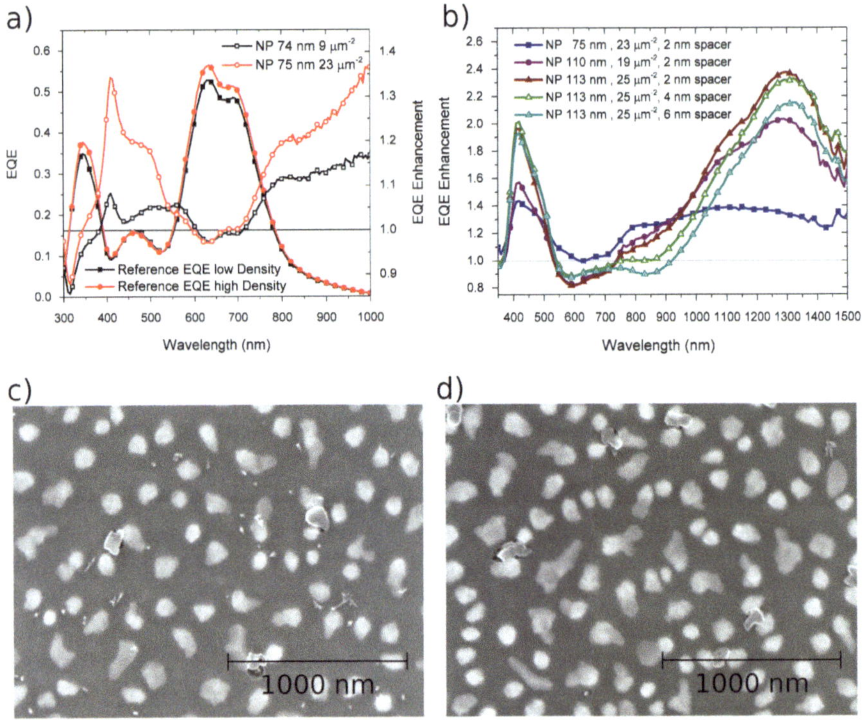

Fig. 7 Effect of particle surface density and interparticle distance. (**a**) External quantum efficiency (EQE) (*filled symbols*) of the reference devices and EQE enhancement (*open symbols*) of plasmonic samples comprising different particle surface densities but comparable particle diameters. (**b**) EQE enhancement spectra of samples including high particle density for different particle diameters and different particle-to-absorber distances. Scanning electron microscopy scan (**c**) of 6 nm of silver and (**d**) of 8 nm of silver deposited at substrate temperatures of 75 °C on a dewetting layer of V_2O_5, resulting in nanoparticle diameters of 110 nm and 113 nm, respectively

Within the plasmonic devices the short-circuit currents are improved for all devices. The EQE in the main ZnPc absorption band is broadened, which is most prominent in the thinnest devices (Fig. 6d). EQE enhancement for all wavelengths is found in the thinnest devices up to a BHJ thickness of 25 nm, as presented in Fig. 6f, where the EQE is normalized to the EQE of the reference device with corresponding BHJ thickness.

Figure 6f also reveals that the photocurrent benefits most in the thin BHJ devices, where, for example, a 30 % relative increase is observed in the 15 − nm sample. This can be understood in the sense of a larger contribution of near-field coupling in those devices as a larger fraction of the near field penetrates the absorber layer. While the photocurrent enhancement in the thick samples mainly results from the short-wavelength region associated with reflection at the particles, long-wavelength near-field enhancement is the dominating contribution in thin BHJ samples, where the near field extends to a larger fraction of the absorber layer and ZnPc peak absorption leaves room for significant improvement.

However, in plasmonic solar cells a small reduction of the open-circuit voltage to 0.50 V and a general reduction of the FF reduce the overall efficiency gain in these devices. With plasmonic photocurrent enhancement, this reduced FF can be overcompensated by using thinner samples, which show better electrical characteristics and improved photocurrents. The optimized absorber thickness of plasmonic devices is at a BHJ thickness between 20 − −25 nm, as suggested by Fig. 6c.

While the observed worsening of the electrical behavior at constant BHJ thickness associated with particle processing may be addressed in further engineering, plasmonic enhanced absorption allowing the use of thinner absorber layers is a tool to improve both the photocurrent and FF in OSCs. This also allows material consumption to be economized [12]. Absorption enhancement in thin layers could become an important feature when the standard thick BHJ design relying on fullerenes as acceptor molecules is challenged by alternative acceptor molecules, which often require the use of thin, flat heterojunctions to achieve good electrical properties. Promising candidates for flat heterojunction devices that use thin, long-wavelength-absorbing flat layers reach PCEs exceeding 8 % [52].

8 Interparticle Distance

A huge absorption enhancement connected to only a moderate EQE enhancement motivates the question of whether reduced particle density could improve the performance of plasmonic devices. Possibly a relevant part of the strong absorption is lost within the particles upon multiple scattering.

Varying the particle density with a fixed particle size is a more elaborate task than varying the particle diameter at a given density. We chose to investigate the effect by depositing particles at different temperatures, resulting in different particle surface densities. According to the assumed particle densities, we deposited different amounts of silver to get comparable particle diameters in devices with different particle densities. EQE spectra of the reference devices for the respective set of samples together with the EQE enhancement for comparable particle diameters but different particle surface densities are presented in Fig. 7a.

The long-wavelength EQE enhancement for mean particle diameters around 75 nm seems to scale with the particle surface density, as it roughly doubles when the particle surface density is increased from 9 to 23 μm^{-2}. The short-wavelength enhancement is much less pronounced in the low-density samples, suggesting that the selective mirror effect depends on particles reflecting collectively in an antenna array–like way.

While varying the surface density of the particles, we so far have not entered a regime where an increased density of the particles reduced device performance. Studies examining the optimum particle density for particles in front of the absorber layer report optima of 45 and 18 μm^{-2} for 67 − nm-diameter silver spheres [28] and 55 − nm-side-length silver on gold core-shell nanocubes [26], respectively.

However, in the high-particle-density samples we observe enhanced EQE for wavelengths longer than 900 nm, which we assign to coupling particles. To exclude that this is an artifact of the low signal level at long wavelength, we remeasured this feature with a more sensitive EQE setup and an additional InGaAs long-wavelength reference diode. EQE enhancement spectra of high-particle-density samples with different particle diameters and a variation of the particle-to-absorber distance are presented in Fig. 7. A very-long-wavelength EQE enhancement is found that is larger for the samples containing larger particles and higher particle densities. Enhancement of the photocurrent response of a factor significantly larger than 2 is found around 1300 nm for samples containing nanoparticles with 113 nm and surface densities of 25 μm^{-2}. This enhancement in the region of absorption of a charge-transfer state between ZnPc and C_{60} [53] has a negligible contribution to the overall photocurrent but demonstrates the advantages of examining plasmonic phenomena in OSCs because the devices themselves can work as sensitive detectors. As an explanation of the very-long-wavelength features, we suggest that particles couple [20, 21] or form-fused particles [42] that show redshifted resonances and increased scattering efficiency. This is more likely for high-particle-density samples and reduced interparticle spacing, which decreases with increasing particle diameter. The SEM scan of a large-diameter and high-density sample containing 8 nm of silver deposited at 75 °C, presented in Fig. 7d, indeed reveals several closely spaced particles but also fused, elongated, and sharp-edged particles. With an increasing spacer layer thickness between particles and absorber, this supposed coupling particle enhancement around 1300 nm drops on larger length scales than the enhancement around 850 nm that was assigned to the single-particle resonance. This can be rationalized with the near field extending to larger distances or with more complicated scattering directivity.

A determination of the interparticle distance in the disordered structure is difficult. Considering that a relatively small number of coupling particle pairs or fused particles can make up for an EQE enhancement that is much larger than that of a much larger number of single particles leads to the assumption that coupling particles have scattering efficiencies that are significantly higher than those of single particles. This is also connected to the question of whether some distances of coupling particles or fused elongated particles [54], also called nanowires in some publications [43], make up a large portion of the EQE enhancement [42].

One way to achieve efficient plasmonic enhancement could be to apply smaller coupling particles which have resonance at more relevant shorter wavelength. This could be a way to reduce dissipation losses but also to reduce overall particle sizes, which can simplify device integration. In fact, one of the most efficient single-junction OSC devices today employs clusters of closely spaced gold particles that have been templated by silver triangles [55]. Such assembled structures can support patterned collective plasmonic modes [56]. A high scattering efficiency and the occurrence of so-called hot-spots might be the reason for the very effective plasmonic enhancement at low cluster surface densities reported in [55].

9 Conclusion

We showed that vacuum thin film processing of organic devices and plasmonic nanoparticles offers versatile ways to investigate plasmonic phenomena and increase photocurrent production in OSCs. We reported how the parameters particle size and dielectric environment influence losses inside plasmonic particles using an analytical model. We discussed how length scales for near-field enhancement, reflection at the particles, and particle surface density influence the EQE response of plasmonic solar cells.

Technological steps involved in the deposition of the particles in small-molecule OSCs so far to some extent have impaired the diode characteristics of the plasmonic device with respect to reference devices of the same thickness. Nevertheless, we emphasize the potential of using plasmonic photocurrent enhancement in thinner plasmonic devices with improved transport properties but comparable photocurrent. With respect to the optimized reference device with a BHJ thickness of 30 nm, a reduction in the BHJ thickness to 20 nm in the optimized plasmonic device leads to an increase in the FF from 49 to 52 % and an enhancement of the short-circuit photocurrent density yet from 8.4 to 8.7 mA/cm^2.

Increasing the particle density led to coupling or elongated particles with strong photocurrent enhancement at very long wavelength, which points to the potential of plasmonic structures with more complicated shapes such as particle assemblies [55, 57], triangles [32, 58], and nanocubes [26].

The optical considerations for photocurrent enhancement were also successfully applied for other organic absorber materials, of which long-wavelength-absorbing materials are best suited. However, the introduction of additional layers for particle formation in proximity to the absorber layer led to a decreased device performance because of worse electrical characteristics. The small specific length scales of near-field enhancement require further studies on the electrical behavior, interdiffusion, and purification of materials involved to find suitable material combinations for the very thin blocking and dewetting layers between particles and absorbers.

Scattering effects that may work as a wavelength-dependent mirror are easier to implement, however, because the particles do not need to be in close proximity to the absorber layer. After all, plasmonic effects have been shown to work on length scales below 8 nm for near-field coupling and a quarter of a wavelength for thin-film optic effects with fixed phase relation, which renders vacuum-processed small-molecule OSCs with absorber layer in the 40 – nm range the solar cell technology that might profit most from plasmonic enhancement.

Acknowledgments We gratefully acknowledge L.M. Eng, M.C. Gather, and M. Riede for fruitful discussions, T. Günther and A. Wendel for sample preparation, S. Goldberg for SEM images, and the DFG for funding via SPP 1355.

References

1. Krebs FC, Espinosa N, Hösel M, Søndergaard RR, Jørgensen M (2014) 25th anniversary article: rise to power – OPV-based solar sprks. Adv Mater 26(1):29–39. doi:10.1002/adma.201302031. http://dx.doi.org/10.1002/adma.201302031
2. You J, Dou L, Yoshimura K, Kato T, Ohya K, Moriarty T, Emery K, Chen CC, Gao J, Li G, Yang Y (2013) A polymer tandem solar cell with 10.6% power conversion efficiency. Nat Commun 4:2411-1–2411-10. http://dx.doi.org/10.1038/ncomms2411
3. Zhang Q, Kan B, Liu F, Long G, Wan X, Chen X, Zuo Y, Ni W, Zhang H, Li M, Hu Z, Huang F, Cao Y, Liang Z, Zhang M, Russell TP, Chen Y (2015) Small-molecule solar cells with efficiency over 9%. Nat Photonics 9(1):35–41. http://dx.doi.org/10.1038/nphoton.2014.269
4. Meerheim R, Körner C, Leo K (2014) Highly efficient organic multi-junction solar cells with a Thiophene based donor material. Appl Phys Lett 105(6):063306-1-063306-4. doi:http://dx.doi.org/10.1063/1.4893012. http://scitation.aip.org/content/aip/journal/apl/105/6/10.1063/1.4893012
5. Haijun Fan XZ (2015) Development of small-molecule materials for high-performance organic solar cells. Sci China Chem 58(6):922-936. doi:10.1007/s11426-015-5418-6. http://chem.scichina.com:8081/sciBe/EN/Y2015/V58/I6/922
6. Donges A (1998) The coherence length of black-body radiation. Eur J Phys 19(3):245-249. http://stacks.iop.org/0143-0807/19/i=3/a=006
7. Selzer F, Weiß N, Kneppe D, Bormann L, Sachse C, Gaponik N, Eychmüller, A, Leo K, Müller-Meskamp L (2015) A spray-coating process for highly conductive silver nanowire networks as the transparent top-electrode for small molecule organic photovoltaics. Nanoscale 7:2777-2783. doi:10.1039/C4NR06502F. http://dx.doi.org/10.1039/C4NR06502F
8. Selzer F, Weiß N, Bormann L, Sachse C, Gaponik N, Müller-Meskamp L, Eychmüller A, Leo K (2014) Highly conductive silver nanowire networks by organic matrix assisted low-temperature fusing. Org Electron 15(12):3818-3824. doi:http://dx.doi.org/10.1016/j.orgel.2014.09.030. http://www.sciencedirect.com/science/article/pii/S1566119914004224
9. Schubert S, Müller-Meskamp L, Leo K (2014) Unusually high optical transmission in Ca:Ag blend films: high-performance top electrodes for efficient organic solar cells. Adv Funct Mater 24(42):6668-6676. doi:10.1002/adfm.201401854. http://dx.doi.org/10.1002/adfm.201401854
10. Schubert S, Hermenau M, Meiss J, Müller-Meskamp L, Leo K (2012) Oxide sandwiched metal thin-film electrodes for long-term stable organic solar cells. Adv Funct Mater 22(23):4993-4999. doi:10.1002/adfm.201201592. http://dx.doi.org/10.1002/adfm.201201592
11. Schubert S, Meiss J, Müller-Meskamp L, Leo K (2013) Improvement of transparent metal top electrodes for organic solar cells by introducing a high surface energy seed layer. Adv Energy Mater 3(4):438-443. doi:10.1002/aenm.201200903. http://dx.doi.org/10.1002/aenm.201200903
12. Ferry VE, Munday JN, Atwater HA (2010) Design considerations for plasmonic photovoltaics. Adv Mater 22(43):4794-4808. doi:10.1002/adma.201000488. http://dx.doi.org/10.1002/adma.201000488
13. Atwater HA, Polman A (2010) Plasmonics for improved photovoltaic devices. Nat Mater 9(3):205-213. doi:10.1038/nmat2629. http://www.nature.com/doifinder/10.1038/nmat2629
14. Stratakis E, Kymakis E (2013) Nanoparticle-based plasmonic organic photovoltaic devices. Mater Today 16(4):133-146. doi:http://dx.doi.org/10.1016/j.mattod.2013.04.006. http://www.sciencedirect.com/science/article/pii/S1369702113001065
15. You J, Li X, Xie FX, Sha WEI, Kwong JHW, Li G, Choy WCH, Yang Y (2012) Surface plasmon and scattering-enhanced low-bandgap polymer solar cell by a metal grating back electrode. Adv Funct Mater 2(10):426-432. doi:10.1002/aenm.201200108. http://dx.doi.org/10.1002/aenm.201200108
16. Sykes ME, Barito A, Amonoo JA, Green PF, Shtein M (2014) Broadband plasmonic photocurrent enhancement in planar organic photovoltaics embedded in a metallic nanocavity. Adv Funct Mater 4(11):1301937-1–1301937-9. doi:10.1002/aenm.201301937. http://dx.doi.org/10.1002/aenm.201301937

17. Hayashi S, Kozaru K, Yamamoto K (1991) Enhancement of photoelectric conversion efficiency by surface plasmon excitation: a test with an organic solar. Solid State Commun 79(9):763-767. doi:10.1016/0038-1098(91)90792-T. http://www.sciencedirect.com/science/article/pii/003810989190792T
18. Morfa AJ, Rowlen KL, Reilly TH, Romero MJ, van de Lagemaat J (2008) Plasmon-enhanced solar energy conversion in organic bulk heterojunction photovoltaics. Appl Phys Lett 92(1):013504-1–013504-4. doi:10.1063/1.2823578
19. Numazawa S, Ranjan M, Heinig KH, Facsko S, Smith R (2011) Ordered Ag nanocluster structures by vapor deposition on pre-patterned SiO2. J Phys Condens Matter 23(22):222203-1–222203-5. doi:10.1088/0953-8984/23/22/222203. http://iopscience.iop.org/0953-8984/23/22/222203
20. Rivera TP, Lecarme O, Hartmann J, Inglebert R, Peyrade D (2009) Spectroscopic studies of plasmonic interactions in colloidal dimers fabricated by convective-capillary force assembly. Microelectron Eng 86(4–6):1089-1092. doi:http://dx.doi.org/10.1016/j.mee.2008.11.015. http://www.sciencedirect.com/science/article/pii/S0167931708005212
21. Lecarme O, Pinedo-Rivera T, Berton K, Berthier J, Peyrade D (2011) Plasmonic coupling in nondipolar gold colloidal dimers. Appl Phys Lett 98(8):083122-1–083122-3. doi:http://dx.doi.org/10.1063/1.3560456. http://scitation.aip.org/content/aip/journal/apl/98/8/10.1063/1.3560456
22. Cordeiro J, Zelsmann M, Honegger T, Picard E, Hadji E, Peyrade D (2015) Table-top deterministic and collective colloidal assembly using videoprojector lithography. Appl Surf Sci 349(0):452-458. doi:http://dx.doi.org/10.1016/j.apsusc.2015.04.223. http://www.sciencedirect.com/science/article/pii/S0169433215011046
23. Müller-Meskamp L, Schubert S, Roch T, Eckhardt S, Lasagni AF, Leo K (2015) Transparent conductive metal thin-film electrodes structured by direct laser interference patterning. Adv Eng Mater. pp. 1215-1219. doi:10.1002/adem.201400454. http://dx.doi.org/10.1002/adem.201400454
24. Wang J, Lee YJ, Chadha AS, Yi J, Jespersen ML, Kelley JJ, Nguyen HM, Nimmo M, Malko AV, Vaia RA, Zhou W, Hsu JWP (2013) Effect of plasmonic Au nanoparticles on inverted organic solar cell performance. J Phys Chem C 117(1):85-91. doi:10.1021/jp309415u. http://dx.doi.org/10.1021/jp309415u
25. Li X, Choy WCH, Huo L, Xie F, Sha WEI, Ding B, Guo X, Li Y, Hou J, You J, Yang Y (2012) Dual plasmonic nanostructures for high performance inverted organic solar cells. Adv Mater (Weinheim Ger) 24(22):3046-3052. doi:10.1002/adma.201200120. http://dx.doi.org/10.1002/adma.201200120
26. Baek SW, Park G, Noh J, Cho C, Lee CH, Seo MK, Song H, Lee JY (2014) Au@Ag core-shell nanocubes for efficient plasmonic light scattering effect in low bandgap organic solar cells. ACS Nano 8(4):3302-3312. doi:10.1021/nn500222q. http://dx.doi.org/10.1021/nn500222q
27. Sachse C, Müller-Meskamp L, Bormann L, Kim YH, Lehnert F, Philipp A, Beyer B, Leo K (2013) Transparent, dip-coated silver nanowire electrodes for small molecule organic solar cells. Org Electron 14(1):143-148. doi:http://dx.doi.org/10.1016/j.orgel.2012.09.032. http://www.sciencedirect.com/science/article/pii/S1566119912004582
28. Baek SW, Noh J, Lee CH, Kim B, Seo MK, Lee JY (2013) Plasmonic forward scattering effect in organic solar cells: a powerful optical engineering method. Sci Rep 3:1726-1–1726-7. http://dx.doi.org/10.1038/srep01726
29. Jung K, Song HJ, Lee G, Ko Y, Ahn K, Choi H, Kim JY, Ha K, Song J, Lee JK, Lee C, Choi M (2014) Plasmonic organic solar cells employing nanobump assembly via aerosol-derived nanoparticles. ACS Nano 8(3):2590-2601. doi:10.1021/nn500276n. http://pubs.acs.org/doi/abs/10.1021/nn500276n
30. Wu B, Wu X, Guan C, Fai Tai K, Yeow EKL, Jin Fan H, Mathews N, Sum TC (2013) Uncovering loss mechanisms in silver nanoparticle- blended plasmonic organic solar cells. Nat Commun 4:3004-1–3004-7. http://dx.doi.org/10.1038/ncomms3004
31. Fung DDS, Qiao L, Choy WCH, Wang C, Sha WEI, Xie F, He S (2011) Optical and electrical properties of efficiency enhanced polymer solar cells with Au nanoparticles in a PEDOT-PSS

layer. J Mater Chem 21(41):16349-16356. doi:10.1039/C1JM12820E. http://pubs.rsc.org/en/content/articlelanding/2011/jm/c1jm12820e

32. Yao K, Salvador M, Chueh CC, Xin XK, Xu YX, deQuilettes DW, Hu T, Chen Y, Ginger DS, Jen AKY (2014) A general route to enhance polymer solar cell performance using plasmonic nanoprisms. Adv Funct Mater 4(9):1400206-1–1400206-7. doi:10.1002/aenm.201400206. http://dx.doi.org/10.1002/aenm.201400206
33. Lilliu S, Alsari M, Bikondoa O, Emyr Macdonald J, Dahlem MS (2015) Absence of structural impact of noble nanoparticles on P3HT:PCBM blends for plasmon-enhanced bulk-heterojunction organic solar cells probed by synchrotron GI-XRD. Sci Rep 5:10633-1–10633-10. http://dx.doi.org/10.1038/srep10633
34. Xue M, Li L, Tremolet de Villers BJ, Shen H, Zhu J, Yu Z, Stieg AZ, Pei Q, Schwartz BJ, Wang KL (2011) Charge-carrier dynamics in hybrid plasmonic organic solar cells With Ag nanoparticles. Appl Phys Lett 98:253302-1–253302-3. doi:10.1063/1.3601742
35. Niesen B, Rand BP, Van Dorpe P, Cheyns D, Tong L, Dmitriev A, Heremans P (2013) Plasmonic efficiency enhancement of high performance organic solar cells with a nanostructured rear electrode. Adv Energy Mater 3(2):145-150. doi:10.1002/aenm.201200289. http://dx.doi.org/10.1002/aenm.201200289
36. Volmer M, Weber A (1926) Keimbildung in Übersättigten Gebilden. Zeitschrift für Physikalische Chemie 119(4):277-301
37. Jägeler-Hoheisel T, Schwab T, Schubert S, Riede M, Leo K (2014) Verfahren zum herstellen eines organischen bauelementes sowie organisches bauelement. Patent Application. https://patentscope.wipo.int/search/en/WO2014094741
38. Metois JJ, Spiller GDT, Venables JA (1982) Lead on graphite: equilibrium shape, crystal growth, melting and the early stages of oxidation. Philos Mag A 46(6):1015-1022. doi:10.1080/01418618208236947. http://dx.doi.org/10.1080/01418618208236947
39. Jägeler-Hoheisel T, Selzer F, Riede M, Leo K (2014) Direct electrical evidence of plasmonic near-field enhancement in small molecule organic solar cells. J Phys Chem C 118(28):15128-15135. doi:10.1021/jp5025087. http://dx.doi.org/10.1021/jp5025087
40. Meiss J, Merten A, Hein M, Schuenemann C, Schäfer S, Tietze M, Uhrich C, Pfeiffer M, Leo K, Riede M (2012) Fluorinated zinc phthalocyanine as donor for efficient vacuum-deposited organic solar cells. Adv Funct Mater 22(2):405-414. doi:10.1002/adfm.201101799. http://doi.wiley.com/10.1002/adfm.201101799
41. Moroz A (2009) Depolarization field of spheroidal particles. J Opt Soc Am B 26(3):517-527. doi:10.1364/JOSAB.26.000517. http://josab.osa.org/abstract.cfm?URI=josab-26-3-517
42. Gunnarsson L, Rindzevicius T, Prikulis J, Kasemo B, Käll M, Zou S, Schatz GC (2005) Confined plasmons in nanofabricated single silver particle pairs: experimental observations of strong interparticle interactions. J Phys Chem 109(3):1079-1087. doi:10.1021/jp049084e. http://dx.doi.org/10.1021/jp049084e
43. Kuwata H, Tamaru H, Esumi K, Miyano K (2003) Resonant light scattering from metal nanoparticles: practical analysis beyond rayleigh approximation. Appl Phys Lett 83(22):4625-4627. doi:http://dx.doi.org/10.1063/1.1630351. http://scitation.aip.org/content/aip/journal/apl/83/22/10.1063/1.1630351
44. Rakić AD, Djurišić AB, Elazar JM, Majewski ML (1998) Optical properties of metallic films for vertical-cavity optoelectronic devices. Appl Opt 37(22):5271-5283. doi:10.1364/AO.37.005271. http://ao.osa.org/abstract.cfm?URI=ao-37-22-5271
45. Schünemann C, Wynands D, Wilde L, Hein MP, Pfützner S, Elschner C, Eichhorn KJ, Leo K, Riede M (2012) Phase separation analysis of bulk heterojunctions in small-molecule organic solar cells using zinc-phthalocyanine and C_{60}. Phys Rev B 85:245314. doi:10.1103/PhysRevB.85.245314. http://link.aps.org/doi/10.1103/PhysRevB.85.245314
46. Zengin G, Johansson G, Johansson P, Antosiewicz TJ, Käll M, Shegai T (2013) Approaching the strong coupling limit in single plasmonic nanorods interacting with J-aggregates. Sci Rep 3:3074-1–3074-8. http://dx.doi.org/10.1038/srep03074

47. Spinelli P, Polman A (2012) Prospects of near-field plasmonic absorption enhancement in semiconductor materials using embedded Ag nanoparticles. Opt Express 20(S5):A641-A654. doi:10.1364/OE.20.00A641. http://www.opticsexpress.org/abstract.cfm?URI=oe-20-105-A641
48. Rand BP, Peumans P, Forrest SR (2004) Long-range absorption enhancement in organic tandem thin-film solar cells containing silver nanoclusters. J Appl Phys N Y 96(12):7519-7526. doi:10.1063/1.1812589. http://link.aip.org/link/JAPIAU/v96/i12/p7519/s1&Agg=doi
49. Standridge SD, Schatz GC, Hupp JT (2009) Distance dependence of plasmon-enhanced photocurrent in dye-sensitized solar cells. J Am Chem Soc 131(24):8407-8409. doi:10.1021/ja9022072. http://dx.doi.org/10.1021/ja9022072
50. Härtling T, Reichenbach P, Eng LM (2007) Near-field coupling of a single fluorescent molecule and a spherical gold nanoparticle. Opt Express 15(20):12806-12817. doi:10.1364/OE.15.012806. http://www.opticsexpress.org/abstract.cfm?URI=oe-15-20-12806
51. Kraner S, Widmer J, Benduhn J, Hieckmann E, Jägeler-Hoheisel T, Ullbrich S, Schütze D, Radke KS, Cuniberti G, Ortmann F, Lorenz-Rothe M, Spoltore D, Meerheim R, Vandewal K, Koerner C, Leo K (2015) Influence of side groups on the performance of infrared absorbing aza-BODIPY organic solar cells. Phys Status Solidi (a) pp. 2747-2753. doi:10.1002/pssa.v212.12
52. Cnops K, Rand BP, Cheyns D, Verreet B, Empl MA, Heremans P (2014) 8.4% efficient fullerene-free organic solar cells exploiting long-range exciton energy transfer. Nat Commun 5:4406-1–4406-6. http://dx.doi.org/10.1038/ncomms4406
53. Scholz R, Luschtinetz R, Seifert G, Jägeler-Hoheisel T, Körner C, Leo K, Rapacioli M (2013) Quantifying charge transfer energies at donor–acceptor interfaces in small-molecule solar cells with constrained DFTB and spectroscopic methods. J Phys Condens Matter 25(47):473201-1–473201-21. http://stacks.iop.org/0953-8984/25/i=47/a=473201
54. Teulle A, Bosman M, Girard C, Gurunatha KL, Li M, Mann S, Dujardin E (2015) Multimodal plasmonics in fused colloidal networks. Nat Mater 14(1):87-94. http://dx.doi.org/10.1038/nmat4114.
55. Park HI, Lee S, Lee JM, Nam SA, Jeon T, Han SW, Kim SO (2014) High performance organic photovoltaics with plasmonic-coupled metal nanoparticle clusters. ACS Nano 8(10):10305-10312. doi:10.1021/nn503508p. http://dx.doi.org/10.1021/nn503508p
56. Jägeler-Hoheisel T, Cordeiro J, Lecarme O, Cuche A, Girard C, Dujardin E, Peyrade D, Arbouet A (2013) Plasmonic shaping in gold nanoparticle three-dimensional assemblies. J Phys Chem C 117(44):23126-23132. doi:10.1021/jp406410k. http://pubs.acs.org/doi/abs/10.1021/jp406410k
57. Wang DH, Park KH, Seo JH, Seifter J, Jeon JH, Kim JK, Park JH, Park OO, Heeger AJ (2011) Enhanced power conversion efficiency in PCDTBT/PC70BM bulk heterojunction photovoltaic devices with embedded silver nanoparticle clusters. Adv Energy Mater 1(5):766-770. doi:10.1002/aenm.201100347. http://dx.doi.org/10.1002/aenm.201100347
58. Viarbitskaya S, Teulle A, Marty R, Sharma J, Girard C, Arbouet A, Dujardin E (2013) Morphology-induced redistribution of surface plasmon modes in two-dimensional crystalline gold platelets. Nat Mater 12(5):426-432. http://dx.doi.org/10.1038/nmat3581.

Synthesis of Conjugated Polymers with Complex Architecture for Photovoltaic Applications

Anton Kiriy and Frederik C. Krebs

Contents

Abstract A common approach to bulk heterojunction solar cells involves a "trial-and-error" approach in finding optimal kinetically unstable morphologies. An alternative approach assumes the utilization of complex polymer architectures, such as donor–acceptor block copolymers. Because of a covalent preorganization of the donor and acceptor components, these materials may form desirable morphologies at thermodynamic equilibrium. This chapter reviews synthetic approaches to such architectures and shows the first photovoltaic results.

A. Kiriy (✉)
Department of Polymer Structures, Leibniz-Institut für Polymerforschung Dresden e.V., Hohe Str. 6, 01069 Dresden, Germany
e-mail: kiriy@ipfdd.de

F.C. Krebs
DTU ENERGY, Department of Energy Conversion and Storage, Technical University of Denmark, Frederiksborgvej 399, Building 111, Room S05, 4000 Roskilde, Denmark
e-mail: frkr@dtu.dk

K. Leo (ed.), *Elementary Processes in Organic Photovoltaics*,
Advances in Polymer Science 272, DOI 10.1007/978-3-319-28338-8_15

Keywords Catalyst-transfer polycondensation • Covalent preorganization • Donor-acceptor block copolymer • Grafting-from • Grafting-through • Grafting-to • Hairy particles • Morphology • PCBM • Self-assembly • Suzuki chain-growth polycondensation

1 Introduction

Controlling nanoscale morphology is crucial to the performance of polymer bulk heterojunction (BHJ) solar cells (SCs) [1]. Recent studies of polymer–polymer [1] and polymer–fullerene [2] blend devices have confirmed the importance of the nanoscale phase separation of the binary electron-donor and electron-acceptor components into bicontinuous networks. Device efficiency is optimized when the component domains are (1) small enough for efficient migration of excitons to an interface but (2) large enough to promote efficient long-range charge separation. Several non–poly(3-hexylthiophene-2,5-diyl) (P3HT)-based SCs were reported, which outperform P3HT/PCBM (phenyl-C_{61}-butyric acid methyl ester) SCs but are worse than expected from their intrinsic potential. These results demonstrate that optimal intrinsic electronic characteristics alone are not a sufficient factor to provide a good photovoltaic (PV) performance, because the morphology and donor–acceptor (D–A) interface structure as well as the molecular orientation also play extremely important roles. It was demonstrated that trapping of electrons, difficulty in dissociating electron–hole pairs, and geminate and bimolecular recombinations are dominant loss channels [3].

In general, spin coating of various polymer/PCBM mixtures from highly volatile solvents leads to blends with kinetically frozen morphologies with rather ultimately intermixed components (Fig. 1, top). These morphologies, favored for the charge-separation process, are, however, hardly suitable for PV devices because they promote the recombination of generated charges. As a rule, long-term annealing of kinetically frozen blends induces macroscopic phase separation close to the point of thermodynamic equilibrium. The resulting morphologies are also undesirable for PV devices because they have a reduced interface between the donor and acceptor components, which, in turn, hampers the charge-separation process. Optimal for PV devices, morphologies with a characteristic ~10-nm domain size are situated somewhere between the two extremes. Meanwhile, significant research efforts were undertaken for the development of universal approaches to such morphologies. Good results were achieved by applying thermal [4] or solvent annealing [5–7]; however, the annealing time must be adjusted for each particular blend. Another promising approach involves high-boiling-point selective solvent additives for one of the components [8–10]. However, this method is also not universal and needs judicious adjustment. As such, blend morphology optimizations usually involve time-consuming trial-and-error efforts, which do not guarantee success at the end. In addition, PV devices that rely upon kinetically frozen morphologies suffer from low operational stability at elevated temperatures.

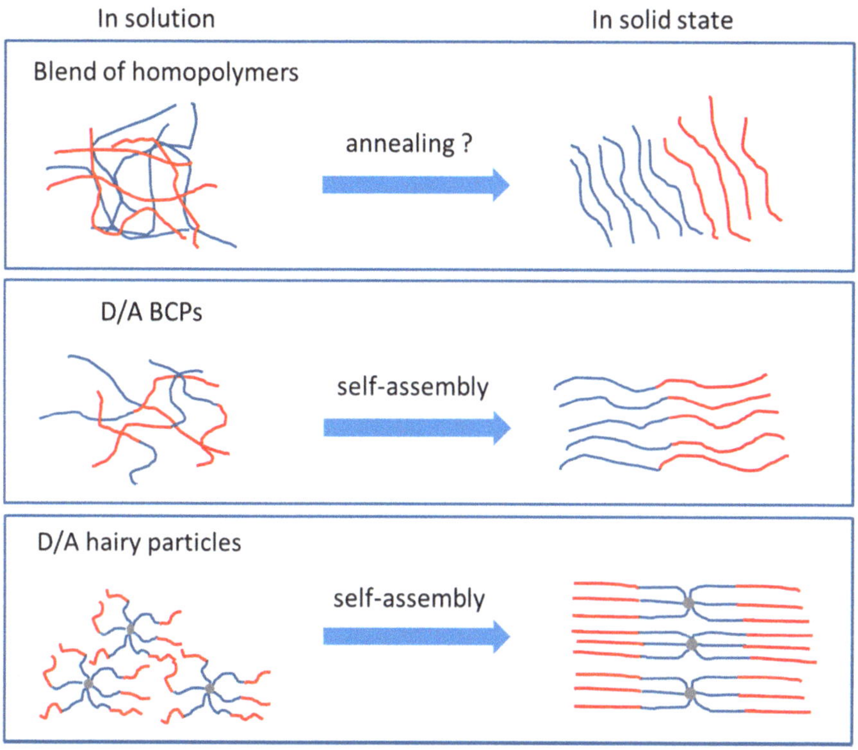

Fig. 1 Schematic representation of a "blend" (*top*) vs. a "covalent preorganization" (*middle* and *bottom*) approach for achieving an "ideal" morphology for all-polymer solar cells. *Top*: a commonly used approach utilizing blends of donor and acceptor materials; self-assembly of block copolymers (*middle*) and hairy particles (*bottom*)

An alternative approach that may help in solving the morphological problem is the application of more complex polymer architectures, the self-assembly properties of which may be encoded into their structure. These architectures may include D–A block copolymers (BCPs) or hairy particles containing donor and acceptor constituents. In BCPs, the donor and acceptor components are covalently attached to one another, which makes macrophase separation impossible (Fig. 1, middle). Instead, a microphase separation under certain conditions may lead to nanomorphologies (lamellar or columnar) desirable for PV devices. On the other hand, hairy particles contain BCP chains grafted to the core so that they appear to be preorganized parallel to one another and stretched, which should favor their crystallization and therefore improve charge-separation and charge-transport properties (Fig. 1, bottom).

However, traditional synthetic tools developed during the past two decades are hardly applicable for controlled synthesis of different architectures of semiconductive polymer (unlike the synthesis of conventional nonconjugated polymers such as polyolefines); therefore, new chemistry methods must be developed to implement the covalent preorganization idea. This chapter reviews recent progress in the development of controlled methods to π-conjugated polymers and describes the synthesis of different polymer architectures relevant to photovoltaics and their application in SCs. A significant potion of this review highlights the progress of Kiriy's and Krebs' groups within SPP 1355, "Elementary Processes in Photovoltaic Devices."

2 BCPs for Solar Cells

One possible approach to direct a delicate self-organization of matter at nanoscale implies the usage of BCPs. Because of an incompatibility between different blocks, BCPs prefer composition-dependent nanophase separation into lamellae, cylinders, or bicontinuous networks [11]. These kinds of organization may be beneficial in BHJ devices if such BCPs comprise state-of-the-art donor and acceptor components [12].

2.1 BCPs with Pendant Semiconductor Groups

Recently, several fully functional D–A BCPs, having pendant fullerene or perylene-diimide (PDI) acceptors and a P3HT donor, have been prepared. Such BCPs were synthesized using one of the following two approaches: (1) synthesis of BCPs using monomers with acceptor pendant units; and (2) synthesis of BCP precursors, followed by the postmodification to incorporate acceptor pendant units. Thelakkat et al. [13–15] reported the synthesis of a series of polymers with triphenylamine and PDI pendant groups by using nitroxide-mediated polymerization (NMP). The resulting BCPs exhibited mostly cylindrical morphologies, such as wire-like or worm-like structures, as revealed by transmission electron microscopy (TEM) measurements.

In addition, the synthesis of the crystalline–crystalline BCP poly(3-hexylthiophene)-block-poly(perylene-bisimide acrylate) (P3HT-b-PDIA, Fig. 2) by NMP using the PDI-based monomer was reported [16]. End-functionalized P3HT was prepared by Grignard metathesis (GRIM) polymerization and then the PDI block was grown from the P3HT macroinitiator. Thus-prepared BCPs exhibited cylindrical morphologies with PDI domains embedded in the P3HT matrix or lamellar morphologies, depending on the ratio of the two blocks [17]. However, the PV performance of these BCPs was below expectations. Although the exact reasons for a modest PV performance are still not fully understood and are in debate, the improper mutual arrangement of the donor and acceptor blocks is likely to play a role. In addition, the presence of an insulating spacer between the donor and acceptor groups may prohibit a direct, through-the-chain charge-separation process.

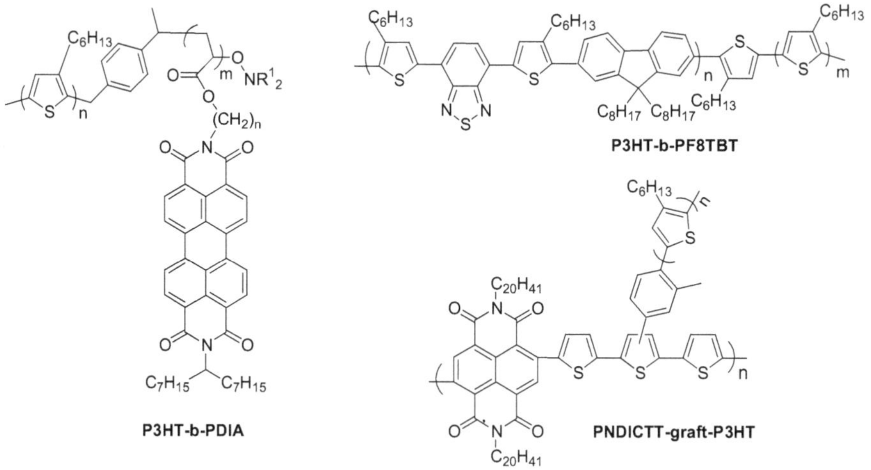

Fig. 2 Examples of different architectures of semiconductive polymers: block copolymer (BCP) with the first conjugated block and the second block having pendant acceptor units (P3HT-*b*-PDIA) [16], all-conjugated BCP (P3HT-*b*-PF8TBT) [28], graft copolymer (PNDICTT-*graft*-P3HT) [35]

A few approaches to fullerene-containing polymers were also developed. Fréchet et al. [18, 19] prepared norbornene-based monomers having P3HT and fullerene substituents for the ring-opening metathesis polymerization. With these monomers Fréchet's group prepared brush-type macromolecules with high fullerene contents up to 50 wt% C60. Lee et al. [20] used polymer-analogous transformations according to which [6,6]-phenyl-C_{61}-butyric acid was attached under Steglich conditions to polymeric hydroxyethyl methacrylate. In another work, alkyne-functionalized fullerenes were clicked to azide groups in BCPs [21].

A new synthetic method for the covalent linking of fullerenes to polymers was introduced by our group [22]. The Bingel reaction was used to prepare bromine-functionalized fullerene building blocks that could be covalently linked to hydroxyl groups of copolymers by the cesium carbonate–promoted Williamson ether synthesis. Subsequently, BCPs with a second block based on styrene and hydroxystyrene or hydroxyethyl methacrylate were synthesized with a P3HT–TEMPO macroinitiator through NMP. Fullerene derivatives were linked to these polymers in a controlled manner and D–A BCPs with high fullerene contents of nearly 50 wt% were achieved. These fullerene-based BCPs were successfully used for morphological control of P3HT/PCBM blends and their thermodynamic stabilization [23]. At the same time, these fullerene-based BCPs did not show an improvement in the efficiency of PV devices, presumably because of the strong aggregation behavior of fullerene-based blocks.

It should be noted that using BCPs having fullerene pendant groups did not result in a substantial increase in the efficiency of PV devices as compared to physical polymer:PCBM blends. However, the use of such BCPs as compatibilizing additives significantly improved the long-term and thermal stability of the corresponding blends [18–21].

2.2 All-Conjugated BCPs

All-conjugated BCPs, which have no nonconjugated bridge between the blocks, is a promising alternative to BCPs having pendant semiconducting units because of the possibility of a direct, in-chain photoinduced charge-separation process. Geng et al. [24] demonstrated that all-conjugated D–A block cooligomers in which a short conjugated oligomer donor is directly attached to the electron acceptor are very promising materials for SCs, showing a relatively high power conversion efficiency (PCE). This result demonstrates the potential of the BCP approach in optoelectronics. However, to realize this potential, all-conjugated D–A BCPs composed of state-of-the-art subunits are needed. Such BCPs can be prepared (1) by coupling of two π-conjugated polymers containing complementary reactive end groups ("grafting-to" approach), and (2) by performing step-growth synthesis of one block in the presence of a properly end-functionalized second block, which serves as an end-capper ("grafting-through" approach). Alternatively, BCPs can be synthesized by (3) polymerization of the second block from the end group of the first block, the latter acting as a macroinitiator ("grafting-from" approach), or (4) by sequential addition-polymerization of two monomers.

Grafting-to Approach An approach to all-conjugated BCPs that implies a coupling of polymers by their end functions was recently implemented by Scherf's group [25]. Its attractive feature is that many standard routes to semiconductive polymers (e.g., step-growth Stille, Suzuki, and Yamamoto polycondensations) can be applied in the synthesis of all-conjugated BCPs. However, this method frequently suffers from an insufficiently high coupling efficiency, which affects the purity of the resulting BCPs especially when the two polymers have high molecular weights [26]. Moreover, an excess amount of one of the polymers is required to increase the coupling efficiency, which requires tedious purification steps.

As a representative example illustrating this approach, all-conjugated D–A–D triblock copolymers poly(3-octylthiophene)-*b*-poly(quinoxaline)-*b*-poly(3-octylthiophene) (P3OT-*b*-PQ-*b*-P3OT) were synthesized in three steps by Woody et al. [27]. The synthesis involved the preparation of a monofunctional telechelic P3OT donor block having a single α-bromothienyl end group, the synthesis of a PQ acceptor block bearing a complementary boronate ester functionality at both ends, and the coupling of the two polymer chains under Suzuki cross-coupling conditions. In this study, the authors used an excess of one of the blocks so that the synthesis required purification of desirable BCPs from unreacted materials. Further investigation of these BCPs revealed an efficient fluorescence quenching of the polymers in the solid state, supporting the existence of the electron transfer.

Recently, Sommer et al. [28] reported the preparation of an all-conjugated D–A diblock copolymer poly(3-hexylthiophene)-block-poly((9,9-dioctylfluorene)-2,7-diyl-alt-[4,7-bis(3-hexylthien-5-yl)-2,1,3-benzothiadiazole]-2′,2″-diyl) (P3HT-*b*-PF8TBT) (Fig. 2) by a similar approach. Particularly, PF8TBT was synthesized via the Suzuki polycondensation followed by in situ end-capping with P3HT-Br. In order to maximize the yield of the BCP, an excess of P3HT-Br was used.

Thus-synthesized BCPs were fractionated using preparative gel permeation chromatography (GPC) to remove unreacted P3HT. However, no microphase separation was observed, as a result of the high miscibility of the two blocks.

Another example relevant to photovoltaics was reported by Greenham et al. [29], who described the preparation of an all-conjugated D–A diblock copolymer poly(3-hexylthiophene)-block-poly[2,7-(9,9-dialkyl-fluorene)-alt-5,5-(4′,7′-di-2-thienyl-2′,1′,3′-benzothiadiazole)] (P3HT-*b*-PFTBTT). In their method, a well-defined P3HT with one terminal bromine functionality was first prepared by GRIM polymerization. The acceptor block, PFTBTT, was synthesized via the Stille cross-coupling procedure using the asymmetric-type monomer with the α-stannyl and γ-bromo end groups. After the reaction, the P3HT-Br was added to the reaction as a macromolecular end-capper. The product synthesized by this approach, however, is a mixture of approximately 53% diblock copolymer, 42% PFTBTT homopolymer, and 5% P3HT homopolymer.

P3HT-*b*-PFTBTT was used as surfactant in all-polymer SCs. The SC based on the P3HT/PFTBTT homopolymer blend showed a PCE of 1.50% after annealing at 130 °C, while the PCEs of the devices with ternary blends of P3HT/PFTBTT/P3HT-*b*-PFTBTT fabricated under the same conditions was 1.13% and 0.76% for the 17% and 40% contents of P3HT-*b*-PFTBTT, respectively. However, after annealing at 200 °C for 10 min, the PCE of the device with the homopolymer blends significantly decreased to 0.61%. In contrast, the PCEs of the devices based on the ternary blends all increased, 1.17% and 0.98% for the 17% and 40% blends of P3HT-*b*-PFTBTT, respectively, which mostly resulted from the increase in the short-circuit current and fill factor. These results demonstrated that the all-conjugated BCPs can be used as surfactants in all-polymer SCs as a new approach toward control of the crystallinity and nanostructure of photovoltaic polymer blends.

Grafting-Through Approach Compared to previously reported methods, the grafting-through approach is similar but technically simpler. Typically, end-functionalized P3HT-Br and the acceptor monomers are added at the same time and copolymerized in a one-pot process. In this case, the donor block acts as the end-capper. With this method, the all-conjugated D–A BCP P3HT-*b*-poly(fluorene-co-benzothiadiazole) was synthesized in a Suzuki polymerization process from fluorene and benzothiadiazole monomers and P3HT-Br [30]. Extensive purifications (solvent extraction and column chromatography) were required to remove the homopolymer contaminations [31].

Recently, the all-conjugated D–A BCP, P3HT-*b*-poly(naphthalene bisimide)-*b*-P3HT (P3HT-*b*-PNBI-*b*-P3HT) was synthesized by Nakabayashi et al. [32] under Yamamoto coupling conditions. In this work, P3HT-Br was prepared with different molecular weights, and it was then allowed to be copolymerized with the NBI-Br_2 monomers to produce P3HT-*b*-PNBI with different P3HT segment compositions. Surprisingly, the triblock copolymer product obtained was pure from byproducts such as P3HT and diblock copolymer P3HT-*b*-PNBI-*b*-P3HT.The potential application of P3HT-*b*-PNBI-*b*-P3HT as acceptor materials in all-polymer SCs in combination with P3HT has also been investigated. Unfortunately, only a modest PCE of 1.2% was found in optimized blends.

In another work of Verduzco et al. [33], GRIM and Suzuki coupling polymerizations were used to synthesize high-molecular-weight P3HT-*b*-PFTBT. For the single-active component devices utilizing P3HT-*b*-PFTBT, the average PCE of 2.7% ($V_{oc} = 1.14$ V, $J_{sc} = 5.0$ mA/cm^2, and the fill factor, FF, = 0.45) was achieved after thermal annealing. X-ray scattering data indicated that P3HT-*b*-PFTBT was self-assembled into lamellar microdomains with a domain spacing of approximately 9 nm, which is comparable to the exciton diffusion length. In contrast, the devices based on the blends of P3HT and PFTBT exhibited a maximum PCE of only 1.0%. In the case of blends, long annealing times led to a drop in performance caused by a macroscopic phase separation [34].

Macroinitiator or "Grafting-From" Approach The methods considered above were used to successfully produce several types of all-conjugated D–A BCPs. However, the generation of homopolymer byproducts is sometimes inevitable. Another drawback of these methods is that, to vary the composition of the donor and acceptor blocks, P3HT-Br with different molecular weights should be individually prepared in advance. To address these issues, the macroinitiator (grafting-from) method has recently been developed by Higashihara [35] to synthesize all-conjugated graft copolymers comprised of an acceptor backbone and donor graft chains. In this work, chloro-toluene groups able to form initiator sites were introduced in the backbone of then-type polymer PNDICTT, followed by externally initiated catalyst-transfer polycondensation of the thiophene monomers to produce a series of all-conjugated graft copolymers, PNDICTT-graft-P3HT (Fig. 2). It was shown that the number of P3HT-grafted chains and their molecular weight could be varied by the macroinitiator structure and reaction conditions. A microphase separation between the P3HT and PNDICTT domains was observed by TEM and grazing-incidence wide-angle X-ray scattering.

The grafting-from method was also used to synthesize all-conjugated D–A–D BCPs. Particularly, an acceptor, di-bromo-terminated poly{[N, N′bis(2-octyldodecyl)-naphthalene-1,4,5,8-bis(dicarboximide)-2,6-diyl]-alt-5,5′-(2,2′-bithiophene)} (PNDITh)-Br_2, was synthesized via the Stille coupling polycondensation, and then P3HT blocks were grown from both ends of the PNDITh-Br_2 via a catalyst-transfer polycondensation to produce an all-conjugated ABA-type BCP P3HT-*b*-PNDI-*b*-P3HT. In this case, the chain length and composition of P3HT in P3HT-*b*-PNDI-*b*-P3HT can be tailored simply by varying the feed molar ratio of the thiophene monomers to the macroinitiators. The TEM results indicated that a well-defined, thermally stable lamellar morphology was formed between the P3HT and PNDITh domains. Unfortunately, the photovoltaic properties of these copolymers were not reported [36].

Chain-Extension Approach The most straightforward route to BCPs implies sequential chain-growth polymerization of different monomers. Several all-conjugated BCPs were prepared via the chain-extension manner after discovery of catalyst-transfer polycondensations of AB-type monomers by McCullough's and Yokozawa's groups [37, 38]. However, a scope of chain-growth polycondensations is mostly limited to the preparation of simple yet electron-rich polymers, such as polythiophenes, polyphenylenes, polypyrroles, and polyfluorenes [39]. At the same time, more complex π-conjugated polymers, such as those comprising electron-deficient building blocks (such as rylenediimides, diketopyrrolopyrroles, isoindigos) and fused electron-rich building blocks (such as dithienosiloles), are needed for real optoelectronic applications. Particularly, naphthalene diimide (NDI) and PDI copolymers are currently evolving as an intriguing class of electron-conducting materials [40]. Bisthiophene-NDI copolymer P(NDI2OD-T2) (also known as and commercially available as ActivInk™ N2200) was reported to yield high electron mobilities up to 0.8 cm^2/Vs. Difficulties in the chain-growth preparation of such polymers are likely to originate from the increased size of monomers and their more complex polarization pattern, which impede an *intra*molecular catalyst transfer (underlying the chain-growth mechanism) and promote step-growth propagation. Copolymerization of monomers having a different nature (e.g., electron-rich with electron-deficient) via the chain-growth mechanism is an even more difficult task because in this case an appropriate common catalyst must be found that supports the *intra*molecular catalyst-transfer process for both monomers [41]. Furthermore, successful block copolymerization requires compatibility of the two polymerization processes (i.e., the presence in the reaction mixture of the first monomer, or its byproducts, must not prohibit polymerization of the second monomer). Finally, efficient transfer of the catalyst from the first block to the second one must be achieved [42]. As a result, examples of all-conjugated D–A copolymers (random, gradient, BCPs) with both units polymerized on the chain-growth manner, yet relevant to applications in SCs, are scarce in the literature. In their attempts to prepare complex architectures of D–A copolymers relevant to PV applications, Kiriy's group contributed substantially to the development of chain-growth polycondensation of rylene-based monomers (see below, Sect. 4).

3 "Hairy Particles"

We supposed that "hairy" particles—nanoscale objects in which polymer chains are attached by their end points to a particle core and stretched away from it to form spherical brushes—would be ~~such~~ an interesting alternative to conventional linear conjugated polymers for applications in SCs. An inherent advantage of this approach is that the chains formed by grafting from the surface appear to be preorganized parallel to one another and more or less stretched because of the known "brush" effect that will favor their controlled crystallization inside the domain (see inset in Fig. 1, bottom).

The grafting-to method to prepare hairy conjugated particles was used, however, with very limited success. Fréchet's and Emrick's groups [43] grafted end-capped P3HT and other conjugated polymers to CdSe quantum dots. However, the grafting density of grafted polymers and the purity of the functionalized quantum dots are usually low. The grafting-from strategy for the preparation of conjugated polymer brushes and hairy particles was pioneered by Kiriy's group [44–46]. In general, surface-initiated polymerization, or the "grafting-from" approach—the method of growing polymer chains selectively from functionalized particles via the one-by-one addition of monomers to surface-immobilized initiators—is one of the most powerful synthetic approaches to hairy particles with variable composition and a tunable grafting density, which requires the chain-growth polymerization mechanism. After the chain-growth mechanism for the preparation of P3HT was demonstrated by McCullough's and Yokozawa's groups, the grafting-from preparation of P3HT became feasible. However, in its "classical" form, because of the requirement for a monomer–monomer initiation stage, the Kumada polycondensation does not allow surface tethering. Recently, we found a method to prepare a library of highly efficient surface-immobilized Ni initiators supported by nitrogen- and phosphorous-based bidentate ligands that allows for the initiation of polymerization into regioregular P3HT with controlled molecular weight and narrow dispersity [47–50]. This novel initiating method was successfully applied in selective grafting of regioregular P3HT from submicron and nanoscale organosilica particles (Fig. 3).

The hairy P3HT nanoparticles were successfully applied in BHJ SCs. A significant dependence of PV performance on both the regioregularity of P3HT and on its molecular weight was previously reported; however, the influence of P3HT's architecture was not investigated in prior works. We demonstrated that the nano-P3HT particles, in combination with PCBM, showed a reasonably good PV performance for relatively large-area devices with an active area of 1 cm^2 (PCE = 1.8–2.3%, J_{sc} = 5.5–6.8 mA/cm^2, V_{oc} = 0.60–0.62 V; FF = 55–58%; Fig. 4).

Although the PV performance of the nano-P3HT particles is lower than that for the best-reported P3HT-PCBM devices, the observed performance for the nano-P3HT particles when used in a BHJ with PCBM is similar to what is obtained with linear P3HT, with a rather high molecular weight (M_W = 36,600 g/mol)

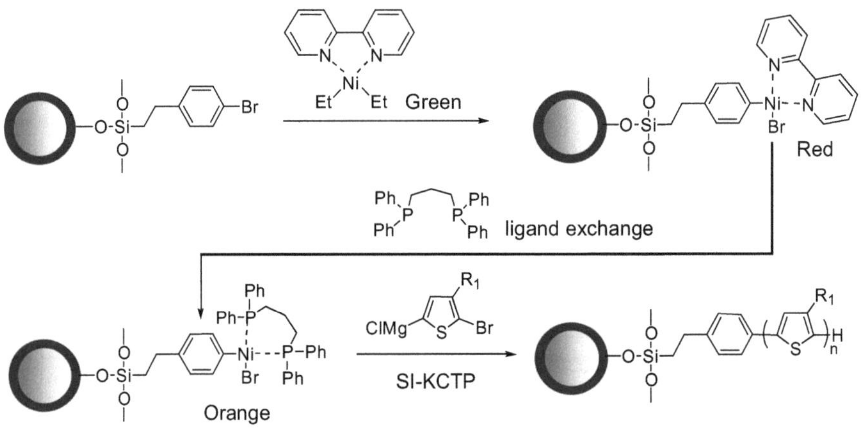

Fig. 3 Schematic representation of the preparation of hairy P3HT nanoparticles by the grafting-from approach [47]

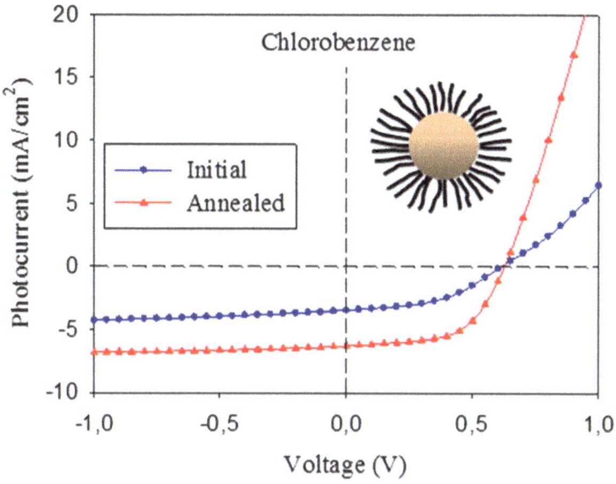

Fig. 4 *I–V* curves of photovoltaic device utilizing P3HT hairy nanoparticles:PCBM as the active layer before (*blue circles*) and after (*red triangle*) thermal annealing (130 °C for 5 min) [47]

for relatively large-active-area devices of 1 cm^2. On the other hand, it is even somewhat better than the performance of previously reported small-area devices that utilized relatively low-molecular-weight P3HTs of $M_W < 13{,}000$ g/mol. This result demonstrates that the developed surface-initiated Kumada catalyst-transfer polycondensation (SI-KCTP) procedure at least does not affect the optoelectronic properties of P3HTs. Although more studies are necessary to comprehensively establish the influence of the polymer architecture on the PV performance, it becomes clear that the molecular weight of linear P3HT constituents of branched P3HT structures must be a subject of further optimization.

Large-polymer SC modules were also prepared using roll-to-roll coating of inks based on preorganized P3HT grafted onto SiO_2 nanoparticles. The device performance was comparable to similar devices prepared using regioregular P3HT [51]. It was found possible to use chlorobenzene as the single solvent for the ink comprising the nanoparticles with grafted P3HT and PCBM. The devices were laminated prior to testing with a 25-μm polyethylene terephthalate (PET)-based overlaminating foil using an acrylic adhesive to protect the printed layers mechanically. These devices required a short annealing period (<10 min) under the solar simulator before reaching the maximum performance, which was typically around 0.9% under 1 sun illumination and 1.2% under about 1/3 sun. The main reason for the higher efficiency at a lower incident light intensity is an increase in the fill factor, which is indicative of extraction problems at a high incident light intensity. Comparison with devices based on regioregular P3HT shows a slightly lower performance, with PCEs of around 1.8% and 2.1%, respectively, for incident light intensities of 1000 and 393 W/m^2. This lower performance can be rationalized by the size of the SiO_2-P3HT nanoparticles, which was significantly lower than what is currently viewed as the correct scale for the nanomorphology.

The fact that we did not obtain clear-cut benefits in PV performance with the nano-P3HT particles compared to the benchmark P3HT/PCBM-based devices is not terribly disappointing because the P3HT/PCBM-based devices are already largely optimized, with their performance approaching the theoretical limit for this particular blend. We therefore believe that a possible impact of the covalent preorganization would be much more pronounced with systems that exhibit much less self-organization propensity into desired ordered ensembles, such as low-bandgap polymers. On the other hand, the hairy particles used in that work (i.e., nanometer cores) are obviously not optimal in their size. We suggest that they are too small to show properties substantially deviating from the properties of conventional linear P3HT. We therefore assumed that hairy particles with a core size of 20–50 nm would be more attractive for PV applications. To verify this assumption, we prepared a series of hairy particles having end-grafted P3HT chains with systematically varied grafting density and investigated their PV properties. To this end, near-monodisperse polymer nanoparticles (which served as a support for polymer brushes, further designated as NPs) with a 5–60-nm diameter were prepared via the miniemulsion copolymerization of divinylbenzene and bromostyrene (polyDVB-co-SBr). In addition, organosilica NPs with diameters of 5 and 30 nm were also prepared by a SI-KCTP that was developed in the Dresden group (Fig. 3). It was demonstrated that the grafting density can be adjusted by varying the diethylbipyridylnickel ($Ni(bipy)Et_2$) quantity used for the immobilization of Ni.

Applying excessive amounts of $Ni(bipy)Et_2$ leads to the highest grafting density of P3HT chains after the graft-polymerization, whereas lowering its quantity is an efficient tool to reduce it. The grafting density up to 0.3 chains/nm^2 was achieved at the maximal initiator loading. It was found that the optical properties greatly depended on the grafting density. Ultraviolet–visible (UV–Vis) spectroscopy and fluorescence of the P3HT hairy particles reflected strong aggregation occurring even in normally good solvents for P3HT. Thus, P3HT degrafted from the hairy NPs exhibits the usual optical properties for P3HT ($\lambda max = 450$ nm and $\lambda em = 560$ nm). In sharp contrast, the dispersion of the hairy P3HT in chloroform has a violet color and a strongly redshifted and structured absorption and fluorescence spectra (absorption: $\lambda max = 518$, 555, 620 nm; fluorescence: $\lambda em = 573$, 670, 730 nm). These spectral features reflect a significant planarization of the conjugated backbone and/or interchain aggregation, resulting in the efficient delocalization of the π-conjugated electrons.

An effective grafting density can also be varied by the chain length of the grafted P3HT; however, the solubility of the resulting hairy P3HT NPs was found to be an important concern, especially for larger NPs. Thus, at the highest grafting density and a core diameter of 40 nm, only the hairy P3HT NPs having relatively small P3HT grafts [with the degree of polymerization (DP) < 30] were soluble in chlorinated solvents; the NPs with larger grafts were not soluble (dispersible), even at elevated temperatures. We believe that the low solubility of the hairy P3HT NPs was caused by the strong aggregation of P3HT chains occurring at high grafting densities. To overcome the solubility problem, we reduced the diameter of the supporting core to about 5–20 nm. In this case, well-dispersible hairy P3HT

NPs with P3HT grafts up to DP ~ 100 were prepared and further used for PV characterizations.

Photovoltaic properties of the hairy P3HT NPs were tested in Krebs' laboratory. It was found that hairy P3HT NPs prepared from NPs with a diameter of 40 nm and larger were poorly dispersible at the single-particle level. As a result, they formed inhomogeneous layers and had a poor PV device performance. In this respect, smaller nanoparticles are more promising materials. Because of a higher surface curvature of these NPs, tethered P3HT chains are close to each other only near the grafting points but quickly become separated as one moves away from the grafting point. We found that only those hairy NPs with 5–10-nm-diameter organic cores showed a reasonable PV performance. In addition, to ensure processability of the hairy NP, it was important to limit the DP of P3HT grafts up to a maximum of 100. By testing a series of P3HT NPs, we determined an optimal diameter and grafting density. Particularly, the highest PCE of 1.7% ($J_{sc} = 7$ mA/cm^2, $V_{oc} = 0.60$ V, FF = 41%) was obtained for P3HT NPs grown from a 10-nm-diameter core and having a DP of P3HT grafts of 100. These NPs outperformed samples of linear P3HT that exhibited a PCE of 1.5%. The reason for the generally lower performance for all samples studied compared to state-of-the-art values for P3HT is the elongated geometry of the cells, which results in a higher serial resistance. An additional source of the low performance is the low shunt resistance of the devices. It is important to note that although absolute values of PCEs obtained in this work were lower than state-of-the-art values for P3HT, the hairy P3HT NPs with larger P3HT grafts outperformed references of linear P3HT samples. A clear trend that is seen is that the hairy NPs having longer grafts perform better than those with smaller grafts. Thus, this is a good starting point to explore low-bandgap polymers having a hairy nanoparticle architecture.

4 Toward Complex Architectures of D–A π-Conjugated Copolymers

As discussed above, the implementation of the chain-growth polycondensation of complex monomers comprising alternating electron-deficient and electron-rich building blocks is a very important yet challenging task.

4.1 *Poly(fluorene-co-benzothiadiazoles), PF8BT*

poly[(9,9-dialkylfluorenyl-2,7-diyl)-co-(1,4-benzo-(2,10,3)-thiadiazole)] (PF8BT), having narrow polydispersities, controlled molecular weights, and well-defined end groups, which was synthesized in a collaboration of Huck's and Kiriy's groups, is perhaps the first example of a D–A copolymer prepared in the chain-growth manner (Fig. 5) [52].

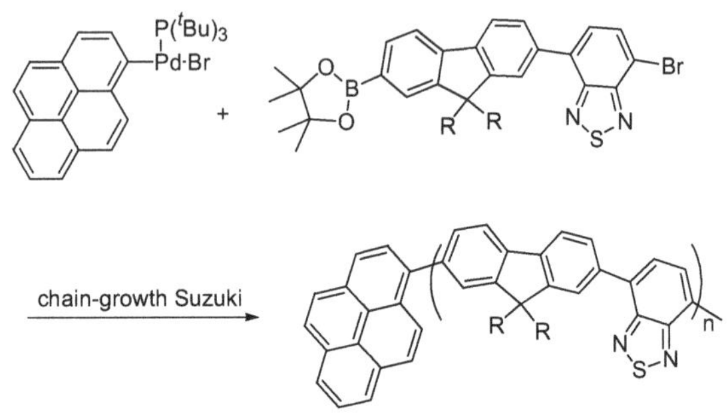

Fig. 5 Chain-growth Suzuki polycondensation [52]

Chain-growth Suzuki polycondensation, which proceeds with AB-type monomers having organoboron and halogen groups in the same monomer molecule, was used in this case. To ensure the (intramolecular) catalyst-transfer mechanism, a "stickiness" of the catalyst to the polymerized π-conjugated chain needed to be provided. Pd complex ligated by a bulky and electron-rich P^tBu_3 ligand was found to be the best catalytic system to date for chain-growth Suzuki polycondensation. Because of the high bulkiness of the ligand, it efficiently supports coordination-unsaturated Pd(0)/P^tBu_3 species formed in the reductive elimination step, which further associates with the polymer backbone and then transfers intramolecularly to the carbon–halogen bond rather than diffuses into the bulk solution. GPC measurements showed that the molecular weight shifted to high values unimodally during polymerization, while maintaining narrow polydispersity. An essential prerequisite for successful polymerization control was found to be the right choice of the supporting ligand for the Pd catalyst (phosphine having bulky, electron-rich substituents), the correct base choice (CsF and crown-ether as phase-transfer catalyst), and a proper polymerization temperature (0 °C). Thus, with this method, it becomes feasible to polymerize both electron-rich and electron-deficient monomers.

4.2 *Chain-Growth Synthesis of P(NDI2OD-T2) and Its Semiconducting Properties*

In 2011, Kiriy's group reported the chain-growth preparation of the NDI-based polymer P(NDI2OD-T2). P(NDI2OD-T2) is a promising polymer for applications in n-type transistors and as the electron-acceptor component in SCs (Fig. 6) [53].

Particularly, we developed a nickel-catalyzed chain-growth polymerization of a highly unusual paramagnetic, anion-radical monomer, Br-TNDIT-Br/Zn. This monomer was prepared by the reaction of activated Zn powder and Br-TNDIT-Br. Interestingly, this reaction did not lead to the usual zinc-organic compound Br-TNDIT-ZnBr in which Zn is inserted between C and Br but rather to a charge-transfer complex. This structural feature follows from the fact that the acidic workup of the product resulted in a quantitative recovering of Br-TNDIT-Br but not

Fig. 6 Chain-growth polycondensation of naphthalene diimide–based anion-radical–type monomer [53]

of Br-TNDIT-H, the protonation product of Br-TNDIT-ZnBr. Despite an unusual monomer structure, the polymerization of Br-TNDIT-Br/Zn proceeded quickly at room temperature and followed the chain-growth mechanism leading to D–A alternating copolymer P(NDI2OD-T2) with a controlled molecular weight and relatively low PDIs of 1.4–1.7. The mechanism of this polymerization was studied in detail through experimental methods and density functional theory calculations. Particularly, we found that polymerization involves Ni(0) intermediates [54], which in the literature are considered key intermediates in catalyst-transfer polycondensations but have never been observed experimentally [39].

It must be noted that the newly developed Ni-catalyzed polymerization provided access to NDI-based polymers with molecular weights limited to approximately 100,000 g/mol. Because high molecular weight P(NDI2OD-T2) could be especially attractive for optoelectronic applications in SCs (as the acceptor component) and in n-type transistors, potentially more active polymerization catalysts were also tested. Very recently, we found that the Negishi chain-growth polycondensation of a Zn-organic AB-type monomer catalyzed by Pd complexes ligated by bulky electron-rich P^tBu_3 occurs extremely fast, in mild conditions, at extremely low catalyst loadings (down to ppm concentrations) and leads to a high molecular weight. The polymerization resulted in poly[2,7-(9,9-bis(2-ethylhexyl)fluorene], PF2/6 [55]. Such low catalyst loadings are attractive for the preparation of electronic-grade materials free from metal impurities and preferable in terms of cost-efficiency. Unprecedentedly high turnover numbers (TON) above 70,000 and turnover frequencies (TOF) up to 280 s^{-1} were observed, which are among the highest values reported in open literature for transition-metal catalyzed cross-coupling polycondensations. In contrast, the corresponding step-growth polycondensation of functionally related AA/BB-type monomers was much less efficient and provided two orders of magnitude lower TONs and TOFs. The much higher polymerization rate and longer catalyst lifetime observed in the chain-growth polymerization are to the result of faster and safer intramolecular (vs. intermolecular) catalyst-transfer processes underlying the chain-growth mechanism.

It was further found that Pd/P^tBu_3 is a remarkably efficient catalyst to polymerize Br-TNDIT-Br/Zn into electronic-grade P(NDI2OD-T2) with a molecular weight in excess of 1000 kg/mol within minutes at room temperature and at rather low catalyst concentrations [56]. The high molecular weight was proven not only by a standard

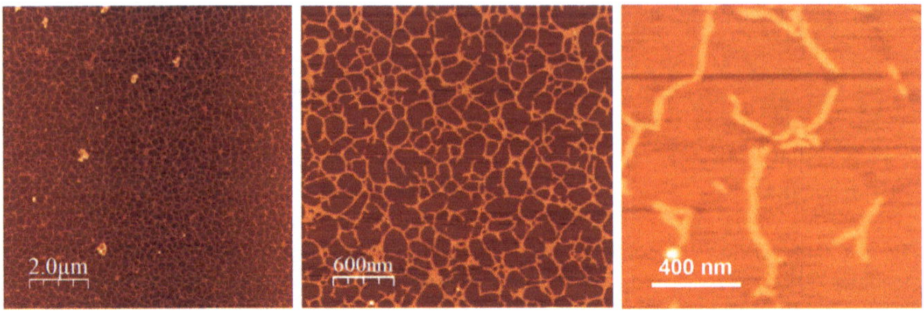

Fig. 7 Spin coating of P(NDI2OD-T2) with relatively high molecular weight ~250 kg/mol on Si wafers from solutions in $CHCl_3$ at 0.01-g/L concentrations form uniform large-area networks of molecular fibers (*left* and *middle* images); the same polymer deposited from a highly diluted solution (0.001 g/L) forms few-molecule aggregates (*right*) [57]

high-temperature GPC method but also by a direct visualization of P(NDI2OD-T2) molecules adsorbed from highly diluted solutions (Fig. 7, right).

In general, obtaining good charge-transport properties of semiconductors are an important prerequisite for the optimal performance of PV devices. P(NDI2OD-T2) was shown to be a promising acceptor component in SCs in combination with various polymeric donors, but the control of morphology P(NDI2OD-T2)-based blends is still a major concern. Chain-growth preparation of P(NDI2OD-T2) paves the way to complex architectures of P(NDI2OD-T2), such as BCPs and hairy particles; therefore, knowledge of the charge-transport properties of P(NDI2OD-T2) prepared by the newly developed protocol is important. To this end, we compared the semiconducting performance of P(NDI2OD-T2) obtained by chain-growth tin-free polymerization and commercially available P(NDI2OD-T2) prepared by Pd-catalyzed step-growth Stille polycondensation (N2200). We found that thus-prepared P(NDI2OD-T2) exhibits field-effect electron mobilities of up to ~0.31 cm^2/Vs, similar to the Stille-derived N2200 control polymer (up to ~0.33 cm^2/Vs) [56].

It is generally accepted that for the fabrication of efficient photovoltaic devices and thin-film transistors, π-conjugated polymers with a high molecular weight are desirable as they frequently show superior charge-transport, morphological, and film-forming properties. As a range of P(NDI2OD-T2) with different molecular weights became available, it was possible to verify this assumption with the example of a high-performance n-type polymer. It was found that high-MW P(NDI2OD-T2) indeed possesses better film-forming properties than low-MW P(NDI2OD-T2), as evident from the simplicity of preparing ultrathin, down to submonolayer P(NDI2OD-T2) films that uniformly cover large-area substrates forming networks of fibers (Fig. 7) [57]. Interestingly, low-molecular-weight P(NDI2OD-T2), when deposited at the same conditions (spin coating from diluted solutions), forms islands of disconnected fibers. To verify how the morphology of P(NDI2OD-T2) affects its electrical properties, different molecular-weight samples ($M_W = 23$, 72, and 250 kg/mol) were used to fabricate field-effect transistors (FETs) with different

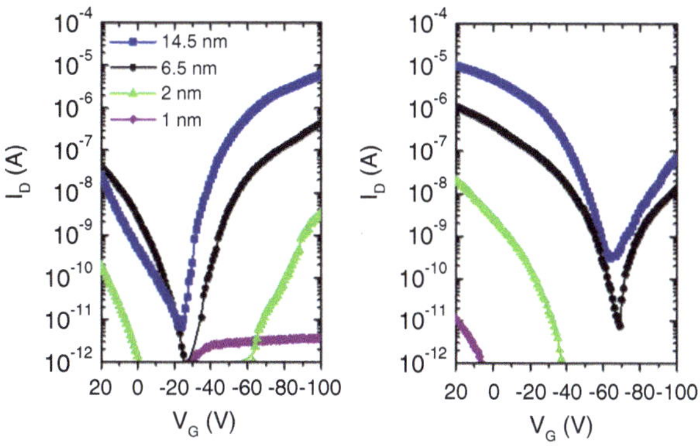

Fig. 8 Representative n-channel and p-channel transfer characteristics measured in ambient conditions for P(NDI2OD-T2), molecular weight ~250 kg/mol, field-effect transistors with the indicated thicknesses [57]

semiconductor layer thicknesses. We found that FETs with semiconductor layer thicknesses from 50 nm to ~15 nm exhibit similar electron mobilities (μ) of 0.2–0.45 cm^2/Vs (representative transfer characteristics are provided in Fig. 7). Reducing the active film thickness led to decreased μ-values; however, FETs with ~5- and ~2-nm-thick P(NDI2OD-T2) films still exhibited substantial μ-values of 0.02–0.01 cm^2/Vs and $\sim 10^{-4}$ cm^2/Vs, respectively (Fig. 8). Interestingly, the lowest molecular-weight sample (P-23, $M_W \approx 23$ kg/mol and PDI = 1.9) exhibited higher μ than the highest molecular-weight one (P-250, $M_W \sim 250$ kg/mol and PDI = 2.3) measured for thicker devices (50–15 nm). This is rather unusual behavior since typically charge carrier mobility increases with the molecular weight, where improved grain-to-grain connectivity normally enhances transport events. We attribute this result to the high crystallinity of the lowest molecular-weight sample, as confirmed by differential scanning calorimetry and X-ray diffraction studies, where this factor may (over)compensate for other effects. In other words, a higher crystallinity inherent to lower-molecular-weight P(NDI2OD-T2) samples is a more important factor for improved charge transport than increased molecular contour length, which, in general, facilitates the conduction of charges through single chains.

In another work (collaboration within SPP 1355), a large-scale alignment of over several centimeters was achieved by high-temperature rubbing of P(NDI2OD-T2) films and was further improved by postrubbing annealing. Oriented and crystalline P(NDI2OD-T2) films were used to probe the anisotropy of electron transport within organic field-effect transistors (OFETs) in top gate geometry. High mobilities up to 0.1 cm^2/Vs were measured along the rubbing direction, that is, along the polymer chains, and appear to be one tenfold lower in the perpendicular direction [58].

4.3 Chain-Growth Preparation and Optoelectronic Properties of PPDIT2

PDI-based alternating main chain copolymers are another intriguing class of electron-conducting materials (in addition to NDI-based polymers) with excellent charge-transport properties. PDI-containing copolymers show a widely tunable optical bandgap and relatively high electron affinities, which makes them well suited to be used as the electron-accepting component in organic SCs [59]. Recently, Hashimoto et al. [60] reported the synthesis of several PDI copolymers. Combined with a polythiophene donor polymer, these polymers reached a PCE above 2%. During the past few years, the efficiency of all-polymer SCs improved significantly by varying the donor and acceptor components. Very recently, Jenekhe et al. [61] reported an impressive PCE of 6.3% for all-polymer SCs utilizing a random copolymer comprising NDI and PDI units. Because NDI- and PDI-based homopolymers show much a worse performance than NDI/PDI -copolymers, the result of Jenekhe's group demonstrates a significant capacity of the "chemical engineering" approach in optimizing the SCs' performance. Typically, the synthesis of rylene-diimide–based alternating copolymers, including copolymers reported by Jenekhe et al., involves conventional step-growth polycondensations that provide only a moderate control over molecular weight, molecular weight distribution, end-group functionality, and polymer architecture. On the other hand, chain-growth polymerizations are potentially more straightforward for the preparation of well-defined, multifunctional polymer architectures as a result of the one-by-one addition of monomers from defined initiators. As discussed in the previous sections, Kiriy's group developed the chain-growth polycondensation of NDI-based polymers, and this approach was recently extended to the polymerization of Zn-activated 2,6-bis(2-bromothien-5-yl)perylene-1,4,5,8-tetracarboxylic-N,N'-bis(2-octyldodecyl) diimide, Br-TPDIT-Br [62].

In order to provide the chain-growth catalyst-transfer mechanism, asymmetric AB-type monomers having metalorganic (nucleophilic) and halide (electrophilic) functions in the same molecule must be involved in polycondensation. A route was found in our previous works to produce AB-type monomers containing electron-deficient naphthalene-diimide–based dihalide, Br-TNDIT-Br, via its activation with Zn [48]. We demonstrated that this approach is also suitable for the activation of PDI-based monomer precursor Br-TPDIT-Br. In particular, we found that Br-TPDIT-Br reacts with activated Zn powder within minutes at room temperature, resulting in a Br-TPDIT-Br/Zn complex soluble in THF [62]. Titration experiments with iodine revealed the one-to-one stoichiometry of the Br-TNDIT-Br/Zn complex, irrespective of whether an equimolar amount or an excess of Zn was added. The acidic workup of the Br-TPDIT-Br/Zn complex resulted in quantitative recovering of Br-TPDIT-Br but not of Br-TPDIT-H. This indicates that the organo-zinc compound Br-TPDIT-ZnBr was not formed under these conditions because otherwise hydrolysis of Br-TPDIT-ZnBr should lead to Br-TPDIT-H. Electron spin-resonance measurements of the Br-TNDIT-Br/Zn complex reveal its paramagnetic

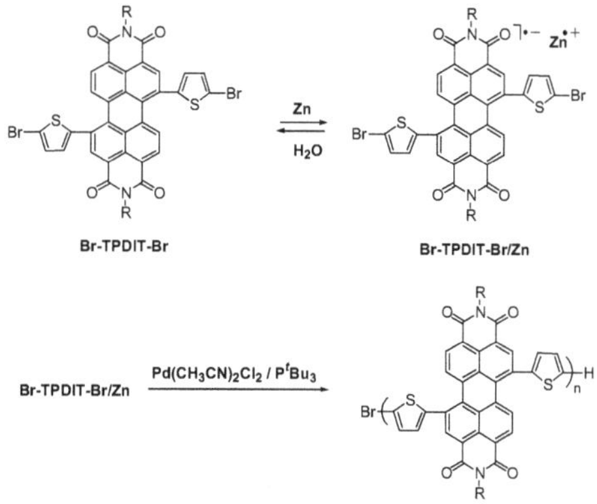

Fig. 9 Synthesis of the Br-TPDIT-Br/Zn radical-anion monomer and its polymerization into PPDIT2 [60]

character. Thus, single electron transfer from Zn to the electron-deficient Br-TPDIT-Br occurs, which leads to Br-TPDIT-Br/Zn having a radical-anion character (Fig. 9).

As such, Br-TPDIT-Br and its NDI-based analogue behave similarly in the presence of Zn. However, despite similarities in the Br-TPDIT-Br/Zn and Br-TNDIT-Br/Zn structures and the fact that Br-TNDIT-Br/Zn smoothly polymerized in the presence of Ni, various Ni catalysts were found to be inactive to polymerize Br-TPDIT-Br/Zn. Therefore, the more active Pd/P^{t}Bu$_3$catalyst was used in this work. Pd/P^{t}Bu$_3$–based catalyst was prepared in situ by mixing 1 equivalent of $Pd(CH_3CN)_2Cl_2$ and 1 equivalent of P^{t}Bu$_3$. Adding Br-TPDIT-Br/Zn to the freshly prepared catalyst at room temperature led to the formation of a dark-blue polymer within several hours. The polymerization course was monitored at [Br-TPDIT-Br/Zn]/[Pd/P^{t}Bu$_3$] ratios of 50/1 and 100/1. As was seen from GPC data of crude reaction mixtures, as the polymerization time increases, the peak corresponding to polymeric products increases while the monomer peak decreases. Peaks corresponding to oligomeric products (between the monomeric and polymeric peaks) are much smaller for all polymerization times. Such behavior is characteristic of the chain-growth propagation mechanism (i.e., one-by-one addition of monomer molecules to growing chains) rather than step-growth propagation for which the monomer should be consumed early in the polymerization. It is noteworthy that the molecular weight of the resulting PPDIT2 increased only modestly with the monomer consumption, presumably as a result of chain-termination reactions that limit the molecular weight at the $M_n \approx 25$-kg/mol level. It is obvious, however, that the polymerization is far from being living (controlled) even though it follows the chain-growth mechanism. Indeed, in living polymerizations, every initiator species polymerizes one chain so that the formation of PPDIT2 with DP = 50 and 100 is expected for the monomer/catalyst ratios of 50/1 and 100/1, respectively, instead of the observed DP ≈ 25 for both catalyst loadings. In addition, the polydispersity index ($PD = M_W/M_n$) is relatively large ($PD = 1.5$–1.8) and it increases as the monomer conversion increases.

Field-Effect Transistor and Photovoltaic Characterizations The performance of PPDIT2 synthesized by the newly developed chain-growth method was further tested and compared with the performance of the same polymer obtained by traditional step-growth Stille coupling polycondensation (designated as $PPDIT2_{control}$). It was previously demonstrated that this polymer exhibits typical n-type semiconductor behavior and provides an electron mobility of $\sim 10^{-3}$ cm^2/Vs when measured in ambient conditions. Instead, PPDIT2 synthesized by the chain-growth method showed a saturated electron mobility of 5×10^{-3} cm^2/Vs. This is a slightly higher mobility value than previously reported for the polymer prepared by step-growth polymerization [59].

The PV performance of PPDIT2 ($M_n \approx 25$ kg/mol, $PD = 1.7$) was further tested. PPDIT2 was used as the electron-accepting component, while regioregular P3HT served as electron donor. Two solvent systems were explored to optimize the devices. The first system is a one-to-one solvent mixture of xylene (Xy) and the high-boiling-point solvent 1-chloronaphthalene (CN), which was recently used to optimize blends of P3HT and P(NDI2OD-T2) [60]. The preparation of the active layer includes a high-temperature drying step of the wet films directly after spin coating. In the second recipe dichlorobenzene (DCB) was used as solvent, which has shown to be an appropriate solvent for blends of P3HT and PPDIT2 [61]. Solar cells were built in the inverted device configuration, with a solution-processed electron-extracting zinc oxide bottom contact and a hole-extracting molybdenum trioxide top contact (for details, see ahead). In accordance with former results, we found that the inverted device structure behaves significantly better than the normal configuration, which has been assigned to the strong vertical composition gradient formed in P3HT:PPDIT2 blends [63].

Figure 10 displays the current density–voltage characteristics of PPDIT2-based SCs under illumination with simulated sunlight. The devices show comparable characteristics when the active layer is dried at high temperature, while the cells prepared from DCB show a slightly higher short-circuit current density (J_{SC}) than those cast from the Xy:CN mixture. We obtain a maximum PCE of 1.3%. On the other hand, slow drying under DCB vapor results in a drastic decrease in the J_{SC} and overall performance, similar to what was observed for P3HT:P(NDI2OD-T2) blends.

For the best-performing cell, the external quantum efficiency (EQE) is also displayed in Fig. 10. The spectra reach a maximum of 32%, 50% higher than the best SCs prepared from the NDI derivative. Also shown is the active layer absorption, which is calculated from the measured optical density of the blend. To account for the reflectance at the top electrode, we assumed that light passes the active layer two times. The good match between absorption and EQE demonstrates that both polymers contribute equally to the photocurrent.

The performance of the PPDIT2 copolymer was further tested against the $PPDIT2_{control}$ copolymer, which was synthesized by a standard step-growth

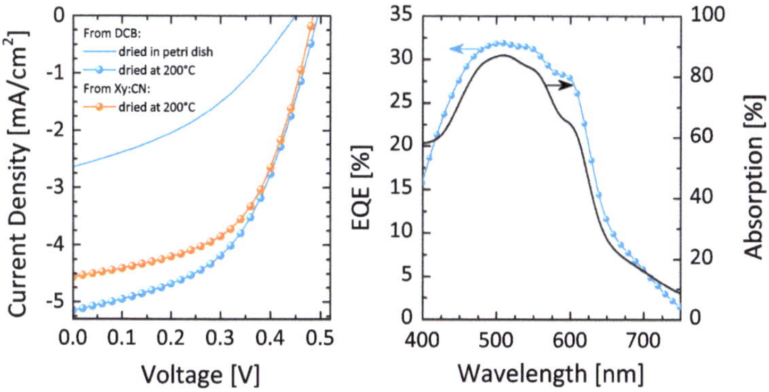

Fig. 10 (*Left*) Current density–voltage characteristics under AM 1.5G illumination at 100 mW/cm^2 of inverted solar cells prepared from a 1.5:1 mixture of P3HT and PPDIT2. Blends prepared from Xy:CN were dried at 200 °C (*orange spheres*), while dichlorobenzene (DCB)-cast blends were either slowly dried in a Petri dish at room temperature (*solid blue line*) or dried quickly at 200 °C (*blue spheres*, *right*). External quantum efficiency (EQE, *blue spheres*) of the 200 °C–dried, DCB-cast solar cell. Also shown is the active layer absorption (*black line*) [62]

Table 1 Solar cell parameters for the devices discussed in the chapter (J_{SC}, short-circuit current density; FF, fill factor; V_{OC}, open-circuit voltage; PCE, power conversion efficiency)

Blend	Solvent	Drying (°C)	J_{SC} (mA/cm^2)	FF (%)	V_{OC}	PCE (%)
P3HT:PPDIT2	DCB	25(Petri dish)	2.64	39	0.45	0.5
P3HT:PPDIT2	DCB	200	5.14	51	0.49	1.3
P3HT:PPDIT2	1:1 Xy:CN	200	4.55	55	0.48	1.2
P3HT:PPDIT2$_{control}$	1:1 Xy:CN	200	4.37	61	0.53	1.4

polycondensation. The SC parameters are given in Table 1. Compared to the best-performing device in Fig. 10, the PPDIT2$_{control}$-based device shows a lower J_{SC} but a higher fill factor and open-circuit voltage, which results in an overall similar PCE of 1.4%. We note that in contrast to the PPDIT2$_{control}$-prepared SCs, PPDIT2-based devices were not fully optimized, as the subject of this chapter was to provide the proof of principle rather than performing a detailed analysis and optimization of the SC performance. Nevertheless, we conclude that the performance of the PPDIT2 copolymer is comparable, whether it was synthesized by a step-growth or by a chain-growth method. This demonstrates that the newly designed method produces high-performance polymers that can readily be applied in functional devices.

4.4 *Chain-Growth Synthesis of NDI-Based BCPs*

As mentioned in previous sections, all-conjugated BCPs and other architectures such as hairy particles comprising electron-rich and electron-deficient blocks may find potential applications in thin-film devices such as ambipolar transistors and all-polymer SCs. It is well known in the case of nonconjugated (and electrically nonconductive) polymers that the chain-growth, better-"living" polymerizations are the most efficient tool for the preparation of such block-like materials. To implement this approach in the case of conjugated polymers, the following general prerequisites must be fulfilled: (1) Polymerization of both electron-rich and electron-deficient monomers must involve a chain-growth mechanism; (2) both monomers must be polymerizable by the same catalyst and under the same conditions; (3) successful block copolymerization requires compatibility of the two polymerization processes (i.e., the presence in the reaction mixture of the first monomer, or its byproducts, must not prohibit polymerization of the second monomer); (4) finally, efficient transfer of the catalyst from the first block to the second one must be achieved.

Recently, we reported the chain-growth Negishi polycondensation of electron-rich AB-type fluorene-based monomer into PF2/6 in the presence of Pd/P^tBu_3 catalyst [55]. In addition, we found that the same catalyst under the same conditions polymerizes electron-deficient NDI- and PDI-based monomers, as discussed in previous sections. These results were a good starting point for the development of the chain-growth block copolymerization process.

For the block copolymerization, the anion-radical naphthalene diimide-dithiophene–based (Br-TNDIT-Br/Zn) and zinc-organic fluorene-based (Br-F2/6-ZnBr) monomers were prepared as reported previously and copolymerized in a single batch in the presence of Pd/P^tBu_3 (Fig. 11). Surprisingly, although the two monomers were simultaneously added to the copolymerization, the polymerization led to a block-like copolymer rather than to a random copolymer or to a mixture of

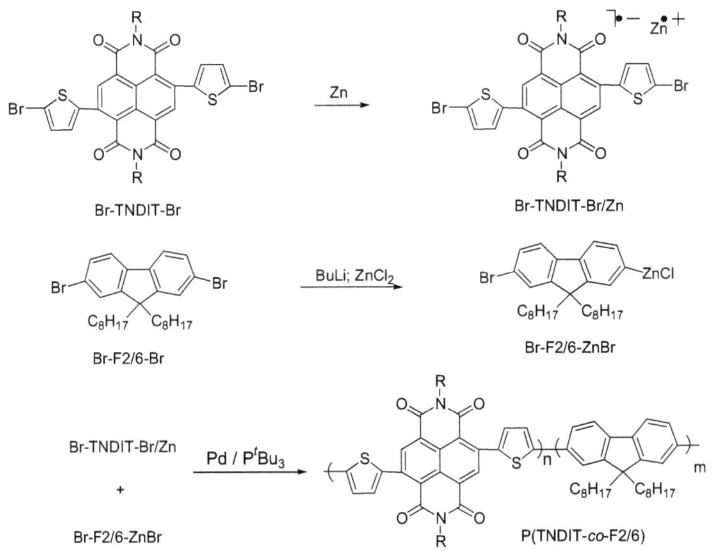

Fig. 11 One-pot synthesis of all-conjugated block-like copolymer P(TNDIT-co-F2/6) [64]

homopolymers, as was evident from NMR, GPC, and AFM data [63]. The block-like copolymer is formed because the fluorenic monomer polymerizes much faster, yet the resulting PF2/6 homopolymer is able to initiate polymerization of the second monomer acting as macroinitiator [64].

Fluorescence-quenching experiments were performed to additionally verify whether the product obtained upon copolymerization of Br-TNDIT-Br/Zn and Br-F2/6-ZnBr is a copolymer or a physical mixture of the corresponding homopolymers P(NDI2OD-T2) and PF2/6. To this end, UV-Vis absorption and fluorescence spectra of the copolymer solution were compared with the spectra of a solution of a physical mixture of PF2/6 and P(NDI2OD-T2) homopolymers. Although solutions of the copolymer and of the physical mixture display approximately the same absorption intensities, the fluorescence intensity of the copolymer is about one order of magnitude lower than one of the mixtures. The fluorescence quenching is due to energy and/or electron transfer from PF2/6 donor to the PTNDIT acceptor. Since the energy- and electron-transfer processes may proceed when the donor and acceptor units are located close to each other, the quenching occurring in solution confirms the covalent connection of the F2/6 and TNDIT blocks. Although the investigated copolymerization does not involve a purely living propagation mechanism and the resulting product is not a well-defined BCP, this result is an important step toward a general protocol for preparing all-conjugated D–A BCPs for optoelectronic applications.

5 Conclusions and Outlook

The most widely utilized "trial-and-error" approach in the fabrication of solution-processed organic SCs involves a codeposition of donor and acceptor components from common solvents followed by the thermal or solvent annealing of resulting blends to achieve one of many kinetically frozen, microphase-separated morphologies, aiming at the best PV performance. As the optimal morphology is situated somewhere between undesirable morphologies—ultimately intermixed and macrophase-separated blends—and because its position is a priori unknown, the search for the optimal morphology is a tedious task. An alternative approach assumes the use of more complex polymer architectures, such as D–A BCPs or hairy particles. As a result of a covalent preorganization of the donor and acceptor components, these materials may, in principle, form desirable morphologies at thermodynamic equilibrium. This chapter reviewed some of the synthetic approaches to complex polymer architectures as well as first results in their utilization in PV devices. It should be noted that to date both semiconducting BCPs and hairy particles have shown a rather modest PV performance. Although the exact reasons for such disappointing results are unknown, we believe that the absence of high-fidelity methods for the synthesis of complex polymer architectures with precisely variable parameters (e.g., the kind and number of blocks, their size, molecular weight, end groups, etc.) is the most probable cause for the modest PV performance, which requires fine-tuning. Therefore, significant efforts in the authors' group were undertaken to develop controlled chain-growth catalyst-transfer polycondensation of D–A building blocks, which was highlighted in this contribution.

Acknowledgments We gratefully acknowledge support from DFG within SPP 1355, "Elementary Processes of Organic Photovoltaics," grant number KI-1094/4.

References

1. Mandoc MM, Veurman W, Sweelssen J, Koetse MM, Blom PWM (2007) Appl Phys Lett 91:073518
2. Kim JY, Lee K, Coates NE, Moses D, Nguyen T-Q, Dante M, Heeger AJ (2007) Science 317:222
3. Shuttle CG, Hamilton R, Nelson J, O'Regan BC, Durrant JR (2010) Adv Funct Mater 20:698
4. Ma W, Yang C, Gong X, Lee K, Heeger AJ (2005) Adv Funct Mater 15:1617
5. Park SH, Roy A, Beaupré S, Cho S, Coates N, Moon JS, Moses D, Leclerc M, Lee K, Heeger AJ (2009) Nat Photonics 3:297
6. Li G, Yao Y, Yang H, Shrotriya V, Yang G, Yang Y (2007) Adv Funct Mater 17:1636
7. Liu J, Chen L, Gao B, Cao X, Han Y, Xie Z, Wang L (2013) J Mater Chem A 1:6216
8. Peet J, Soci C, Coffin RC, Nguyen TQ, Mikhailovsky A, Moses D, Bazan GC (2006) Appl Phys Lett 89:252105
9. Liang Y, Xu Z, Xia J, Tsai S-T, Wu Y, Li G, Ray C, Yu L (2010) Adv Mater 22, E135
10. Peet J, Kim JY, Coates NE, Ma WL, Moses D, Heeger AJ, Bazan GC (2007) Nat Mater 6:497
11. Yassara A, Miozzoa L, Girondaa R, Horowitz G (2013) Prog Polym Sci 38:791
12. Segalman RA, McCulloch B, Kirmayer S, Urban JJ (2009) Macromolecules 42:9205
13. Sommer M, Lindner SM, Thelakkat M (2007) Adv Funct Mater 17:1493
14. Linder SM, Hüttner S, Chiche A, Thelakkat M, Krausch G (2006) Angew Chem Int Ed 45:3364
15. Hüttner S, Sommer M, Chiche A, Krausch G, Steiner U, Thelakkat M (2009) Soft Matter 5:4206
16. Sommer M, Lang AS, Thelakkat M (2008) Angew Chem Int Ed 47:7901
17. Lohwasser RH, Gupta G, Kohn P, Sommer M, Lang AS, Thurn-Albrecht T, Thelakkat M (2013) Macromolecules 46:4403
18. Ball ZT, Sivula K, Fréchet JMJ (2006) Macromolecules 39:70
19. Zachary KS, Ball T, Watanabe N, Fréchet JMJ (2006) Adv Mater 18:206
20. Lee JU, Cirpan A, Emrick T, Russell TP, Jo WH (2009) J Mater Chem 19:1483
21. Dong X-H, Zhang W-B, Li Y, Huang M, Zhang S, Quirk RP, Cheng SZD (2012) Polym Chem 3:124
22. Heuken M, Komber H, Voit B (2012) Macromol Chem Phys 213:97
23. Heuken M, Komber H, Erdmann T, Senkovskyy V, Kiriy A, Voit B (2012) Macromolecules 45:4101
24. Bu LJ, Guo XY, Yu B, Qu Y, Xie ZY, Yan DH, Geng YH, Wang FS (2009) J Am Chem Soc 131:13242
25. Tu G, Li H, Forster M, Heiderhoff R, Balk LJ, Scherf U (2006) Macromolecules 39:4327
26. Scherf U, Gutacker A, Koenen N (2008) Acc Chem Res 41:1086 and references herein
27. Woody KB, Leever BJ, Durstock MF, Collard DM (2011) Macromolecules 44:4690
28. Sommer M, Komber H, Huettner S, Mulherin R, Kohn P, Greenham NC, Huck WTS (2012) Macromolecules 45:4142
29. Mulherin RC, Jung S, Huettner S, Johnson K, Kohn R, Sommer M, Allard S, Scherf U, Greenham NC (2011) Nano Lett 11:4846

30. Verduzco R, Botiz I, Pickel DL, Kilbey SM II, Hong K, Dimasi E, Darling SB (2011) Macromolecules 44:530
31. Botiz I, Schaller RD, Verduzco R, Darling SB (2011) J Phys Chem C 115:9260
32. Nakabayashiand K, Mori H (2012) Macromolecules 45:9618
33. Lin YH, Smith KA, Kempf CN, Verduzco R (2013) Polym Chem 4:229
34. Guo C, Lin YH, Witman MD, Smith KA, Wang C, Hexemer A, Strzalka J, Gomez ED, Verduzco R (2013) Nano Lett 13:2957
35. Wang J, Lu C, Mizobe T, Ueda M, Chen WC, Higashihara T (2013) Macromolecules 46:1783
36. Wang J, Ueda M, Higashihara T (2013) ACS MacroLett 2:506
37. Iovu MC, Sheina EE, Gil RR, McCullough RD (2005) Macromolecules 38:8649
38. Miyakoshi R, Yokoyama A, Yokozawa T (2005) J Am Chem Soc 127:17542
39. Kiriy A, Senkovskyy V, Sommer M (2011) Macromol Rapid Commun 32:1503
40. Yan H, Chen Z, Zheng Y, Newman C, Quinn JR, Dötz F, Kastler M, Facchetti A (2009) Nature 457:679
41. Erdmann T, Back J, Tkachov R, Ruff A, Voit B, Ludwigs S, Kiriy A (2014) Polym Chem 5:5383
42. Tkachov R, Senkovskyy V, Komber H, Sommer J-U, Kiriy A (2010) J Am Chem Soc 132:7803
43. Liu J, Tanaka T, Sivula K, Alivisatos AP, Fréchet JMJ (2004) J Am Chem Soc 126:6550
44. Senkovskyy V, Khanduyeva N, Komber H, Oertel U, Stamm M, Kuckling D, Kiriy A (2007) J Am Chem Soc 129:6626
45. Khanduyeva N, Senkovskyy V, Beryozkina T, Horecha M, Stamm M, Uhrich C, Riede M, Leo K, Kiriy A (2009) J Am Chem Soc 131:153
46. Beryozkina T, Boyko K, Khanduyeva N, Senkovskyy V, Horecha M, Oertel U, Simon F, Komber H, Stamm M, Kiriy A (2009) Angew Chem Int Ed 48:2695
47. Senkovskyy V, Tkachov R, Beryozkina T, Komber H, Oertel U, Horecha M, Bocharova V, Stamm M, Gevorgyan SA, Krebs FC, Kiriy A (2009) J Am Chem Soc 131:16445
48. Senkovskyy V, Sommer M, Komber H, Tkachov R, Huck W, Kiriy A (2010) Macromolecules 43:10157
49. Senkovskyy V, Senkovska I, Kiriy A (2012) ACS Macro Lett 1:494
50. Komber H, Senkovskyy V, Tkachov R, Johnson K, Kiriy A, Huck WTS, Sommer M (2011) Macromolecules 44:9164
51. Krebs FC, Senkovsky V, Kiriy A (2010) IEEE J Sel Top Quantum Electron 16:1821
52. Elmalem E, Kiriy A, Huck WTS (2011) Macromolecules 44:9057
53. Senkovskyy V, Tkachov R, Komber H, Sommer M, Heuken M, Voit B, Huck WTS, Kataev V, Petr A, Kiriy A (2011) J Am Chem Soc 131:19966
54. Senkovskyy V, Tkachov R, Komber H, John A, Sommer J-U, Kiriy A (2012) Macromolecules 45:7770
55. Tkachov R, Senkovskyy V, Beryozkina T, Boyko K, Bakulev V, Lederer A, Sahre K, Voit B, Kiriy A (2014) Angew Chem Int Ed 53:2402
56. Tkachov R, Karpov Y, Senkovskyy V, Raguzin I, Zessin J, Lederer A, Stamm M, Voit B, Beryozkina T, Bakulev V, Zhao W, Facchetti A, Kiriy A (2014) Macromolecules 47:3845
57. Karpov Y, Zhao W, Raguzin I, Beryozkina T, Bakulev V, Al-Hussein M, Häußler L, Stamm M, Voit B, Facchetti A, Tkachov R, Kiriy A (2015) ACS Appl Mater Interfaces 7:12478
58. Tremel K, Fischer FSU, Kayunkid N, Di Pietro R, Tkachov R, Kiriy A, Neher D, Ludwigs S, Brinkmann M (2014) Adv Energy Mater 4:1301659
59. Chen Z, Zheng Y, Yan H, Facchetti A (2009) J Am Chem Soc 131:8
60. Zhou E, Cong J, Wei Q, Tajima K, Yang C, Hashimoto K (2011) Angew Chem 50:2799
61. Hwang Y-J, Earmme T, Courtright BAE, Eberle FN, Jenekhe SA (2015) J Am Chem Soc 137:4424

62. Liu W, Tkachov R, Komber H, Senkovskyy V, Schubert M, Neher D, Zhao W, Facchetti A, Kiriy A (2014) Polym Chem 5:3404
63. Schubert M, Dolfen D, Frisch J, Roland S, Steyrleuthner R, Stiller B, Chen Z, Scherf U, Koch N, Facchetti A, Neher D (2012) Adv Energy Mater 2:369
64. Tkachov R, Komber H, Rauch S, Lederer A, Oertel U, Häußler L, Voit B, Kiriy A (2014) Macromolecules 47:4994

Electronic Properties of Interfaces with Oligo- and Polythiophenes

Petra Tegeder

Contents

Abstract Using energy- and femtosecond time-resolved two-photon photoemission spectroscopy and second harmonic generation to investigate interfaces with oligo- and polythiophenes provides important parameters such as energetic positions of transport levels, lifetimes of excitonic states, and charge-transfer times across donor–acceptor interfaces. They are essential for designing organic material–based optoelectronic devices.

Keywords Exciton binding energies • Exciton dynamics • Second harmonic generation • Transport levels • Two photon photoemission

1 Introduction

Semiconducting materials based on organic molecules are being explored for a number of fascinating applications in optoelectronic devices [1–4]. For device performance the properties of interfaces between the inorganic substrate and the

P. Tegeder (✉)
Physikalisch-Chemisches Institut, Ruprecht-Karls-Universität Heidelberg, Im Neuenheimer Feld 253, 69120 Heidelberg, Germany
e-mail: tegeder@uni-heidelberg.de

K. Leo (ed.), *Elementary Processes in Organic Photovoltaics*,
Advances in Polymer Science 272, DOI 10.1007/978-3-319-28338-8_16

organic material as well as between the active layers (donor–acceptor) are crucial. Here the orientation of the molecules is of particular importance, especially for the light emission/absorption and charge-transport properties. Intermolecular interactions compete with molecule–substrate interactions, defining both the adsorption geometry and the arrangement of the molecular film and accordingly the electronic structure of the system. The electronic structure at interfaces between organic semiconductors and inorganic substrates is the most fundamental issue to understand the functionalities of organic films [5–7]. Measuring the energetic positions of molecular electronic states (transport levels) with respect to the Fermi level of the (metallic) substrate yields quantitative information on the barriers for electron (hole) injection at the molecule–metal interface. Additionally, excitonic states play an important role; in particular, their binding energy is one of the key parameters as well as the formation, relaxation, and decay dynamics of excitons at interfaces (e.g., donor–acceptor interface) and within the molecular films govern the physics of optoelectronic devices.

Because of the great influence of the molecular layer structure on the electronic properties of organic semiconductors, the adsorption, growth, and electronic structure of thin films of various organic molecules on different noble metal surfaces have been studied extensively [8–16]. Whereas the energy-level alignment of states due to the highest occupied molecular orbital (HOMO) has been widely studied using ultraviolet photoemission spectroscopy (UPS) [2], spectroscopic information on unoccupied molecular states, for example, the lowest unoccupied molecular orbital (LUMO), is rather limited. With UPS and inverse photoemission (IPS) transport levels have been determined for a few organic compounds [17, 18]. It has been shown that two-photon photoemission (2PPE) is a very powerful tool not only to elucidate the energetic positions of unoccupied electronic states [19–29] but also to determine exciton binding energies [30]. Investigations on the dynamics of exciton formation and decay/dissociation within organic semiconductors or at interfaces, for instance, between a donor and acceptor compound have been conducted by time-resolved optical spectroscopy ranging from femtosecond (fs) to millisecond time domains. Also, nonlinear optical (NLO) spectroscopy, namely second harmonic generation (SHG), has been utilized to study the exciton decay [31]. Additionally, fs time-resolved 2PPE allowed the exciton decay to be followed in C_{60} films epitaxially grown on noble metals [30] and thin PTCDA films [32], and to probe charge-transfer (CT) excitons on a surface of organic semiconductors, namely, pentacene and tetracene, respectively [33, 34].

Unsubstituted oligothiophenes, such as α-sexithiophene (6T) and α-octithiophene (8T), have been widely investigated because they serve as model systems for the structurally less controllable polythiophene. 6T has been used in organic field-effect transistors [35, 36] and in bulk heterojunction photovoltaic cells [37]. Several surface science studies focusing on both the adsorption behavior on noble metal surfaces and occupied band structures have been reported (see, e.g., [17, 38–43]). Oligothiophene derivatives, in particular dicyanovinyl (DCV)-substituted oligothiophenes, have been utilized in vacuum-processed small-molecule organic solar cells [44–47]. Using these low-bandgap donor materials,

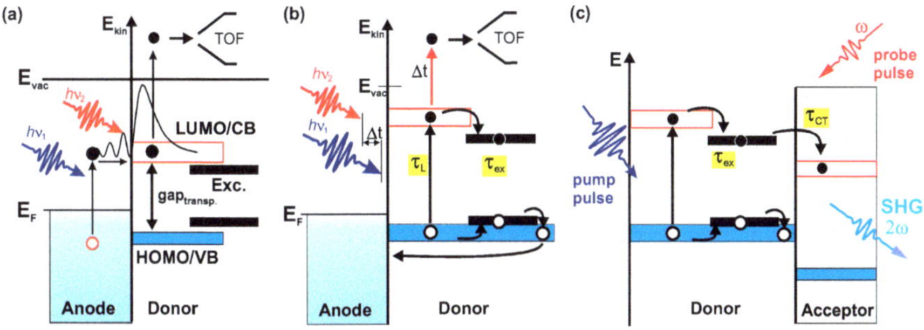

Fig. 1 (**a**) Scheme of the interface between an electrode and donor material, relevant parameters such as work functions, energetic position of occupied and unoccupied electronic and excitonic states (HOMO, LUMO positions; conduction and valence band positions), transport levels and gaps, as well as exciton binding energies can be determined with two-photon photoemission (2PPE) spectroscopy. (**b**) Femtosecond time-resolved 2PPE is utilized to study the charge carrier, exciton formation, and decay dynamics. (**c**) Femtosecond time-resolved second harmonic generation (SHG) is used to investigate the charge-transfer dynamics across the donor–acceptor interface

Meerheim et al. [48] obtained promising efficiencies of up to 8.3 % for a single junction and up to 9.7 % for a triple-junction cell. DCV-substituted oligothiophenes combine an electron-rich donor backbone and two terminal electron-poor acceptor moieties. On the other hand, semiconducting polymers have attracted large interest because of their (opto)electronic properties and accordingly their application in organic field-effect transistors and organic photovoltaic devices. One of the most prominent representatives of this class of material is the poly(3-hexylthiophene) (P3HT; see, e.g., [49] and references therein). Numerous studies dealing with its morphology, electronic properties, and device performance exist; it is well established that the molecular packing and morphology of P3HT films are crucial for the optical and electronic properties and therefore also for device performance.

In this work we used 2PPE spectroscopy (see Fig. 1a) to gain insights into the electronic structure, that is, energetic positions of occupied, unoccupied electronic and excitonic states of several oligothiophenes adsorbed on Au(111). We determined the energetic position of several affinity levels as well as ionization potentials originating from LUMOs and HOMOs, thus providing transport levels and gaps. Additionally, excitonic states have been identified. In the case of sexithiophene adsorbed on Au(111), the whole life cycle from the creation to decay of excitons has been investigated by fs time-resolved 2PPE (see Fig. 1b). Fs time-resolved SHG is used to investigate the CT dynamics across a donor–acceptor interface (see Fig. 1c), namely the P3HT/C_{60} interface. SHG is a well-known second-order NLO technique [50–53]. It is surface selective because, as a second-order process in the dielectric field, it is forbidden under the electric-dipole approximation in a medium with centrosymmetry. At a surface or interface the inversion symmetry is broken. As a result of the unique symmetry selection rule of the second-order NLO process, this technique is intrinsically surface sensitive, in contrast to absorption spectroscopy.

2 Results and Discussion

In the following we first present case studies on the electronic structure of unsubstituted oligothiophenes as well as DCV-substituted oligothiophenes adsorbed on an Au(111) substrate. Then we introduce time-resolved experiments, which allow one to investigate the exciton formation and decay dynamics in optically excited 6T and the CT dynamics across the P3HT/C_{60} interface.

2.1 *Electronic Structure*

Before presenting and discussing the results obtained from 2PPE, we want to point out the possible excitation mechanisms in 2PPE. The transient population of unoccupied molecular electronic states dominantly occurs either via a photoinduced substrate to molecule electron transfer or through intramolecular excitation, for example, HOMO–LUMO transition. In the former case a negative ion resonance is created; thus, we obtain the electron affinity of the respective molecular state (transport state). Occupied electronic states are ionized (as in UPS), creating a positive ion resonance; therefore, the ionization potential is measured. From the resulting energetic positions one can determine transport gaps. The intramolecular excitation is certainly needed for exciton formation.

2.1.1 Unsubstituted Oligothiophenes: Sexithiophene and Octithiophene

An exemplary one-color (pump and probe photon have the same energy) 2PPE spectrum of 2 monolayer (ML) 6T adsorbed on Au(111) recorded at $h\nu = 4.3$ eV shows several peaks (see Fig. 2b). In order to identify whether the peaks originate from occupied initial, unoccupied intermediate, or final states in the 2PPE process, their energetic position has to be measured as a function of photon energy [55]. On the basis of such photon energy–dependent measurements, the peak at 8.18 eV can be assigned to the $n = 1$ image potential state (IPS), which exhibits an energetic position of 3.88 eV with respect to E_F (corresponding to ≈ 0.7 eV below the vacuum level, E_{vac}). The other two 2PPE features are associated with photoemission from occupied electronic states. They can be assigned to the 6T-derived HOMO and HOMO-1 located at −0.8 and -1.5 eV below E_F. For comparison, UPS studies on thick 6T films observed the HOMO level between -1.2 and -1.0 eV [17, 56]. Figure 2c shows a two-color (pump and probe photon have different photon energies) 2PPE spectrum of 0.8 ML 6T recorded at $h\nu_1 = 3.0$ eV and $h\nu_2 = 4.4$ eV. Besides the IPS, further peaks are observed which can be assigned to unoccupied electronic states. The peaks at 6.52 and 5.6 eV in the 2PPE spectrum both originate from states probed by the 4.4- eV photons; thus, they are located at 2.12 and 1.2 eV above E_F, respectively. We assign the peak at 2.1 eV to the LUMO and the one at

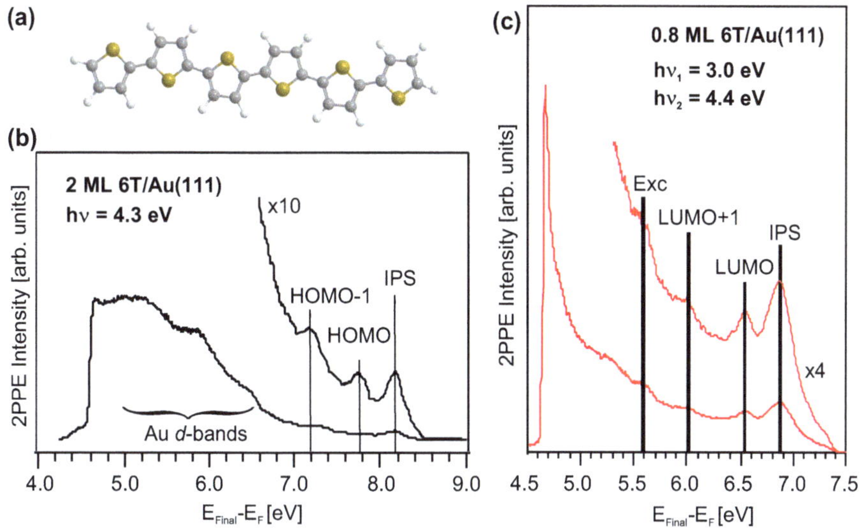

Fig. 2 (**a**) Chemical structure of the α-sexithiophene (6T). (**b**) One-color 2PPE spectrum of 2 monolayer (ML) 6T/Au(111) taken at a photon energy of 4.3 eV. (**c**) Two-color 2PPE spectrum of 0.8 ML 6T/Au(111) recorded with 3.0- and 4.4- eV photons. The spectra are displayed as a function of the final state energy above the Fermi level, $E_{\text{Final}} - E_F = E_{\text{kin}} + \Phi$ (with Φ the work function). [Reproduced from [54]]

1.2 eV to an excitonic (Exc) state (see ahead). On the other hand, the feature seen at 6.01 eV is probed by the 3- eV photons; accordingly it possesses an energetic position of 3.0 eV with respect to E_F and is attributed to the LUMO+1 state [54].

Coverage-dependent measurements [54] show a pronounced photoemission intensity increase from the excitonic state with rising coverage. In contrast, the intensities of the peaks attributed to the LUMO and LUMO+1 drop with increasing coverage. In order to understand this behavior, one has to consider the excitation mechanisms possible in 2PPE (see above). We believe that the LUMO and LUMO+1 level are transiently populated by an optically induced interfacial electron transfer from the gold substrate to the molecules. This interpretation is consistent with the observed intensity decrease of the LUMO and LUMO+1 peaks with increasing 6T coverage. With increasing film thickness, the electronic coupling between the 6T molecules and the metal surface is expected to decrease and accordingly the transition dipole moment for electron-transfer excitation from the substrate to the adsorbate. On the other hand, the proposed excitonic state is visible at any film thickness because of the intramolecular excitation process. In the first layer, quenching may occur to a certain extent by the stronger interaction with the metallic substrate. A similar behavior of 2PPE peak intensities resulting from molecular electronic and exciton states as a function of film thickness has been observed for C_{60} adsorbed on Cu(111) [57] and Au(111) [30]. Note that shifting of energetic positions of the electronic states as a function of coverage (at least up to

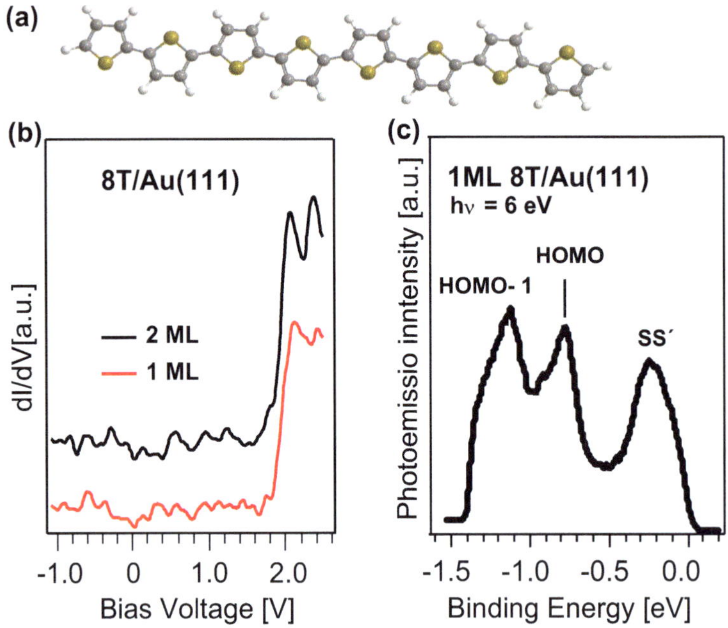

Fig. 3 (**a**) Chemical structure of the α-octithiophene (8T). (**b**) dI/dV spectroscopic data taken at a monolayer and bilayer coverage of 8T on Au(111). (**c**) Photoemission data taken with 6-eV photons at a coverage of 1 ML 8T

3 ML) as known for other organic–electrode interfaces [22, 30, 57] is not found here (see ahead).

To determine the electronic structure of 8T/Au(111), we performed scanning tunneling spectroscopy (STS) and UPS measurements. Figure 3a shows STS data of the 8T/Au(111) recorded at a coverage of 1 and 2 ML, respectively. The dI/dV spectrum of 1 ML taken at the center of 8T molecules shows a double-peak structure at a positive sample bias with maxima at 2.0 and 2.4 V. We attribute these features as arising from the 8T-derived LUMO and LUMO+1, respectively. At 2 ML 8T the energetic position of the unoccupied molecular electronic states with respect to the Fermi level is the same as for 1 ML, but the LUMO and LUMO+1 peaks are nicely resolved. The separation of 0.4 eV between the two peaks nicely fits the energy difference determined on a thin 8T film by using electronic high-resolution electron energy loss spectroscopy [58, 59]. At both 1 and 2 ML the narrow peaks and the constant energetic positions observed in the STS data reveal that the adsorbate–substrate interaction and accordingly the electronic coupling between the 8T molecules and the metal surface are comparatively weak. This can be attributed to the adsorption geometry. The 8T molecules are arranged in a tilted geometry with respect to the surface plane, which allowed for intermolecular $\pi-\pi$ interaction [60]. With such an adsorption geometry, already the adsorbate–substrate interaction in the first layer and, accordingly, the electronic coupling are not strong. Note that the electronic states determined with STS are transport states, since electron tunneling into unoccupied states leads to the creation of negative ion resonances; tunneling out of occupied states leads to the generation of positive ion resonances.

The dI/dV spectrum of 1 ML does not exhibit any clear peak structure up to −1.0 V, a region where we would expect to observe the HOMO of 8T. However, it is known that STS of occupied electronic states of small molecules adsorbed on a metal surface can be affected by the high local density of states (LDOS) of the metal at E_F. When applying a negative bias voltage between the sample and the STM tip, it is normally more likely to fill the empty states of the tip with electrons coming from occupied electronic states at E_F of the metal rather than from molecular orbitals (e.g., the HOMO) because of the large difference in LDOS. Thus, in order to gain information about occupied electronic states of 8T on Au(111), we performed photoemission experiments using 6- eV photons. Figure 3b shows photoemission data of the 8T/Au(111) recorded at a coverage of 1.00 ML. The peak labeled SS$'$ at a binding energy of −0.3 eV is the surface state of the adsorbate-covered Au(111) surface [23]. In addition, two peaks are observed at binding energies of −0.8 and −1.1 eV, which we attribute to photoemission from the HOMO and HOMO-1, respectively.

Figure 4 summarizes the energetic positions and assignments of the 6T- and 8T-derived electronic states observed at a coverage of 1 ML 6T and 8T on Au(111), respectively. The work function at a 1 ML coverage of 6T is 4.55 eV, while for 8T it is 4.6 eV; that is, a decrease of 0.85 eV for 6T and 0.8 eV for 8T compared to the bare Au(111) surface is observed. With the first affinity level at 2.45 eV associated with the LUMO and the ionization potential of 5.35 eV (HOMO), 6T possesses a transport gap of 2.9 eV. The corresponding gap for 8T is 2.7 eV; namely, the gap in 8T is smaller than in 6T, which is in agreement with the prediction [59, 61]. In both systems excitonic states have been identified, which are clearly visible in 2PPE at higher coverages (data not shown). At multilayers (above 6 ML) the spectra

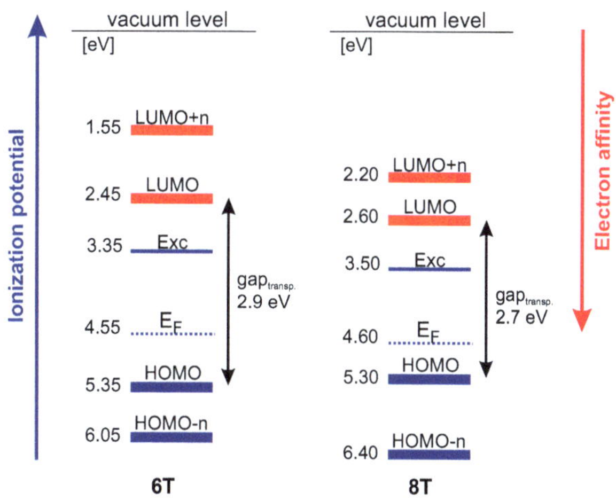

Fig. 4 Energy-level diagram of 1 ML 6T and 8T adsorbed on Au(111), respectively. The *blue levels* are the ionization potentials (*left axis*) and the *red ones* are the electron affinities (*right axis*); E_F denotes the Au(111) Fermi level

are dominated by photoemission from these states and they possess considerable lifetimes (see Chap. 2.2). For 6T and 8T the energetic position for the excitonic state is 0.9 eV below the affinity level corresponding to the LUMO.

2.1.2 DCV-Substituted Oligothiophenes: DCV-Dimethyl-Pentathiophene and DCV-Sexithiophene

Using STS and 2PPE, we obtained a comprehensive picture of the energetic positions of the adsorbate-derived unoccupied and occupied electronic states of the DCV-substituted oligothiophenes, namely DCV-dimethyl-pentathiophene (DCV5T-Me_2) [62] and DCV-sexithiophene (DCV6T) [63] adsorbed on Au(111), respectively. We also studied the coverage-dependent energetic positions of molecular states in DCV5T-Me_2. The energy levels and their assignments are summarized in Fig. 5. Starting with the DCV5T-Me_2: At 1 ML, with the LUMO level (affinity level) at 3.8 eV and the HOMO (ionization potential) at 5.2 eV we obtain a transport gap of 1.4 eV. For comparison, the DCV5T-Me_2 in solution possesses oxidation and reduction potentials of −3.75 and −5.66 eV measured with an electrochemical technique of cyclic voltammetry corresponding to an energy gap of 1.91 eV [47], 0.5 eV larger than the gap measured in our study. This clearly indicates that the presence of the metallic substrate leads to a decrease in the gap size because of

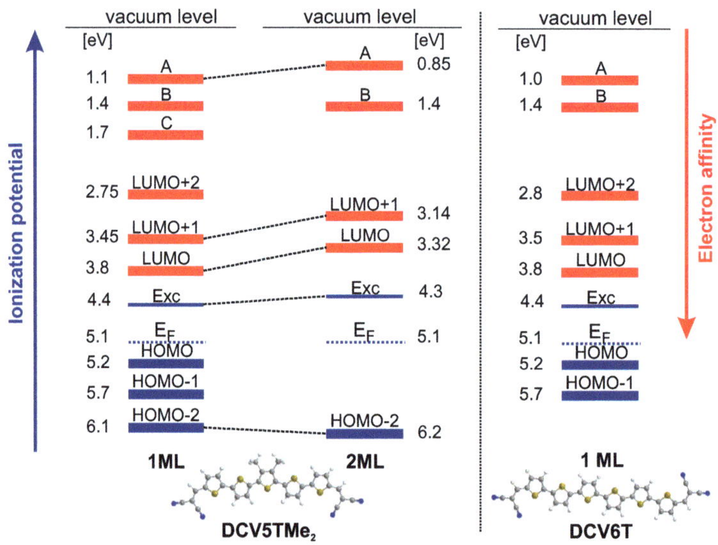

Fig. 5 Energy-level diagram of 1 and 2 ML DCV5T-Me_2 as well as 1 ML DCV6T adsorbed on Au(111). The *blue levels* are the ionization potentials (*left axis*) and the *red ones* are the electron affinities (*right axis*); E_F denotes the Au(111) Fermi level

screening of molecular charge states [64]. The excitonic state is located 0.6 eV below the affinity level associated with the LUMO.

To clarify the origin of the higher-lying unoccupied states labeled A, B, and C, we performed angular-resolved 2PPE measurements (data not shown). Both states A and B show no dispersion (the signal of state C was too weak to resolve it). Additionally, they are not pinned to the vacuum level. Thus, we can exclude that they arise from an IPS, since IPSs possess a free-electron–like dispersion and are pinned to the vacuum level. Other possibilities are (1) interface states (ISs), (2) CT excitons [33, 34] on the surface of the DCV5T-Me_2 with the image band serving as an electron acceptor, or (3) higher-lying molecule-derived states. Let's consider possibility (1): ISs have been observed at organic–metal interfaces, for instance at the interface of an imine-derivative at Au(111) [23]. There two delocalized unoccupied electronic states have been identified, which have been proposed to originate from a modified image potential at the interface. At the perylene–tetracarboxylic acid dianhydride (PTCDA) on Ag(111) and Ag(100) interface [65–67], respectively, an interface state generated from a Shockley resonance has been suggested, where the organic overlayer induced energetic shifts of a Shockley resonance. However, interface states disperse and do not reproduce the absence of angular dispersion found for states A and B. In both cases (2) and (3) one would expect no dispersion of the respective states, as we find here. Photoemission from an exciton (case 2) does not show dispersion since the photoionization process destroys the exciton (quasi-particle) [68]. Dispersion of molecular states (case 3) is highly improbable in flat-lying molecular layers. Hence, possibilities (2) and (3) have to be taken into account.

Going to higher coverages, that is, above 1 ML, we observe pronounced shifts in energy for several molecular states, most dominant for the LUMO and LUMO+1 levels, where a shift of $\approx$ 0.5 eV is found. This effect is attributed to an electronic decoupling from the metallic substrate. In contrast, for the 8T/Au(111) no energetic shift of the LUMO and LUMO+1 in the first and second layers was found [69] (see above). In the present system DCV5T-Me_2 molecules adsorb in a flat configuration; hence, there is a stronger coupling with metal states. The first layer then acts as an isolating layer, resulting in electronically decoupled molecular states in the second layer. The comparison between the two systems DCV5T-Me_2/Au(111) and 8T/Au(111) nicely depicts the effect of the adsorption geometry on the electronic structure of the adsorbate–substrate system (structure–property relationship).

For 1 ML DCV6T, with the LUMO level at 3.8 eV and the HOMO at 5.2 eV, we obtain a transport gap of 1.4 eV. In solution DCV6T has oxidation and reduction potentials of −3.87 and −5.43 eV. Thus, an energy gap of 1.56 eV is found [46], which is only 0.16 eV larger than the gap measured in our study. This may indicate that the interaction between the DCV6T molecules and the metal substrate is weaker compared with the DCV5T-Me_2/Au(111) system (see above). For DCV6T we also observed higher-lying unoccupied states (labeled A and B), which exhibit the same properties as found for the DCV5T-Me_2. For both molecular compounds an excitonic state has been identified; they are located 0.6 eV below the affinity levels associated with the LUMOs. Note that for higher coverages (in the multilayer regime) these states possess considerable lifetimes, as observed with time-resolved 2PPE (data not shown here).

2.2 *Excited States Dynamics*

2.2.1 Exciton Formation and Decay Dynamics in 6T

Some time-resolved measurements ranging from femtoseconds to milliseconds exist to elucidate the decay dynamics of excitons in 6T films with thicknesses about 50–350 nm [70–72]. Here we study exciton formation, relaxation, and decay at 6T/Au(111) interfaces and within thin films using fs time-resolved 2PPE. All measurements have been conducted at a substrate temperature of $\approx$ 90 K. In the low-coverage regime (1–2 ML), the lifetimes for electrons in molecular excited states, namely the LUMO and LUMO+1, possess values below the limit of our experimental resolution ($\tau < 10$ fs). These ultrashort lifetimes point toward a strong electronic coupling between 6T and Au(111), allowing for the efficient back transfer of electrons to the metal substrate. At metallic surfaces such short lifetimes of excited states, that is, in the range of fs, have been found for several adsorbates [23, 26, 30, 73]. In contrast, the exciton exhibits a pronounced lifetime, as can clearly be seen in Fig. 6a for 30 ML 6T/Au(111). In this false-color, two-dimensional representation, the 2PPE intensity at particular energy with respect to the Fermi level, E_F, as a function of time delay (Δt) between the two laser pulses $h\nu_1 = 3.0$ eV and $h\nu_2 = 4.3$ eV is displayed. The cross-correlation (XC) curve for the energy region of the excited state is shown in Fig. 6b. The solid line corresponds to a fit used to determine lifetimes. The fit model contains a $sech^2$ function representing the laser pulse duration convoluted to the response function of the intermediate state. The response function is biexponential, yielding the indicated lifetime values τ_1 and τ_2, which correspond to two different states [74].

Returning to the time-resolved 2PPE spectra shown in Fig. 6a, a fast energetic stabilization of electrons (dashed line) leading to the formation of the Frenkel exciton is clearly seen. In addition, the excitonic state stabilizes on a slower timescale (dotted line) [74]. The first fast process, the stabilization around 0.9 eV, occurs within 100±20 fs (see Fig. 6a). We interpret this observation as follows: Optical excitation with 3.0- eV photons leads to an intramolecular HOMO–LUMO transition, followed by the decay of the polarization and formation of the exciton population (correlated electron-hole pair), which occurs on a femtosecond timescale [75]. The transition of the electron from a bonding (the HOMO) to an antibonding state (the LUMO) is a very efficient source for the stimulation of molecular vibrations; therefore, the exciton binding energy of 0.9 eV dissipates to molecular vibrations on a timescale of the inverse of a vibrational quantum (≤ 100 fs, for instance, the C-S-C ring deformation mode at 700 cm^{-1} [76],

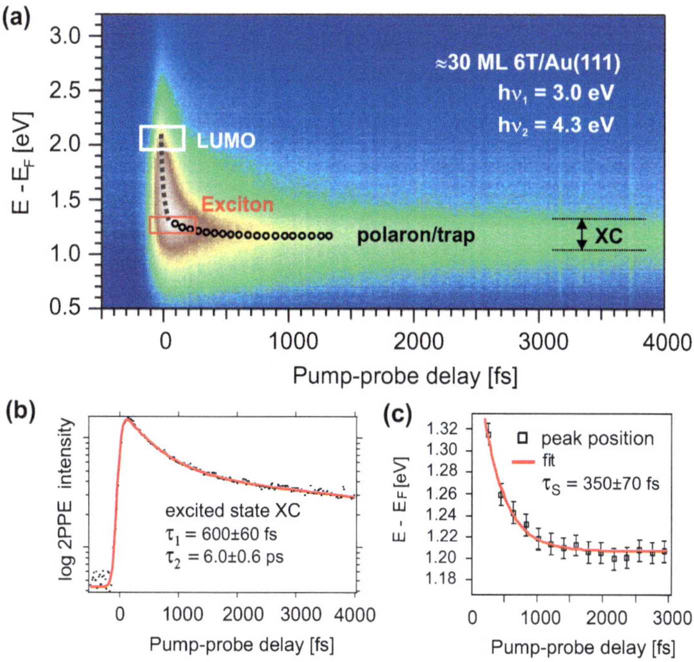

Fig. 6 Two-dimensional spectrum of time-resolved 2PPE measurements recorded with $h\nu_1 = 3.0\,\text{eV}$ and $h\nu_2 = 4.3\,\text{eV}$. The *dashed line* in (**a**) indicates the energetic stabilization of electrons leading to the formation of the exciton, and the *dotted line* represents the energetic stabilization of electrons caused by the formation of polarons or at defect sites trapped electrons (see text). (**b**) Cross-correlation (XC) traces of the 2PPE intensity integrated over the excited-state peak intensities [XC energy range in (**a**)] with a biexponential fit that gives the indicated lifetimes. (**c**) Maximum peak intensity of the exciton state as a function of pump-probe delay, yielding the indicated energy stabilization time because of polaron formation or localization at defect sites. [Adapted from [74]]

which corresponds to a period of 48 fs). On the other hand, coupling of the exciton to molecular vibrations (electron–nuclear interaction) can lead to the generation of a polaron. This process may be responsible for the observed energetic stabilization of around 0.1 eV, which proceeds on a timescale of 350±70 fs (see Fig. 6c). In addition to the decrease in electron energy, this kind of localization resulting from the polarization in the nuclear coordinates should be accompanied by a time-dependent disappearance of parallel dispersion [77, 78]. However, the photoemission of the electron, which is intrinsic to 2PPE, breaks the exciton, and thus we cannot measure the dispersion of this excitonic state. In addition to polaron formation responsible for the energetic stabilization, the electron may be trapped in a defect site [79]. Localization by defects has been shown to occur on a 100- fs timescale in particular systems [80]. Distinguishing between electrons trapped in a defect site and a self-localizing polaron is difficult because of the similar timescales for both formation processes.

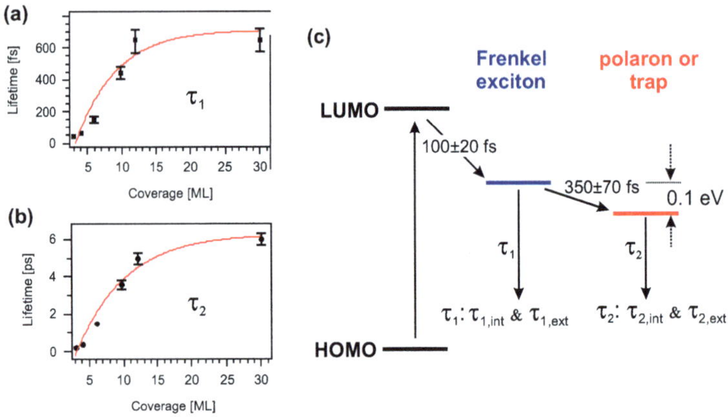

Fig. 7 The dependence on film thickness of the transient lifetime (**a**) τ_1 and (**b**) τ_2. (**c**) Formation and decay processes of a Frenkel exciton in sexithiophene adsorbed on Au(111). τ_{int} indicates the intrinsic (for bulk 6T) and τ_{ext} the external (distance-dependent) lifetime (see text). [Reproduced from [74]]

In order to gain further insights into the exciton decay, we investigated the dynamics as a function of 6T coverage. As shown in Fig. 7a and b, both lifetime values τ_1 and τ_2 exhibit a clear dependence on layer thickness: τ_1 increases from 40±10 fs at 3 ML to 650±65 fs at 30 ML, and τ_2 rises from 200±20 fs (3 ML) to 6±0.6 ps (30 ML). The solid lines are fits to a simple exponential function. The asymptotic values are 650±50 fs for τ_1 and 6.3±0.6 ps for τ_2, respectively. A coverage-dependent decay time indicates that the exciton decays via two relaxation channels, an intrinsic and a distance-dependent channel [30]. The intrinsic decay rate corresponds to the decay in the bulk material and the distance-dependent (external) rate to quenching by the metal substrate. Accordingly for 6T/Au(111) the coverage dependence of both decay times indicates that each of them contains at least two relaxation channels, an intrinsic and an external channel. The intrinsic decays for bulk 6T are related to the asymptotic values of 650±50 fs for τ_1 and 6.3±0.6 ps for τ_2 (Fig. 7a, b) [74].

Figure 7c summarizes the exciton formation and decay processes as determined in our time-resolved photoemission study. We attribute the two lifetime values τ_1 and τ_2 obtained from the biexponential decay (see Fig. 6) to originate from the decay of the singlet Frenkel exciton (τ_1) and the polaron or defect (trap) state (τ_2), respectively. As discussed above, both values contain an intrinsic (τ_{int}, for bulk 6T) and external (τ_{ext}, coverage-dependent decay to metal states) decay. Before addressing the coverage-dependent lifetime, we discuss the intrinsic lifetime of the Frenkel exciton and the polaron/defect state in bulk 6T by comparing them with time-resolved studies known from the literature. Most studies [70–72] determined the dynamical processes after optical excitation of higher-lying electronic states compared to the formation and decay of the lowest (first) Frenkel exciton state investigated here. In addition, almost all of these experimental results revealed dynamic processes on much longer timescales. For the singlet Frenkel exciton, we observe a lifetime of $\tau_{1,\mathrm{int}} = 650 \pm 50\,\mathrm{fs}$. In comparison, a value of $\approx$ 1 ps has been found for the decay in 6T films using picosecond fluorescence spectroscopy [71]. The lifetime of the polaron/defect state is $\tau_{2,\mathrm{int}} = 6.3 \pm 0.6\,\mathrm{ps}$, thus one order of magnitude longer than the first exciton state. The coverage- or distance-dependent lifetime (τ_{ext}) is associated with the back transfer of electrons to the metallic substrate (substrate-mediated quenching). As the 6T coverage increases, the probability density of the excited state (exciton, polaron, or trap) is further away from the metal–molecule interface. As a result, its coupling (wavefunction mixing) to the Au surface decreases and the electron-transfer rate decreases [30].

2.2.2 Charge-Transfer Dynamics Across the P3HT/C_{60} Interface

Organic photovoltaic cells (OPVs) based on semiconducting polymers or molecules offer potential advantages over inorganic modules, such as lightweight, low-cost fabrication and flexibility [81]. One of the most prominent donor–acceptor (D–A) systems is P3HT:PCBM (phenyl-C61-butyric acid methyl ester) [82]. However, the widespread reported efficiencies of devices based on this prototypal blend [81] are all well below the values reached by state-of-the-art inorganic multijunction solar cells [83], indicating a demand for further investigations into the underlying ultrafast photoinduced dynamics at the D–A interface. In particular, the role of excess energy from above-bandgap excitation currently experiences revision from generally being considered to be lost in waste heat [31]. On the one hand, recent findings indicate the crucial role of hot excitons: (1) for populating the CT states [84]; (2) to overcome the poorly screened Coulomb barrier in organic semiconductors and avoid trapping in relaxed CT ground states localized at the interface [85]; and (3) to enhance the probability of charge dissociation out of delocalized CT states [86]. Furthermore, although theoretical modeling is generally difficult because of the large system sizes [85], wavefunction-based *ab initio* calculations recently revealed that the experimentally observed delocalization of excitonic states accompanying the high energetic excitation [84] is crucial for the stabilization of the CT states [87]. On the other hand, for the ultrafast charge separation into band-like acceptor states in blends of small molecules and polymers with fullerenes, respectively, no indication for the need of thermal excess energy has been found [88]. Moreover, no correlation between the excitation energy and the external quantum efficiency of a variety of OPV model systems has been found in recent extensive studies [89, 90].

We investigate the CT across the P3HT/C_{60} interface, which constitutes a central process in the light-to-current conversion in OPVs (see Fig. 8). By applying time-resolved SHG (TR-SHG), we investigate this process with a temporal resolution below 50 fs and excitation energy resolved.

The photoinduced dynamics at the D–A interface of the P3HT/C_{60} model system can be expected to be very similar to those in P3HT:PCBM blends, as outlined in the following. To obtain high-performance OPVs based on the bulk

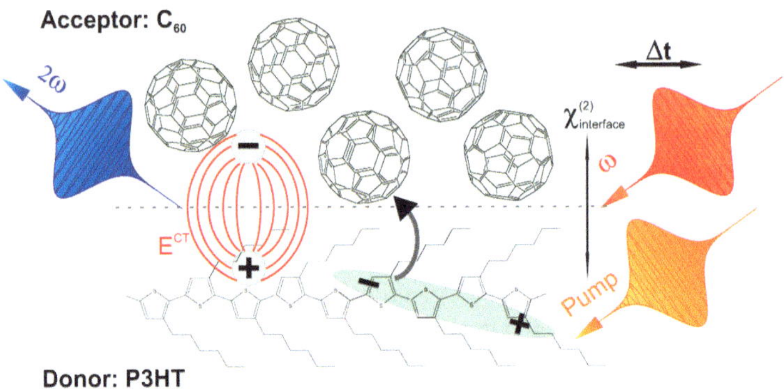

Fig. 8 Schematic illustration of the time-resolved second harmonic generation experiment. A pump pulse generates hot, delocalized excitons in the semiconducting polymer P3HT (*bottom*). It thus induces a distortion in the interfacial electron density to which the second-order nonlinear susceptibility $\chi^{(2)}$ probed in SHG measurements is correlated. The subsequent formation of interfacial electric fields caused by charge transfer to the electron acceptor C_{60} (*top*) results in electric field–enhanced second harmonic generation (EFISH). Varying the pump-probe delay Δt allows us to monitor the temporal evolution of the exciton population. [Adapted from [98]]

heterojunction concept, an annealing step must be seen as crucial to achieve [91]: (1) a phase separation that results in the formation of domains of donor and acceptor molecules [92, 93] with a large interfacial area [94]; (2) the crystallization of the polymer phase and hence improved transport properties [95, 96]; and (3) low-ohmic contacts between the active layer and the electrodes and accordingly a reduced series resistance [91]. Regarding the comparability of the photoinduced dynamics in our model system and in the bulk heterojunctions, the first two points are of crucial importance. First, we chose vacuum evaporation of the not-soluble C_{60} instead of spin coating of PCBM to reduce diffusion of acceptor molecules into the donor layer. Second, it has been demonstrated that neat P3HT of high regioregularity as investigated here exhibits the characteristic vibronic pattern in the absorption profile associated with the π-stacking-mediated self-organized lamella superstructure [97] and accordingly inheres high transport capabilities even without prior annealing [95], in contrast to the blends.

As illustrated in Fig. 8, TR-SHG probes the second-order NLO properties of a medium described by the susceptibility tensor $\chi^{(2)}$ (or its molecular equivalent, the first hyperpolarizability β) [53]. The quadratic dependency of the SHG signal on the strong electric field of the probing femtosecond laser pulse results in the intrinsic interface and surface sensitivity of the method: For homogeneous samples, only the symmetry-breaking interfaces contribute to the SHG signal [51]. As the $\chi^{(2)}$-tensor reflects the electronic structure of the system, the SHG probe is sensitive to the dynamic distortion of the interfacial electron density, induced by the excitation of electron–hole pairs by a pump laser beam. In the D–A heterojunction the energetic relaxation is in competition with the CT across the interface, which is accompanied by the formation of an electric field between the adjacent materials.

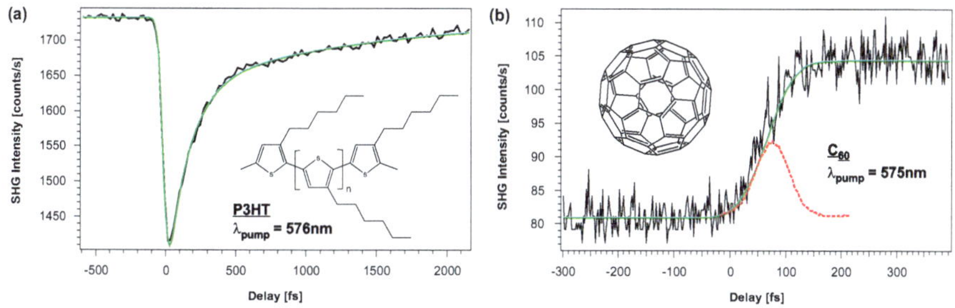

Fig. 9 (**a**) Time-resolved second harmonic generation (TR-SHG) data of pristine P3HT. At time zero, the pump beam induces a pronounced decrease in the nonlinear optical response, which returns to the initial level within a few picoseconds. The fit (*green*) is the iterative convolution of the cross correlation with a biexponential function, yielding $\tau_1 = 154 \pm 6$ fs and $\tau_1 = 1370 \pm 80$ fs. (**b**) TR-SHG data of pristine C_{60}. The initial excitation (cross correlation shown in *red*) results in an increasing TR-SHG signal level, which remains constant for at least 20 ps. [Reproduced from [98]]

The corresponding amplification of the TR-SHG signal is known as electric field–enhanced second harmonic generation (EFISH) [99].

Figure 9a shows the TR-SHG data of pristine P3HT. The pump-induced excitation at time zero leads to a decrease in the SHG signal amplitude, which returns to the initial level within a few picoseconds. The data are fitted (green curve) by iteratively convoluting the Gaussian profile of the pump-probe cross correlation with a biexponential function yielding $\tau_1 = 154 \pm 6$ fs and $\tau_1 = 1370 \pm 80$ fs. The changes in the NLO response are attributed to the excitation and subsequent relaxation and localization of hot singlet excitons [98]. To elucidate why the relaxation within the donor occurs with two time constants, we investigated the sample using spectroscopic SHG and ultraviolet–visible absorption measurements (data not shown; see [98]). We found that optical excitation (579 and 619 nm) into the the S_2 band yielded a biexponential decay, whereas when we used a pump wavelength of 652 nm (excitation into the S_1 band), only the fast component was observed. Accordingly, the fast-relaxation channel is associated with the hot exciton relaxation in the energetically lower S_1 band. The slow-decay channel needs to be associated with the S_2 band. For pristine C_{60} (see Fig. 9b), the TR-SHG signal amplitude rises during the excitation, as indicated by the cross correlation. The SHG signal remains at the high level within the 20- ps time window of our measurement, indicating the population of a long-living excited state. Because this wavelength is insufficient for the excitation of triplet excitonic states and singlet transitions are optically dipole forbidden, the change in the NLO response is attributed to the population of long-living charged polarons [98].

Figure 10a shows the TR-SHG data of the C_{60}/P3HT sample. The red dashed curve is given as a visual guide. It constitutes the sum of the TR-SHG data of both pristine samples (cf. Fig. 9), which should reflect the expected qualitative observation if the presence of the acceptor does not influence the induced dynamics in the donor, and vice versa. While the initial signal decrease at short delays resembles

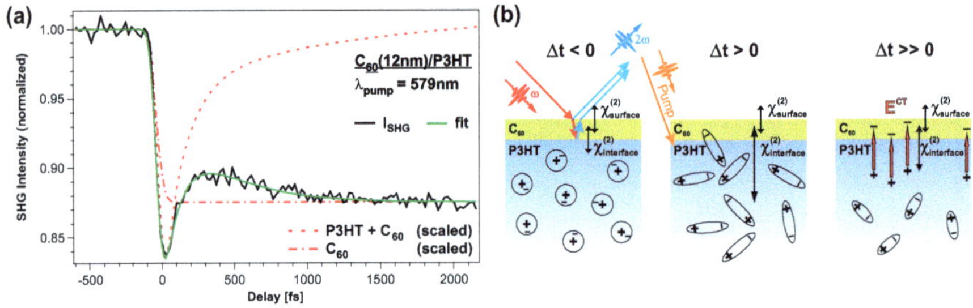

Fig. 10 (**a**) Time-resolved second harmonic generation (TR-SHG) data of C_{60}/P3HT differ significantly from the sum of the pristine sample data (*red dashed line*). The low signal level at large delays originates from the charge-transfer–induced field enhancement. The *red dash-dotted line* is a visual guide to indicate that the charge-transfer state is not directly populated by the pump beam. (**b**) $\Delta t < 0$: SHG probes the steady-state $\chi^{(2)}$-tensor. $\Delta t > 0$: Distortion in the interfacial electron density caused by highly delocalized hot excitons. $\Delta t \gg 0$: Energetic relaxation and localization of hot excitons and simultaneous electric field formation because of the charge transfer. [Reproduced from [98]]

the dynamics observed for pristine P3HT, the long-term temporal evolution of the signal reflects a completely different behavior, as the SHG signal amplitude does not return to the initial level. It remains constant at the low level for the 50- ps time window of our measurement. It is important to note that the signal level at large delays cannot originate from a state that is directly populated by the pump beam as was the case for pristine C_{60}. This is qualitatively illustrated by the red dash-dotted line given as a visual guide, which was obtained by multiplying the normalized C_{60} signal by a factor of -0.5.

As the topmost C_{60} layer is only 12 nm thin and the 579- nm excitation wavelength is within the optical gap of the acceptor [100], the pump beam initially predominantly excites hot, delocalized singlet excitons inside the donating P3HT layer. In the presence of the electron acceptor, an additional interface-derived relaxation pathway besides the fast thermalization channel within P3HT exists, namely, the CT to C_{60}. Hence, the local maximum in the SHG signal of the D–A sample observed at a pump-probe delay of approximately 350 fs denotes the turning point between the two nonequilibrium contributions to the TR-SHG signal, that is, (1) the decreasing distortion of the interfacial electron density by hot delocalized excitons (Fig. 10b, $\Delta t > 0$) and (2) the field enhancement from the CT across the interface (Fig. 10b, $\Delta t \gg 0$). Note that the "field enhancement" results in a decrease in the TR-SHG signal amplitude, as the signal change ΔI_{CT} is proportional to the electric field (vide infra), which in the present case points out of the sample, in contrast to the more commonly chosen opposite geometry [31, 85]. The green curve in Fig. 10a results from the iteratively convoluting fit, which incorporates the thermalization of hot excitons and the delayed population of the CT state out of the initially excited singlet excitons [98]. It yields $\tau_{P3HT} = 146 \pm 8$ fs, identical to the localization within P3HT, and $\tau_{CT} = 320 \pm 20$ fs for the population of the CT state. The latter time constant fits well with other reported subpicosecond

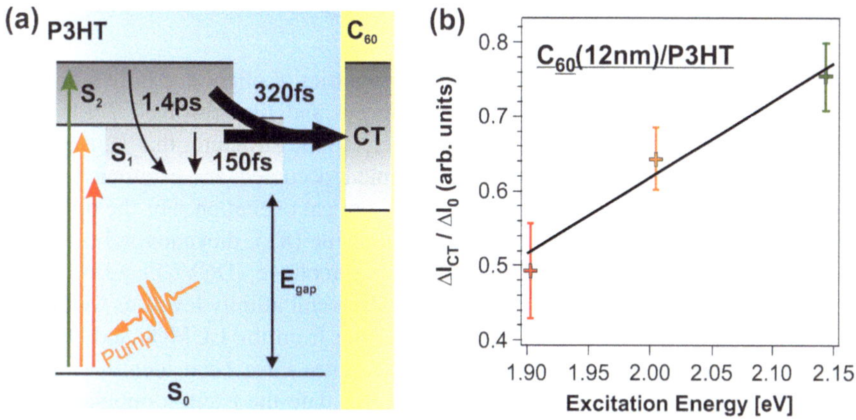

Fig. 11 (**a**) The charge-transfer state becomes populated out of the S_1 band of the polymer. The process is enhanced if the excitation energy is sufficient to involve the S_2 band in addition. (**b**) The charge-transfer yield $\Delta I_{CT}(\lambda)/\Delta I_0(\lambda)$ correlates with the excitation wavelength (ΔI_0 as the second harmonic generation signal amplitude changes at short delays and ΔI_{CT} at long delays). It rises by approximately 50 % when the pump photon energy is increased from 1.90 to 2.14 eV. [Reproduced from [98]]

population times of CT states in systems such as P3HT:PCBM [101], P3HT blends, and bulk heterojunctions with silicon [102], PCPDTBT:PCBM [86], or copper-phthalocyanine bilayers with C_{60} and C_{70} [85].

In a next step, the photon energy of the pump beam is varied to excite the polymer energetically above (576 nm), in overlap with (618 nm), and below (652 nm), the $S_0 \rightarrow S_1$ absorption maximum centered at 610 nm (data not shown here; see [98]). The dynamics induced by the intermediate photon energy are identical ($\tau_{P3HT} = 146 \pm 7$ fs, $\tau_{CT} = 320 \pm 20$ fs) as for the 579- nm excitation. For the 652- nm excitation, the CT is slightly faster ($\tau_{P3HT} = 154 \pm 26$ fs, $\tau_{CT} = 249 \pm 66$ fs). From these results two important conclusions can be derived. First, the CT is also operative upon the 652- nm excitation although the S_2 band of the polymer is not populated. In conjunction with the above finding of the direct, ultrafast CT population out of the S_2 band for higher excitation energies, the corresponding conclusion is depicted in Fig. 11a: The CT state becomes populated out of the S_1 band of the polymer, and the process is enhanced when the excess excitation energy is sufficient to populate the S_2 band in addition. Second, this results in an increased CT yield as a function of the excitation wavelength, as shown in Fig. 11b. The CT yield increases by approximately 50 % when the pump photon energy is altered from 1.90 to 2.14 eV. Not only do these numbers confirm the qualitatively evident competition between energetic relaxation and localization of hot excitons and the CT (cf. Fig. 10): They also show that the balance between these two channels progressively shifts in favor of the CT-state population, with increasing excess energy at the higher energetic excitations.

3 Conclusions and Outlook

In this work, we have studied the electronic properties of oligo- and polythiophenes at interfaces and within the films using energy- and time-resolved 2PPE spectroscopy as well as SHG. This includes the electronic structure, that is, the determination of the energetic positions of unoccupied, occupied, and excitonic states as well as the dynamics after excited states after optical excitation. For the oligothiophene derivatives sexithiophene (6T), octithiophene (8T), dicyanovinyl-dimethyl-pentathiophene (DCV5T-Me_2), and DCV-sexithiophene (DCV6T) adsorbed on Au(111), respectively, the energetic position of several affinity levels as well as the ionization potentials (transport levels) originating from the LUMOs and HOMOs have been determined. In addition, excitonic states have been identified for all molecular systems. For 6T we were able to elucidate the exciton population and decay dynamics using femtosecond time-resolved 2PPE. Thus, demonstrating the 2PPE promises to provide a link between optical and electron spectroscopy. SHG, an intrinic interface-sensitive NLO method, allowed us to gain insights into the CT dynamics at the P3HT/C_{60} interface. The deexcitation of hot singlet excitons in the conduction bands of the polymer into localized excitonic states is observed. In the presence of the electron acceptor, the ultrafast population of a CT state is identified as the dominating relaxation channel. Interestingly, the CT yield correlates with the excitation wavelength and rises with increasing excess energy.

While 2PPE has been used so far to study the electronic properties of organic molecule–metal interfaces, future studies will concentrate on interfaces between the active layers (donor–acceptor), thus allowing us not only to elucidate the energetic positions of occupied and unoccupied molecular states but also to examine the CT dynamics. In combination with the complementary method fs TR-SHG, this will enable us to gain a comprehensive picture of CT processes across D–A interfaces.

Acknowledgments It is a pleasure to thank all coworkers who made this work possible and carried out the experiments, especially Lea Bogner, Michael Schulze, and Erwan Varene. I am grateful for the fruitful collaboration with Peter Bäuerle, Katharina J. Franke, Dieter Neher, Nacho Pascual, Yan Pennec, and their coworkers. Funding by the Deutsche Forschungsgemeinschaft through the priority program SPP 1355 (project: TE 479/1) has been essential for this work.

References

1. Katz HE, Huang J (2009) Thin-film organic electronic devices. Annu Rev Mater Res 39:1–92
2. Koch N (2007) Organic electronic devices and their functional interfaces. ChemPhysChem 8:1438–1455
3. Günes S, Neugebauer H, Sariciftci NS (2007) Conjugated polymer-based organic solar cells. Chem Rev 107:1324–1338
4. Walzer K, Maennig B, Pfeiffer M, Leo K (2007) Organic devices based on electrically doped transport layers. Chem Rev 107:1233–1271

5. Ishii H, Sugiyama K, Ito E, Seki K (1999) Energy level alignment and interfacial electronic structures at organic/metal and organic/organic interfaces. Adv Mater 11:605–625
6. Braun S, Salaneck, WR, Fahlman M (2009) Energy-level alignment at organic/metal and organic/organic interfaces. Adv Mater 21:1450–1472
7. Oehzelt M, Koch N, Heimel G (2014) Organic semiconductor density of states controls the energy level alignment at electrode interfaces. Nat Comm 5:4174
8. Umbach E, Sokolowski M, Fink R (1996) Substrate-interaction, long-range order, and epitaxy of large organic adsorbates. Appl Phys A Mater Sci Process 63:565–576
9. Schreiber F (2000) Structure and growth of self-assembling monolayers. Prog Surf Sci 65:151–257
10. Witte G, Wöll C (2004) Growth of aromatic molecules on solid substrates for applications in organic electronics. J Mater Res 19:1889–1916
11. Love JC, Estroff LA, Kriebel JK, Nuzzo RG, Whitesides GM (2005) Self-assembled monolayers of thiolates on metals as a form of nanotechnology. Chem Rev 105:1103–1169
12. Barth JV (2007) Molecular architectonic on metal surfaces. Annu Rev Phys Chem 58:375–407
13. Heath JR (2009) Molecular electronics. Annu Rev Mater Res 39:1–23
14. Bombis C, Weigelt S, Knudsen MM, Nørgaard M, Busse C, Lægsgaard E, Besenbacher F, Gothelf KV, Linderoth TR (2010) Steering organizational and conformational surface chirality by controlling molecular chemical functionality. ACS Nano 4:297–311
15. Gooding JJ, Ciampi S (2011) The molecular level modification of surfaces: from self-assembled monolayers to complex molecular assemblies. Chem Soc Rev 40:2704–2718
16. Otero R, Gallego JM, Vázquez de Parga AL, Martín N, Miranda R (2011) Molecular self-assembly at solid surfaces. Adv Mat 23:5148–5176
17. Hill IG, Kahn A, Soosb ZG, Pascal RA Jr (2000) Charge-separation energy in films of π-conjugated organic molecules. Chem Phys Lett 327:181–188
18. Krause S, Casu MB, Schöll A, Umbach E (2008) Determination of transport levels of organic semiconductors by UPS and IPS. New J Phys 10:085001–16
19. Zhu X-Y (2004) Electronic structure and electron dynamics at molecule-metal interfaces: implications for molecule-based electronics. Surf Sci Rep 56:1–83
20. Lindstrom CD, Zhu X-Y (2006) Photoinduced electron transfer at molecule-metal interfaces. Chem Rev 106:4281–4300
21. Tegeder P, Hagen S, Leyssner F, Peters MV, Hecht S, Klamroth T, Saalfrank P, Wolf M (2007) Electronic structure of the molecular switch tetra-tert-butyl-azobenzene adsorbed on Ag(111). Appl Phys A 88:465–472
22. Blumenfeld ML, Steele MP, Monti OLA (2010) Near- and far-field effects on molecular energy level alignment at an organic/elecrode interface. J Phys Chem Lett 1:145–148
23. Hagen S, Luo Y, Haag R, Wolf M, Tegeder P (2010) Electronic structure and electron dynamics at an organic molecule/metal interface: interface states of tetra-tert-butyl-imine/Au(111). New J Phys 12:125022 (17pp)
24. Leyssner F, Hagen S, Óvári L, Dokić J, Peters MV, Hecht S, Saalfrank P, Klamroth T, Tegeder P (2010) Photoisomerization ability of molecular switches adsorbed on Au(111): comparison between azobenzene and stilbene derivatives. J Phys Chem C 114:1231–1239
25. Bronner C, Schulze G, Franke KJ, Pascual JI, Tegeder P (2011) Switching ability of nitrospiropyran on Au(111): electronic structure changes as a sensitive probe during a ring-opening reaction. J Phys Condens Matter 23:484005
26. Bronner C, Schulze M, Hagen S, Tegeder P (2012) The influence of the electronic structure of adsorbate-substrate complexes on the photoisomerization ability. New J Phys 14:043032
27. Bronner C, Leyssner F, Stremlau S, Utecht M, Saalfrank P, Klamroth T, Tegeder P (2012) Electronic structure of a sub-nanometer wide bottom-up fabricated graphene nanoribbon: end states, band gap and dispersion. Phys Rev B 86:085444
28. Bronner C, Utecht M, Haase A, Saalfrank P, Klamroth T, Tegeder P (2014) Electronic structure changes during the surface-assisted formation of a graphene nanoribbon. J Chem Phys 140:024701

29. Bronner C, Haase A, Tegeder P (2015) Image potential states at chevron-shaped graphene nanoribbons/Au(111) interfaces. Phys Rev B 91:045428
30. Dutton G, Quinn DP, Lindstrom CD, Zhu X-Y (2005) Exciton dynamics at molecule-metal interfaces: C_{60} /Au(111). Phys Rev B 72:045441
31. Chan W-L, Ligges M, Jailaubekov A, Kaake L, Miaja-Avila L, Zhu X-Y (2011) Observing the multiexciton state in singlet fission and ensuing ultrafast multielectron transfer. Science 334:1541–1545
32. Marks M, Sachs S, Schwalb CH, Schöll A, Höfer U (2013) Electronic structure and excited state dynamics in optically excited PTCDA films investigated with two-photon photoemission. J Chem Phys 139:124701
33. Muntwiler M, Yang Q, Tisdale WA, Zhu X-Y (2008) Coulomb barrier for charge separation at an organic semionductor interface. Phys Rev Lett 101:196403
34. Zhu X-Y, Yang Q, Muntwiler M (2009) Charge-transfer excitons at organic semiconductor surfaces and interfaces. Acc Chem Res 42:1779–1787
35. Katz HE (1997) Organic molecular solids as thin film transistor semiconductors. J Mater Chem 7:369–376
36. Fichou D (2000) Structural order in conjugated oligothiophenes and its implications on opto-electronic devices. J Mater Chem 10:571–588
37. Sakai J, Taima T, Saito K (2008) Efficient oligothiophene: fullerene bulk heterojunction organic photovoltaic cells. Org Electron 9:582–590
38. Kiguchi M, Entani S, Saiki K, Yoshikawa G (2004) One-dimensional ordered structure of α-sexithienyl on Cu(110). Appl Phys Lett 84:3444–3446
39. Koch N, Heimel G, Wu J, Zojer E, Johnson RL, Brédas J-L, Müllen K, Rabe JP (2005) Influence of molecular conformation on organic/metal interface energetics. Chem Phys Lett 413:390–395
40. Kiel M, Duncker K, Hagendorf C, Widdra W (2007) Molecular structure and chiral separation in α-sexithiophene ultrathin films on Au(111): low-energy electron diffraction and scanning tunneling microscopy. Phys Rev B 75:195439
41. Grobosch M, Knupfer M (2007) Charge-injection barriers at realistic metal/organic interfaces: metals become faceless. Adv Mater 19:754
42. Yokoyama T, Kurata S, Tanaka S (2006) Direct identification of conformational isomers of adsorbed oligothiophene on Cu(100). J Phys Chem B 110:18130
43. Kakudate T, Tsukamoto S, Nakaya M, Nakayama T (2006) Initial stage of adsorption of octithiophene molecules on Cu(111). Surf Sci 605:1021–1026
44. Mishra A, Uhrich C, Reinold E, Pfeiffer M, Bäuerle P (2011) Synthesis and characterization of acceptor-substituted oligothiophenes for solar cell applications. Adv Energy Mater 1:265–273
45. Ziehlke H, Fitzner R, Körner C, Gresser R, Reinold E, Bäuerle P, Leo K, Riede M (2011) Side chain variations on a series of dicyanovinyl-terthiophenes: a photoinduced absorption study. J Phys Chem A 115:8437–8446
46. Fitzner R, Reinold E, Mishra A, Mena-Osteritz E, Ziehlke H, Körner C, Leo K, Riede M, Weil M, Tsaryova O, Weiß A, Uhrich C, Pfeiffer M, Bäuerle P (2011) Dicyanovinyl-substituted oligothiophenes: structure-property relationships and application in vacuum-processed small-molecule organic solar cells. Adv Funct Mater 21:897–910
47. Fitzner R, Mena-Osteritz E, Mishra A, Schulz G, Reinold E, Weil M, Körner C, Ziehlke H, Elschner C, Leo K, Riede M, Pfeiffer M, Uhrich C, Bäuerle P (2012) Correlation of π-conjugated oligomer structure with film morphology and organic solar cell performance. J Am Chem Soc 134:11064–11067
48. Meerheim R, Körner C, Leo K (2014) Highly efficient organic multi-junction solar cells with a thiophene based donor material. Appl Phys Lett 105:063306
49. Ludwigs S (ed) (2014) P3HT revisited—from molecular scale to solar cell devices. Advances in polymer science, vol 265. Springer, Berlin, Heidelberg
50. Shen YR (1984) The principles of nonlinear optics. Wiley, New York

51. Shen YR (1989) Surface properties probed by second-harmonic and sum-frequency generation. Nature 337:519–525
52. Heinz TF (1991) Second-order nonlinear optical effects at surfaces and interfaces. Elsevier, Amsterdam
53. McGilp JF (1996) A review of optical second-harmonic and sum- frequency generation at surfaces and interfaces. J Phys D Appl Phys 29:1812–1821
54. Varene E, Martin I, Tegeder P (2011) Optically induced inter- and intrafacial electron transfer probed by two-photon photoemission: electronic states of sexithiophene on Au(111). J Phys Chem Lett 2:252–256
55. Tegeder P (2012) Optically and thermally induced molecular switching processes at metal surfaces. J Phys Condens Matter 24:394001 (34pp)
56. Knupfer M, Liu X (2006) Interface electronic properties of oligothiophenes: the effect of chain length and chemical substituents. Surf Sci 600:3978–2981
57. Dutton G, Zhu X-Y (2002) Unoccupied states in C_{60} thin films probed by two-photon photoemission. J Phys Chem B 106:5975–6981
58. Oeter D, Egelhaaf HJ, Ziegler Ch, Oelkrug D, Göpel W (1994) Electronic transitions in α-oligothiophene thin films. comparison of ultraviolet/visible absorption spectroscopy and high resolution electron energy loss spectroscopy investigations. J Chem Phys 101:6344–6352
59. Egelhaaf HJ, Oelkrug D, Oeter D, Ziegler Ch, Göpel W (1995) HREELS and UV/VIS spectroscopic studies on the electronic structure of oligothiophene thin films. J Mol Struct 384:405–408
60. Varene E, Bogner L, Meyer S, Pennec Y, Tegeder P (2012) Coverage-dependent adsorption geometry of octithiophene on Au(111). Phys Chem Chem Phys 14:691–696
61. Zade SS, Bendikov M (2006) From oligomers to polymer: convergence in the HOMO-LUMO gaps of conjugated oligomers. Org Lett 8:5243–5246
62. Bogner L, Yang Z, Corso M, Fitzner R, Bäuerle P, Franke KJ, Pascual JI, Tegeder P (2015) Electronic structure and excited states dynamics in a dicyanovinyl-substituted oligothiophene on Au(111). Phys Chem Chem Phys 17:27118–27126
63. Bogner L, Yang Z, Corso M, Fitzner R, Bäuerle P, Franke KJ, Pascual JI, Tegeder P (2016, in preparation) Electronic structure and exciton dynamics in optically excited dicyanovinyl-sexithiophene on Au(111)
64. Fernández Torrente I, Franke KJ,Pascual JI (2008) Spectroscopy of C_{60} single molecules: the role of screening on energy level alignment. J Phys Condens Matter 20:184001
65. Schwalb CH, Sachs S, Marks M, Schöll A, Reinert F, Umbach E, Höfer U (2008) Electron lifetime in a Shockley-type metal-organic interface state. Phys Rev Lett 101:146801
66. Galbraith MCE, Marks M, Tonner R, Höfer U (2014) Formation of an organic/metal interface state from a shockley resonance. J Phys Chem Lett 5:50–55
67. Marks M, Schöll A, Höfer U (2014) Formation of metal-organic interface states studied with 2PPE. J Electron Spectrosc Relat Phenom 195:263–271
68. Weinelt M, Kutschera M, Fauster T, Rohlfing M (2004) Dynamics of exciton formation at the Si(100) c(4×2) surface. Phys Rev Lett 92:126801.
69. Varene E, Pennec Y, Tegeder P (2011) Assembly and electronic structure of octithiophene on Au(111). Chem Phys Lett 515:141–145
70. Lanzani G, Frolov SV, Lane PA, Vardeny ZV, Nisoli M, De Sivestri S (1997) Transient spectroscopy of frenkel and charge transfer excitons in α-sexithienyl films. Phys Rev Lett 79:3066–3069
71. Watanabe K, Asahi T, Fukumura H, Masuhara H, Hamano K, Kurata T (1997) Ultrafast decay dynamics of excited and charged states in α-sexithienyl film as revealed by femtosecond transient absorption and picosecond fluorescence spectroscopy. J Phys Chem B 101:1510–1519
72. Urbasch G, Giessen H, Murgia M, Zamboni R, Mahrt RF (2000) Femtosecond differential transmission spectroscopy of α-sexithienyl thin film at low temperature. J Phys Chem B 104:6536–6540

73. Yang A, Shipman ST, Garrett-Roe S, Johns J, Strader M, Szymanski P, Muller E, Harris CB (2008) Two-photon photoemission of ultrathin film ptcda morphologies on Ag(111). J Phys Chem C 112:2506
74. Varene E, Bogner L, Bronner C, Tegeder P (2012) Ultrafast exciton population, relaxation, and decay dynamics in thin oligothiophene films. Phys Rev Lett 109:207601
75. Koch SW, Kira M, Khitrova G, Gibbs HM (2006) Semiconductor excitons in new light. Nat Mat 5:523–531
76. Louarn G, Buisson JP, Lefrant S, Fichou D (1995) Vibrational studies of a series of α-oligothiophenes as model systems of polythiophene. J Phys Chem 99:11399–11404
77. Johns JE, Muller EA, Frechet JMJ, Harris CB (2010) The origin of charge localization observed in organic photovoltaic materials. J Am Chem Soc 132:15720–15725
78. Miller AD, Bezel I, Gaffney KJ, Garrett-Roe S, Liu SH, Szymanski P, Harris CB (2002) Electron solvation in two dimensions. Science 297:1163–1166
79. Kaake LG, Barbara PF, Zhu X-Y (2010) Intrinsic charge trapping in organic and polymeric semiconductors: a physical chemistry perspective. J Phys Chem Lett 1:628–635
80. Bovensiepen U, Stähler J, Gahl C, Bockstedte M, Meyer M, Baletto F, Scandolo S, Zhu X-Y, Rubio A, Wolf M (2009) A dynamic landscape from femtoseconds to minutes for excess electrons at ice-metal interfaces. J Phys Chem C 113:979–988
81. Dang MT, Hirsch L, Wantz G (2011) P3HT:PCBM best seller in polymer photovoltaic research. Adv Mater 23:3597–3602
82. Dennler G, Scharber MC, Brabec CJ (2009) Polymer-fullerene bulk-heterojunction solar cells. Adv Mater 21:1323–1338
83. Green MA, Emery K, Hishikawa Y, Warta W, Dunlop ED (2012) Solar cell efficiency tables (version 39). Prog Photovolt Res Appl 20:12–20
84. Chen K, Barker AJ, Reish ME, Gordon KC, Hodgkiss JM (2013) Broadband ultrafast photoluminescence spectroscopy resolves charge photogeneration via delocalized hot excitons in polymer: fullerene photovoltaic blends. J Am Chem Soc 135:18502–18512
85. Jailaubekov AE, Willard AP, Tritsch JR, Chan W-L, Sai N, Gearba R, Kaake LG, Williams KJ, Leung K, Rossky PJ et al (2013) Hot charge-transfer excitons set the time limit for charge separation at donor/acceptor interfaces in organic photovoltaics. Nat Mater 12:66–73
86. Grancini G, Maiuri M, Fazzi D, Petrozza A, Egelhaaf H-J, Brida D, Gerullo G, Lanzani G (2013) Hot exciton dissociation in polymer solar cells. Nat Mater 12:29–33
87. Borges I Jr, Aquino AJA, Köhn A, Nieman R, Hase WWL, Chen LX, Lischka H (2013) Ab initio modeling of excitonic and charge-transfer states in organic semiconductors: the PTB1/PCBM low band gap system. J Am Chem Soc 135:18252–18255
88. Gélinas S, Rao A, Kumar A, Smith SL, Chin AW, Clark J, van der Poll TS, Bazan GC, Friend RH (2014) Ultrafast long-range charge separation in organic semiconductor photovoltaic diodes. Science 343:512–516
89. Vandewal K, Albrecht S, Hoke ET, Graham KR, Widmer J, Douglas JD, Schubert M, Mateker WR, Bloking JT, Burkhard FG et al (2014) Efficient charge generation by relaxed charge-transfer states at organic interfaces. Nat Mater 13:63–68
90. Albrecht S, Vandewal K, Tumbleston JR, Fischer FSU, Douglas JD, Fréchet JMJ, Ludwigs S, Ade H, Salleo A, Neher D (2014) On the efficiency of charge transfer state splitting in polymer:fullerene solar cells. Nat Mater 26:2533–2539, 2014.
91. Ma W, Yang C, X Gong X, Lee K, Heeger AJ (2005) Thermally stable, efficient polymer solar cells with nanoscale control of the interpenetrating network morphology. Adv Funct Mater 15:1617–1622
92. Campoy-Quiles M, Ferenczi T, Agostinelli T, Etchegoin PG, Kim Y, Anthopoulos TD, Stavrinou PN, Bradley DDC, Nelson J (2008) Morphology evolution via self-organization and lateral and vertical diffusion in polymer:fullerene solar cell blends. Nat Mater 7:158–164
93. Xu Z, Chen L-M, Chen M-H, Li G, Yang Y (2009) Energy level alignment of poly(3-hexylthiophene): [6,6]-phenyl c61 butyric acid methyl ester bulk heterojunction. Appl Phys Lett 95:013301

94. Kim Y, Cook S, Tuladhar SM, Choulis SA, Nelson J, Durrant JR, Bradley DDC, Giles M, McCulloch I, Ha C-S et al (2006) Strong regioregularity effect in self-organizing conjugated polymer films and high-efficiency polythiophene:fullerene solar cells. Nat Mater 5:197–203
95. Mauer R, Kastler M, Laquai F (2010) The impact of polymer regioregularity on charge transport and efficiency of P3HT:PCBM photovoltaic devices. Adv Funct Mater 20:2085–2092
96. Agostinelli T, Lilliu S, Labram JC, Campoy-Quilles M, Hampton M, Pires Rawle EJ, Bikondoa O, Bradley DDC, Anthopoulos D (2011) Real-time investigation of crystallization and phase-segregation dynamics in P3HT:PCBM solar cells during thermal annealing. Adv Funct Mater 21:1701–1708
97. Chirvase D, Parisi J, Hummelen JC, Dyakonov V (2004) Influence of nanomorphology on the photovoltaic action of polymer-fullerene composites. Nanotechnology 15:1317–1323
98. Schulze M, Hänsel M, Tegeder P (2014) Hot excitons increase the donor/acceptor charge transfer yield. J Phys Chem C 118:28527–28534
99. Bouhelier A, Beversluis M, Hartschuh A, Novotny L (2003) Field second-harmonic generation induced by local field enhancement. Phys Rev Lett 90:013903–1–013903–4, 2003.
100. Janner A-M, Eder R, Koopmans R, Jonkman HT, Sawatzky GA (1995) Excitons in C_{60} studied by temperature-dependent optical second-harmonic generation. Phys Rev B 52:17158–17164
101. Guo J, Ohkita H, Benten H, Ito S (2010) Charge generation and recombination dynamics in poly(3-hexylthiophene)/fullerene blend films with different regioregularities and morphologies. J Am Chem Soc 132:6154–6164
102. Herrmann D, Niesar S, Scharsich C, Köhler A, Stutzmann M, Riedle E (2011) Role of structural order and excess energy on ultrafast free charge generation in hybrid polythiophene/Si photovoltaics probed in real time by near-infrared broadband transient absorption. J Am Chem Soc 133:18220–18233

Impact of Charge Carrier Mobility and Electrode Selectivity on the Performance of Organic Solar Cells

Annika Spies, Jeneke Reinhardt, Mathias List, Birger Zimmermann, and Uli Würfel

Contents

Abstract Low charge carrier mobilities as often observed for photoactive materials of organic solar cells have significant impact on their performance. They cause accumulation of charge carriers which can be described quantitatively by a nonohmic transport resistance in the framework of an analytical model. Further addressed in this work is surface recombination stemming from insufficient electrode selectivity which is another factor limiting the performance of organic solar cells.

Keywords Charge carrier mobility • Power conversion efficiency • Surface recombination

A. Spies • J. Reinhardt • U. Würfel (✉)
Freiburg Material Research Center (FMF), Stefan-Meier-Str. 21, 79104 Freiburg, Germany

Fraunhofer Institute for Solar Energy Systems ISE, Heidenhofstr. 2, 79110 Freiburg, Germany
e-mail: uli.wuerfel@fmf.uni-freiburg.de

M. List • B. Zimmermann
Fraunhofer Institute for Solar Energy Systems ISE, Heidenhofstr. 2, 79110 Freiburg, Germany

K. Leo (ed.), *Elementary Processes in Organic Photovoltaics*,
Advances in Polymer Science 272, DOI 10.1007/978-3-319-28338-8_17

1 Recombination and Transport Properties of Organic Solar Cells

Organic solar cells (OSCs) have triggered considerable academic and industrial research because of their beneficial properties. The possibility of a low-cost roll-to-roll production and a high degree of flexibility make them promising candidates to contribute to the future energy supply.

However, the organic materials used in OSCs are inherently more disordered than their inorganic, mostly crystalline counterparts and possess a higher degree of impurities. Thus, the mobility of charge carriers is significantly lower and the open-circuit voltage is severely limited by nonradiative recombination pathways. To understand the underlying elementary processes in more detail, further research has to be conducted. In this chapter we address some of the main limiting factors and describe design guidelines how to improve the performance of OSCs.

The charge carrier mobility μ is the parameter that relates the motion of charges either to an electric field or to diffusion, which is expressed by

$$\begin{aligned} J_{\mathrm{e,h}}(x) &= e\left[n_{\mathrm{e,h}}(x)\mu_{\mathrm{e,h}}E(x) - z_{\mathrm{e,h}}D_{\mathrm{e,h}}\frac{\partial n_{\mathrm{e,h}}(x)}{\partial x}\right] \\ &= e\mu_{\mathrm{e,h}}\left[n_{\mathrm{e,h}}(x)E(x) - z_{\mathrm{e,h}}\frac{k_{\mathrm{B}}T}{e}\frac{\partial n_{\mathrm{e,h}}(x)}{\partial x}\right], \end{aligned} \tag{1}$$

where e is the elementary charge, z the charge number, which is $+1$ for holes and -1 for electrons, x the spatial coordinate, n the charge carrier density, E the electric field, D the diffusion constant, k_{B} Boltzmann's constant, and T the temperature. The mean drift velocity v_{D} of the charges in the respective electric field is given by

$$v_{\mathrm{D}} = \mu E. \tag{2}$$

As mentioned previously, the mobility depends strongly on the order and purity of the material. It can reach values up to 1–10 $\mathrm{cm^2/Vs}$ for organic crystals, but photoactive materials used in OSCs often possess significantly lower values in the range of 10^{-5}–10^{-3} $\mathrm{cm^2/Vs}$ [1–5]. This can be explained by the fact that the materials applied in OSCs are rather amorphous and therefore show a temperature-activated hopping-like transport [6, 7]. An overview of the mobilities of some semiconductors is given in Table 1. The mobilities of photoactive materials of state-of-the-art OSCs are orders of magnitudes lower than those of crystalline silicon [13, 14].

Charge carrier mobilities are often determined using time-of-flight, space charge limited current (SCLC), or organic field-effect transistor (OFET) measurements [11, 15, 16]. However, these techniques need a sample separate from the OSC, the properties of which can differ substantially from those of a photovoltaic device. The time-of-flight technique requires large film thicknesses with a high optical density. This is in contrast to the optimization process for a real photovoltaic device, where

Table 1 Charge carrier mobilities μ at $T = 300$ K for different semiconductors

Material	Mobility (μ, cm^2/Vs)	Method	References
C8-BTBT (single crystal)	43 (holes)	OFET	[8]
Pentacene (single crystal)	40	OFET	[9, 10]
P3HT:PC_{61}BM	3×10^{-3} (electrons)	SCLC	[11]
PffBT4T-2OD:P_{61}BM	$1.5 - 3.0 \times 10^{-2}$ (holes)	SCLC	[12]
PCDTBT:PC_{71}BM	5×10^{-5} (eff. mob.)	CELIV	[13]
Crystalline silicon	1350, 480 (electrons, holes)	Pulsed field method	[14]

a compromise between absorption (thick layers) and transport (thin layers) has to be found. In addition, it was shown that the charge carrier mobility critically depends on the morphology, which itself is affected by the layer thickness [17–20].

In OFET measurements the motion of charges between laterally arranged electrodes is measured, which hardly resembles the operating conditions of sandwiched-like device architectures with the transport direction perpendicular to the electrodes and transport distances orders of magnitudes (typically a factor of at least 100) shorter than in a typical OFET device [16]. An alternative to these techniques is the measurement of SCLCs from electron- and hole-only devices [11]. Here, the photoactive layer is applied between two electron or hole contacts, and therefore the film thickness can equal that in an OSC. Nevertheless, it is not possible to use the same selective materials as in the OSC because the latter requires an electron- and a hole-selective contact, and thus the morphology can still differ.

In this regard, a valuable technique was introduced by Juška et al. [21, 22] in 2000, namely charge extraction by linearly increasing voltage (CELIV). This method offers the advantage that a working solar cell can be used. Here, charge carriers are excited by a short light pulse (photo-CELIV). Subsequently, a linearly increasing voltage in the reverse direction is applied and the current response is measured over time. Note that we stick to the convention that a voltage in blocking or reverse direction of a diode has a negative sign. The displacement current arising from the geometrical capacitance of the solar cell is constant over time as the derivative of the voltage is constant. In addition, free-charge carriers are extracted under the formation of a depletion region. This produces an additional current response peak, with the position of the maximum (and its absolute value) depending strongly on the charge carrier mobility.

The photo-CELIV technique has frequently been used to investigate the field and time dependence of the charge carrier transport in various systems [5, 23–27]. Although the CELIV technique enables the mobility of a complete solar cell to be determined, it can also lead to erroneous results [28, 29].

In general, it is questionable whether a single value for the charge carrier mobility can describe the fundamental properties of dispersive transport that relies on statistical distributions.

Also, the assumption of Juška's approach of a uniform sheet of charge carrier density extending across the device is not valid in the case of trapped charge carriers

[30]. In this case the sweeping out of charges will not leave a region fully depleted of charges. Rather, the charge sheet will become distorted and spread out by charge carriers that are trapped and released [31]. Thus, the mobility becomes a function of time and will depend strongly on the applied voltage ramp, leading to the fact that the mobility can easily vary by two orders of magnitude during the measurement.

Additionally, the CELIV measurement reveals a value for the effective mobility, and we must keep in mind that trapped charge carriers have a mobility of zero and only free-charge carriers are mobile [31]. Moreover, dispersive transport due to energetic and structural disorder may yield incorrect mobility values [26, 32]. However, it was recently observed that for high efficient solar cells made of poly[3,6-dithiophene-2-yl-2,5-di(2-octyldodecyl)-pyrrolo[3,4-*c*]pyrrole-1,4-dione-*alt*-naphthalene] (PDPP-TNT):[6,6]-phenyl-C_{71}-butyric-acid-methyl-ester ($PC_{71}BM$) (PDPP-TNT:$PC_{71}BM$), the mobility is time independent on a timescale relevant for extraction [27]. These authors state that the thermalization processes occur on shorter timescales than the carrier transit time. However, contradictory results were found for prototypical poly[2,3-bis-(3-octyloxyphenyl)quinoxaline-5,8-diyl-*alt*-thiophene-2,5-diyl] ((TQ1):$PC_{71}BM$) blends [33, 34]. Here, the authors concluded that the charge carrier motion and thus the mobility under operational conditions are highly dispersive and orders of magnitudes faster than under near-equilibrium conditions [33]. This shows that depending on the applied materials, there may be substantially different transport properties. In addition, more practical problems can arise from the fact that the number of free-charge carriers can be too low and the signal is not detectable by using a light pulse to generate free-charge carriers. In this case the injection-CELIV (i-CELIV) technique provides a straightforward alternative where charge carriers are injected via forward bias. But as the distribution of charge carriers in the device is rather unknown and further is a strong function of both the applied voltage and the mobility, the authors found that it is not practical to determine the charge carrier mobility of OSCs by i-CELIV [35].

Another prerequisite is that the layer thickness has to be adjusted such that the time constant of the resistor-capacitor circuit is shorter than the carrier transit time; therefore, it is not suitable for all devices [36].

In addition to the development of organic semiconductors with improved absorption, morphology, or beneficial alignment of their energy levels, recent experimental and theoretical work shows that the performance of different polymer:fullerene blends are still largely affected by inefficient charge extraction due to low mobilities and high recombination [37, 38].

In general, the *J–V* characteristics of a solar cell are often described in terms of the Shockley equation:

$$J(V) = J_0 \left[\exp\left(\frac{eV}{n_{\mathrm{id}} k_{\mathrm{B}} T} \right) - 1 \right] - J_{\mathrm{gen}}, \tag{3}$$

where J_0 is the dark generation current density, V is the voltage, n_{id} is the ideality factor, and J_{gen} is the photogenerated current density. The Shockley equation gives the current with respect to voltage as the sum of a photogenerated current density J_{gen} and the recombination current density $J_{\mathrm{rec}}(V) = J_0\left[\exp\left(eV/n_{\mathrm{id}}k_{\mathrm{B}}T\right) - 1\right]$. This recombination current strongly depends on the recombination mechanism present in the cell, which is usually expressed by the ideality factor n_{id}. The recombination rate R is determined by the charge carrier densities of electrons (n_{e}) and holes (n_{h}), its reaction order β, and the corresponding recombination coefficient k_{rec}:

$$R = k_{\mathrm{rec}}(n_{\mathrm{e}}n_{\mathrm{h}})^{\beta/2}. \tag{4}$$

As we are interested in the photogenerated electrons and holes, we will only regard their energetic transport levels. The electron-transport level (ETL) corresponds to the lowest unoccupied molecular orbital (LUMO) of the acceptor and the hole-transport level (HTL) to the highest occupied molecular orbital (HOMO) of the donor. The corresponding quasi-Fermi levels are referred to as $E_{\mathrm{F,ETL}}$ and $E_{\mathrm{F,HTL}}$. This model is called the effective semiconductor (ES) approach, which is often used in the literature [39–41]. In other words, the ES is a hybrid semiconductor that represents the donor–acceptor junction via effective parameters. Thus, for the recombination the current density follows:

$$J_{\mathrm{rec}} = ek_{\mathrm{rec}}\int_0^d (n_{\mathrm{e}}n_{\mathrm{h}})^{\beta/2}\mathrm{d}x = ek_{\mathrm{rec}}n_{\mathrm{i}}^{\beta}\int_0^d \left(\exp\left[\left(E_{\mathrm{F,ETL}} - E_{\mathrm{F,HTL}}\right)/k_{\mathrm{B}}T\right]\right)^{\beta/2} dx, \tag{5}$$

where n_{i} is the intrinsic charge carrier density of the ES and d is the thickness of the photoactive layer. The term $e\,k_{\mathrm{rec}}\,n_{\mathrm{i}}^{\beta}\,d = J_0$ equals the dark current generation, and therefore the splitting of the quasi-Fermi levels can be directly correlated to the number of charge carriers that recombine.

For infinitely large conductivities there are no gradients of the quasi-Fermi levels throughout the photoactive layer, and thus

$$E_{\mathrm{F,ETL}}(x) - E_{\mathrm{F,HTL}}(x) = eV. \tag{6}$$

Under the assumption of infinite conductivities, the application of Eq. (6) in Eq. (5) gives the Shockley equation, Eq. (1). If one recombination mechanism is dominant, the ideality factor n_{id} can be related to the reaction order β of the recombination process:

$$n_{\mathrm{id}} = 2/\beta. \tag{7}$$

Therefore, for direct bimolecular recombination $n_{\mathrm{id}} = 2$, for Shockley–Read–Hall recombination $n_{\mathrm{id}} = 1$, and for Auger recombination $n_{\mathrm{id}} = 2/3$ [42, 43].

The idealization of infinitely large conductivities was reasonable at the time when Shockley and coworkers derived their expression. It was in good agreement with their observations because at that time diodes were solid-state devices made of inorganic materials, such as Ge, Si, or GaAs, with high charge carrier mobilities.

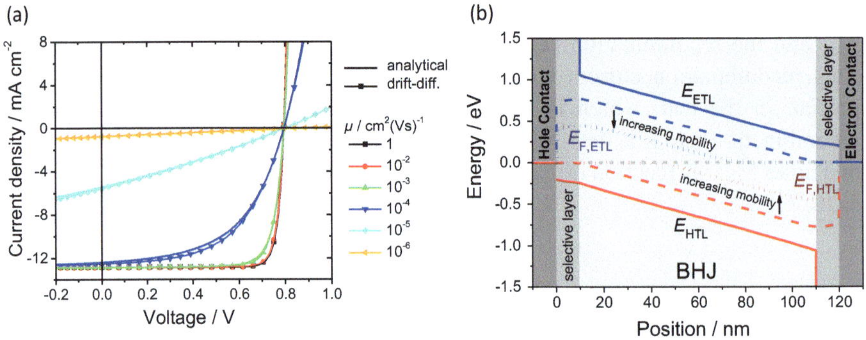

Fig. 1 (**a**) *J–V* curves for different mobilities. The *solid lines* represent the results calculated with the analytical model, and the *lines with symbols* show the results of the full drift-diffusion simulations. (**b**) The energy-level diagram for a mobility of $\mu = 10^{-6}$ cm/Vs (*dashed lines*) and $\mu = 1$ cm^2/Vs (*dotted lines*) under short-circuit conditions is illustrated. The *dashed lines* depict the quasi-Fermi-level splitting, which decreases for increasing mobilities and should be zero for an ideal case of infinitely large mobilities

As described previously, this assumption no longer holds for disordered organic materials with relatively poor transport properties, which leads to an additional voltage drop caused by a transport resistance R_{tr} [44]. The transport resistance is a nonohmic resistance and it depends strongly on the charge carrier density and thus on the working point of the solar cell. The influence on the *J–V* characteristics is shown in Fig. 1a. It becomes clear that a lower mobility first affects the fill factor (FF) and then also the J_{SC}. Additionally, the forward current density is significantly reduced for lower mobilities.

In case of poor transport properties, the transit time of the charge carriers becomes too long to be extracted at the electrodes, and hence they accumulate in the photoactive layer. This is illustrated in Fig. 1b, which depicts energy-level diagrams for $\mu = 10^{-6}\ \mathrm{cm^2/Vs}$ and $\mu = 1\ \mathrm{cm^2/Vs}$ under short-circuit conditions. Here, it becomes obvious that the quasi-Fermi-level splitting is large; as mentioned previously, this correlates with the recombination rate R. Consequently, this violates the main condition of Eq. (6) as the recombination current cannot be described by $J_{rec}(V) = J_0\ [\exp(eV/n_{id}k_BT) - 1]$ as $V_{ext} = 0$ would yield $J_{rec}(V = 0) = 0$, which is obviously a wrong result.

We can conclude that for finite mobilities in the voltage range $V_{ext} \leq V_{OC}$, the quasi-Fermi-level splitting is always larger than the externally applied V_{ext}. These results are consistent with experimental results from Schiefer et al. as well as Albrecht et al. [45–47]. To describe the extent of the additional recombination caused by the transport resistance, first we calculate J_{rec} from $J_{rec}(V) = J(V) + J_{gen}$. Next, we define an internal voltage V_{int} by

$$eV_{int} = k_BT \cdot \ln \frac{J_{rec}}{J_0}, \tag{8}$$

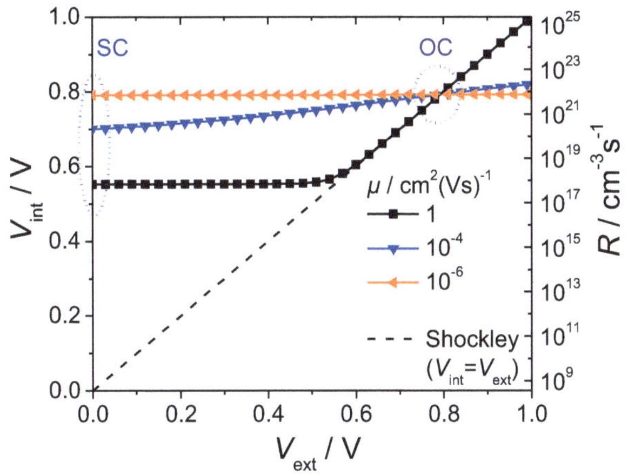

Fig. 2 Correlation among V_{int}, V_{ext}, and the recombination rate R for different mobilities. The diagram shows that V_{int} differs from V_{ext} at every working point of the solar cell except at V_{OC}. The *dashed line* corresponds to $V_{int} = V_{ext}$ and represents the case where the Shockley equation can successfully be applied

where eV_{int} represents the quasi-Fermi-level splitting that yields the same J_{rec} if the gradients of the quasi-Fermi levels are constant throughout the photoactive layer. Figure 2 shows the correlation among V_{int}, the recombination rate R, and V_{ext}. It becomes clear that the recombination rate depends exponentially on V_{int} (as expected) but not on V_{ext}. Furthermore, for external short-circuit conditions, that is, $V_{ext} = 0$ V, the internal voltage is still considerably high even for a high mobility of $\mu = 1$ cm^2/Vs. In contrast, V_{int} equals V_{ext} for all mobilities at open circuit where the net current is zero (transport negligible) and all generated charge carriers recombine.

In principle, it can be distinguished between a constant ohmic resistance of the circuitry, called R_{cir} in the following, and a nonohmic transport resistance R_{tr} of the photoactive layer originating from low mobilities, as previously introduced. If both types of series resistances are present, it becomes difficult to differentiate unambiguously between them. However, it is known that at high forward voltages the transport resistance becomes rather small (due to very high charge carrier densities and thus conductivities), which allows an estimation of the constant R_{cir}.

The determination of the transport resistance gives useful insights into the potential limitations of a solar cell, and methods to determine R_{tr} will be described in detail in the following.

2 Determination of the Transport Resistance in Organic Solar Cells

The Suns–V_{OC} method is a measurement technique to reveal pseudo-J–V curves without the contribution from series resistances and hence only reflects generation and recombination. It was introduced by Sinton and Cuevas [48] in 2000 based on the method of Wolf and Rauschenbach, who measured J_{SC}–V_{OC} values under various light intensities [49].

For inorganic solar cells, the Suns–V_{OC} method is a widely used characterization tool due to its fast and facile determination of the transport resistance. Schiefer et al. [45] demonstrated the applicability of this method on OSCs. In comparison to the steady-state measurement of J_{SC}–V_{OC} values, the Suns–V_{OC} method uses a slowly decaying flashlight while the V_{OC} is measured and the intensity is recorded by a reference cell. Consequently, a pseudo-J–V curve is created that reflects only the generation and recombination processes without the impact of any resistance that is in series (either from the circuitry, R_{cir}, or photoactive layer, R_{tr}) because the measurement is at open circuit with zero net current. An implied current density is calculated assuming a linear relationship between the current density and the light intensity.

The condition of a quasi-steady state requires that the recombination of excess charge carriers is faster than the variation in the flashlight intensity. However, this can be checked separately by comparing a transient Suns–V_{OC} measurement with some J_{SC}–V_{OC} values measured under steady-state conditions at different intensities. If these values coincide with the transient Suns–V_{OC} curve, the assumption of quasi–steady-state conditions is verified and the faster Suns–V_{OC} technique can be applied.

Hence, for continuity, Eq. (9) follows:

$$\frac{\partial \Delta n_{\mathrm{e/h}}(x,t)}{\partial t} = G(x,t) - R(x,t) + \frac{1}{e}\frac{\partial J_{\mathrm{e/h}}(x,t)}{\partial x}, \tag{9}$$

with the assumption of steady-state conditions:

$$\frac{\partial \Delta n_{\mathrm{e/h}}(x,t)}{\partial t} = 0, \tag{10}$$

$$\frac{1}{e}\frac{\partial J_{\mathrm{e/h}}(x,t)}{\partial x} = G(x,t) - R(x,t)\,. \tag{11}$$

Because the voltage is measured under open-circuit conditions, the average generation rate therefore equals the average recombination rate (excluding cases where surface recombination is present):

$$\frac{1}{d}\int_0^d G(x)dx = \overline{G} = \frac{1}{d}\int_0^d R(x)\,dx = \overline{R}. \tag{12}$$

During a Suns–V_{OC} measurement the open-circuit voltage is measured but not the current density. As a consequence, the current density has to be calculated from the light intensity by assuming a linear relationship. The so-called implied current density J_{imp} (quasi–steady-state conditions) is determined from the current density at 1 sun illumination intensity of a standard J–V-curve measurement and the respective light intensity of the flashlight, which was tracked by a reference cell:

$$J_{imp} = J_{gen}\big|_{1\ sun}\left(1 - I_{flash}\right). \tag{13}$$

As a result, we get J–V values for several orders of magnitudes of light intensity and thus also of recombination, which mimic the different working points of a solar cell. Most important is the fact that the resulting pseudo J–V curve does not have any contribution from series resistances—in contrast to a standard J–V measurement.

The difference in voltage at a respective current density is the voltage drop arising from the series resistance:

$$R_S(J) = \frac{V_{JV}(J) - V_{suns}(J)}{J} = R_{cir} + R_{tr}(J), \tag{14}$$

where V_{JV} is the voltage of the standard J–V curve and V_{suns} is that from the pseudo J–V curve. The conductivity of the photoactive layer rises in the forward direction because of the increase in charge carrier density, whereas the resistance arising from the circuitry is not affected. Therefore, for high forward currents the resistance approaches the value of R_{cir} and can be subtracted, yielding the R_{tr} from the photoactive layer as a function of voltage/current.

However, for cases where transient effects are not negligible, the more time-consuming method of recording V_{OC} values by varying the light intensity of a steady-state light source (e.g., with filters) is a possible alternative and from a physical point of view equal to the decaying flashlight. An example is shown in Fig. 3.

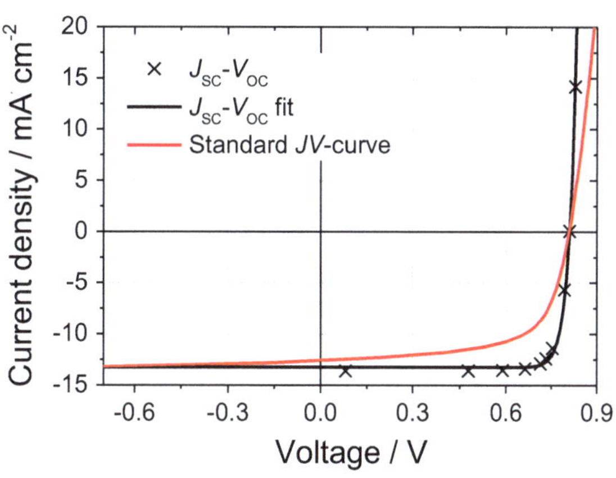

Fig. 3 Comparison of a standard J–V curve measured under an intensity of 1 sun and the respective J_{SC}–V_{OC} measurement. The J_{SC}–V_{OC} data were obtained using a light-emitting diode light source and neutral density filters to ensure steady-state conditions

3 Analytical Model for Efficiency Predictions or Extraction of Specific Parameters

After successfully demonstrating the applicability of the Suns–V_{OC} method on OSCs, Schiefer et al. [46] further determined the intrinsic and the injection-dependent charge carrier density with an analytical approach and the Suns–V_{OC} method. Therefore, they used the assumption of constant quasi–Fermi-level gradients that are related to the current density $J_{e/h}$ and the electrical conductivity $\sigma_{e/h}$ of the charge carriers:

$$\mathrm{grad}\,E_{\mathrm{F,ETL/HTL}} = e\frac{J_{\mathrm{e/h}}(x)}{\sigma_{\mathrm{e/h}}(x)} = \text{const.} = \frac{eV_{\mathrm{tr}}}{d}. \tag{15}$$

Assuming further that the mobilities for holes and electrons are balanced ($\mu_e = \mu_h = \mu$), it follows

$$\frac{J_{\mathrm{e,h}}(x)}{n_{\mathrm{e,h}}(x)} = \text{const.} \tag{16}$$

If the recombination is homogeneous in the device and surface recombination can be excluded, the voltage drop over the transport resistance V_{tr} can be described by the average quantities:

$$V_{\mathrm{tr}} = \frac{d}{e\mu}\left(\frac{\overline{J_{\mathrm{e}}}}{\overline{n_{\mathrm{e}}}} + \frac{\overline{J_{\mathrm{h}}}}{\overline{n_{\mathrm{h}}}}\right), \tag{17}$$

with $\overline{J_e} = \overline{J_h} = J/2$. Hence, with $\overline{n_e} = \overline{n_h} = n_{av}$ it follows that

$$V_{\mathrm{tr}} = \frac{Jd}{2e\mu n_{\mathrm{av}}}. \tag{18}$$

As the Suns–V_{OC} measurement yields R_{tr}, the average density of mobile charge carriers results in

$$n_{\mathrm{av}}(J) = \frac{d}{2e\mu R_{\mathrm{tr}}(J)}. \tag{19}$$

The measured voltage in the Suns–V_{OC} method equals the quasi–Fermi-level splitting because it measures the voltage under open-circuit conditions; thus, $eV_{suns} = eV_{int} = E_{F,ETL} - E_{F,HTL}$. Hence the intrinsic charge carrier density can easily be calculated using the Boltzmann description of excess charge carriers:

$$n_e n_h = n_i^2 \exp\left(eV_{suns}/k_B T\right), \tag{20}$$

$$n_i = \frac{d}{2e\mu R_{tr}(J)\exp\left(eV_{suns}/2k_B T\right)}. \tag{21}$$

The investigation of an exemplary poly[[2,60-4,8-di(5-ethylhexylthienyl)–benzo[1,2-*b*;3,3-*b*]dithiophene][3-fluoro-2[(2-ethylhexyl)carbonyl]–thieno[3,4-b]-thiophenediyl]] (PBDTT-FTTE:PC_{71}BM; commercial name PCE-10) OSC gave an average mobile charge carrier density of $n_{av} = 6 \times 10^{16}\text{cm}^{-3}$ under open-circuit conditions at 1 sun's illumination intensity (see Fig. 3). The intrinsic charge carrier density was determined to be about $n_i = 9 \times 10^9\text{cm}^{-3}$. In order to exclude any influence from transient effects, the pseudo *J*–*V* values were recorded under steady-state conditions by varying the light intensity of a light-emitting diode (LED) using neutral density filters. Note that in Eqs. (19) and (21) the mobility has a huge influence on the resulting charge carrier densities. As for OSCs, it is not possible to determine the mobilities very precisely: There remains a certain degree of uncertainty for n_{av} and n_i.

Additionally, to evaluate the accuracy of this approach, full drift-diffusion simulations were carried out which revealed that also for realistic scenarios, that is, inhomogeneous generation and unbalanced charge carrier mobilities, the method delivers a reasonably good approximation of the charge carrier densities as shown in [45].

Moreover, this approach offers the possibility to predict efficiency potentials while explicitly taking charge carrier transport into account [44]. To do so, the intrinsic charge carrier density and the charge carrier mobility are used as parameters in an analytical model to calculate *J*–*V* curves.

Assuming again constant gradients of the quasi-Fermi levels (15), the relationship of the internal voltage V_{int} and the external voltage V_{ext} are expressed as

$$V_{ext} = V_{int} + JR_{tr} = V_{int} + J\frac{d}{\sigma_e + \sigma_h}. \tag{22}$$

The conductivity is approximated with the effective charge carrier mobility and the intrinsic charge carrier density as parameters:

$$\sigma = \sigma_e + \sigma_h = 2e\mu_{eff} n_i \exp\left(eV_{int}/2k_B T\right). \tag{23}$$

To calculate the total current density, first we calculate the recombination current with respect to the quasi–Fermi-level splitting and, as such, the V_{int}:

$$J_{rec}\left(V_{int}\right) = edk_{rec} n_e n_h = edk_{rec} n_i^2 \exp\left(eV_{int}/k_B T\right) = J_0 \exp\left(eV_{int}/k_B T\right). \tag{24}$$

With that, it follows for the total current:

$$J\left(V_{\text{int}}\right)=J_0\left[\exp\left(eV_{\text{int}}/k_{\text{B}}T\right)-1\right]-J_{\text{gen}}. \tag{25}$$

Equation (25) resembles the well-known Shockley equation, but it relates the total current density to the internal voltage rather than to the externally applied voltage. This is important because only the internal voltage gives information about the actual recombination and with that about the mobility. Applying (23) and (25) into (22), we get an expression that describes a J–V curve analytically, explicitly taking the charge carrier transport into account:

$$V_{\text{ext}}\left(V_{\text{int}}\right)=V_{\text{int}}+J\left(V_{\text{int}}\right)\frac{d}{2e\,\mu_{\text{eff}}\,n_{\text{i}}\exp\left(eV_{\text{int}}/2k_{\text{B}}T\right)}. \tag{26}$$

The excellent agreement between the analytical approach and the full drift-diffusion simulations carried out with the software TCAD Sentaurus from Synopsis is shown in Fig. 1a [50].

The achievable power conversion efficiency is calculated as a function of the bandgap of the donor $E_{\text{gap,D}}$ and is depicted in Figs. 4 and 5b. A variable offset Δ_{LL} was used to simulate the dependency of the overall efficiency on the LUMO level distance of donor and acceptor, shown in Fig. 5a. To calculate the photogenerated current density, we used the bandgap of the donor $E_{\text{gap,D}}$ via $J_{\text{gen, SQ}}\left(E_{\text{gap,D}}\right)=e\int_{E_{\text{gap,D}}}^{\infty}J_{\gamma,\text{ AM1.5G}}\left(\hbar\omega\right)\,\text{d}\hbar\omega$. Accordingly, we assumed that all photons with $\hbar\omega\geq E_{\text{gap,D}}$ generate electron–hole pairs. The dark-generated current density was determined by the black body spectrum at $T=300$ K of the effective semiconductor with the effective bandgap $E_{\text{gap,ES}}$ as the distance between the HOMO of the donor and the LUMO of the acceptor: $J_{0,\text{SQ}}\left(E_{\text{gap,ES}}\right)=\frac{e}{4\pi^2\hbar^3c}\int_{E_{\text{gap,ES}}}^{\infty}\frac{(\hbar\omega)^2}{\exp\left[\hbar\omega/k_{\text{B}}T\right]-1}\text{d}\hbar\omega$.

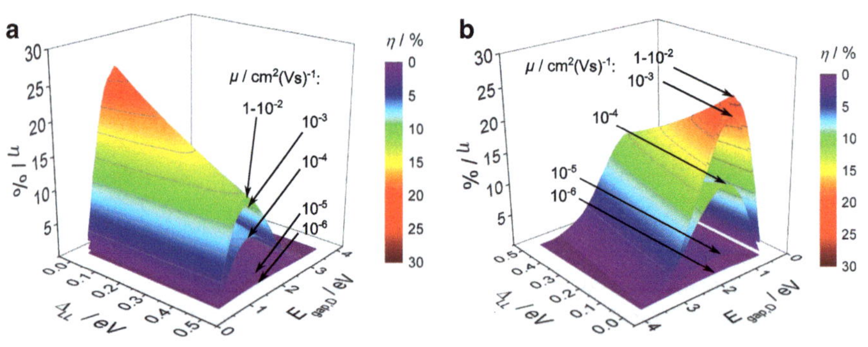

Fig. 4 Achievable power conversion efficiencies as a function of the lowest unoccupied molecular orbital (LUMO) offset, (Δ_{LL}), and the bandgap of the donor $E_{\text{gap,D}}$. For a better visualization, the three-dimensional plot is shown from two different perspectives, (**a**) and (**b**)

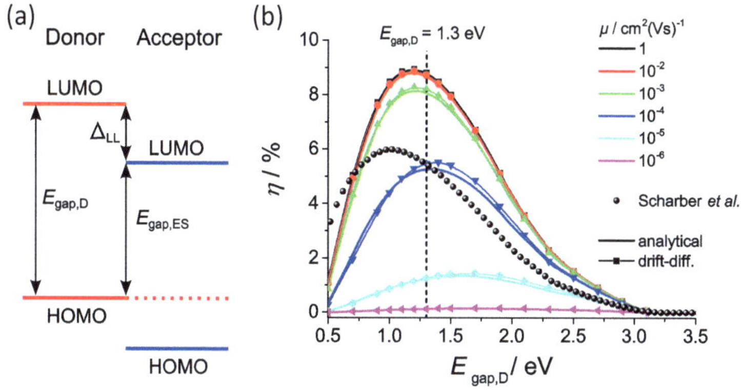

Fig. 5 (**a**) Scheme of energy levels involved in the photocurrent $J_{gen,SQ}$ and dark generation current $J_{0,SQ}$ calculation. (**b**) Achievable power conversion efficiencies with the parameter set used for the *J–V* curves in Fig. 1 for $E_{G,abs} = 1.3$ eV and $\Delta_{LL} = 0.78$ eV

Hence, we make use of the assumption that the absorption via the charge-transfer state hardly contributes to the overall photogeneration process, yet the recombination is dominated by the effective semiconductor gap, which agrees with experimental results [51–53]. Additionally, an amplification factor was used that expresses the fact that in OSCs most charge carriers recombine through additional nonradiative recombination pathways. It was adjusted in accordance with the parameter set of Fig. 1, that is, $\alpha J_{0,\,SQ}\left(E_{gap,ES}\right) \stackrel{!}{=} J_0\left(E_{gap,ES}\right) = ek_{rec}\,n_i^2 d \rightarrow \alpha = 5564$.

Therefore, for cases where transport limitation and the energetic loss for the exciton dissociation are negligible, $\mu = 1\ \text{cm}^2/\text{Vs}$ and $\Delta_{LL} = 0.0$ eV, the efficiency is limited by nonradiative recombination, which otherwise should reach a maximum of above 33 % according to the Shockley–Queisser limit [54].

Other models such as the frequently used model from Scharber et al. [55] do not take charge carrier transport into account but rather use a constant value of the FF for their calculations. As the FF strongly depends on the transport properties, this will give imprecise results, which is illustrated in Fig. 5b for a set of parameters that give the *J–V* curves of Fig. 1 for $E_{gap,D} = 1.3$ eV.

These examples show that the transport resistance gives valuable information to calculate efficiency limitations or to extract intrinsic parameters of the solar cell by fitting *J–V* curves or extracting information from Suns–V_{OC} measurements. On the other hand, analytical descriptions are only valid for a certain parameter range and give misleading results if, for instance, the depletion of charge carriers at interfaces or surface recombination at the electrodes is the dominating performance limitation. This leads to an inhomogeneous distribution of charge carriers and with that to a violation of Eq. (15).

However, the exact determination of the charge carrier distribution within the photoactive layer is hardly possible. Average charge carrier densities are often determined with extraction methods or other techniques, but the actual spatial distribution remains unknown [46, 56]. In this regard, drift-diffusion simulations are a powerful tool to investigate the effect of charge carrier inhomogeneity as it evolves, for example, in the case of surface recombination at the electrodes. This effect is not only of phenomenological interest, yet it is a crucial factor that limits the solar cell performance, especially when it comes to the substitution of expensive electrode metals such as silver or gold with cheaper and less noble metals or when alternatives to the frequently used indium tin oxide (ITO) electrode are used. In the following the effects of surface recombination will be addressed in more detail.

Under open-circuit conditions there is no net current and all generated charge carriers recombine. Nevertheless, it is possible that charge carriers flow internally from the bulk toward the electrodes. This is the case when the electrodes are not ideally selective and the "wrong" charge carrier, for example, holes, can arrive at the electron contact and vice versa, which is defined as surface recombination and is a parasitic loss mechanism. It becomes especially challenging when producing ITO-free OSCs [57].

One possibility to prevent surface recombination is the application of metal electrodes with work function $W_{F,m}$ lower or higher in energy with respect to the ETL or HTL of the photoactive layer. This is indicated in Fig. 6b, where green (highly selective) or red (poorly selective) gradients indicate the selectivity of the contacts depending on the metal work function. The higher the absolute energetic distance between the ETL or HTL and $W_{F,m}$, the better is the selectivity. In case of the electron contact, the position of the $W_{F,m}$ needs to be closer to the vacuum level than to the ETL. In equilibrium conditions (dark, 0 V) there is only one Fermi distribution with no gradient for the ETL and HTL, and hence the net current must be zero. As the metal work functions are energetically different, an electric potential

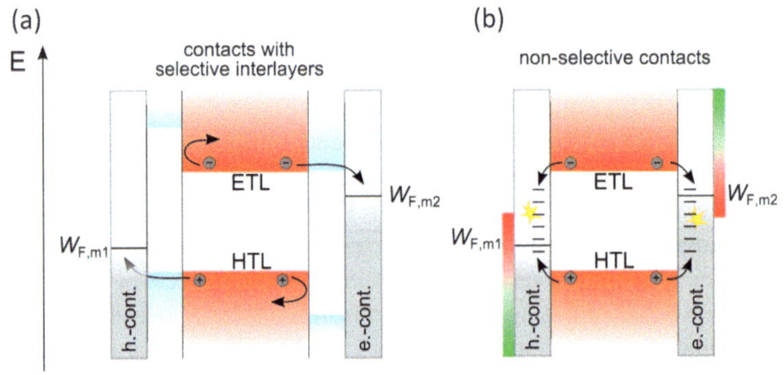

Fig. 6 (**a**) Schematic illustration of a device that possesses selective interlayers and where surface recombination is successfully suppressed. (**b**) Schematic illustration to show the effects of surface recombination. The metal electrodes depict a continuum of accessible states where nonradiative recombination is facilitated. The gradients from *green* to *red* indicate the degree of surface recombination that results in case the Fermi energy of the applied metal is energetically aligned with respect to the semiconductor

difference results between the electron and hole contact which is commonly referred to as "built-in voltage." As a consequence of the constant Fermi energy throughout the device, electrons will accumulate at the electron contact and holes at the hole contact, respectively [58].

Consequently, the holes will recombine and deplete in the vicinity of the electron contact, which, in turn, will tremendously decrease their conductivity, whereas the conductivity of electrons is increased by orders of magnitude due to their accumulation and vice versa in the vicinity of the hole contact.

Thus, under operational conditions almost no holes will reach the electron contact, whereas the transport of electrons is facilitated and the selectivity of the electrode is thus ensured. This concept works well for the simulation of OSCs; however, as a result of the formation of interface states at the metal–semiconductor interface, it is scarcely practical and only a few cases have been reported where this concept successfully resulted in highly selective metal electrodes such as calcium [59].

Surface recombination leads to two effects that limit the device performance: on the one hand, it leads to a loss of charge carriers because it offers an additional nonradiative recombination pathway; on the other hand, it leads to an internal voltage drop close to the surface, which lowers the overall V_{OC} measured externally at the contacts. This is illustrated schematically in Fig. 7.

The lower voltage comes with a reduction in the conductivity of the majority charge carriers. As the contact is getting less selective, the surface recombination current is increased and due to the poor conductivity it is accompanied by gradients of the quasi-Fermi levels which is described by

$$J_{e/h} = \frac{1}{e}\sigma_{e/h} \cdot \nabla E_{F,ETL/HTL}. \tag{27}$$

The impact of surface recombination with respect to the *J*–*V* curve (especially V_{OC}, FF, and forward current density) and electroluminescence was investigated by Reinhardt et al. [57]. Devices with the same photoactive layer composite but varied

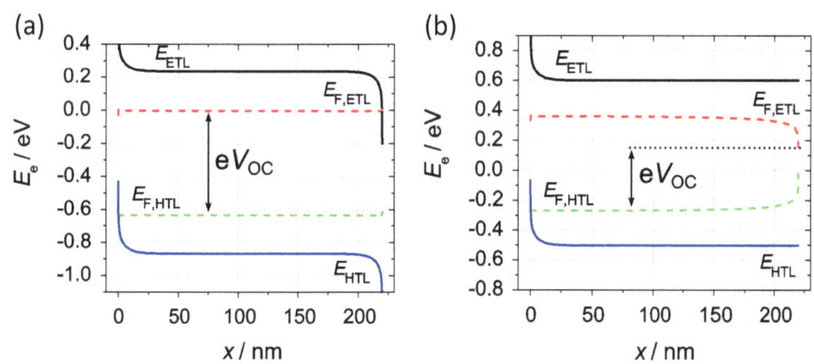

Fig. 7 Energy-level diagram of a device (**a**) with no surface recombination and (**b**) with strong surface recombination

contact materials revealed that the observed lower V_{OC}, reduced electroluminescence signal, and significant decrease in the forward current density can be fully attributed to surface recombination at the electrodes. These results are particularly important for the fabrication of ITO-free inverted devices, as explained previously. With these characterization techniques at hand, Reinhardt et al. investigated three different photoactive layers in the inverted and standard configurations, which differ in the direction of the extracted current. It was shown that depending on the HOMO position of the respective polymer, the inverted devices are severely affected by surface recombination, which lowers the overall device performance.

Further, the influence of charge carrier mobility on the degree of surface recombination has been investigated by several groups [60–62]. They showed that there is an optimum mobility for OSCs if surface recombination is present, which seems to be counterintuitive. A high mobility leads to an increase in the depletion of charge carriers, which will reduce the overall device performance although the transport properties are enhanced. Hence, a lower mobility is beneficial for the voltage as the transport to the electrodes is significantly reduced. On the other hand, lower mobilities lead to a decrease in FF and with that to a lower overall power conversion efficiency. Therefore, it is always a trade-off between good transport properties leading to a high FF and low mobility to suppress surface recombination if present.

In order to improve the device performance with respect to contact selectivity, new materials as well as new device architectures have been developed. One possibility to prevent the approach of minority charge carriers to the electrode is the application of wide-bandgap semiconductors, so-called window layers. State-of-the-art OSCs have thin interlayers of materials such as ZnO, TiO_x, MoO_3, NiO, or V_2O_5 [57, 63–65]. Alternatively, thin charge carrier selective dipole layers can be applied in front of the electrode. These layers have been identified to introduce a vacuum-level shift, which results in the buildup of a balancing space charge region that increases the conductivity of majority charge carriers while decreasing the conductivity of minority charge carriers in the vicinity of the contact. Depending on the orientation of the dipole, it results in a higher selectivity of the contact, which barely depends on the applied metal and its work function. This enables the application of less noble metals as electrode materials, which can lead to a reduction in the overall production cost.

References

1. Ebenhoch B et al (2015) Org Electron 22:62
2. Baumann A et al (2012) Adv Mater 24:4381
3. Lv A et al (2012) Adv Mater 24:2626
4. Mas-Torrent M et al (2004) J Am Chem Soc 126:984
5. von Hauff E, Dyakonov V, Parisi J (2005) Sol Energy Mater Sol Cells 87:149
6. Bässler H (1993) Phys Stat Sol 175:15
7. Coehoorn R, Pasveer WF, Bobbert PA, Michels MAJ (2005) Phys Rev B 72:155206

8. Yuan Y et al (2014) Nat Commun 5:3005
9. Jurchescu OD, Baas J, Palstra TTM (2004) Appl Phys Lett 84:3061
10. Jurchescu OD, Popinciuc M, van Wees BJ, Palstra TTM (2007) Adv Mater 19:688
11. Mihailetchi VD et al (2006) Adv Funct Mater 16:699
12. Liu Y et al (2014) Nat Commun 5:5293
13. Clarke TM et al (2012) Org Electron 13:2639
14. Ludwig GW, Watters RL (1956) Phys Rev 101:1699
15. Geens W et al (2004) Thin Solid Films 451–452:498
16. Park JH, Jung EH, Jung JW, Jo WH (2013) Adv Mater 25:2583
17. Martens T et al. (2003) In: Kafafi ZH (ed) Organic photovoltaics III. International symposium on optical science and technology. Proceedings of SPIE, Seattle, vol 4801, 07 July 2002, p 40
18. Kline RJ, McGehee MD (2006) J Macromol Sci Polym Rev 46:27
19. Oklobia O, Shafai TS (2014) Sol Energy Mater Sol Cells 122:158
20. van Duren JKJ et al (2004) Adv Funct Mater 14:425
21. Juška G, Arlauskas K, Viliunas M (2000) Phys Rev Lett 84
22. Juška G, Arlauskas K, Viliunas M, Genevicius K (2000) Phys Rev B 62:16235
23. Albrecht S et al (2012) J Phys Chem Lett 3:640
24. Schubert M et al (2013) Phys Rev B 87
25. Mozer AJ et al (2005) Phys Rev B 72
26. Homa B, Andersson M, Inganäs O (2009) Org Electron 10:501
27. Philippa B et al (2015) Org Electron 16:205
28. Juška G et al. (2011) Phys Rev B 84
29. Bange S, Schubert M, Neher D (2010) Phys Rev B 81
30. MacKenzie RCI, Kirchartz T, Dibb GFA, Nelson J (2011) J Phys Chem C 115:9806
31. Hanfland R et al (2013) Appl Phys Lett 103:63904
32. Philippa B et al (2014) Sci Rep 4:5695
33. Melianas A et al (2014) Adv Funct Mater 24:4507
34. Kim Y, Yeom HR, Kim JY, Yang C (2013) Energy Environ Sci 6:1909
35. Armin A et al (2012) Appl Phys Lett 101:83306
36. Neukom MT, Reinke NA, Ruhstaller B (2011) Sol Energy 85:1250
37. Foster S et al (2014) Adv Energy Mater 4:1400311
38. Bartesaghi D et al (2015) Nat Commun 6:7083
39. Stelzl FF, Würfel U (2012) Phys Rev B 86:75315
40. Stelzl FF (2013) PhD Thesis, Albert-Ludwigs-Universität Freiburg im Breisgau
41. Wagenpfahl A et al (2010) Phys Rev B 82
42. Shockley W, Read W (1952) Phys Rev 87:835
43. Sinton RA, Swanson RM (1987) IEEE Trans Electron Devices 34:1380
44. Würfel U, Neher D, Spies A, Albrecht S (2015) Nat Commun 6:6951
45. Schiefer S, Zimmermann B, Glunz SW, Würfel U (2014) IEEE J Photovoltaics 4:271
46. Schiefer S, Zimmermann B, Würfel U (2014) J Appl Phys 115:44506
47. Albrecht S et al (2014) J Phys Chem Lett 5:1131
48. Sinton RA, Cuevas A (eds) (2000) A quasi-steady-state open-circuit voltage method for solar cell characterization. In: Proceedings of the 16th European photovoltaic solar energy conference, Glasgow, UK
49. Wolf M, Rauschenbach H (1963) Adv Energy Conversion 3:455
50. Synopsis, TCAD Sentaurus: Sentaurus device user guide, release H-2013.03 (2013)
51. Vandewal K et al (2009) Nat Mater 8:904
52. Tvingstedt K et al (2009) J Am Chem Soc 131:11819
53. Vandewal K et al (2010) Phys Rev B 81:125204
54. Shockley W, Queisser HJ (1961) J Appl Phys 32:510
55. Scharber MC et al (2006) Adv Mater 18:789
56. Kniepert J, Schubert M, Blakesley JC, Neher D (2011) J Phys Chem Lett 2:700
57. Reinhardt J et al (2014) Adv Energy Mater 4:1400081
58. Würfel U, Cuevas A, Würfel P (2015) IEEE J Photovoltaics 5:461

59. Peters CH et al (2011) Adv Energy Mater 1:491
60. Deibel C, Wagenpfahl A, Dyakonov V (2008) Phys Status Solidi RRL 2:175
61. Kirchartz T, Pieters BE, Taretto K, Rau U (2009) Phys Rev B 80:35334
62. Mandoc MM, Koster LJA, Blom, PWM (2007) Appl Phys Lett 90:133504
63. Seßler M, Saeed A, Kohlstädt M, Würfel U (2014) Org Electron 15:1407
64. Ratcliff EL et al (2013) Adv Energy Mater 3:647
65. Ratcliff EL, Zacher B, Armstrong NR (2011) J Phys Chem Lett 2:1337

Index

K. Leo (ed.), *Elementary Processes in Organic Photovoltaics*,
Advances in Polymer Science 272, DOI 10.1007/978-3-319-28338-8

MIX
Papier aus verantwortungsvollen Quellen
Paper from responsible sources
FSC® C105338

If you have any concerns about our products,
you can contact us on
ProductSafety@springernature.com

In case Publisher is established outside the EU,
the EU authorized representative is:
Springer Nature Customer Service Center GmbH
Europaplatz 3, 69115 Heidelberg, Germany

Printed by Libri Plureos GmbH
in Hamburg, Germany